Advanced Ground-Penetrating Radar (GPR) Technologies and Applications

Advanced Ground-Penetrating Radar (GPR) Technologies and Applications

Guest Editors

Xuan Feng
Haoqiu Zhou
Zejun Dong

MDPI

Basel • Beijing • Wuhan • Barcelona • Belgrade • Novi Sad • Cluj • Manchester

Guest Editors
Xuan Feng
College of Geo-Exploration Science and Technology
Jilin University
Changchun
China

Haoqiu Zhou
College of Geo-Exploration Science and Technology
Jilin University
Changchun
China

Zejun Dong
College of Geo-Exploration Science and Technology
Jilin University
Changchun
China

Editorial Office
MDPI AG
Grosspeteranlage 5
4052 Basel, Switzerland

This is a reprint of the Special Issue, published open access by the journal *Remote Sensing* (ISSN 2072-4292), freely accessible at: https://www.mdpi.com/journal/remotesensing/special_issues/U62OQDK631.

For citation purposes, cite each article independently as indicated on the article page online and as indicated below:

Lastname, A.A.; Lastname, B.B. Article Title. *Journal Name* **Year**, *Volume Number*, Page Range.

ISBN 978-3-7258-7396-8 (Hbk)
ISBN 978-3-7258-7397-5 (PDF)
https://doi.org/10.3390/books978-3-7258-7397-5

Contents

About the Editors vii

Donato D'Antonio
Investigation of Bored Piles Under Deep and Extensive Plinth Foundations: Method of Prospecting and Mapping with Pulse Georadar
Reprinted from: *Remote Sens.* **2025**, *17*, 3228, https://doi.org/10.3390/rs17183228 1

Yuwei Chen, Beizhen Bi, Pengyu Zhang, Liang Shen, Chaojian Chen, Xiaotao Huang and Tian Jin
Weather-Resilient Localizing Ground-Penetrating Radar via Adaptive Spatio-Temporal Mask Alignment
Reprinted from: *Remote Sens.* **2025**, *17*, 2854, https://doi.org/10.3390/rs17162854 25

Feifei Hou, Yu Zhang, Jian Dong and Jinglin Fan
End-to-End Model Enabled GPR Hyperbolic Keypoint Detection for Automatic Localization of Underground Targets
Reprinted from: *Remote Sens.* **2025**, *17*, 2791, https://doi.org/10.3390/rs17162791 42

Hang Zhang, Zhijie Ma, Xinyu Fan and Feifei Hou
Synergistic Multi-Model Approach for GPR Data Interpretation: Forward Modeling and Robust Object Detection
Reprinted from: *Remote Sens.* **2025**, *17*, 2521, https://doi.org/10.3390/rs17142521 70

Yu Liu, James Irving and Klaus Holliger
Estimating Subsurface Geostatistical Properties from GPR Reflection Data Using a Supervised Deep Learning Approach
Reprinted from: *Remote Sens.* **2025**, *17*, 2284, https://doi.org/10.3390/rs17132284 90

Qingqi Xu, Qinghua Liu and Shan Ouyang
GPR-Based Leakage Reconstruction of Shallow-Buried Water Supply Pipelines Using an Improved UNet++ Network
Reprinted from: *Remote Sens.* **2025**, *17*, 2174, https://doi.org/10.3390/rs17132174 106

Yao Xiao, Guangyue Liu, Guojie Hu, Defu Zou, Ren Li, Erji Du, et al.
Mapping Mountain Permafrost via GPR-Augmented Machine Learning in the Northeastern Qinghai–Tibet Plateau
Reprinted from: *Remote Sens.* **2025**, *17*, 2015, https://doi.org/10.3390/rs17122015 128

Hui Cheng, Yonghui Zhao and Kunwei Feng
Subsurface Cavity Imaging Based on UNET and Cross–Hole Radar Travel–Time Fingerprint Construction
Reprinted from: *Remote Sens.* **2025**, *17*, 1986, https://doi.org/10.3390/rs17121986 145

Congde Lu, Senguo Cao, Xiao Wang, Guanglai Jin, Siqi Wang and Wenlong Cai
OEM-HWNet: A Prior Knowledge-Guided Network for Pavement Interlayer Distress Detection Based on Computer Vision Using GPR
Reprinted from: *Remote Sens.* **2025**, *17*, 1554, https://doi.org/10.3390/rs17091554 161

Hanqing Qiao, Minghe Zhang and Maksim Bano
Harris Hawks Optimization for Soil Water Content Estimation in Ground-Penetrating Radar Waveform Inversion
Reprinted from: *Remote Sens.* **2025**, *17*, 1436, https://doi.org/10.3390/rs17081436 183

Ding Yang, Cheng Guo, Raffaele Persico, Yajie Liu, Handing Liu, Changjin Bai, et al.
Adaptive Variational Mode Decomposition and Principal Component Analysis-Based Denoising Scheme for Borehole Radar Data
Reprinted from: *Remote Sens.* **2025**, *17*, 525, https://doi.org/10.3390/rs17030525 . . . . **207**

Mingze Wu, Qinghua Liu and Shan Ouyang
Two-Stage GPR Image Inversion Method Based on Multi-Scale Dilated Convolution and Hybrid Attention Gate
Reprinted from: *Remote Sens.* **2025**, *17*, 322, https://doi.org/10.3390/rs17020322 . . . . **232**

Yuanjin Fang, Feng Yang, Xu Qiao, Maoxuan Xu, Liang Fang, Jialin Liu and Fanruo Li
Cross-Line Fusion of Ground Penetrating Radar for Full-Space Localization of External Defects in Drainage Pipelines
Reprinted from: *Remote Sens.* **2025**, *17*, 194, https://doi.org/10.3390/rs17020194 . . . . **250**

Satoshi Ebihara, Raiki Masui, Koki Koyama, Yuki Tsujikawa and Yuto Nishida
Effect of the Outer Pipe on Reducing Direct Coupling of the Thin Borehole Radar Probe in Thick Water-Filled Borehole
Reprinted from: *Remote Sens.* **2025**, *17*, 100, https://doi.org/10.3390/rs17010100 . . . . **272**

Jie Xu, Qifeng Lai, Dongyan Wei, Xinchun Ji, Ge Shen and Hong Yuan
The Ground-Penetrating Radar Image Matching Method Based on Central Dense Structure Context Features
Reprinted from: *Remote Sens.* **2024**, *16*, 4291, https://doi.org/10.3390/rs16224291 . . . . **297**

Tina Wunderlich, Bente S. Majchczack, Dennis Wilken, Martin Segschneider and Wolfgang Rabbel
What Is Beyond Hyperbola Detection and Characterization in Ground-Penetrating Radar Data?—Implications from the Archaeological Site of Goting, Germany
Reprinted from: *Remote Sens.* **2024**, *16*, 4080, https://doi.org/10.3390/rs16214080 . . . . **321**

Jialin Liu, Xiaosong Tang, Feng Yang, Xu Qiao, Fanruo Li, Suping Peng, et al.
Study on the Identification Method of Planar Geological Structures in Coal Mines Using Ground-Penetrating Radar
Reprinted from: *Remote Sens.* **2024**, *16*, 3990, https://doi.org/10.3390/rs16213990 . . . . **341**

Dae Wook Park, Han Eung Kim, Kicheol Lee and Jeongjun Park
Development of a Background Filtering Algorithm to Improve the Accuracy of Determining Underground Cavities Using Multi-Channel Ground-Penetrating Radar and Deep Learning
Reprinted from: *Remote Sens.* **2024**, *16*, 3454, https://doi.org/10.3390/rs16183454 . . . . **361**

About the Editors

Xuan Feng

Xuan Feng received his B.S. degree in applied geophysics and his M.S. and Ph.D. degrees in geo-exploration and information technology from Jilin University, Changchun, China, in 1996, 1999, and 2002, respectively. From 2003 to 2005, he held a postdoctoral position. From 2005 to 2006, he served as a Research Assistant at Tohoku University, Sendai, Japan. From 2006 to 2008, he was an Associate Professor at Jilin University and has been a professor there since 2008. From 2014 to 2016, he worked as a Visiting Scientist at the Massachusetts Institute of Technology (MIT), Cambridge, MA, USA. From 2017 to 2024, he served as the Deputy Dean of the College of Geo-Exploration Science and Technology and has been the Dean of the college since 2025. His research interests include signal processing technology of ground-penetrating radar (GPR) and seismic exploration, as well as the joint inversion of GPR and seismic exploration. In the field of GPR, he has developed a full-polarization GPR system and established the theory of underground target detection and data analysis technology for full-polarization GPR. In the field of seismic exploration, he discovered the short-term memory characteristics of rock nonlinear viscoelasticity and established a new phenomenological model of rock nonlinear viscoelasticity. In the direction of joint inversion of GPR and seismic exploration, he developed a cross-gradient joint inversion algorithm, which improved the imaging accuracy.

Haoqiu Zhou

Haoqiu Zhou, PhD, is a Tang Aoqing Young Scholar and Associate Professor at Jilin University. He received his doctoral degree in June 2023 from the College of Geo-exploration Science and Technology, Jilin University. In July 2023, he was selected for the Dingxin Scholar Support Program of Jilin University and joined the College of Geo-exploration Science and Technology, Jilin University, as a faculty member in July 2025. His research focuses on full-polarimetric ground-penetrating radar (FP GPR) data analysis and planetary radar data interpretation. In the field of full-polarimetric GPR, he has developed high-precision target analysis techniques, including polarimetric interference suppression, multi-polarization data fusion imaging, and high-precision target recognition using multi-polarization attributes. In lunar and Martian geological studies based on planetary radar data, he has developed dedicated processing and analysis methods using fully polarimetric techniques, yielding new insights into lunar and Martian regolith structures, impact crater formation, lava tube formation, and other geological processes. He is the principal investigator of a National Natural Science Foundation of China (NSFC) for Young Scholars and a China Postdoctoral Science Foundation General Program. He was also selected for the National Funded Postdoctoral Fellowship. He has published dozens of SCI papers in journals such as GRL, JGR Planets, IEEE TAP, IEEE TGRS, and ApJ. He received the Excellent Master's and Doctoral Dissertation Awards of Jilin Province.

Zejun Dong

Zejun Dong, PhD, is a Tang Aoqing Young Scholar, Associate Professor, and PhD Supervisor at Jilin University. She received her doctoral degree from the College of Geo-exploration Science and Technology, Jilin University, in June 2022. In July 2022, she was selected for the Dingxin Scholar Support Program of Jilin University. Since September 2024, she has been working at the College of Geo-exploration Science and Technology, Jilin University. Her research interests include engineering and environmental geophysics, full-polarimetric ground-penetrating (FP-GPR) radar data processing and interpretation, and planetary radar data processing and interpretation. She developed the induced field rotation theory at rough interfaces and the polarization rotation theory in anisotropic media for FP-GPR data, solved the problem of FP-GPR signal analysis under conditions of rough interfaces and anisotropic media, and realized ultra-high-precision detection of small underground targets. She was awarded the Excellent Doctoral Dissertation by the Chinese Geophysical Society. Prof. Dong is the principal investigator of several projects, including the National Natural Science Foundation of China (Youth Fund), a sub-task of the National Key Science and Technology Project for Deep Earth Resources, a Scientific Research Project of the Department of Education of Jilin Province, and a Postdoctoral Talent Funding Project for Retaining Talents in Jilin Province. She has published dozens of SCI papers in journals, such as IEEE TAP, IEEE TGRS, including an ESI highly cited paper (top 1%).

remote sensing

MDPI

Article

Investigation of Bored Piles Under Deep and Extensive Plinth Foundations: Method of Prospecting and Mapping with Pulse Georadar

Donato D'Antonio

Independent Researcher, Vicolo III San Nicola n.2, 86013 Gambatesa, Italy; info@tellurad.it; Tel.: +39-3791128093

Highlights:

What are the main findings?

- Transillumination velocity tests on the massive plinth provided permittivity values between 5.9 and 6.7; CMP velocity tests on natural soil provided permittivity values between 9.4 and 13.9.
- Backscattering field analysis highlights pile anomalies even in individual radargrams; acquisitions on exposed faces of the excavation reveal unmistakable pile anomalies, in slices, when combined in the form of perimeter tomography.

What is the implication of the main finding?

- It is possible to study piles beneath massive foundations using the transillumination ground-penetrating radar technique. The results indicating the precise position and geometry of the piles are dependent on the availability of trench excavation to below the top of the plinth.
- A simulated prospection of piles under massive plinths, if conducted from ground level, returns results that are very difficult to decode if one does not possess adequate a priori knowledge.

Abstract

Ground-penetrating radar surveys on structures have a wide range of applications, and they are very useful in solving engineering problems: from detecting reinforcement, studying concrete characteristics, unfilled joints, analyzing brick elements, detecting water content in building bodies, and evaluating structural deformation. They generally pursued small investigation areas with measurements made in direct contact with target structures and for small depths. Detecting deep piles presents specific challenges, and surveys conducted from the ground level may be unsuccessful. To reach great depths, medium-low frequencies must be used, but this choice results in lower resolution. Furthermore, the pile signals may be masked when they are located beneath massive reinforced foundations, which act as an electromagnetic shield. Finally, GPR equipment looks for differences in the dielectric of the material, and the signals recorded by the GPR will be very weak when the differences in the physical properties of the investigated media are modest. From these weak signals, it is difficult to identify information on the differences in the subsurface media. In this paper, we are illustrating an exploration on plinth foundations, supported by drilled piles, submerged in soil, extensive, deep and uninformed. Pulse GPR prospecting was performed in common-offset and single-fold, bistatic configuration, exploiting the exposed faces of an excavation around the foundation. In addition, three velocity tests were conducted, including two in common mid-point and one in zero-offset transillumination, in order to explore the range of variation in relative dielectric permittivity in the investigated media. Thanks to the innovative survey on the excavation faces, it is possible to perform profiles perpendicular to the strike direction of the interface. The electromagnetic backscattering analysis approach allowed us to extract the weighted average frequency attribute section. In it, anomalies emerge in the presence of drilled piles with four piles with an estimated diameter of 80 cm.

Keywords: ground-penetrating radar; pile; common mid point; transillumination; plynth; backscattering field analisys; weighted average frequency, signal attributes

1. Introduction

The study described herein fits into the broad field of applications of ground-penetrating radar (GPR) and, in particular, that of human-made structures. The recent developments [1] in the methods of acquisition, data processing [2] and presentation of results have strengthened a tougher confidence in the use of this tool [3,4]. GPR has been successfully used to study beams, columns, slabs, bridges, tunnels, reinforcing bars, cracks and cracking in concrete, subsidence, stratigraphy, voids and cavities and foundations [5–7]. We describe a method of geophysical prospecting on bored piles, the verification of their number, and the estimation of their geometry under extended foundations on plinth, on which they are constrained, under abutment foundation. The prospecting was carried out in May 2023 at the Liscione Bridge, built between 1967 and 1974, in the Monte Arcano district, on the SS 647 Biferno valley floor, in the province of Campobasso, Molise. An excavation carried out adjacent to an abutment, functional for the maintenance of the drainage around the foundation, uncovered the plinth and surrounding soil and provided a unique opportunity to access the seabed walls and verify the presence and quantity of piles drilled below the plinth. The work performed is important because it allows us to shed light on relatively recent human-made structures for which design drawings are no longer available. Our work represents a turning point, leaving behind all the problems of interpretative ambiguity through the innovative lateral scanning of the excavation fronts. Indirectly investigating, with geophysical methods, the bored piles below the foundations is difficult from the ground surface, especially if the depth between the head of the piles and the ground level is considerable, as in this case, which is 4 m. Especially if it is not possible to access all sides of the pylon, on the surface, the relationships between the geometric variables (Figure 1) and the displacement of the antennas must be favorable; otherwise, the signals coming from the pylon will not be recorded.

For example, assuming a good contrast in dielectric permittivity, f should be very small compared to D; conversely, if f is large, then F should be very small; similarly, D should be much larger than d; E should be at least twice F; and α (and the form of radiation pattern) for low permittivity soils is high and decreases more and more as the permittivity increases [8]. In the following paragraphs a simulation of a direct investigation on the surface is proposed. Surveys are successful when both the transmitter and receiver antennas are sensitive to the target region. For each antenna, this region is defined by the refraction focusing peak, and it depends on the critical refraction angle of the air–earth interface. An estimate of the optimum antenna separation is given by

$$S = \frac{2d}{\sqrt{\epsilon - 1}} \quad (1)$$

where ϵ is the relative permittivity and d is the approximate depth of the target. It should be noted, however, that depth resolution can be increased by the decreasing of the antenna separation. GPR studies of this kind are very recent [5,9], and the need for extraordinary maintenance for aging civil works is giving a boost to prospecting and research with GPR aimed at these applications.

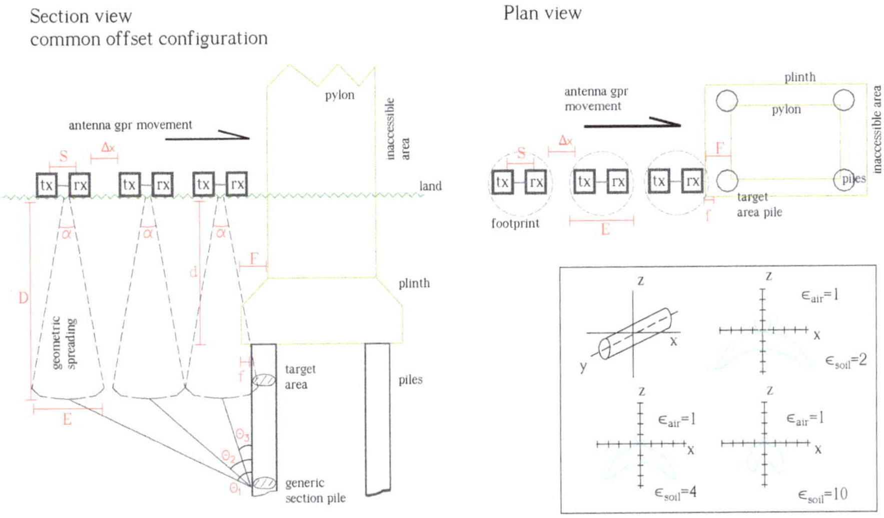

Figure 1. Critical geometric variables, in red, that determine the success rate of a GPR survey of a deep pile beneath a plinth and affect vertical and horizontal resolution; D is the penetration depth, d is the depth of the pile head, E is the energy diameter, α is the opening angle of the illumination cone, θ is the angle of attack between the generic target section of the pile and the base of the footprint, F is the distance of the edge of the plinth from the pile, f is the distance of attachment of the pile measured from the edge of the plinth, Δx is the station spacing that controls aliasing, and S is the antenna separation. The theoretical variation in the radiation pattern [10] with the dielectric permittivity is shown in the box; the cylinder is the antenna orientation.

2. Materials and Methods

Ground-penetrating radar detects dielectric discontinuities and hence makes it possible to investigate objects that are buried in visually opaque media. The dielectric wave has a lot of information that can be related to the parameters significant to hydrology, soil mechanics, and material property characterizations [11]. Moreover, the dielectric wave is also efficient for quantifying the properties of the host materials and the nature and size of the buried superficial bodies as well [12]. The method is based on the transmission of electromagnetic waves through the subsurface to characterize terrains with changes in their electromagnetic properties. The electromagnetic wave propagation is governed by Maxwell's equations [13]. The way that GPR responds to each material is governed by two physical properties of the material: electrical conductivity and dielectric constant. The relative dielectric permittivity (RDP) is a descriptive number that indicates, among other things, how fast radar energy travels through a material. In materials of low magnetic susceptibility, velocity v and attenuation α are related to the RDP ϵ and loss tangent, δ, by the expressions [14]

$$v = \frac{c}{\sqrt{\epsilon}} \tag{2a}$$

$$\alpha = \frac{\omega \cdot \sqrt{k}}{2 \cdot c} \cdot tg\delta \tag{2b}$$

where $\omega = 2\pi f$ is the angular frequency, f is the frequency and c is the velocity of the

EM wave in the empty space (30 cm/ns). There are some main parameters to define for common-offset, single-fold pulse radar reflection survey [15].

Operating frequency: a simple rule of thumb for your choice is

$$log_{10}f = -0.95\,log_{10}z + 6.15 \tag{3}$$

where f is the operating frequency, and z is the required depth of investigation;

time window

$$tw = 1.3\,\frac{2z}{v} \tag{4}$$

sampling interval

$$t = 100\,\frac{6}{f} \tag{5}$$

where t is the time in ns and f is the center frequency in Mhz;

vertical resolution can be taken as one-quarter of the wavelength [16] of the EM wave [17]

$$\lambda = \frac{v}{f} \tag{6}$$

where t is the time in ns and f is the center frequency in Mhz;

horizontal resolution measures for the horizontal resolution of the signal (so-called first Fresnel zone) that can be achieved with a given antenna frequency and in a specified material by

$$r = \sqrt{\frac{\lambda z}{2} + \frac{\lambda^2}{16}} \tag{7}$$

where z is the depth nd λ is the wavelenght;

radius of footprint of the elliptical illumination cone of GPR into the ground is [18]

$$A = \frac{\lambda}{4} + \frac{D}{\sqrt{\epsilon + 1}} \tag{8}$$

where λ is the center frequency wavelength of the EM wave, D is the depth from the ground surface to the reflection surface, and ϵ is the relative dielectric permittivity of the material from the ground to the surface depth (D).

Electrical properties control how electromagnetic waves travel through a material; dielectric permittivity primarily controls wave speed; and conductivity determines the signal attenuation. The relationship between the GPR (transmitter, receiver), buried target distance and the subsurface is described in detail by the radar equation. It is useful both for determining the maximum depth to which the radar can "see" (radar range) and for designing the system itself. Furthermore, it shows that GPR measurements ultimately detect reflected or scattered energy. This ability is governed by the Fresnel coefficient; Fresnel reflection coefficients quantify the amplitude of reflected and transmitted signals at boundaries. The ratio of the reflected-to-incident signal amplitudes is the reflection coefficient; the ratio of the transmitted-to-incident signal amplitudes is the transmission coefficient. Reflection coefficients depend on the angle of incidence, the polarization of the incident field, and the velocity contrast. At the investigated site, the pulsed GPR geophysical prospecting method was used to obtain high-resolution subsurface information. Pulsed GPR surveys use the broadband antennas (UWB) to emit and receive electromagnetic signals from the subsurface. These signals are recorded in the radargrams [13]. The advantages of UWB technology for GPR are as follows:

- It offers high vertical resolution (also known as the down-range resolution), which is closely related to the operating bandwidth [19];

- Covers a frequency range between an octave and a decade by using short-time pulses,

$$\Delta V = \frac{c}{2 \times \Delta f} \tag{9}$$

where Δv is the vertical resolution of the system, c is the velocity of an electromagnetic signal in vacuum, and Δf is the bandwidth used of the system [20].

The radargram is the visual representation of the data recorded by the GPR in a distance-time cross-section. It is composed of a series of successive traces (A-scans), called B-scans. The A-scan is the result of a single radar measurement on an oscilloscope and provides amplitude–time record of a single measurement over a target. At boundaries where the electromagnetic properties of the subsurface change abruptly, radio wave signals can undergo transmission, reflection, and refraction. In this way, the objects present can be studied through signal variations generated by contrasts in their intrinsic physical properties. These variations are defined as signal "anomalies" and they are related to the relative dielectric permittivity (ϵ) and electrical conductivity (σ).

In pulsed GPR, the shape of the transmitted pulse is calibrated to obtain a Gaussian-type spectral distribution [16], where the central value represents the characteristic frequency of the antenna, which corresponds to the dominant frequency of the pulse. It determines the resolution and the maximum exploration depth characteristics. The working hypothesis is based on the question of whether it is possible to reveal the presence of bored piles hidden by a plinth and surrounding soil with GPR. The use of common-offset and single-fold scans on the sides of each exposed face of the foundation could show anomalies if there are contrasts in dielectric permittivity and/or electrical conductivity [14]. To verify this, first we need to explore the site-specific relationship between these variables, depending on the spatial geometry and acquisition method [21]. The sediments encompassing the piles were wet, and we know that the permittivity and conductivity of rocks and soils depend to a large extent on their water content [22]. Both permittivity and conductivity also depend on other properties of the investigated material, in particular on its porosity and clay content, such that the application of such empirical laws should be site- and scale-dependent [23]. The influence of electrical conductivity on the attenuation of the GPR wave can only be neglected in dry soils that do not contain clay particles. In the radargrams, the major signal reflections are related to the dielectric permittivity of the materials involved rather than variations in electrical conductivity [24]. If the contrasts associated with these parameters are weak then the study of the electromagnetic signal and its attributes will be necessary in order to see if the objects can be detected [25]. The attributes from which the best information can be extracted are as follows [26–28]:

- Arc length of time window is defined as the expanded length of the waveform curve within the analysis time window; it is a joint attribute that combines amplitude and frequency characteristics and can be used to distinguish the phase characteristics of GPR records, such as between strong amplitude and high frequency, between strong amplitude and low frequency, between weak amplitude and high frequency and between weak amplitude and low frequency.
- Product of instantaneous amplitude and cosine of instantaneous phase that strengthens the amplitudes of wave peaks and troughs and inverts all trough amplitudes into apparent peak amplitudes.
- Slope of instantaneous frequency is defined as the change rate of instantaneous frequency over time within the analysis time window. It highlights changes in the local frequency, so it can more effectively reflect differences in thin layers, and is effective for dividing edge phases.
- Slope of reflection strength is defined as the change in instantaneous reflection intensity (i.e., instantaneous amplitude) with reflection time. If the reflection intensity remains

throughout the analysis time window interval, the slope value will approach zero. If the reflection intensity gradually increases from the top to the bottom of the time window, the slope value is positive; on the contrary, the slope value is negative. The attribute can reflect the vertical distribution characteristics and interbedding conditions of the stratum.

2.1. Geometry of the Site, B-Scans, and GPR Velocity Tests

In situ, the bridge pylon rests on a trapezoidal plinth. The plinth in turn is constrained to deep bored piles. Geologically, the area consists of upper Pleistocene terraced alluvial soils. The lithologies are mixed-grained: silt, sand, clay and gravel in varying proportions, with the sandy component prevailing; moisture in the exposed soils was evident. From a stratigraphic point of view, the succession of soils on the site is given by the following:

- An 80 cm vegetal cover, starting from the ground level;
- The silty–sandy–clayey soils up to the excavation depth.

The plinth was 10 m long, 8 m wide and 2.3 m high, while the underlying soil was 1.3 m thick to the bottom of the excavation. The base of the plinth was at a depth of about 4 m above the ground level.

2.2. GPR Instrumentation and Acquisition Parameters

The survey was carried out in the spring using a radar system from "Radar System Inc." (Riga, Latvia) with a time range up to 2000 ns, a frequency of pulse repetition of transmitter 115 kHz, 320 scans per second, 1024 samples per scan and stacking of traces up to 128. The antennas used, transmitter and receiver separated, are 200 Mhz, 300 Mhz and 500 Mhz, respectively, and a connecting cable of length 30 m (Figure 2). The antennas are bow-tie, ground-coupled, shielded type. GPR raw data were processed using PRISM 2.54, Georadar-Expert 2.13, Matlab 2020, Seismic Unix 40. This combination of tools enabled sophisticated data analysis and visualization, enhancing the quality of the survey results. Preliminarily, the propagation of the electromagnetic wave was simulated qualitatively in the time domain to understand how different center frequency antennas could perform, as well as to get an idea of how the A-scan might look like in the given conditions [29]. The positioning was achieved with metric measurements of the entire site and plinth geometry in order to obtain a 3D reconstruction (Figure 3). The scans were measured and marked using some spray references.

Figure 2. Excavation of the foundation for drainage maintenance and acquisition steps with the antennas, control unit, and connection cable reel.

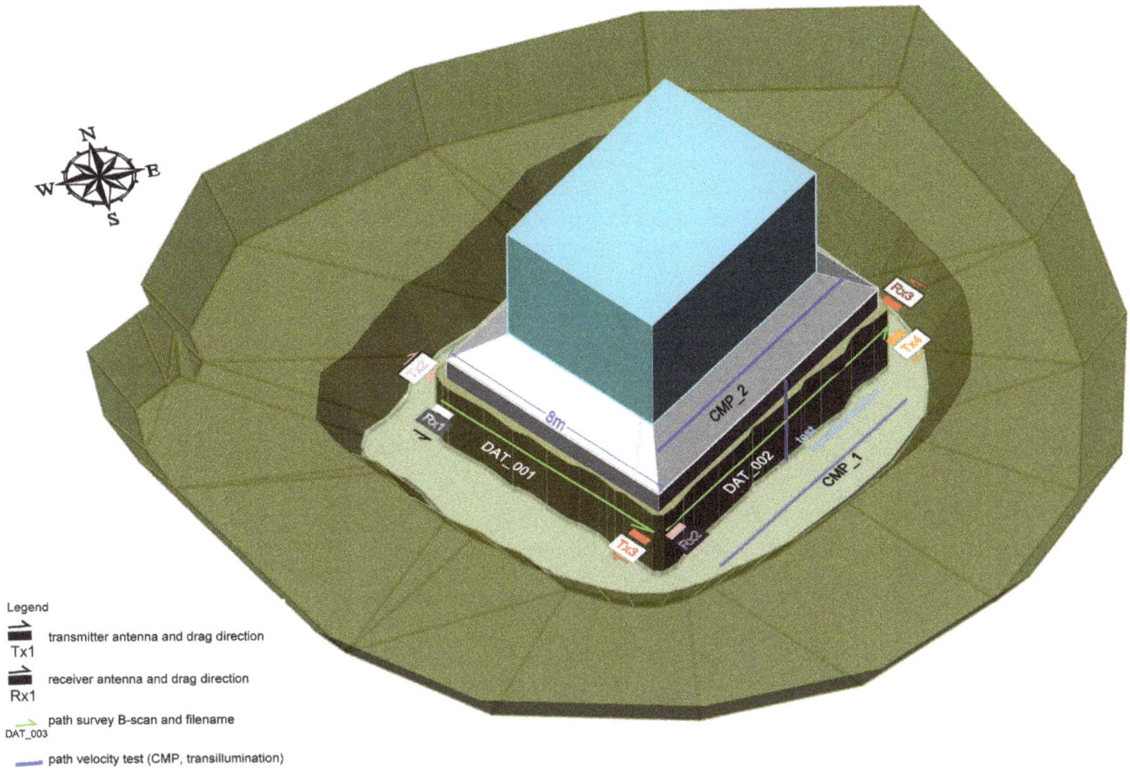

Figure 3. 3D view of the site: The stations of the 2 CMP tests, the transillumination test, and the 4-sided B-scans are shown with the location of the transmitting and receiving antennas, for each acquisition, drag direction and filename.

2.3. Direct In Situ Velocity Tests: CMP and Zero-Offset Transillumination

For the calibration of the propagation speed of the electromagnetic wave in the medium, the following in situ velocity tests had to be carried out:

- One test of the common mid point on the excavated bottom;
- One test of the common mid point on the exposed plinth;
- One test transillumination zero-offset on the two faces of the short side of the foundation.

The two CMPs were performed by symmetrically moving 2 antennas with a constant step, a 500 Mhz transmitter and a 300 Mhz receiver, along the same direction and each in opposite directions starting from a common midpoint (Figure 4). The CMP method is suitable for building structures [30].

In the CMP test the constraint is that the process requires more time to survey a site, since the transmitter and receiver must be moved independently from each other for every scan [30].

To determine the velocity in the crossed medium and so the relative permittivity, the following trigonometric relation can be applied:

$$vt_n = 2\sqrt{(x - x_0)^2 + (\frac{vt_0}{2})^2} \tag{10}$$

where the time of flight of the signal t_n is the time the signal needs from the transmitter and back to the receiver, while t_0 is the time to and from the target at the coordinate of the minimum distance (x_0). From the survey schematic, we see that if the interface is approximately flat, the point of reflection is the same for all readings. As a result, the signal from the reflected wave in the radargram should form a hyperbola (Figure 5). The advantage of the CMP test is that the receiver does not become saturated while receiving the reflected signal [30]. The radargrams of the CMP acquisitions can be seen in the following figure, wherein the gray box of each are calculated the results of the velocities and permittivities, shown in detail in Table 1. The zero-offset transillumination test between the transmitting and receiving antennas was possible by having 4 vertical faces, with the grounds to be tested, exposed. The antenna used for receiving was the 300 Mhz and transmitting the 200 Mhz. The center-band frequency was 195 Mhz and the wavelength in the medium was 0.4 m, which was optimal to avoid near-field problems since the two opposite faces of the short side stood 8 m apart, which is much more than 1.5 times the wavelength. The 2 antennas were moved from top to bottom with point acquisition every 10 cm.

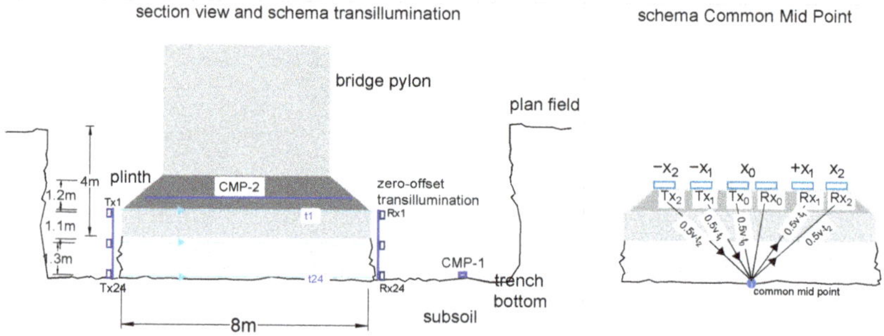

Figure 4. Geometry of the acquisition area, CMP tests, and zero-offset transillumination.

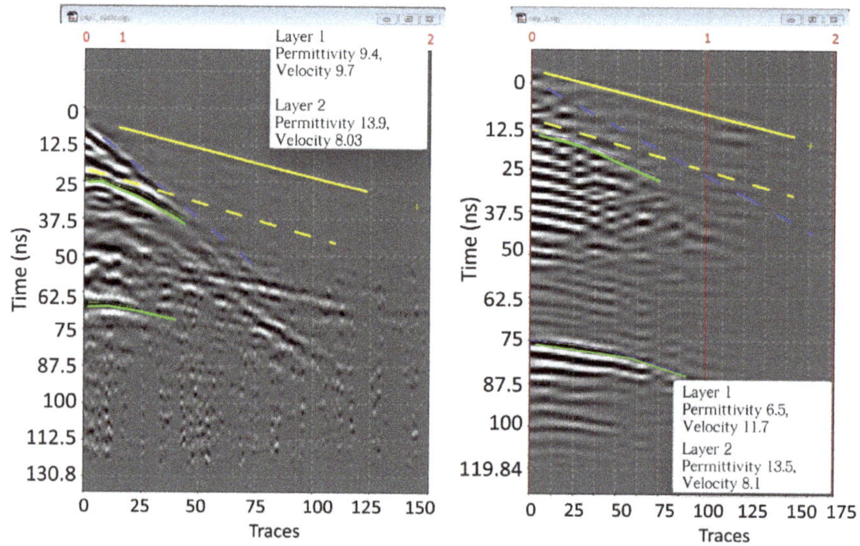

Figure 5. CMP1 radargrams on the left and CMP2 on the right. The free airwave has been highlighted with the solid yellow line, the dashed yellow line instead of the critically refracted airwave, the blue dashed line for the ground wave and the lines in green for the refracted waves.

Table 1. Velocity and dielectric permittivity of the tested layers.

	CMP1 on Trench		CMP2 on Plynth	
	Layer		Layer	
	1	2	1	2
Thickness (m)	0.96	1.68	1.1	2.47
Layer velocity (cm/ns)	9.79	8.04	11.74	8.15
Permittivity	9.4	13.94	6.53	13.56
Error [RMS]	0.11	0.02	0.32	0.001

The distance measured at the top and bottom of the excavation was always 8 m on the short side. A total of 24 "illuminations" were made. The waves received and the traces of each measurement point are shown in Figure 6. To calculate the airwave, the length of the perimeter path on the ground between the two antennas is required. This is 11 m when the antennas are on the top of the plinth and 16 m when the antennas are on the base of the trench. The first arrival in these B-scans can be identified as the first significant change in amplitude. The period of small or no response in each trace is the time elapsed by the pulse between the transmitting and receiving antenna on the other side of the exposed face. In the radargram, through appropriate trace gain, the entire portion from 70 ns onwards takes on much greater prominence than the first part; these signals have higher amplitude than the first arrivals of the wave in the air. This is because for the signal that travels only in air, the transmission and reflection coefficients do not vary since the permittivity is always the same.

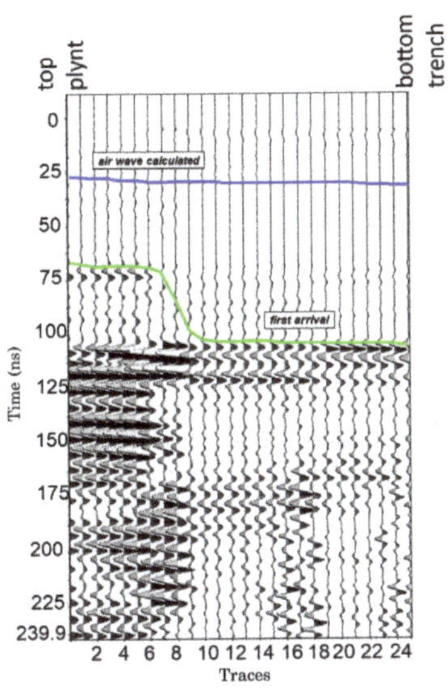

Figure 6. Wave arrivals in the transmission test. In blue are the calculated potential arrivals of the wave in the air, and in green, the first arrivals of the direct wave.

2.4. Acquisition B-Scan GPR

Four B-scan profiles were acquired in single-fold mode. The 200 Mhz transmitting antenna is placed on one side of the trench, at the base of the excavation, while the 300 Mhz receiving antenna is placed on the opposite side and on the same reference line with respect to the cardinal points (Figure 7). The GPR step is 0.01 m and the spatial separations of the transmitting and receiving units are as follows:

- For 200 mhz, $\Delta x = 70$ cm;
- For 300 mhz, $\Delta x = 68$ cm;
- For 500 mhz, $\Delta x = 45$ cm.

The antennas were used to acquire data sequentially, where each one transmits and receives simultaneously, then operating the data fusion.

The antennas used are dipolar and radiate with a preferred polarity (parallel endfire). The antennas are normally oriented so that the electric field is polarized parallel to the long axis or strike direction of the target. The antenna's illumination cone is extended perpendicularly to the exposed face of the trench where the antenna is placed (Figure 8).

In this way, the transmitted electromagnetic energy passes through the adjacent layers it encounters. In this case, all the terrain and objects along the radiation pattern are struck: a first layer of soil, then the concrete plinth, then the soil, then the concrete plinth again, and so on. The antenna orientation dictates the direction of the electric field (E) in the subsurface. *E-field* direction, in turn, dictates how well a target will couple to the excitation field and generate a measured reflected or scattered signal. Antennas are copolar and normally oriented to the target, which is not equidimensional.

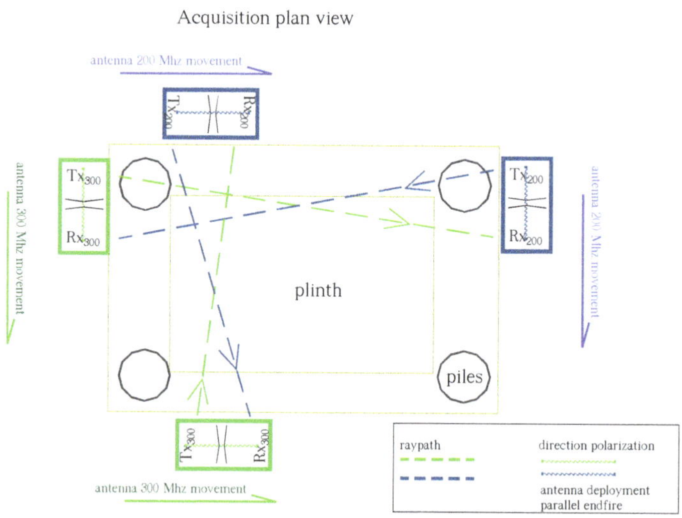

Figure 7. Plane view acquisition geometry of the B-scans on the exposed faces of the trench: relationship between the antenna deployment, polarization direction, raypath, and target. The arrow indicates the direction of the transmitting signal. The electric field is assumed to be aligned along the antenna's long axis.

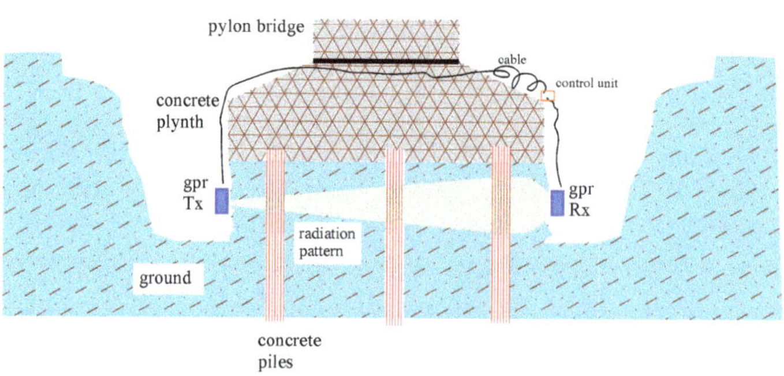

Figure 8. Schematic of B-scan acquisitions with separate Tx and Rx antennas placed on opposite sides of each side of the bottom trench, the long cable connecting to the control unit.

The acquisition parameters are grouped in Table 2 in the first 4 columns; in the remaining ones are the values estimated using $f = 300$ Mhz, $\epsilon_{mean} = 12$, and $\sigma_{mean} = 0.5$ mS/m.

Table 2. Acquisition parameters in bistatic configuration.

Time Window (ns)	Stack Traces	Sample	Realtime Filter	Wavelength Medium (m)
200	4	512	no	0.27
1/2λ; 1/4λ	**Fresnel zone foot (m)**	$f_{Nyquist}$ **Mhz**	**Loss tangent**	**Q-estimation**
0.14; 0.07	2.2; 1.1 [1]	1280	2.24	30

[1] First, number is semimajor axes and second is semiminor axis.

The acquired radargrams are, here, grouped in the form of horizontally compressed wiggle plots (Figure 9). The wiggle plot is a collection of single traces with no color intensity, and the visual aspect focuses on the shape and pattern of the data, on the deviation in positive and negative values from a baseline such as the reference polarity. The polarities, positive and negative, of the reflection provide much information about the type of material from which it comes. A change in polarity can help in the interpretation of data by indicating a transition from a metallic to a plastic object, for example.

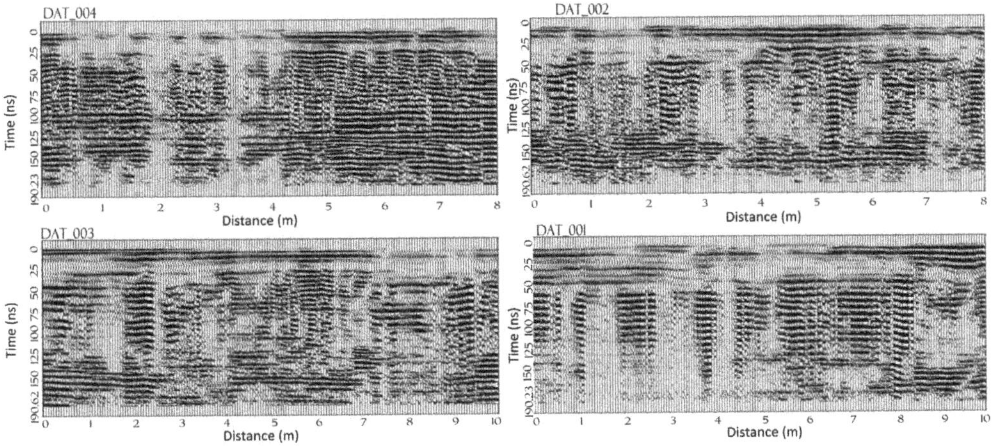

Figure 9. Radargrams from DAT_001, bottom right, to DAT_004, top left. The wiggle plot shows the combination of a series of A-scan.

2.5. Processing GPR Data

In order to generate GPR signals with adequate signal-to-noise ratio (SNR) and to extract meaningful information from the radargram, GPR data processing is implemented. The following data processing helps to reduce clutter or eliminate unwanted noise in the data set:

- Time-zero correction to ensure that all reflections are correctly aligned by setting the airwave and direct wave of the trace at the first break point (or the first negative lobe) to a particular time-zero position.
- Background signal removal that is present at the beginning of a radar signal known as the direct wave, which is due to the coupling of the antenna with the vertical face of the trench. This part of the signal is considered as unwanted noise or clutter.

- Exponential gain to equalize the amplitude of the emitted wave, which suffers a significant attenuation along the medium.
- Ormsby bandpass filter along a trace for low-frequency interference and signal's high-frequency components suppression. The algorithm used comprises three steps: application of direct FFT (fast Fourier transform) for transition from the time domain into the frequency domain of low-frequency and high-frequency trace spectrum components suppression and application of reverse FFT for transition from the frequency domain into the time domain; the cutoff frequencies are from 30 to 110 Mhz (in low frequency) and 230 to 340 Mhz in high frequency.
- Time-depth conversion should be used for restructuring the initial time profile into a depth profile in compliance with the velocity areas.

2.6. Backscattering Electromagnetic Field (BSEF) Analysis

There are situations, even very common ones, in which the geophysical characteristics of the subsurface change fluidly, without jumps. This is due to the diffuse nature of the adjacent layers. In these situations, it is difficult to distinguish the variations or the reflections [31] from the GPR signal recordings with visual analysis of the radargrams. The only solution is to manually determine, as is well known in geophysical techniques, the velocity characteristics of the various discontinuities in the radargrams based on the diffracted waves and combine the areas with similar velocity values into layers. The diffracted waves are formed following reflections from point-like objects in the subsurface based on their cross-sections affected by the energy. The cross-section is a measure of the effective area that a scatterer objects into the path of the incident radar signal. The incident radar wavefront energy per unit area multiplied by the cross-sectional area determines the energy the scatterer extracts from the incident wave.

The energy-extracted signal can be absorbed or re-radiated in any direction. Scattering describes deviations in the paths of electromagnetic waves due to localized non-uniformities, which are less than 1/4 the wavelength of the radio wave signal; backscatter gain measures the amount of energy re-radiated back in the direction of the incident signal. Backscatter gain and cross-sectional area are either computed from numerical modeling or measured for standard geometrical shapes in laboratories. Some simple geometries yield relatively compact analytical backscatter gain formulas. The cross-sectional area is a function of the true geometrical cross-section of an object as well as the contrast in electrical properties. The backscattered gain is primarily controlled by the geometrical attributes of the object. When an electromagnetic wave impinges on the interface between two media with different dielectric properties, a part of its energy is backscattered [32,33] and the remainder is transmitted into the lower medium.

The directions of the incident, reflected and transmitted waves are related to each other by Snell's law. The radiation pattern in the real case of inhomogeneous media becomes increasingly diffuse, and backscattering tends to increase. Part of the transmitted energy is scattered at the inhomogeneities and can pass through the upper medium to the surface [31]. Scattering, which is called volume scattering, is weaker than surface scattering because it causes a redistribution of energy in the transmitted wave in multiple directions, resulting in energy loss. The total loss that an electromagnetic wave experiences in a physical medium is the sum of the loss due to diffusion and the loss due to conduction [34]. The radar backscatter of an object, as mentioned by Toropainen [35], is measured in decibels (dB) using Equation (11) (the equation is used in the routine for automatic calculation of BSEF analysis sections):

$$\sigma_{cp} = \sigma_s \perp [\frac{1 - e^{-2\sigma_t d}}{2\sigma_t} + 2[\Gamma_2]^2 d e^{-2\sigma_t d} + \frac{|\Gamma_2|^4}{2\sigma_t} e^{-2\sigma_t d}(1 - e^{-2\sigma_t d})] \tag{11}$$

where the first term represents the backscatter of the incident wave; the second term represents the forward scattered wave; the third term represents the reduced incident wave's backscatter; Γ_2 represents the voltage reflectance coefficient at the lower surface; σ_t represents the cross-section of the material; σ_s represents the co-polarized cross-section of the material, and their angular variation is according to Fresnel reflectivity of two types of polarizations; and d is the distance traveled by the signal. With these reflection patterns, recorded in the radargrams, the properties of the investigated materials can be extracted directly through dedicated processing.

From this premise, it seems natural to use the backscatter field [36] of electromagnetic waves to determine the physical structure (electric and magnetic) of the propagating medium [37]. One of the advantages of BSEF analysis is the ability to study the subsurface, whose physical (electric and magnetic) characteristics change vertically and uniformly, without sudden jumps, and they cannot form reflections. In such soils, often, only high-frequency noise and various types of interference are present on the GPR profile, provided that the medium does not contain local objects. Local subsurface objects are defined as those whose linear dimensions are comparable to the wavelength of the GPR antenna pulse and whose physical characteristics differ from those of their encompassing medium [33]. They become a source of diffracted waves, whose kinematic and dynamic characteristics contain information about the properties of the subsurface. With this analysis, the attribute sections are obtained [36] to study the subsurface even in the absence of reflections from the layer boundaries. The backscattered wavefield attributes of the GPR profile are quantitative characteristics of a recording. Based on the wavefield attributes, subsurface parameters such as permittivity, humidity, electrical resistivity, conductivity, amplitude, frequency, phase, attenuation, etc., are calculated. In our processing, the attribute section Weighted average frequency was studied, which is the weighted average frequency [33] of the spectrum of the reflected signal, calculated by the formula,

$$F_{wa} = \frac{\sum A_j F_j}{\sum A_j} \tag{12}$$

where A_j is the spectral amplitude at the frequency F_j measured in MHz.

3. Results

Survey results are represented in the following order: velocity tests, bistatic scans, and common-offset single-fold measurements.

Velocity Test Results

The results of the two CMP velocity tests, shown in Table 3, demonstrate that the velocity of the EM wave undergoes a clear variation propagating through the plinth and then into the encompassing soil. The velocity of the soil outside the plinth also shows values consistent with the lithologies present. In addition, the clay component of the soils and their moisture did not result in a lowering in propagation velocity. In the transillumination test, first arrivals, distances and depths of travel of antennas were measured (Table 3). Knowing the horizontal separation between antennas and the travel time of the radar energy at each point, the velocity and relative dielectric permittivity can be calculated. The velocity measurements, at each point, thus represented as a function of antenna depth, show the trend in the tested material, any gradient or whether there are jumps (Figure 10). In this case, a dip between 70 cm and 100 cm depth is noted (measurements started from above), which corresponds to the interface between the plinth and the underlying soil on which it rests. The conducted velocity tests were used to explore the dielectric permittivity field across the media to be investigated and used them for BSEF analysis.

Table 3. Zero-offset transillumination test results.

Trace	Depth (cm)	Distance (cm)	Time (ns)	Velocity (cm/ns)	ϵ
1	0	800	65.27	12.26	5.98
2	10	800	66.86	11.97	6.28
3	20	800	67.44	11.86	6.39
4	30	800	66.86	11.97	6.28
5	40	800	66.86	11.97	6.28
6	50	800	66.86	11.97	6.28
7	60	800	67.44	11.86	6.39
8	70	800	69.2	11.56	6.73
9	80	800	85.04	9.41	10.15
10	90	800	96.18	8.32	12.99
11	100	800	97.94	8.17	13.47
12	110	800	97.94	8.17	13.47
13	120	800	98.52	8.12	13.63
14	130	800	99.11	8.07	13.79
15	140	800	99.7	8.02	13.96
16	150	800	100.87	7.93	14.29
17	160	800	105.87	7.56	15.74
18	170	800	104.46	7.66	15.32
19	180	800	105.35	7.59	15.59
20	190	800	104.9	7.63	15.45
21	200	800	105.66	7.57	15.68
22	210	800	105.75	7.56	15.71
23	220	800	106.57	7.51	15.95
24	230	800	106.92	7.48	16.05

During the GPR investigation of the soil under the foundation, four GPR profiles were recorded. Analysis of each of these B-scan GPR profiles produced no results and no visual evidence of amplitude anomalies that could be directly interpreted from radargrams and correlated with the target under investigation (Figures 11 and 12). In the two radargrams reproduced here (Figures 11 and 12), the areas (red boxes) with some of the most evident reflections have been highlighted, which are distributed along the entire length.

The average spectrum is very narrow, and the peaks with the highest spectral density are at 141 Mhz and 134 Mhz. The attenuation curves fluctuate within about ten decibels from zero. We would expect to find sharp, narrow, well-defined signal discontinuities at the beginning and end, or hyperbolas for the piles, but there is no evidence of it. It became necessary to combine these 4 profiles into a single set.

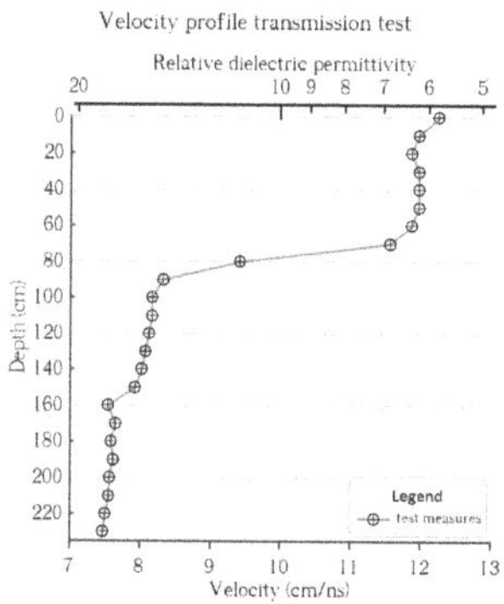

Figure 10. Graph of velocity and dielectric permittivity ϵ (in logarithmic scale) as a function of measurement depth.

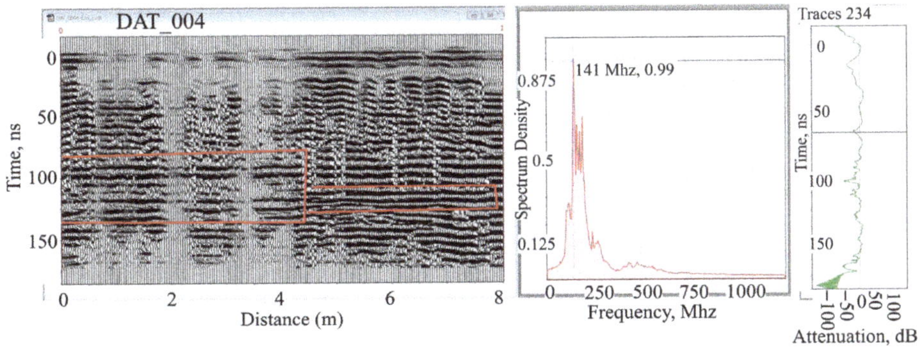

Figure 11. Processed radargram DAT_004, with relative attenuation curve of trace 234 and spectrum, where the spectral density peak at 141 MHz is highlighted.

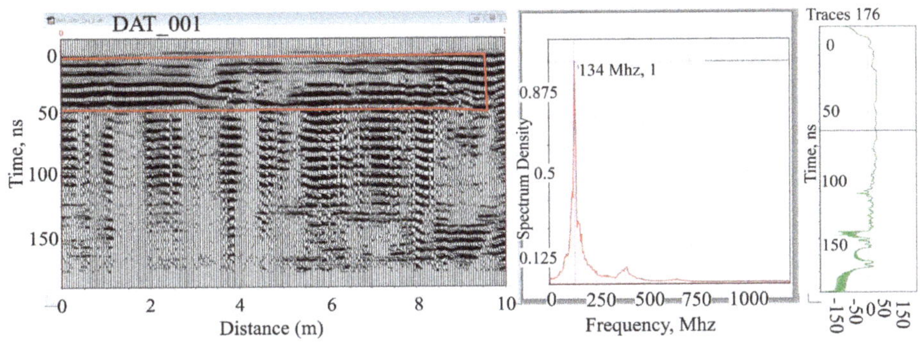

Figure 12. Processed radargram DAT_001, with relative attenuation curve of trace 176 and spectrum, where the spectral density peak at 134 MHz is highlighted.

In general, this can be achieved easily if the GPR profiles have the pulses propagating in one direction. In this particular case, however, the antenna radiation propagates in the same plane but in different directions as the antennas move along a perimeter. With the help of BSEF analysis, the sections of the four radargrams of the attribute "Weighted Average Frequency" were first combined to obtain a perimeter tomography. In GPR data processing, A-scan traces are sequentially plotted as B-scan profiles. Often, direct interpretation of each individual profile may be sufficient when the geology or stratigraphy of the subsurface is not complex. In complex cases, analysis of the spatial distribution of the amplitudes of the reflected waves is used, as shown on the left of Figure 13.

However, compared to the Cartesian plane, in this left section of the figure we have many radargrams (B-scans 1 to *n*) where the propagation direction of the electromagnetic signal is always the same. In our case, shown on the right, we have four radargrams whose signal propagates on the same plane in different directions. To combine these signals, it was first necessary to separate the radargrams into pairs with similar propagation directions but with opposite directions. For example, B-scans d1 and d3 are grouped into a first group, and B-scans d2 and d4 are grouped into a second group.

The next step consists of inverting the traces of one of the two radargrams along the t-axis of the B-scan so that the traces of the two radargrams in a group have the time direction pointing in the same direction. This inversion can also concern the x-axis if, for example, d1 was acquired towards the east and d3 towards the west. In fact, the opposite radargrams (d1–d3 or d2–d4), when taken in pairs, cover a positive half-space and a negative half-space with respect to the distance-time axes. Subsequently, the processing is analogous to the extraction of amplitude-averaged time slices from co-polar radargrams, where the same area is illuminated first by a signal polarized in one direction and then by a signal polarized in another direction perpendicular to it. In Figure 14, we can appreciate the results of the BSEF analysis of individual radargrams using the weighted average frequency attribute.

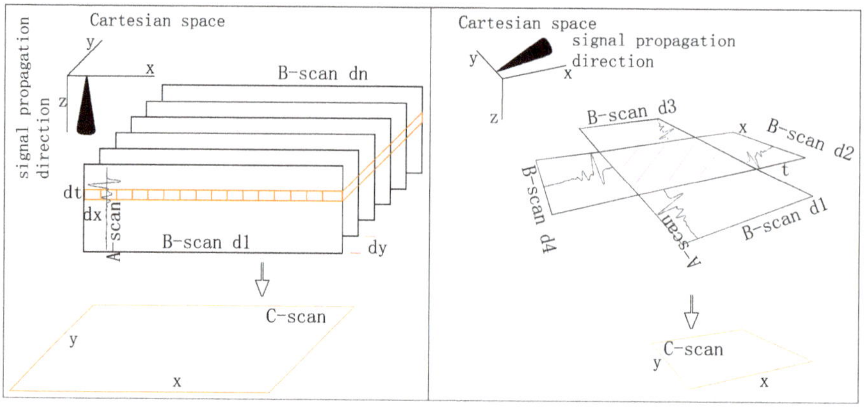

Figure 13. Representation of the construction of horizontal time slices from standard two-dimensional GPR profiles in a grid delineated by x and y values, where dy is the distance between profiles; dx is the distance along the profiles, where the reflected waves are averaged; and dt is the thickness of the slices, measured in nanoseconds (double travel time). The mean of the square of the amplitudes, in the dx, dt windows and located at x, y in the grid, is the parameter displayed in the amplitude anomaly maps.

Discontinuities in the low values (blue shades) can be detected from left to right of each. The combination performed with perimeter tomography allowed us to obtain a section on the XY plane. By appropriately tuning the color scale and the values in the color palette, we arrive at the following Figure 15. In this flat slice, four anomalies were identified, comparable in size to drilled piles with a diameter of 0.8 m. These anomalies are located at the corners of the foundation a little less than 1 m inward. In the following 3D image (Figure 16), the perimeter and horizontal tomographic section are placed in the ground and make the pile anomalies visible (circled with black dotted lines).

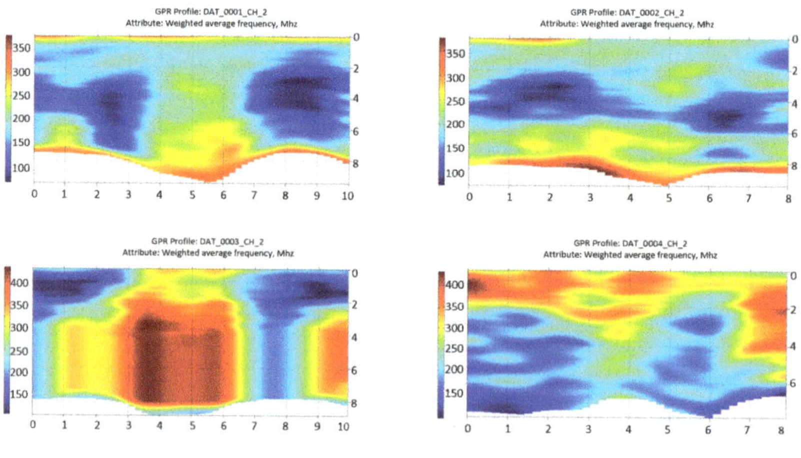

Figure 14. Result of BSEF analysis of each radargram: section of the attribute Weighted average frequency (in Mhz).

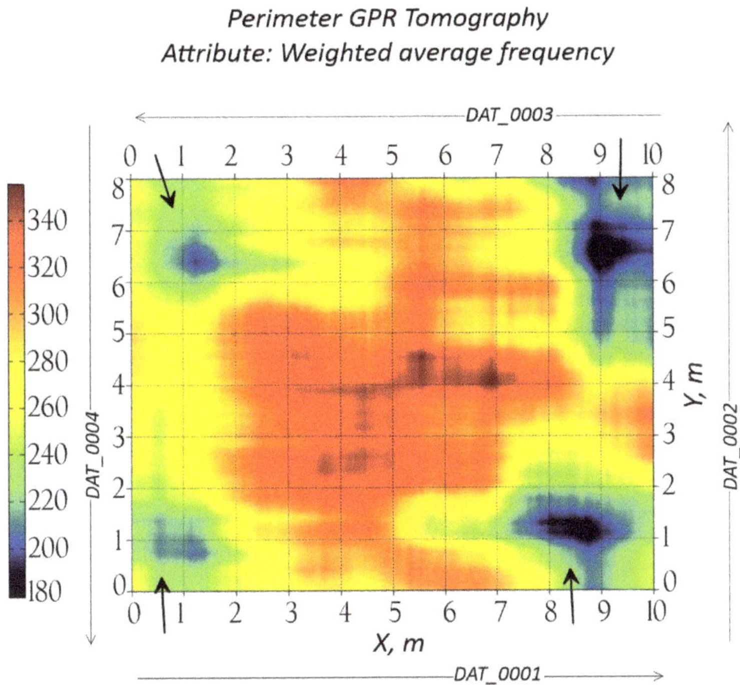

Figure 15. Result of BSEF analysis and combined section of the attribute Weighted average frequency (in Mhz); the arrows highlight the 4 anomalies.

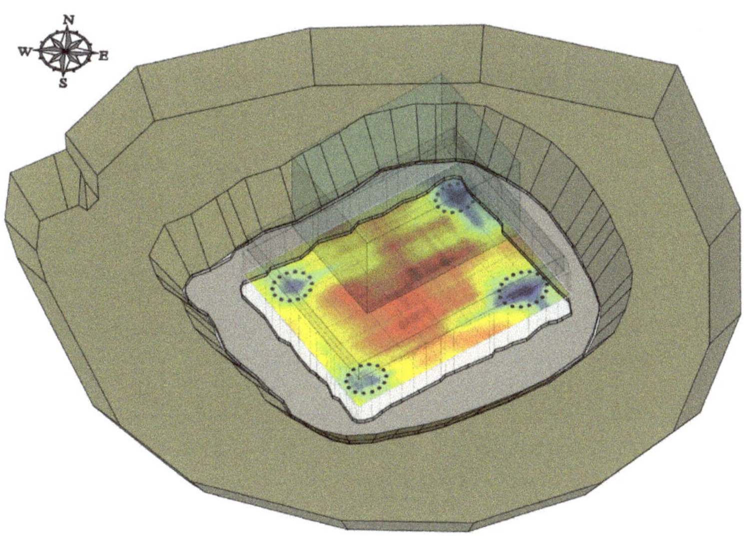

Figure 16. Overlay of the perimeter tomography to the 3D model of the structure investigated below the plinth. The anomalies of the piles have been circumscribed here. The color scale ranges from blue to red. The values with the lowest weighted average frequency are in blue, and the values whit the highest weighted average frequency are in red.

4. Discussion

We conducted a study on the search for bored piles below a large and deep foundation using GPR methods and a medium–low frequency of acquisition. Exploring a foundation plinth, and especially the piles at its base, with GPR, is difficult if approached from ground level (as explained below with the simulation). The geometry between the data acquisition plane, the position of the piles, and the aperture angle of the illumination cone is unfavorable.

The possibility of investigating the target is significantly reduced when the plinths are very thick and placed 4 m below ground level, as was addressed in this study. For an acquisition performed from ground level, the depth to be reached of several meters, as in this case, certainly falls within the potential of GPR, using low frequencies. However, the target to be investigated, the piles, would be located at a considerable distance from the GPR antennas, and both the radiation lobe and the pile are approximately parallel to each other. Furthermore, as the antenna moves toward the structure, the circular cross-section of a pile in the target area would be struck at varying angles. These increasingly smaller angles would be critical, and the returning echoes would not be picked up at the surface. To pursue a rigorous interpretation, we analyzed all aspects related to the survey and the data. The characteristics of the survey are as follows:

- The lowest operating frequency is about 134 Mhz (peak of the spectrum in Figures 11 and 12), and the wavelet pulse width is 0.27 m.
- Due to local physical properties, we know that concrete has low electrical conductivity compared to clayey silts, while velocity tests indicate acceptable dielectric permittivity contrasts because they range from 5.9 to 16.
- The soil investigated is dry and has low permeability, and no features are detected that could cause strong signal dispersion.
- The vertical resolution is about 0.14 m, and the investigation depth is about 8 m.
- The Tx-Rx separation distance is 68 cm.

Data characterizations for visual analysis include the following considerations:

- The first useful signal is measured as early as 15 ns, while the last one is measured at 175 ns.
- The attenuation of the traces is minimal (as evidenced in Figures 11 and 12 on the right); moreover, the reflections show an amplitude ranging from a minimum of −17 to a maximum of 2200.
- In the wavelet signals, one does not find polarity reversals leading to a pile interface, but one finds diffuse reversals along the entire distance axis.
- Unfortunately no hyperbolic features emerge; for direct identification of the pile, one might expect a hyperbola since it is assimilated to a cylinder.
- No point reflectors are detected.
- Linear features emerge (the wavelet offset at a certain time is zero) that might indicate a flat interface or dipping across the entire radargram. We would expect a flat interface only at the progressives where the pile is. Also, there is no obvious reflector below which other reflectors emerge.

Since the propagation and response of electromagnetic waves near the pile and near the plinth is really complex, for the correct interpretation of the radargrams, it was essential to prepare a geophysical model (Figure 17). This is to have a comparison of what might happen to signals recorded with an acquisition conducted from the surface.

In the absence of piles (left of figure), electromagnetic waves are reflected only at the interfaces of layers with different impedance. GPR traces are easily interpreted.

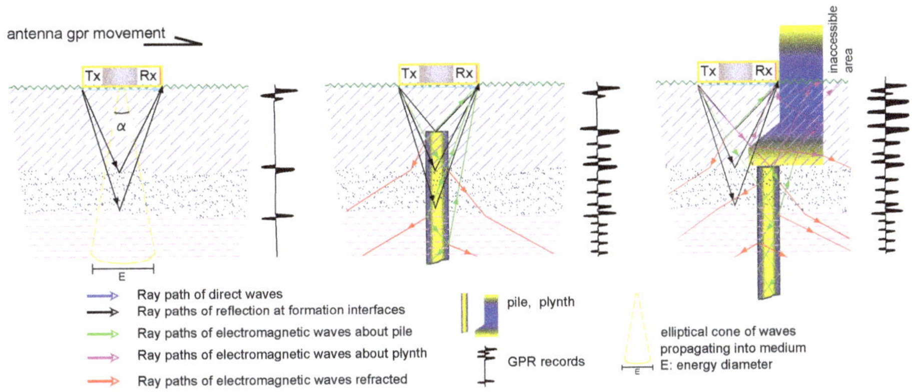

Figure 17. Geophysical modeling of electromagnetic propagation, in various layers of the subsoil (colored hatch), in the presence of foundation piles and plinths using a GPR prospection conducted from the surface.

In the presence of a pile (center of figure), reflections are always formed in the layer interfaces and in more complex reflections. They are both transmitted, refracted and guided waves in the pile and between the pile and the soil layer. Therefore, responses with relatively stable frequency and amplitude are formed in the GPR trace. Still, it is possible to distinguish reflections in the pile top.

In the presence of a plinth and the pile (right part of the figure), in addition to the formation of the waves already described for the pile, additional waves will be formed from refractions and reflections and additional guided waves in the plinth. The energy distribution of electromagnetic waves will now be changed more, which will affect the response of the GPR. In particular, the amplitude becomes weaker (response characteristics as at left) or stronger (response characteristics as at center). The frequency also becomes lower or higher, but for the oscillating waves appearing, we hypothesize that this may be due to the guided wave phenomenon generated by electromagnetic waves within or between the pile, plinth and pylon.

Instantaneous attributes, shown in Figure 18, were extracted for the acquired radargrams with the Hilbert transform: at the top the positive envelope of the traces (instantaneous amplitude) and at the bottom the instantaneous phase. In the instantaneous amplitude section, the most intense envelope traces on the positive axis (black contour) range from a minimum of 80 dB to a maximum of 118 dB, with widths from a minimum of 10 ns to a maximum of 37 ns.

In the phase section, a weak anomaly is identified at about 2 m and 25 ns and a second, but lateral, variation from 7 m to 9 m and from about 0 ns to 100 ns (white boxes). Compared with the BSEF analysis of the weighted average frequency attribute of the individual radargrams (Figure 14), here one would expect informative anomalies on both the instantaneous amplitude section, and the phase section indicating the presence of the piles. Differences in electromagnetic properties between the subsurface, the pile, and the plinth should cause changes in the GPR response. This response should manifest itself in changes in waveform, amplitude, frequency, phase, and related attributes. However, unambiguous and clearly evident variations are visible only in the frequency attribute sections, particularly in the weighted average frequency (Figure 14). To test the applicability of the method, it was necessary to conduct a simulation with scans taken from the surface to verify the response according to the geometric variables (defined in Figure 1) and the parameters of the media involved. Thus a model (Figure 19 on the left) was constructed where on one side there was only one pile (in blue without the overlying basement) and on the other side the pile with the basement. The depth of the pile head is 4 m. The simulation led to the following result (Figure 19 on the center and right): when the values of the geometry and permittivities are identical to those of the known real case, by prospecting from the surface, reflections are obtained. In the synthetic radargram at 300 Mhz, a hyperbola (progressive 2 m) with numerous multiples is formed in the center at the single blue pile; however, the hyperbola is sharp. On the pile side with the basement (progressive 7.5 m), flat reflections are obtained in the first 100 ns with crossed multiples and numerous reverberations on the edges of the structure in the first 90 ns.

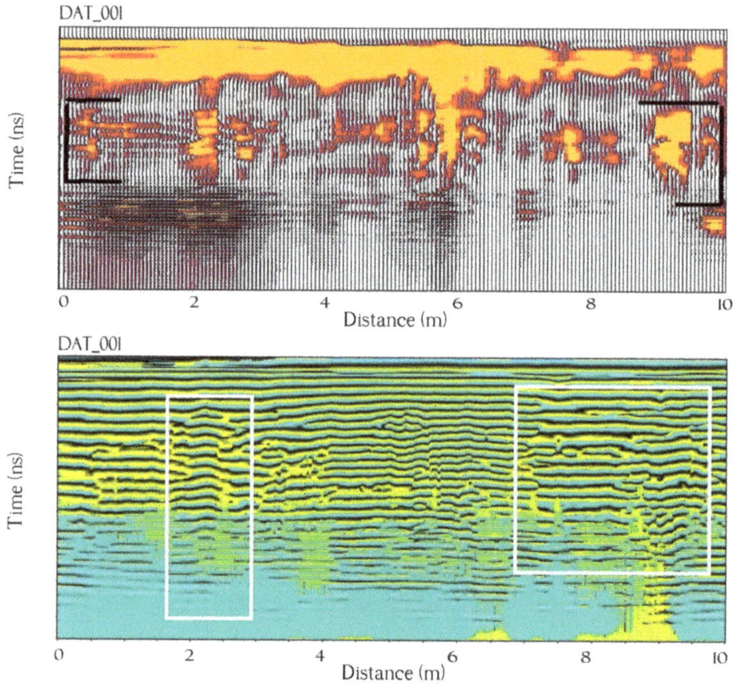

Figure 18. Attribute sections of the instantaneous amplitude and phase of the radargram DAT_001. The boxes highlight the areas of anomaly.

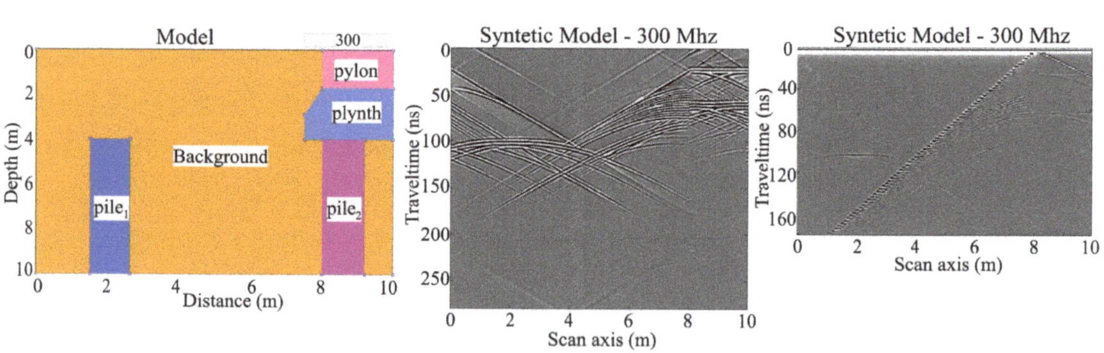

Figure 19. (**Left**) is the initial model; the colors indicate different permittivities and electrical conductivities; in the (**center**) there is the simulation with the TM-mode, Finite Difference, time-domain method [38]; on the (**right**), the simulation with doubled dielectric permittivity ratios and method Split-step 2D [39].

Changes in polarity and amplitude alone do not help to understand the onset of the pile and its development. A strong disturbing element is cross reflections related to guided waves throughout the radargram.

In the synthetic radargram at 300 Mhz, right, the dielectric permittivity ratios between the media have been doubled while leaving other geometric values unchanged. A more defined hyperbola (progressive 2 m) with a single multiple is formed at the single blue pile. On the side of the pile with the basement (progressive 7.5 m), we now obtain two plane reflections in the first 100 ns with a single multiple. On the edges of the structure, however, reverberations always emerge in the first 90 ns. Again, a strong disturbing element here is the cross reflections associated with guided waves throughout the radargram. Changes in polarity and amplitude alone could now only help to discriminate pile beginnings if the construction depths are known a priori. With the scans that were possible on the opposite faces of the excavations, the data are acquired perpendicular to the pile axis, and they do not contain guided wave noise, reverberations and multiples.

Extraction of weighted average frequency attributes and combination in perimeter tomography eliminates any interpretative ambiguity. Ultimately, scans from the surface can provide the presence of the pile even at depths of 4 m if it is without the basement. In the presence of the plinth in the GPR response signals, it can be very complex to extract pile information. The limitation of the automated BSEF analysis method is that the real part of the complex permittivity ϵ' cannot be calculated from the parameters of the diffraction component of the wave field if this component is absent on the GPR profile. The reasons for the absence of a diffraction component may be one of the following:

- Absence of local inhomogeneities whose linear dimensions are comparable to the wavelength of the GPR probing pulse.
- Large step of GPR profiling (aliasing).

In natural and artificial strata, there are always local objects of various scales and natures that are the source of diffracted waves, so even for small permittivity contrasts (e.g., when the difference of $\epsilon < 3$), this method can be used because it counts the presence of local objects that create diffracted waves.

5. Conclusions

In this study, a new approach for GPR detection of large-diameter piles below a deep, massive foundation is presented. The approach involves performing GPR tests on the exposed faces of the perimeter excavation carried out up to 6 m depth, processing the data and, in particular, with BSEF analysis and testing the results by comparing simulated synthetic radargrams with acquisitions from the surface. GPR tests in the field were conducted to evaluate the foundation problem on piles. The pile depth information provided by the GPR evaluation is useful and important in resolving the engineering dispute about the actual number of piles present at the base of the plinth.

The transillumination and CMP acquisition methods were used to estimate the wave velocity in the investigated media. Based on the EM wave velocity, the investigation depths were estimated. For good penetration of the EM signal at great depth, the antenna with low center frequency was used, which is essential to provide high-energy waves in order to detect deep objects. This study presents the application of combined and sequential use in transmission and reception of 300 and 200 Mhz antennas. Single radargrams, after post-processing and extraction of instantaneous amplitude and phase, show anomalies, which cannot be unambiguously interpreted, leaving an ambiguous solution if no a priori knowledge is available. Similarly, simulation tests generate radargrams with anomalies affected by a minimum of guided wave disturbance after doubling the ratios of dielectric permittivity values. However, correct interpretation of the data always requires a priori knowledge of the geometries involved. Conversely, the scans performed below the plinth on which they rest produced the anomalies, but combining the sections in the form of "perimeter tomography". In fact, the scans are performed on the same plane but in different directions.

The BSEF analysis of the radargrams and the study of the weighted average frequency attribute sections allowed the detection of pile anomalies already in the single radargrams. These, in turn, combined in the form of perimeter tomography, clearly showed the presence of four piles below the plinth. The technique of backscattered field analysis made it possible to interpret the data unambiguously, leading to an accurate answer as to the number of existing engineering structures. The assumptions adopted in the analysis made the computational process much easier and more accurate, thanks to the velocity tests performed. GPR prospecting in conjunction with advanced data processing proved to be an effective and reliable technique in providing accurate information on the presence, number and location of large-diameter piles below a massive plinth. The results show a favorable accuracy that emerged with a robust analysis of the data with respect to the chosen detection geometry. Further studies will expand this work and will be aimed at investigating the variation in the minimum and maximum angle between the target and the radiation lobe of the antennas, with respect to which the presence of the piles could still be probed, moving the antennas along the vertical axis.

Funding: This research received no external funding.

Data Availability Statement: No data are associated with this research.

Acknowledgments: I thank Roman R. Denisov for his technical contribution, his computer support and advice, and his checks on data processing. I thank Paolo Spallacci for the opportunity provided to perform the georadar surveys.

Conflicts of Interest: The author declares no conflicts of interest.

Abbreviations

The following abbreviations are used in this manuscript:

BSEF	Backscattering Electromagnetic Field
CMP	Common Mid Point
EM	Electromagnetic
GPR	Ground-Penetrating Radar
RDP	Relative Dielectric Pemittivity
Rx	Receiver
SNR	Signal-to-Noise Ratio
Tx	Transmitter
UWB	Ultra-Wide Band

References

1. Paoletti, V.; D'Antonio, D.; De Natale, G.; Troise, C.; Nappi, R. Large-Depth Ground-Penetrating Radar for Investigating Active Faults: The Case of the 2017 Casamicciola Fault System, Ischia Island (Italy). *Appl. Sci.* **2024**, *14*, 6460. [CrossRef]
2. Nappi, R.; Paoletti, V.; D'Antonio, D.; Soldovieri, F.; Capozzoli, L.; Ludeno, G.; Porfido, S.; Michetti, A.M. Joint Interpretation of Geophysical Results and Geological Observations for Detecting Buried Active Faults: The Case of the "Il Lago" Plain (Pettoranello Del Molise, Italy). *Remote Sens.* **2021**, *13*, 1555. [CrossRef]
3. Alani, A.M.; Morteza, A.; Gokhan, K. Applications of Ground Penetrating Radar (GPR) in Bridge Deck Monitoring and Assessment. *J. Appl. Geophys.* **2013**, *97*, 45–54. [CrossRef]
4. Soldovieri, F.; Persico, F.; Utsi, E.; Utsi, V. The Application of Inverse Scattering Techniques with Ground Penetrating Radar to the Problem of Rebar Location in Concrete. *NDT Int.* **2006**, *39*, 602–607. [CrossRef]

5. D'Antonio, D.; Di Stefano, S.; Faustoferri, A.; De Leo, A. Opi (AQ) Ponte Romano Sul Sangro. *Quad. Archeol. D'Abruzzo* **2014**, *5*, 128–134.
6. Lai, W.L.; Kou, S.C.; Tsang, W.F. Characterization of Concrete Properties from Dielectric Properties Using Ground Penetrating Radar. *Cem. Concr. Res.* **2009**, *39*, 687–695. [CrossRef]
7. Tosti, F.; Slob, E. Determination, by Using GPR of the Volumetric Water Content in Structures, Substructures, Foundations and Soil, Civil Engineering Applications of Ground Penetrating Radar. In *Springer Transaction in Civil and Environmental Engineering*; Springer: Cham, Switzerland, 2015.
8. Morozov, P.; Morozov, F.; Lazarev, M.; Popov, A. Characterization of Antenna Radiation Pattern and Penetration Depth in Ground Penetrating Radar Field Missions. *Remote Sens.* **2023**, *15*, 5452. [CrossRef]
9. Zhou, D.; Zhu, H. Application of Ground Penetrating Radar in Detecting Deeply Embedded Reinforcing Bars in Pile Foundation. *Hindawi Adv. Civ. Eng.* **2021**, *2021*, 4813415. [CrossRef]
10. Diamanti, N.; Annan, P. Characterizing the energy distribution around GPR antennas. *J. Appl. Geophys.* **2013**, *99*, 83–90. [CrossRef]
11. Bradford, J.H. Frequency Dependent Attenuation of GPR Data as a Tool for Material Property Characterization: A Review and New Developments. In Proceedings of the 6th International Workshop on Advanced GPR, Aachen, Germany, 22–24 June 2011; pp. 1–4.
12. Grandjean, G.; Gourry, J.C.; Bitri, A. Evaluation of GPR Technique for Civil-Engineering Applications: Study on a Test Site. *Appl. Geophys.* **2000**, *45*, 141–156. [CrossRef]
13. Annan, A.P. *Ground Penetrating Radar Principles, Procedures and Applications*, Sensor and Software ed.; Sensors & Software Inc.: New York, NY, USA, 2003; Volume 278.
14. Davis, J.L.; Annan, A.P. Ground Penetrating Radar for High-Resolution Mapping of Soil and Rock Stratigraphy. *Geophysics* **1989**, *37*, 531–551.
15. Annan, A.P.; Cosway, S.W. Symposium on the Application of Geophysics to Engineering and Environmental Problems. In *Ground Penetrating Radar Survey Design*; Environmental & Engineering Geophysical Society: New York, NY, USA, 1992.
16. Daniels, D.J. *Ground Penetration Radar*, 2nd ed.; Number 15 in IEE Radar, Sonar and Navigation; Institut of Engineering & Technology: London, UK, 2004.
17. Burger, H.R. *Exploration Geophysics of the Shallow Subsurface*, Englewood Cliffs ed.; Print Hall: Hackensack, NJ, USA, 1992.
18. Conyers, L.B. *Ground-Penetrating Radar for Archaeology*; AltaMira Press: Walnut Creek, CA, USA, 2004.
19. Skolnik, M.I. *Introduction to Radar System*, 3rd ed.; MacGraw-Hill: New York, NY, USA, 2001; Volume 278, pp. 3–5.
20. Wu, Y.; Shen, F.; Liu, R. An ultra-wideband antenna with low dispersion for ground penetrating radar system. *IEEE Sens. J.* **2021**, *21*, 15171–17179. [CrossRef]
21. Lavoue, F. 2D Full Waveform Inversion of Ground Penetrating Radar Data: Towards Multiparameter Imaging from Surface Data. Ph.D. Thesis, Université de Grenoble, Sigillum Universitatis Gratianopolitane, Saint-Martin-d'Hères, France, 2014.
22. Klysz, G.; Balayssac, J.P. Determination of Volumetric Water Content of Concrete Using Ground-Penetrating Radar. *Cem. Concr. Res.* **2007**, *37*, 1164–1171. [CrossRef]
23. Moysey, S.; Knight, R.J. Modeling the field-scale relationship between dielectric constant and water content in heterogeneous systems. *Water Resour. Res.* **2004**, *40*, W03510. [CrossRef]
24. Jonscher, A.K. Dielectric relaxation in solids. *J. Phys. D Appl. Phys.* **1999**, *32*, 57–70. [CrossRef]
25. Olhoeft, G.R. Maximizing the Information Return from Ground Penetrating Radar. *J. Appl. Geophys.* **2000**, *43*, 175–187. [CrossRef]
26. Zhao, W.; Forte, E.; Pipan, M. Texture attribute analysis of GPR data for archaeological prospection. *Pure Appl. Geophys.* **2016**, *173*, 2737–2751. [CrossRef]
27. Khwanmuang, W.; Udphuay, S. Ground-penetrating radar attribute analysis for visualization of subsurface archaeological structures. *Lead Edge* **2011**, *31*, 946–949. [CrossRef]
28. Tronicke, J.; Böniger, U. GPR attribute analysis: There is more than amplitudes. *First Break* **2013**, *31*, 103–108. [CrossRef]
29. Giannopoulos, A. GprMax2D V 1.5 User's Manual. 2002. Available online: https://www.gprmax.com/ (accessed on 10 October 2024)
30. Persico, R. *Introduction to Ground Penetrating Radar: Inverse Scattering and Data Processing*; Wiley-IEEE Press: Hoboken, NJ, USA, 2014.
31. Radzevicius, S.; Daniels, J.J. Ground Penetrating Radar Polarization and Scattering from Cylinders. *J. Appl. Geophys.* **2000**, *45*, 111–125. [CrossRef]
32. Langley, K.A.; Hamrain, S.E. Use of C-band Ground Penetrating Radar to Determine Backscatter Sources within Glaciers. *IEEE Trans. Geosci. Remote Sens.* **2005**, *45*, 1236–1246. [CrossRef]
33. Denisov, R.R. Processing of georadary data in automatic mode. *Sci. Tech. J. Geophys.* **2010**, *4*, 76–80.
34. Ulaby, F.T.; Moore, R.K.; Fung, A.K. Microwave Remote Sensing Active And Passive Volume Ii Radar Remote Sensing and Surface Scattering and Emission Theory. In *Radar Remote Sensing and Surface Scattering And Emission Theory*; Addison-Wesley Publishing Company: Boston, MA, USA, 1982; Volume II.

35. Toropainen, A.P. Measurement of the Properties of Granular Materials by Microwave Backscattering. *J. Microw. Power Electromagn. Energy* **1995**, *30*, 240–245. [CrossRef]
36. González, R.C.; Woods, E.C.; Eddins, S.L. *Digital Image Processing Using MATLAB*; Pearson Prentice Hall: Saddle River, NJ, USA, 2003.
37. Ivan Levyant, V.B.; Petrov Pankratov, S.A. Seismic Exploration Technologies. In *Research of the Characteristics of Longitudinal and Exchange Waves of Retroactive Response Scattering from the Zones of the Cracked Collector*; Moscow Institute of Physics and Technology: Moscow, Russia, 2009; pp. 3–11.
38. Irving, J.; Knight, R. Numerical modeling of ground-penetrating radar in 2-D using MATLAB. *Comput. Geosci.* **2006**, *32*, 1247–1258. [CrossRef]
39. Bitri, A.; Grandjean, G. Frequency–wavenumber modelling and migration of 2D GPR data in moderately heterogeneous dispersive media. *Geophys. Prospect.* **1998**, *46*, 287–301. [CrossRef]

remote sensing

MDPI

Article

Weather-Resilient Localizing Ground-Penetrating Radar via Adaptive Spatio-Temporal Mask Alignment

Yuwei Chen [1], Beizhen Bi [1,*], Pengyu Zhang [1], Liang Shen [2], Chaojian Chen [3], Xiaotao Huang [1] and Tian Jin [1]

1 College of Electronic Science and Technology, National University of Defense Technology, Changsha 410073, China; chenyuwei@nudt.edu.cn (Y.C.); zhangpengyu19@nudt.edu.cn (P.Z.); xthuang@nudt.edu.cn (X.H.); tianjin@nudt.edu.cn (T.J.)
2 Test Center, National University of Defense Technology, Xi'an 710106, China; shenliang16@nudt.edu.cn
3 School of Geosciences and Info-Physics of Central South University, Changsha 410012, China; chenchaojian@csu.edu.cn
* Correspondence: bibeizhen@nudt.edu.cn

Abstract

Localizing ground-penetrating radar (LGPR) benefits from deep subsurface coupling, ensuring robustness against surface variations and adverse weather. While LGPR is widely recognized as the complement of existing vehicle localization methods, its reliance on prior maps introduces significant challenges. Channel misalignment during traversal positioning and time-dimension distortion caused by non-uniform platform motion degrade matching accuracy. Furthermore, rain and snow conditions induce subsurface water-content variations that distort ground-penetrating radar (GPR) echoes, further complicating the localization process. To address these issues, we propose a weather-resilient adaptive spatio-temporal mask alignment algorithm for LGPR. The method employs adaptive alignment and dynamic time warping (DTW) strategies to sequentially resolve channel and time-dimension misalignments in GPR sequences, followed by calibration of GPR query sequences. Moreover, a multi-level discrete wavelet transform (MDWT) module enhances low-frequency GPR features while adaptive alignment along the channel dimension refines the signals and significantly improves localization accuracy under rain or snow. Additionally, a local matching DTW algorithm is introduced to perform robust temporal image-sequence alignment. Extensive experiments were conducted on both public LGPR datasets: GROUNDED and self-collected data covering five challenging scenarios. The results demonstrate superior localization accuracy and robustness compared to existing methods.

Keywords: ground-penetrating radar (GPR) image processing; dynamic time warping; multi-level discrete wavelet transform

1. Introduction

The concept of localizing ground-penetrating radar (LGPR) was first introduced by Cornick et al. in 2016 [1]. Unlike vision-based systems, LGPR operates reliably under varying ground conditions and provides rich subsurface reflections, enabling accurate localization without dependence on stable visual environments credited on GPR's performance [2,3]. As a result, this approach has attracted significant research attention [4–9]. The core principle of LGPR involves matching real-time GPR profiles with georeferenced profiles in a prebuilt map library to identify the best correspondence, thereby achieving re-localization on a prior map. This process can be viewed as an image retrieval task [10].

Consequently, positioning accuracy critically depends on high-precision registration between the prior map and the revisited GPR echoes images.

Since LGPR relies on image matching using priori maps, numerous studies have explored registration techniques for GPR A-scan, B-scan, and sequence images. For instance, ref. [11] proposed localization using GPR A-scans within a known underground map. Further, ref. [12] investigated faster R-CNN-based feature extraction from GPR B-scans, while [13] introduced a phase-symmetric feature descriptor to enhance GPR image registration by leveraging hyperbolic signatures. Nevertheless, ref. [7] demonstrated that single GPR scans from distinct scenes exhibit high similarity, risking false matches between different locations. In computer vision (CV) and LiDAR localization, this issue is typically mitigated via feature vector aggregation [14,15] or Siamese networks [16,17]. However, due to GPR's limited vertical detection range [1,8], direct adoption of these methods degrades LGPR accuracy. To address this, refs. [7,18,19] employed GPR sequence images to improve LGPR precision.

However, LGPR exhibits fundamental distinctions from conventional wheel odometry, inertial navigation, and vision-based systems through its utilization of temporally acquired subsurface reflections instead of real-time measurements [1,10,20]. Specifically, in LGPR applications, the temporal acquisition mode is typically adopted, with subsequent distance normalization enabling the computation of ground-penetrating radar mile positions based on standard channel spacing. This methodology, however, introduces inherent sensitivity to platform movement: velocity fluctuations not only distort the dimensional accuracy of GPR reflection profiles but also compound ranging inaccuracies. While the GROUNDED dataset [10,21] addresses temporal synchronization through Savitzky–Golay filtering of multi-sensor timestamps, persistent misalignments occur in high-frame-rate GPR acquisitions owing to constrained detection ranges, resulting in compromised feature correspondence and degraded positioning performance. In addition, LGPR imposes stringent requirements on vehicle trajectory accuracy. As shown in Figure 1, channel deviations in the GPR array antenna, which occur between prior map collection and the online positioning phase, can alter the target echoes acquired from the same road segment. This variability leads to positioning inaccuracies caused by channel misalignment in the vehicle's travel path. Lane deviations cause misaligned echo images at the same forward position, leading to incorrect matches, akin to viewpoint variations in CV positioning [22]. While LiDAR/CV mitigates such issues via rotation-invariant descriptors (e.g., spherical projection [23,24]), GPR's sparse feature points limit their effectiveness. While the standard DTW method can address some of these issues, it fails to resolve the channel dimension deviation problem inherent in GPR sensors.

To address the critical challenges in LGPR, including GPR data distorted induced by variable carrier velocity and lane offset, positioning mismatches caused by groundwater content variations during inclement weather (e.g., rain/snow) and computational complexity in GPR spatio-temporal feature extraction are investigated. We propose a novel framework employing adaptive alignment and dynamic time warping (DTW) strategies. Our approach generates bias-aware masks to sequentially rectify both channel and temporal misalignments in GPR sequences, followed by calibration of GPR maps. The methodology enables precise comparison between local features in query images and their semantically corresponding representations in reference maps. The system operates through three key phases. The first is that the generated correction masks adjust the GPR query sequences. Notably, our channel dimension calibration framework incorporates an adaptive parameter κ to generate mathematically constrained dimension masks, providing a complete solution to the systematic deviations depicted in Figure 1. The second group of low-frequency channel-dimensional features are extracted from both the map and query GPR time series.

The third adaptive DTW performs spatio alignment of the processed sequences. This integrated approach demonstrates significant improvements in localization accuracy while enhancing the real-time capability of LGPR systems. We present a weather-resilient adaptive spatio-temporal mask alignment algorithm for LGPR, a novel framework with three key contributions:

- We develop an adaptive DTW algorithm that generates a map deviation-aware mask to address channel and temporal dimensional deformations in LGPR acquisition.
- We introduce a channel alignment module employing MDWT-based low-frequency signal enhancement, coupled with time-dimensional DTW local matching, to overcome challenges from both lane deviations and diverse weather conditions.
- Extensive evaluations on both the public GROUNDED dataset and self-collected data demonstrate the superior performance of the proposed method. The framework effectively eliminates positioning mismatches caused by lane deviations while significantly improving robustness under precipitation (rain/snow), surpassing state-of-the-art methods. Visualization results further validate its capability to mitigate channel misalignments and temporal feature distortions, highlighting its exceptional performance.

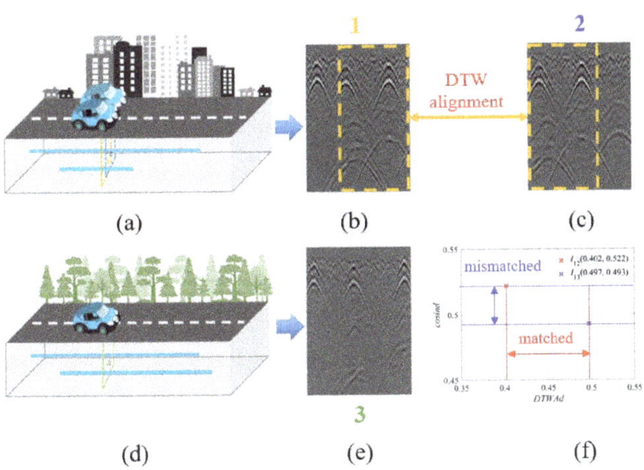

Figure 1. Illustration of lane deviation effects on GPR echo signals. (**a**) Scene of lane deviation during two GPR mapping campaigns. (**b**,**c**) Underground echo images collected from the two deviated lanes in (**a**), (**b**) and (**c**) correspond to conditions 1 and 2 in (**a**), respectively. (**d**) GPR mapping scene at a location distinct from (**a**). (**e**) GPR echo image acquired at (**d**), (**e**) corresponds to condition 3 in (**a**). Despite the spatial proximity of (**b**,**c**), the echo patterns of (**b**,**e**) exhibit higher similarity. (**f**) Quantitative comparison of image distances between GPR echoes: cosine (cosine distance) and align (alignment-based distance).

2. Preliminary

Since our proposed method builds upon DTW and MDWT, this section briefly reviews DTW and MDWT.

2.1. DTW

The dynamic time warping (DTW) algorithm employs a recursive approach to compute an optimal warping path for aligning two temporal sequences, where the path length quantifies their similarity. Initially developed for speech recognition [25,26], DTW has become a standard sequence alignment tool in this field due to its ability to synchronize time series with variable acquisition times and speeds. Subsequently, it has been successfully extended to diverse domains, including data mining [27] and image processing [28], establishing itself as a fundamental distance metric for time series analysis [29]. In visual location recognition, DTW was first adapted to align spatially distributed local features while measuring inter-image distances. For instance, ref. [30] utilized DTW to compare video sequences, effectively addressing temporal scale variations. Inspired by these advances, we propose the first application of DTW to LGPR. Specifically, we perform dynamic time regularization between GPR prior maps and query sequences to mitigate spatio-temporal misalignments. The core computation process is described as follows:

Given two GPR sequences $\mathbf{M} = \{\mathbf{m}_1, \mathbf{m}_2, ..., \mathbf{m}_m\}$ and $\mathbf{Q} = \{\mathbf{q}_1, \mathbf{q}_2, ..., \mathbf{q}_n\}$, the DTW distance is computed through the following three-step process. Firstly, we compute the pairwise Euclid distance matrix $\boldsymbol{\mathcal{D}}$ (or cosine distance), where each element $d_{i,j}$ represents the distance between frame $\mathbf{m}_i$ and $\mathbf{q}_j$:

$$\boldsymbol{\mathcal{D}} = \begin{bmatrix} d_{1,1} & d_{1,2} & \cdots & d_{1,m} \\ d_{2,1} & d_{2,2} & \cdots & d_{2,m} \\ \vdots & \vdots & \ddots & \vdots \\ d_{n,1} & d_{n,2} & \cdots & d_{n,m} \end{bmatrix}, \quad d_{i,j} = \|\mathbf{m}_i - \mathbf{q}_j\|_2. \tag{1}$$

Secondly, using dynamic programming, we compute the accumulated cost matrix $\mathbf{S}$, where each entry $s_{i,j}$ stores the minimum cumulative distance up to (i, j):

$$s_{i,j} = \begin{cases} d_{i,j} & \text{if } (i,j) = (1,1) \\ d_{i,j} + s_{i,j-1} & \text{if } i = 1,\ j \in [2,s] \text{ (H)} \\ d_{i,j} + s_{i-1,j} & \text{if } i \in [2,s],\ j = 1 \text{ (V)} \\ \min \begin{cases} \alpha \cdot d_{i,j} + s_{i-1,j-1} \\ d_{i,j} + s_{i-1,j} \\ d_{i,j} + s_{i,j-1} \end{cases} & \text{if } i,j \in [2,s] \text{ (D)} \end{cases}. \tag{2}$$

The fundamental recursive formulation of DTW computes the minimal cumulative distance $d_{i,j}$ between the initial alignment point $(0,0)$ and current position (i,j), in which $d_{\mathbf{m}_i,\mathbf{q}_j}$ denotes the local distance metric between elements $\mathbf{m}_i$ and $\mathbf{q}_j$ of the respective time series. The minimization operation evaluates three distinct warping path transitions:

- Vertical movement V: progression in the reference sequence;
- Horizontal movement H: advancement in the query sequence;
- Diagonal movement D: simultaneous progression in both sequences.

Thirdly, we find an optimal alignment path in the matrix $\boldsymbol{\mathcal{D}}$ by minimizing the overall distance between sequence $\mathbf{M}$ and $\mathbf{Q}$. The optimal alignment between the two sequences is indicated by the elements of the optimal alignment path $\boldsymbol{W}$.

$$W = \{w_1, w_2, ..., w_k\}. \tag{3}$$

W is derived by back-tracking from (i,j) to $(1,1)$, selecting the direction with the smallest accumulated cost at each step, where K is the length of the warping path ($mas(m,n) <= K < m + l - 1$). The final sequence similarity is provided by the normalized cumulative cost:

$$d'_{\mathbf{m}_i,\mathbf{q}_j} = \frac{\mathbf{S}_{i,j}}{K}, \tag{4}$$

where K is the length of the warping path.

This method ensures precise spatio-temporal alignment for visual place recognition under appearance changes. Emerging research evidence confirms DTW's successful application in GPR data analysis [31], which substantiates the validity of our DTW-based approach to LGPR challenges.

2.2. MDWT

During GPR map and query acquisitions, matching images are typically captured at different times under varying weather conditions. Significantly, rainy and snowy environments induce shallow subsurface humidity variations that subsequently cause dielectric constant fluctuations. These variations critically influence electromagnetic wave time delays, thereby generating substantial image discrepancies. Consequently, to achieve robust positioning performance in adverse weather, more stable information extraction becomes essential for LGPR systems to effectively mitigate matching errors. Importantly, as demonstrated in [5,32], feature points from low-frequency B-scan components exhibit remarkable stability against moisture variations, clearly indicating that low-frequency-based registration offers superior reliability for LGPR applications. Accordingly, prior to channel-dimension alignment, we strategically employ multi-level DWT to enhance low-frequency components, as illustrated in Figure 2, which fundamentally improves weather-resistant matching capability. Consequently, we employ frequency-domain transformation to process GPR images, specifically utilizing the multi-level discrete wavelet transform (MDWT) with Symlet 4 as the wavelet basis to decompose the original image. Notably, this approach systematically enhances low-frequency information while simultaneously attenuating high-frequency components, thereby improving feature stability against environmental variations. The transformed representation effectively preserves structurally significant patterns while suppressing weather-induced noise components that predominantly reside in higher frequency bands. The MDWT operation is formally expressed as follows:

$$\mathbf{X}_{MDWT} = \mathcal{MDWT}\{\mathbf{x}\}, \tag{5}$$

where $\mathcal{MDWT}$ denotes the MDWT operator. This approach systematically enhances low-frequency components while attenuating high-frequency noise, improving environmental robustness. The multi-level enhancement process proceeds as follows:

$$\begin{cases} \mathbf{x}_{LL_1}, \{\mathbf{x}_{LH_1}, \mathbf{x}_{HL_1}, \mathbf{x}_{HH_1}\} = \mathcal{W}_1(\mathbf{x}) & \text{(Level 1)} \\ \mathbf{x}_{LL_2}, \{\mathbf{x}_{LH_2}, \mathbf{x}_{HL_2}, \mathbf{x}_{HH_2}\} = \mathcal{W}_2(\mathbf{x}_{LL_1}) & \text{(Level 2)}. \\ \mathbf{x}_{LL_3}, \{\mathbf{x}_{LH_3}, \mathbf{x}_{HL_3}, \mathbf{x}_{HH_3}\} = \mathcal{W}_3(\mathbf{x}_{LL_2}) & \text{(Level 3)} \end{cases} \tag{6}$$

The enhanced reconstruction is computed through weighted synthesis:

$$\mathbf{x}_{\text{MDWT}} = \sum_{k=1}^{3} \alpha_k \mathcal{W}_k^{-1}(\mathbf{x}_{LL_k}) + \beta \sum_{k=1}^{3} \mathcal{W}_k^{-1}(\mathbf{x}_{HF_k}), \tag{7}$$

where $\mathbf{x}_{HF_k}$ are high-frequency components (LH, HL, and HH) of an image post-MDWT, excluding the low-frequency LL sub-band. α_k enhances level-k low-frequency components ($\alpha_3 > \alpha_2 > \alpha_1 > 1$), and $\beta \in (0, 1)$ suppresses high-frequency noise.

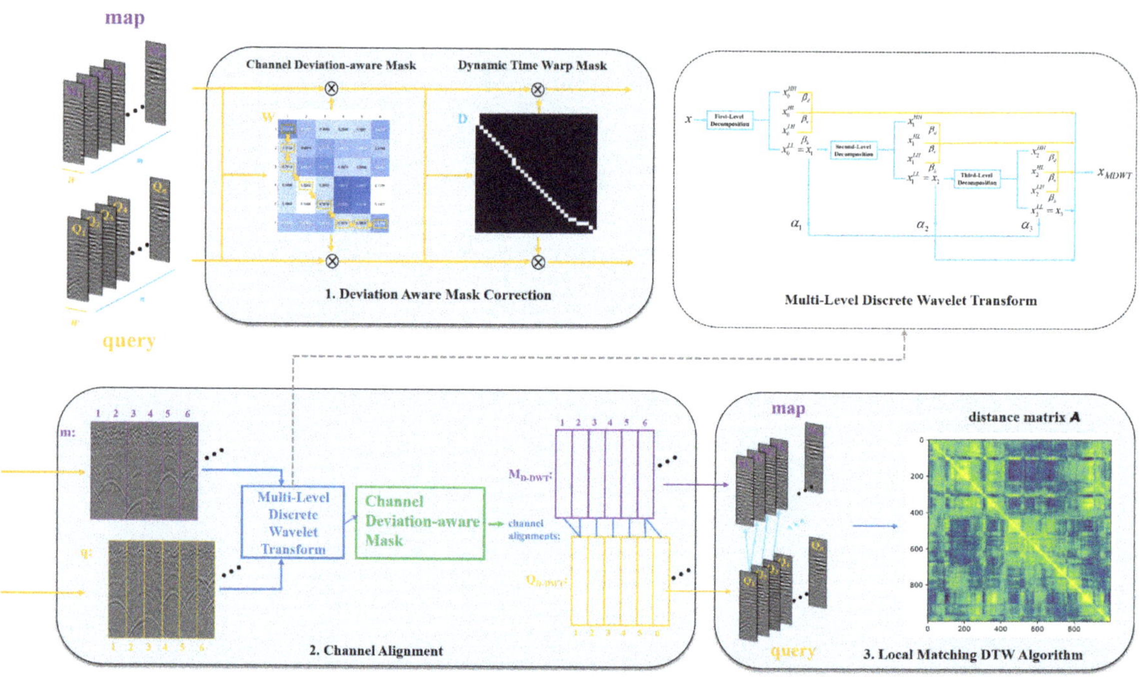

Figure 2. Overview of the proposed framework. The proposed framework comprises three key components: (**1**) Deviation-aware mask correction: Compensates for map distortions induced by platform posture variations and motion speed changes (Section 3.1). (**2**) Channel alignment module: Addresses channel offsets caused by yaw deviations (Section 3.2). (**3**) Local matching DTW algorithm: Generates the alignment distance matrix $\mathcal{A}$ for robust feature matching (Section 3.3); MDWT module: Extracts low-frequency features from aligned channel data to mitigate weather-dependent underground moisture variations (Section 2.2), the blue module illustrates the processing pipeline of the proposed methodology.

3. Methodology

This section presents the proposed framework, which consists of three key components: a deviation-aware mask, channel alignment, and localized feature matching. Figure 2 illustrates the overall pipeline of the proposed framework. The framework works as follows. Firstly, we use DTW to align two segments of raw GPR data collected at different times on the same road section, generating a deviation-aware mask. Secondly, the aligned data enter the channel alignment module, where MLDWT enhances low-frequency features. The processed signal is realigned in the channel dimension using the channel deviation-aware mask, improving alignment and reducing antenna-channel mismatch caused by track offset. Finally, we apply the local matching DTW algorithm. In the following section, we will elaborate on each module in detail.

3.1. Deviation-Aware Mask for GPR Sequence Calibration

To mitigate map deformation artifacts induced by lane deviations or platform velocity fluctuations during GPR data acquisition, we propose a novel two-stage mask calibration framework. The first stage generates a lane deviation mask for precise channel-dimension alignment of dual-channel GPR data. Building upon this, the second stage employs an optimized dynamic time warping (DTW) algorithm to construct a deviation-aware mask that achieves concurrent spatial and temporal registration of GPR sequences, as shown in Figure 2.

Consider two GPR sequences, $\mathbf{M} = \{\mathbf{m}_1, \mathbf{m}_2, ..., \mathbf{m}_m\}$ and $\mathbf{Q} = \{\mathbf{q}_1, \mathbf{q}_2, ..., \mathbf{q}_n\}$, exhibiting both spatial (cross-track) and temporal (along-track) distortions due to channel misalignments and non-uniform acquisition velocities during data collection. In the absence of precise prior information, direct sequence alignment becomes particularly challenging. To address this, we partition each sequence into N localized segments. For channel-dimensional DTW mask generation, we employ cosine distance as the fundamental similarity measure between subsequences, defined as $d_{i,j}$, as shown in Equation (1). A smaller $d_{i,j}$ value indicates higher similarity between subsequences. We introduce an adaptive parameter κ to regulate diagonal shift weighting in our enhanced DTW formulation, yielding the channel-dimensional alignment mask $\boldsymbol{W}$ (detailed in Section 3.2). Subsequently, mask $\boldsymbol{W}$ generates channel-corrected sequences for both reference and query data, enabling temporal alignment through DTW with the Euclidean distance metric:

$$d_{euc} = \sqrt{\sum_{i=1}^{n} (\mathbf{m}_i - \mathbf{q}_j)^2}. \tag{8}$$

Following distance computation and DTW cost matrix generation, we obtain the cumulative DTW distance and perform path backtracking to derive the temporal alignment mask $\boldsymbol{D}$. This mask is then applied to generate the corrected GPR time series. Figure 3 demonstrates the comparative results between the query data and the prior map after offset-aware mask correction, showing significant improvement in alignment accuracy.

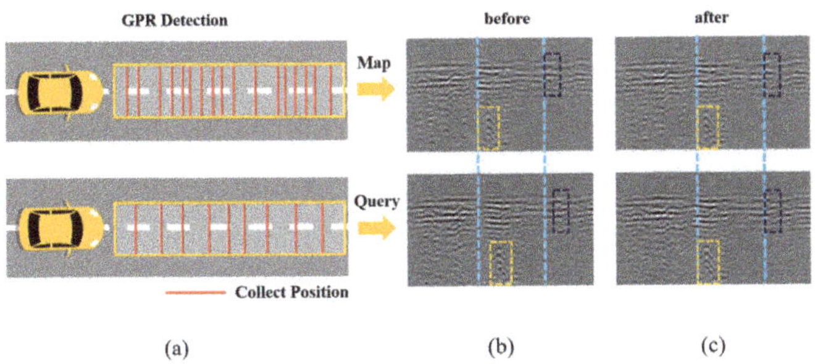

Figure 3. Visualization of deviation-aware mask calibration for GPR sequence alignment. (**a**) GPR signal acquisition positions under varying platform speeds during map and query. (**b**) Time-domain comparison of a channel before calibration, showing significant temporal misalignment (yellow and blue boxes highlight identical features). (**c**) Post-calibration result, demonstrating corrected feature alignment in the B-scan.

3.2. Channel Alignment

Owing to the inherent operational characteristics of GPR systems, vehicles must precisely adhere to the prior map's acquisition trajectory to obtain properly aligned GPR echo images in the antenna channel dimension. As demonstrated in Figure 4, any trajectory deviation causes channel dimension misalignment, introducing viewpoint discrepancies in the acquired B-scans. Direct application of cosine or Euclidean distance metrics under such misalignment conditions inevitably results in matching errors. To address this challenge, we propose an adaptive DTW-based block alignment method for GPR channel-dimensional B-scan registration. This approach computes the alignment distance between GPR time-series pairs while accounting for non-linear temporal deformations, thereby enabling robust similarity measurement.

As illustrated in Figure 4, the images of the GPR sequence map $\mathbf{M} = \{\mathbf{m}_1, \mathbf{m}_2, ..., \mathbf{m}_m\}$ and query sequence $\mathbf{Q} = \{\mathbf{q}_1, \mathbf{q}_2, ..., \mathbf{q}_n\}$ are $\mathbf{m}$ and $\mathbf{q}$, which are further segmented into W channel-wise blocks separately, generating block sequences $\mathbf{m}_W = \{\mathbf{m}_1, \mathbf{m}_2, \ldots, \mathbf{m}_W\}$ and $\mathbf{q}_W = \{\mathbf{q}_1, \mathbf{q}_2, \ldots, \mathbf{q}_W\}$. W must evenly divide the channel dimension in the along-track direction. Based on typical GPR channel dimensions and hardware limitations, we adopt W = 10 as a practical default value. While the choice of W impacts positioning accuracy, its selectable range is primarily determined by the GPR channel configuration. Consequently, the optimal practice is to set W as either the full channel dimension or a proper divisor thereof. Afterwards, they are sent to MDWT separately. The three-level DWT was chosen for optimal GPR signal analysis, balancing time-frequency resolution and computational efficiency. This decomposition depth simultaneously resolves shallow features and deep layers, while avoiding the inadequate resolution of fewer levels or the excessive fragmentation of deeper transforms. The weighting parameters α and β were empirically optimized to 1.5 and 0.2, respectively, effectively suppressing high-frequency noise components while amplifying relevant low-frequency signals in the GPR data.

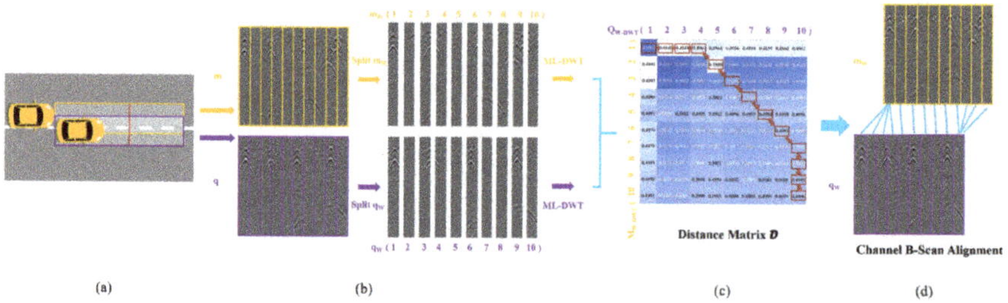

Figure 4. Channel alignment module workflow. (**a**) Yaw-induced channel offset between map and query GPR data. (**b**) Each B-scan image is partitioned into W (W = 10 in Figure 4) uniform sub-regions, which subsequently undergo MDWT decomposition for localized low-frequency feature extraction. (**c**) Cosine distance matrix computation: A $W \times W$ distance matrix $\mathcal{D}$ is constructed from the localized low-frequency feature. (**d**) Channel-dimension alignment: dynamic time warping (DTW) is applied to matrix $\mathcal{D}$ to generate the final channel alignment matrix.

The similarity measurement employs the cosine distance (defined in Section 3.1) between MDWT-processed low-frequency features (detailed in Section 2.2). This metric operates on vectorized B-scan pairs after DWT decomposition, with the distance $d_{\mathrm{i,j}} \in [0, 2]$. This local similarity measure enables pointwise comparison of time series characteristics, where decreasing distance values indicate higher feature similarity. From pairwise block comparisons, we construct a $W \times W$ distance matrix $\mathcal{D}$. The DTW algorithm then identifies the optimal alignment path (The mask) $\mathbf{W}$, which is shown in Equation (3), with path coordinates establishing the temporal alignment between sequences $\mathbf{M}$ and $\mathbf{Q}$. The accumulated lane deviation-aware matrix $\mathbf{S}$ is computed along the optimal path, with each element $s_{i,j}$ representing the minimum cost of the optimal path from $(1, 1)$ to (i, j). The computation of $W_{i,j}$ is provided by Equation (2).

The fundamental recursive formulation of DTW computes the minimal cumulative distance $d_{i,j}$ between the initial alignment point $(0, 0)$ and the current position (i, j), in which $d_{\mathbf{m}_i,\mathbf{q}_j}$ denotes the local distance metric between elements $\mathbf{m}_i$ and $\mathbf{q}_j$ of the respective time series. This recursive computation ensures optimal non-linear alignment while preserving temporal ordering constraints through systematic evaluation of all feasible warping paths. The proposed method introduces an adaptive deformation coefficient κ to the conventional DTW framework, enhancing its capability to accommodate feature variations while maintaining temporal alignment integrity. This modification addresses the limitations of standard DTW in handling non-linear distortions through a dynamically adjusted constraint parameter, as demonstrated in prior work [29]. The adaptive mechanism automatically regulates the warping path flexibility based on local feature characteristics, achieving improved alignment accuracy compared to traditional fixed-constraint approaches.

$$\kappa = \sqrt{1 + \delta \left| I_{\lfloor W/2 \rfloor} - \lfloor W/2 \rfloor \right|} \quad (W = 11 \Rightarrow \lfloor W/2 \rfloor = 5). \tag{9}$$

Here, δ is a predefined constant (typically set to 1), and $I_{\lfloor W/2 \rfloor}$ corresponds to the index of the optimal matching feature $\mathbf{q}_W$ in image $\mathbf{q}$ for the reference feature $\mathbf{m}_W$ in image $\mathbf{m}$. The warping path establishes spatio alignment between the two sequences from their initial positions $(1,1)$ to terminal coordinates (W, W), as illustrated in Figure 4. The exemplar alignment demonstrates that while the path primarily connects semantically equivalent channel-dimensional B-scan pairs (e.g., $\mathbf{m}_1 - \mathbf{q}_3$), it necessarily incorporates some non-corresponding feature matches (e.g., $\mathbf{m}_1 - \mathbf{q}_1$). These incidental alignments arise from trajectory deviations during GPR data acquisition, where the survey vehicle's positional offset introduces content disparities between co-located regions of images $\mathbf{m}$ and $\mathbf{q}$. Crucially, retaining these non-corresponding matches preserves the spatial coherence of local features and ensures metrologically sound distance computation between the B-scan images. To account for path length variability in dynamic time warping alignment, we employ normalized cumulative distance to derive a scale-invariant similarity measure between images $\mathbf{m}$ and $\mathbf{q}$. The normalized distance metric is formulated as Equation (4). This normalization procedure effectively eliminates the dimensional bias introduced by varying warping path lengths while preserving the essential dissimilarity information between the compared B-scan sequences.

The proposed alignment methodology begins by computing a local distance matrix $\mathcal{D}$ for each pair of GPR channel-dimensional B-scans. Subsequently, DTW is systematically applied to derive the cumulative distance matrix, thereby enabling precise temporal alignment of the channel-dimensional images. Notably, during this alignment process, an adaptive parameter κ is dynamically optimized to effectively modulate the diagonal path cost, consequently compensating for non-linear temporal distortions that may arise during data acquisition. Ultimately, as clearly demonstrated in Figure 4, this comprehensive procedure successfully yields a set of optimally aligned channel-dimensional B-scan distance measurements, which are characterized by enhanced accuracy and reliability for subsequent analytical processing.

3.3. Local Matching DTW Algorithm

Finally, we systematically compute pairwise distances between the channel-aligned map and query GPR time series using the local matching DTW algorithm, thereby generating a comprehensive distance matrix $\mathcal{A}$. Crucially, the matching performance is rigorously evaluated by identifying minimum-distance correspondences within matrix $\mathcal{A}$, which conclusively demonstrates the effectiveness of the proposed method under varying environmental conditions. Specifically, the consistent localization accuracy achieved through this approach verifies the method's robustness against moisture-induced signal variations that typically degrade conventional GPR matching performance.

4. Experiment and Results Analysis

In this section, we first describe the data acquisition process and evaluation metrics adopted for the experiments. Subsequently, we conduct an ablation study to rigorously examine the influence of three critical components—namely, the deviation-aware mask (DAM), channel alignment (CA), and multi-level discrete wavelet transform (MDWT)—on LGPR performance under various weather conditions and different lane deviation scenarios. Finally, to comprehensively validate the efficacy of our framework, we compare the proposed method with state-of-the-art (SOTA) approaches, demonstrating its superior robustness and accuracy.

4.1. Data Acquisition and Data Pre-Processing

To rigorously validate the universality of the proposed method, we employ two complementary datasets: the publicly available GROUNDED dataset [21] and a self-collected dataset [5]. Both datasets comprehensively capture multiple traversals of identical routes under varying temporal conditions, with each route being recorded at least three times at different periods. Subsequently, we provide detailed descriptions of these datasets to facilitate reproducibility and demonstrate their suitability for evaluating our method. Finally, Table 1 presents the precise experimental configurations and dataset specifications for reference. In addition, the table summarizes the lengths of the chosen map and query sequences, the sampling interval, and the lane offset measured by RTK-GPS during acquisition.

Table 1. Dataset specifications and parameters

Dataset	Frames (Map/Query)	Interval (cm)	Channel Offset	Tolerance (Frames)
GROUNDED	1000/1000	5	$\geq$3	$\pm$2
Self	900/900	7	0	$\pm$3
Self (Yaw)	660/660	7	2/4	$\pm$3

- GROUNDED dataset: To comprehensively characterize the experimental data, we utilize a unique large-scale LGPR dataset that currently represents the sole publicly available collection employing 11-channel 400 MHz GPR measurements. The dataset systematically incorporates three distinct weather conditions (rain, snow, and clear) while simultaneously encompassing seven diverse road environments, including highways, urban centers, and suburban areas. Furthermore, the dataset specifically addresses multi-lane mapping challenges through a carefully designed acquisition plan, thereby providing a robust testbed for evaluating LGPR performance across varying operational scenarios. In this study, Route5's run56 (sunny) is selected as the map, while run57 (sunny), run90 (rainy), and run57 (snowy) serve as the query sequences.
- Self dataset: To ensure comprehensive data coverage, we employed a multi-channel GPR system to systematically acquire ground-penetrating radar data across various road surfaces in Changsha, China. The system configuration consists of a dual-row uniform antenna array comprising 20 radar channels with a total array width of 1.5 m, operating within a broad frequency range of 100–3000 MHz. While the nominal penetration depth reaches approximately 3 m, we note that actual penetration performance varies significantly depending on subsurface soil characteristics. Furthermore, to guarantee high-resolution data collection, we integrated a distance measurement instrument (DMI) that triggers the GPR system at precisely 7-cm intervals along each survey line. Importantly, our acquisition plan specifically addresses lane deviation challenges by intentionally collecting data with 0–5 channel offsets, while simultaneously ensuring environmental variability through repeated surveys of identical road sections under both sunny and rainy weather conditions. We selected non-yawed trajectories (self) under sunny conditions as the map, while sunny and rainy sequences served as queries. Additionally, we included sunny trajectories with intentional yaw deviations (self yaw), using the non-yawed sequence as the map and trajectories with two-lane and four-lane offsets as queries.

4.2. Valuation Metric

To objectively assess recognition performance, we primarily employ the F_1-score as our evaluation metric, where ground truth is established through frame-level correspondences within each dataset. Specifically, we determine true positive matches by comparing query sequences with result sequences, requiring both the sequence distance to remain below a predetermined threshold and the recognition error to fall within acceptable tolerances (as detailed in Table 1). The best results are highlighted in bold.

4.3. Ablation Study

To systematically evaluate the contribution of each module in the proposed framework, we conduct comprehensive ablation experiments, with comparative results presented in Table 2. The baseline localization using original GPR data shows limited effectiveness, particularly under adverse weather conditions. In the GROUNDED dataset, we observe F_1-scores of only 0.496 for sunny/sunny conditions and approximately 0.2 for sunny/rainy conditions. Similar trends are evident in the Self dataset, where the F_1-score drops from 0.802 for sunny/sunny conditions to 0.285 during sunny/rainy conditions. These results highlight the significant challenges posed by weather-induced subsurface variations in traditional LGPR approaches. Our proposed framework demonstrates substantial improvements across all tested conditions. The adaptive spatio-temporal alignment strategy successfully maintains F_1-scores above 0.9 in both sunny/sunny and sunny/rainy conditions for the GROUNDED dataset, while achieving 0.986 in the sunny/rainy conditions of the Self dataset. This enhanced performance is particularly notable given that rainfall typically causes more severe subsurface dielectric variations than snowfall due to greater moisture infiltration, making rainy conditions theoretically more challenging for GPR-based localization. The methodology also shows remarkable robustness to lane offset scenarios. In the Self dataset with intentional trajectory deviations, the system improves localization accuracy from baseline F_1-scores of 0.78 (two-lane offset) and 0.577 (four-lane offset) to 0.901 and 0.845, respectively. This performance gain underscores the effectiveness of our channel alignment module in compensating for acquisition geometry variations. Notably, the relatively smaller performance gain for sunny/snowy conditions in the GROUNDED dataset (compared to sunny/rainy conditions) may be attributed to platform motion artifacts during data acquisition. As indicated in Table 3, advanced feature extraction techniques could potentially bridge this gap. Future work will explore the integration of learning-based approaches to further enhance localization robustness under such challenging conditions. Specifically, we analyze five distinct configurations:

- Proposed: the complete proposed method (Figure 2), representing our full proposed framework.
- No DAM: the variant excluding DAM calibration, where original dataset data undergo CA after MDWT filtering.
- No CA:a configuration without CA, utilizing cosine distance metrics on DAM calibrated MDWT data for single-image matching.
- No MDWT: an ablated version removing the MDWT module entirely.
- Original data: baseline performance using raw, unprocessed origin data.

This structured ablation approach enables the precise quantification of each component's impact on overall system performance.

The ablation of the DAM module results in a significant degradation of positioning performance in both the GROUNDED and self-collected datasets, with particularly pronounced effects under varying weather conditions. To further investigate the DAM module's influence on the LGPR task, we visualize the maps and queries under different weather conditions before and after DAM module processing, as illustrated in Figure 5.

Table 2. F_1-score comparison of different ablated versions.

Method/Challenge	Grounded Dataset			Self Dataset			
	Sunny/Sunny	Sunny/Snowy	Sunny/Rainy	Sunny/Sunny	Sunny/Rainy	Yaw0/2	Yaw0/4
Proposed	**0.988**	**0.778**	**0.924**	**0.975**	**0.986**	**0.901**	**0.845**
No DAM	0.508	0.223	0.338	0.949	0.344	0.873	0.838
No CA	0.949	0.447	0.771	**0.969**	**0.981**	0.783	0.613
No MDWT	**0.982**	0.667	0.877	0.804	0.767	0.876	0.823
Original Data	0.496	0.201	0.207	0.802	0.285	0.780	0.577

Table 3. F_1-score comparison of SOTA.

Method/Challenge	Grounded Dataset			Self Dataset			
	Sunny/Sunny	Sunny/Snowy	Sunny/Rainy	Sunny/Sunny	Sunny/Rainy	Yaw0/2	Yaw0/4
Proposed	**0.988**	**0.778**	**0.924**	**0.975**	**0.986**	**0.901**	**0.845**
STU-Net	0.621	0.530	0.573	0.742	0.586	0.536	0.265
LFE-MDP	0.933	0.562	0.387	0.913	0.847	0.315	0.243
fast DTW	**0.950**	0.521	0.797	0.804	0.647	0.720	0.435
Savitzky–Golay filter	0.612	0.389	0.536	0.832	0.301	0.784	0.579

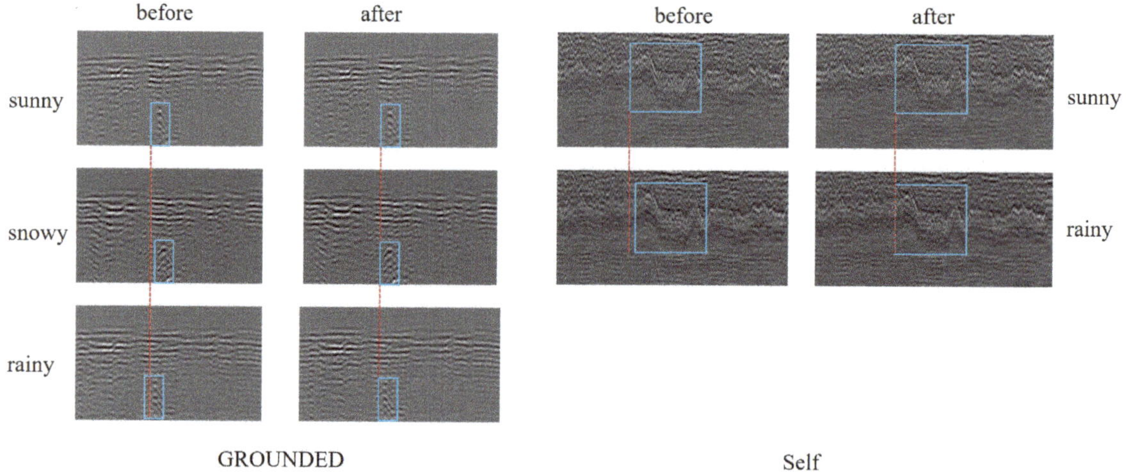

Figure 5. Comparison of fifth-channel B-scans before and after DAM module processing in the time domain. (**Left**) GROUNDED dataset (map: sunny day; query: snowy/rainy day). (**Right**) Self-collected dataset (map: sunny day; query: rainy day). Blue boxes highlight GPR echoes from identical shallow subsurface features, while red dashed lines indicate alignment improvements after DAM processing.

Figure 5 demonstrates that the DAM module effectively corrects the GPR echo query by aligning it with the map's shallow subsurface target echoes in both the GROUNDED and Self datasets. As analyzed in Section 3.1, this approach mitigates GPR map distortions caused by platform speed variations. Notably, despite the visual discrepancies between map and query echoes due to differing subsurface water content under varying weather conditions, the DAM module reliably captures consistent target features. Future work could further enhance the DAM module's robustness in matching target echoes across weather conditions through advanced learning methods.

In Addition, the CA module significantly improves positioning accuracy in both the GROUNDED dataset and the Self (yaw) dataset, but its impact is less pronounced in the standard Self dataset. This discrepancy arises because the standard self-collected dataset maintains strict channel alignment during data acquisition, whereas both the GROUNDED dataset and the yaw-variant dataset exhibit GPR channel misalignment due to lane deviations.

We further evaluate the MDWT module and demonstrate its effectiveness in enhancing the robustness of LGPR tasks under varying weather conditions. On the GROUNDED dataset, the MDWT module improves the F_1-score by 11% for snowy conditions and 5% for rainy conditions. The performance gain is more pronounced in the self-collected dataset, with F_1-score improvements of 10% (sunny) and 20% (rainy). This discrepancy arises because the higher acquisition frequency of the self-collected dataset makes target echoes more susceptible to high-frequency noise interference. To validate this analysis, we provide visualizations of the self-collected dataset and further examine the MDWT module's performance, as illustrated in the Figure 6.

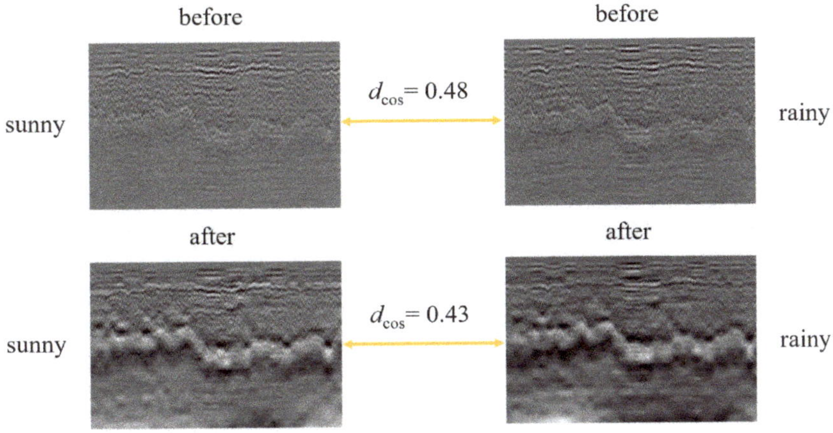

Figure 6. Visualization of map–query pairs in the self-collected dataset: (**left**) map (sunny) and (**right**) query (rainy), showing data before and after MDWT module processing. Corresponding cosine distance metrics between map and query are included for both processing states.

As shown in Figure 6, the MDWT module's enhanced map-query matching capability under both clear and rainy conditions. Post-processing results show, firstly, improved distinctiveness of target feature echoes, secondly, reduced cosine distance between co-located map-query pairs, and thirdly effective compensation for subsurface water content variations. As analyzed before, the performance improvement is particularly significant for our self-collected dataset (operational bandwidth: 100–3000 MHz) compared to conventional 400 MHz systems, since the broader frequency range introduces additional high-frequency noise components that the MDWT module effectively suppresses.

Moreover, we compute the cosine distance matrix for localization using LGPR data ($\mathcal{A}$ Figure 7). In the cosine-distance heatmaps, smaller cosine distances between frames are represented by lighter colors, clearly demonstrating enhanced segment correspondence. Comparative analysis shows the proposed method significantly improves positioning accuracy under rainy conditions, as evidenced by the sharper diagonal pattern in the processed matrix versus the original data. The right sub-figure (proposed) exhibits a markedly clearer diagonal with more distinct light-colored segments, indicating superior inter-segment matching compared to the left sub-figure (original data).

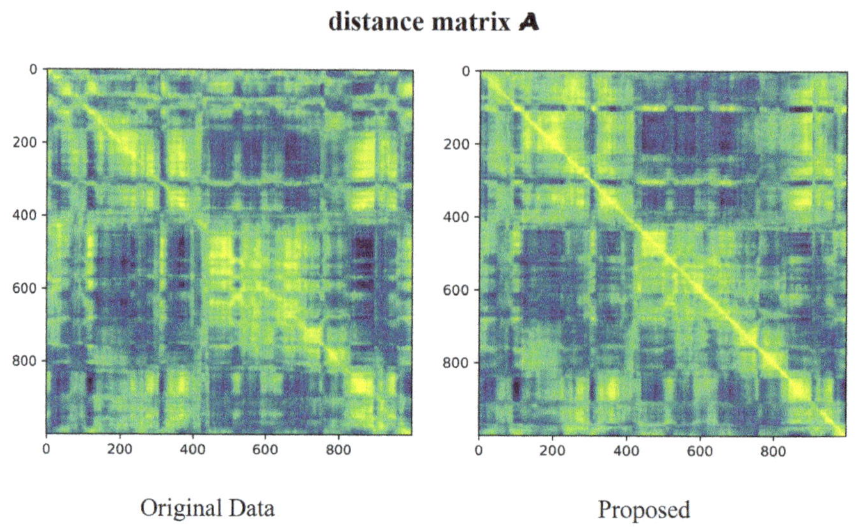

Figure 7. Comparison of cosine distance matrices for local positioning using GPR data: (**left**) raw unprocessed data; (**right**) data processed with the proposed method. Experimental validation conducted using the GROUNDED dataset under sunny mapping and rainy query conditions.

In summary, as evidenced by the F_1-score comparisons in Table 2, the proposed method demonstrates substantial positioning accuracy improvements across all experimental conditions. Specifically, the deviation-aware mask module yields remarkable performance gains, elevating F_1-scores by 0.4–0.5 for both sunny and rainy conditions in the GROUNDED and Self datasets. Similarly, the channel alignment module proves particularly effective for challenging scenarios, improving F_1-score by 0.2–0.3 in rainy/snowy sections of the GROUNDED dataset and lane-offset conditions in our Self data. Furthermore, the MDWT module shows consistent enhancement (ΔF_1-score = 0.2) for multi-weather positioning by effectively mitigating subsurface water content variations, with notably greater impact on our Self dataset. Moreover, baseline comparisons using raw GPR B-scan cosine distances further validate the proposed method's superiority.

4.4. Comparison with the State-of-the-Art Methods

In order to rigorously validate the performance enhancement of our proposed LGPR method, systematic comparisons have been conducted against SOTA approaches encompassing both learning-based feature localization techniques and map alignment preprocessing solutions, including STU-Net [19] as a deep learning architecture for GPR echo feature extraction and enhancement, along with hybrid-driven recovery with low-frequency enhancement (LFE-MDP) [33], in addition to the widely adopted Savitzky–Golay filter [10] for LGPR preprocessing and the conventional fast DTW algorithm [34], all of which were reimplemented and retrained within our unified LGPR framework to ensure fair evaluation. Building upon our previous work [19,33], which demonstrated the superior performance of learning-based methods such as STU-Net and HRLE over alternative learning schemes for LGPR tasks, and considering that the Savitzky–Golay filter represents the current standard preprocessing approach in LGPR applications [7,10,18] while the fast DTW has been successfully applied across various computer vision applications [30,35], the quantitative comparison results presented in Table 3, the best results are highlighted in bold. clearly demonstrate significant performance improvements achieved by our proposed method across all **SOTA**, with particular emphasis on its unique advantages for subsurface sensing applications, including enhanced feature discrimination in challenging environments and improved robustness to signal variations.

Table 3 demonstrates the superior performance of our proposed method for LGPR localization. While both STU-Net and LFE-MDP learning methods show limited improvement over original data positioning in rainy/snowy conditions (in GROUNDED, 32.9%/36.6% and 36.1%/18% accuracy gains, respectively), their performance remains constrained by the inherent limitations of B-scan channel dimension features. Comparative analysis reveals HRLE's better suitability for LGPR tasks (in GROUNDED and Self, 10–20% higher than STU-Net in sunny/sunny), confirming the importance of low-frequency information sensitivity in LGPR processing. However, neither learning-based approach effectively mitigates yaw-induced positioning errors. Furthermore, conventional methods (fast DTW and Savitzky–Golay filter) exhibit significantly lower robustness to weather variations compared to our proposed technique.

The Figure 8 presents precision–recall (PR) curves comparing the proposed method with baseline approaches across three distinct weather conditions in the GROUNDED dataset, quantitatively demonstrating the method's superior robustness to environmental variations.

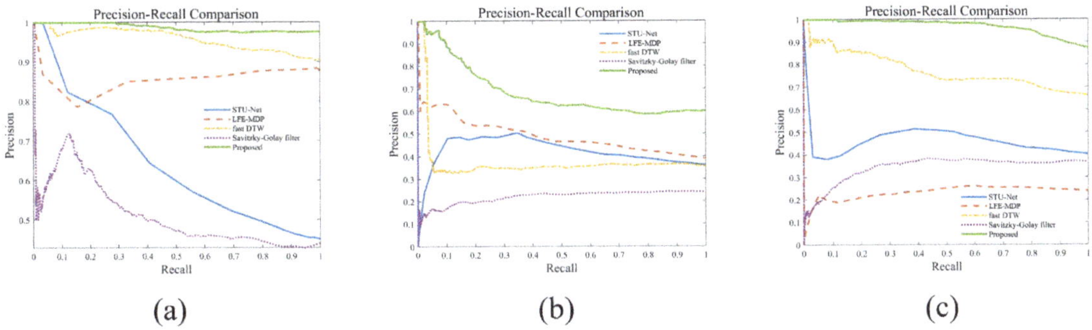

Figure 8. Precision–recall (PR) performance comparison of state-of-the-art methods in the GROUNDED dataset under varying weather conditions: (**a**) sunny, (**b**) snowy, and (**c**) rainy days.

5. Conclusions

We present the proposed method, which fundamentally enhances LGPR accuracy by innovatively adapting DTW principles from audio processing while meticulously accounting for GPR data acquisition characteristics. Specifically, our technical contributions include four key components: the DAM correction module addressing speed variations in GPR platform motion; the CA module resolving channel mismatches induced by vehicle trajectory deviations; MDWT handling feature distortions from groundwater content fluctuations due to climatic variations; a novel local matching DTW algorithm integrating these solutions. Collectively, these innovations establish a robust framework for high-precision LGPR. Extensive experimental validation demonstrates our method's consistent positioning accuracy across diverse weather conditions and channel offset scenarios, while comprehensive benchmarking confirms its superior performance over **SOTA** approaches on multiple datasets.

Author Contributions: Conceptualization, Y.C. and P.Z.; Methodology, Y.C., B.B. and P.Z.; Formal analysis, Y.C.; Writing—original draft, Y.C.; Writing—review & editing, B.B.; Visualization, Y.C.; Supervision, L.S. and C.C.; Funding acquisition, X.H. and T.J. All authors have read and agreed to the published version of the manuscript.

Funding: This research received no external funding.

Data Availability Statement: The raw data supporting the conclusions of this article will be made available by the authors on request.

Conflicts of Interest: The authors declare no conflicts of interest.

References

1. Cornick, M.; Koechling, J.; Stanley, B.; Zhang, B. Localizing ground penetrating radar: A step toward robust autonomous ground vehicle localization. *J. Field Robot.* **2016**, *33*, 82–102. [CrossRef]
2. Xu, L.; Winner, V.; Maurer, H. Gradient-constrained model parametrization in 3-D compact full waveform inversion. *Geophys. J. Int.* **2023**, *232*, 366–397. [CrossRef]
3. Wang, X.; Yu, T.; Feng, D.; Li, B.; Ding, S. Spatiotemporal Optimization of GPR Full Waveform Inversion Based on Super-Resolution Technology. *IEEE Trans. Geosci. Remote Sens.* **2025**, *63*, 5908413. [CrossRef]
4. Ort, T.; Gilitschenski, I.; Rus, D. Autonomous navigation in inclement weather based on a localizing ground penetrating radar. *IEEE Robot. Autom. Lett.* **2020**, *5*, 3267–3274. [CrossRef]
5. Bi, B.; Shen, L.; Zhang, P.; Huang, X.; Xin, Q.; Jin, T. TSVR-Net: An End-to-End Ground-Penetrating Radar Images Registration and Location Network. *Remote Sens.* **2023**, *15*, 3428. [CrossRef]
6. Li, H.; Guo, J.; Song, D. Subsurface Feature-based Ground Robot/Vehicle Localization Using a Ground Penetrating Radar. In Proceedings of the 2024 IEEE International Conference on Robotics and Automation (ICRA), Yokohama, Japan, 13–17 May 2024; IEEE: Piscataway, NJ, USA, 2024; pp. 1716–1722.
7. Zhang, P.; Zhi, S.; Yuan, Y.; Bi, B.; Xin, Q.; Huang, X.; Shen, L. Looking Beneath More: A Sequence-based Localizing Ground Penetrating Radar Framework. In Proceedings of the 2024 IEEE International Conference on Robotics and Automation (ICRA), Yokohama, Japan, 13–17 May 2024; pp. 8515–8521.
8. Zhang, K.; Chi, Y.; Guo, J.; Bai, C. Underground Robot Localization Based on Ground-Penetrating Radar. In Proceedings of the International Conference on Autonomous Unmanned Systems, Singapore, 26–28 October 2022; Springer: Berlin/Heidelberg, Germany, 2022; pp. 3577–3588.
9. Xu, J.; Lai, Q.; Wei, D.; Ji, X.; Shen, G.; Yuan, H. The Ground-Penetrating Radar Image Matching Method Based on Central Dense Structure Context Features. *Remote Sens.* **2024**, *16*, 4291. [CrossRef]
10. Ort, T.; Gilitschenski, I.; Rus, D. GROUNDED: The Localizing Ground Penetrating Radar Evaluation Dataset. In Proceedings of the Robotics: Science and Systems, Virtual, 12–16 July 2021; Volume 2.
11. Skartados, E.; Kargakos, A.; Tsiogas, E.; Kostavelis, I.; Giakoumis, D.; Tzovaras, D. Gpr antenna localization based on a-scans. In Proceedings of the 2019 27th European Signal Processing Conference (EUSIPCO), A Coruña, Spain, 2–6 September 2019; IEEE: Piscataway, NJ, USA, 2019; pp. 1–5.
12. Ni, Z.; Ye, S.; Shi, C.; Pan, J.; Zheng, Z.; Fang, G. A deep learning assisted ground penetrating radar localization method. *J. Electron. Inf. Technol.* **2022**, *44*, 1265–1273.
13. Zhang, P.; Shen, L.; Wen, T.; Huang, X.; Xin, Q. Vector phase symmetry for stable hyperbola detection in ground-penetrating radar images. *IEEE Trans. Geosci. Remote Sens.* **2021**, *60*, 5107912. [CrossRef]
14. Arandjelovic, R.; Gronat, P.; Torii, A.; Pajdla, T.; Sivic, J. NetVLAD: CNN architecture for weakly supervised place recognition. In Proceedings of the IEEE Conference on Computer Vision and Pattern Recognition, Las Vegas, NV, USA, 27–30 June 2016; pp. 5297–5307.
15. Arandjelovic, R.; Zisserman, A. All about VLAD. In Proceedings of the IEEE conference on Computer Vision and Pattern Recognition, Portland, OR, USA, 23–28 June 2013; pp. 1578–1585.
16. Noh, H.; Araujo, A.; Sim, J.; Weyand, T.; Han, B. Large-scale image retrieval with attentive deep local features. In Proceedings of the IEEE International Conference on Computer Vision, Venice, Italy, 22–29 October 2017; pp. 3456–3465.
17. Ma, J.; Zhang, J.; Xu, J.; Ai, R.; Gu, W.; Chen, X. OverlapTransformer: An efficient and yaw-angle-invariant transformer network for LiDAR-based place recognition. *IEEE Robot. Autom. Lett.* **2022**, *7*, 6958–6965. [CrossRef]
18. Zhang, P.; Chen, X.; Chen, Y.; Bi, B.; Xu, Z.; Jin, T.; Huang, X.; Shen, L. EDENet: Echo Direction Encoding Network for Place Recognition Based on Ground Penetrating Radar. In Proceedings of the AAAI Conference on Artificial Intelligence, Vancouver, BC, Canada, 23–27 February 2025; Volume 39, pp. 10067–10075.
19. Chen, Y.; Zhang, P.; Bi, B.; Shen, L.; Jin, T.; Huang, X. Spatial–Temporal U-Net for Localizing Ground-Penetrating Radar. *IEEE Geosci. Remote Sens. Lett.* **2025**, *22*, 3504905. [CrossRef]
20. Stasewitsch, I.; Schattenberg, J.; Frerichs, L. Robust Monte Carlo Localisation Using a Ground Penetrating Radar. In Proceedings of the Iberian Robotics Conference, Sevilla, Spain, 23–25 November 2022; Springer: Berlin/Heidelberg, Germany, 2022; pp. 247–258.

21. Ort, T.; Gilitschenski, I.; Rus, D. GROUNDED: A localizing ground penetrating radar evaluation dataset for learning to localize in inclement weather. *Int. J. Robot. Res.* **2023**, *42*, 901–916. [CrossRef]
22. Lu, F.; Chen, B.; Zhou, X.D.; Song, D. STA-VPR: Spatio-temporal alignment for visual place recognition. *IEEE Robot. Autom. Lett.* **2021**, *6*, 4297–4304. [CrossRef]
23. Yin, H.; Wang, Y.; Ding, X.; Tang, L.; Huang, S.; Xiong, R. 3d lidar-based global localization using siamese neural network. *IEEE Trans. Intell. Transp. Syst.* **2019**, *21*, 1380–1392. [CrossRef]
24. Kim, G.; Kim, A. Scan context: Egocentric spatial descriptor for place recognition within 3d point cloud map. In Proceedings of the 2018 IEEE/RSJ International Conference on Intelligent Robots and Systems (IROS), Madrid, Spain, 1–5 October 2018; IEEE: Piscataway, NJ, USA, 2018; pp. 4802–4809.
25. Deng, J.J.; Leung, C.H. Dynamic time warping for music retrieval using time series modeling of musical emotions. *IEEE Trans. Affect. Comput.* **2015**, *6*, 137–151. [CrossRef]
26. Sakoe, H.; Chiba, S. Dynamic programming algorithm optimization for spoken word recognition. *IEEE Trans. Acoust. Speech Signal Process.* **2003**, *26*, 43–49. [CrossRef]
27. Keogh, E.J.; Pazzani, M.J. Scaling up dynamic time warping for datamining applications. In Proceedings of the Sixth ACM SIGKDD International Conference on Knowledge Discovery and Data Mining, Boston, MA, USA, 20–23 August 2000; pp. 285–289.
28. Maus, V.; Câmara, G.; Appel, M.; Pebesma, E. dtwsat: Time-weighted dynamic time warping for satellite image time series analysis in r. *J. Stat. Softw.* **2019**, *88*, 1–31. [CrossRef]
29. Lu, F.; Chen, B.; Guo, Z.; Zhou, X. Visual sequence place recognition with improved dynamic time warping. In Proceedings of the 2019 IEEE 31st International Conference on Tools with Artificial Intelligence (ICTAI), Portland, OR, USA, 4–6 November 2019; IEEE: Piscataway, NJ, USA, 2019; pp. 1034–1041.
30. Deng, Z.P.; Jia, K.B. A video similarity matching algorithm supporting for different time scales. In Proceedings of the 2008 Eighth International Conference on Intelligent Systems Design and Applications, Kaohsiung, Taiwan, 26–28 November 2008; IEEE: Piscataway, NJ, USA, 2008; Volume 3, pp. 570–574.
31. Wen, J.; Huang, T.; Cui, X.; Zhang, Y.; Shi, J.; Jiang, Y.; Li, X.; Guo, L. Enhancing Image Alignment in Time-Lapse-Ground-Penetrating Radar through Dynamic Time Warping. *Remote Sens.* **2024**, *16*, 1040. [CrossRef]
32. Luo, W.; Lee, Y.H.; Sun, H.H.; Ow, L.F.; Yusof, M.L.M.; Yucel, A.C. Tree roots reconstruction framework for accurate positioning in heterogeneous soil. *IEEE J. Sel. Top. Appl. Earth Obs. Remote Sens.* **2022**, *15*, 1912–1925. [CrossRef]
33. Bi, B.; Shen, L.; Zhang, P.; Chen, Y.; Huang, X.; Jin, T. Hybrid-Driven with Low-Frequency Enhancement for LGPR Map Reconstruction. *IEEE Sens. J.* **2025**, *in press*. [CrossRef]
34. Salvador, S.; Chan, P. Toward accurate dynamic time warping in linear time and space. *Intell. Data Anal.* **2007**, *11*, 561–580. [CrossRef]
35. Chen, A.P.; Lin, S.F.; Cheng, Y.C. Time registration of two image sequences by dynamic time warping. In Proceedings of the IEEE International Conference on Networking, Sensing and Control, Taipei, Taiwan, 21–23 March 2004; IEEE: Piscataway, NJ, USA, 2004; Volume 1, pp. 418–423.

remote sensing

MDPI

Article

End-to-End Model Enabled GPR Hyperbolic Keypoint Detection for Automatic Localization of Underground Targets

Feifei Hou [1], Yu Zhang [1], Jian Dong [2] and Jinglin Fan [3,4,*]

[1] School of Automation, Central South University, Changsha 410017, China; houfeifei@csu.edu.cn (F.H.); 8207221712@csu.edu.cn (Y.Z.)
[2] School of Electronic Information, Central South University, Changsha 410017, China; dongjian@csu.edu.cn
[3] School of Architecture and Art, Central South University, Changsha 410017, China
[4] School of Construct Management, Hunan University of Finance and Economics, Changsha 410205, China
[*] Correspondence: jinglin2021@csu.edu.cn

Abstract

Ground-Penetrating Radar (GPR) is a non-destructive detection technique widely employed for identifying underground targets. Despite its utility, conventional approaches suffer from limitations, including poor adaptability to multi-scale targets and suboptimal localization accuracy. To overcome these challenges, we propose a lightweight deep learning framework, the Dual Attentive YOLOv11 (You Only Look Once, version 11) Keypoint Detector (DAYKD), designed for robust underground target detection and precise localization. Building upon the YOLOv11 architecture, our method introduces two key innovations to enhance performance: (1) a dual-task learning framework that synergizes bounding box detection with keypoint regression to refine localization precision, and (2) a novel Convolution and Attention Fusion Module (CAFM) coupled with a Feature Refinement Network (FRFN) to enhance multi-scale feature representation. Extensive ablation studies demonstrate that DAYKD achieves a precision of 93.7% and an mAP50 of 94.7% in object detection tasks, surpassing the baseline model by about 13% in F1-score, a balanced metric that combines precision and recall to evaluate overall model performance, underscoring its superior performance. These findings confirm that DAYKD delivers exceptional recognition accuracy and robustness, offering a promising solution for high-precision underground target localization.

Keywords: ground-penetrating radar (GPR); hyperbolic characteristic; YOLO; dual-task learning; objection detection; keypoint detection

1. Introduction

Ground-Penetrating Radar (GPR), an advanced non-destructive testing (NDT) technology, is renowned for its high-precision localization, rapid scanning, operational flexibility, and superior detection accuracy [1]. This technique finds broad applications, ranging from soil contamination assessment [2], underground pipeline detection [3], sediment detection [4], and subsurface exploration [5]. By emitting high-frequency electromagnetic waves and analyzing their reflections, GPR enables the accurate identification of subsurface anomalies [6]. Effective target localization enhances the interpretation of GPR data, thereby facilitating the accurate mapping of subsurface features. The advantages of precise localization are substantial. It enables the identification of underground voids, which is crucial for assessing potential hazards such as sinkholes or soil instability [7]. Furthermore, it ensures reliable mapping of utility networks, minimizing excavation risks during construction [8]. Such capabilities make GPR indispensable for infrastructure safety and geotechnical engineering.

Conventional GPR image recognition predominantly depends on template matching and image processing techniques. Liu et al. [9] applied the Sobel operator to detect hyperbolic edges in GPR images, while Li et al. [10] leveraged randomized Hough transforms for automated root identification. Maas et al. [11] adopted the Viola–Jones learning algorithm to constrain regions of interest, thereby reducing computational overhead in Hough transform-based localization. M. Sharafeldin et al. [12,13] deployed a total of 10 electrical resistivity imaging (ERI), 26 shallow seismic refraction (SSR), and 19 GPR survey lines across the Giza Plateau, and performed integrated inversion to build a three-layer subsurface model, accurately characterizing the groundwater aquifer and its water table depth. Luo et al. [14] introduced a Laser Dynamic Deflectometer method that uses vehicle-mounted laser Doppler sensors to capture pavement deflection velocity anomalies as indicators of subsurface cavities for rapid, non-invasive road network screening. Recently, machine learning has revolutionized hyperbolic feature recognition in GPR data. Dou et al. [15] developed a Connected Component Clustering (C3) algorithm to isolate hyperbolas within target regions, whereas Zhang et al. [16] proposed a symmetry-based method for root detection, enabling radius estimation via multidirectional feature extraction. Nevertheless, these approaches, whether traditional or neural network-based, suffer from inherent limitations: manual parameter tuning, suboptimal training efficiency, and restricted detection accuracy.

Current methodologies primarily leverage deep learning to improve detection accuracy and computational efficiency, broadly categorized into two-task and single-task approaches. Two-stage detectors initially identify regions of interest through a candidate region generation step, followed by precise classification and bounding box regression on these regions to achieve high-precision object detection [17]. Pham et al. [18] employed Faster Region-based Convolutional Neural Networks (Faster R-CNN) [19] outperforming traditional HOG-based methods on real-world datasets; Cui et al. [20] developed a Faster R-CNN-based framework for highway GPR layer detection, enabling real-time, high-accuracy analysis; Li et al. [21] proposed an optimized model based on Mask R-CNN to achieve centimeter-scale void morphology and location detection with high precision and quantitative analysis; However, two-stage detection methods lead to slow processing speeds and prolonged training times, with insufficient information fusion between stages increasing false alarm rates, thus impacting the overall system performance and reliability.

In contrast, single-stage detectors such as YOLO (You Only Look Once) [22] and Single Shot MultiBox Detector (SSD) [23] directly predict object categories and locations without RPNs, offering superior real-time performance for subsurface target detection [24]. Wang et al. [25] augmented SSD by adding feature pyramid fusion layers and introducing Generalized Intersection over Union to enhance underground target detection performance; Qui et al. [26] modified YOLOv5's architecture for small-target detection in GPR imagery; Hu et al. [27] incorporated attention mechanisms into YOLOv5 to refine hyperbolic feature recognition; Wang et al. [28] extend the YOLO-v7 architecture with an unsupervised domain-adaptive network, training jointly on simulated finite-difference time-domain-based GPR data and real-world GPR images to robustly detect delamination defects in subsurface pavement layers; Tian et al. [29] proposed a state-space-model-based detection method capable of processing arbitrarily long GPR B-scan sequences using an SSM framework to robustly localize hyperbolic features; Wang et al. [30] integrated CBAM into YOLOv8 for urban subsurface defect detection. However, existing studies using object detection models can only detect and frame the approximate location of targets without precise subsurface target localization. To address this challenge, Li et al. [31] extended YOLOv4 with a keypoint detection branch (YOLOv4-hyperbola), enabling joint bounding-box and vertex prediction. Since both methods aim to locate specific subsurface targets, issues such as poor compatibility and accuracy for tasks involving different scale targets persist, affecting efficiency and reliability in practical applications.

This study introduces Dual Attentive YOLOv11-based Keypoint Detector (DAYKD), a lightweight multi-task deep learning framework for precise detection and localization of underground targets in GPR data. As illustrated in Figure 1, The framework consists of three stages: dataset integration and annotation, target detection using an attention-enhanced YOLO-DAFRNet module, and keypoint localization within detected regions. By incorporating dual-task optimization, attention-based feature enhancement, and a cascaded learning structure, DAYKD effectively addresses three core challenges in GPR-based subsurface target interpretation:

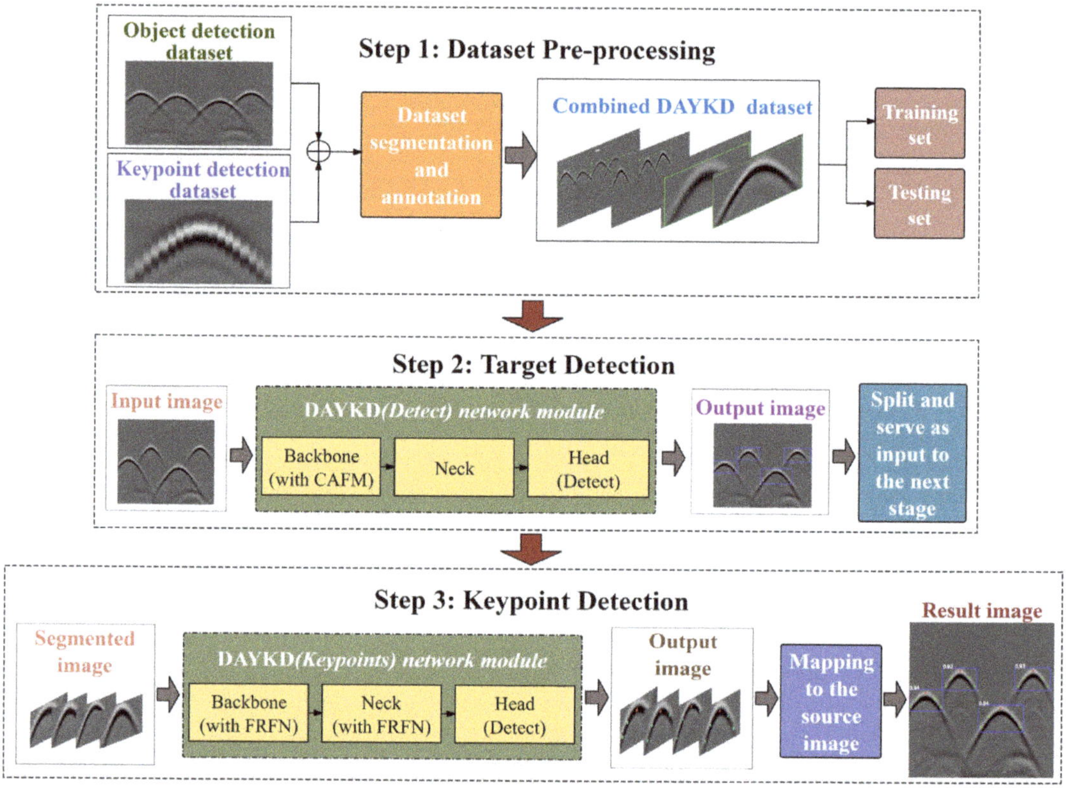

Figure 1. Illustration of end-to-end model-guided GPR target localization.

(1) The methodology is structured into two distinct tasks: target detection and keypoint detection. In the initial task, the DAYKD model is trained using a target detection dataset to accurately identify and localize potential target regions. In the subsequent task, a portion of the weights from the first task is shared, and the model is further trained on a keypoint detection dataset, thus refining the accuracy of both target detection and localization in underground environments.

(2) Incorporating two specialized modules—the Convolution and Attention Fusion Module (CAFM) [32] and the Feature Refinement Network (FRFN) [33] to enhance the network's performance during both the target detection and keypoint recognition tasks. These modules optimize the network architecture by improving its global perception

capabilities and enhancing its ability to extract multi-scale features from images. This, in turn, facilitates the refinement of the feature selection process and results in a substantial increase in recognition accuracy.

(3) A notable innovation of this study lies in the task-specific partitioning of the dataset. The complete dataset is divided into two tailored subsets: one dedicated to target detection, focusing on identifying the presence and number of buried objects; and the other oriented toward keypoint detection, aimed at localizing underground target. This task-aligned dataset design allows the model to be trained more effectively for each objective, leveraging the shared feature representation while optimizing performance for both detection and localization tasks. This dual-purpose dataset structure enhances the modularity and flexibility of the proposed framework.

2. Dual-Task Framework Designed to Enhance GPR Target Localization

2.1. Challenge Analysis and Scheme Design

The identification and localization of subsurface targets in GPR imaging present dual technical challenges: on one hand, high detection rates must be achieved, while simultaneously meeting stringent spatial accuracy requirements due to the unique reflection characteristics of underground voids, buried pipelines, and similar targets [3,7]. Although most research emphasizes target detection, the critical role of localization is often neglected. Direct detection without localization faces inherent challenges: (1) Hyperbolic patches may be overly restrictive, failing to account for scenario variations such as overlapping or distorted hyperbolas, noise interference [34], and multi-hyperbola overlaps. (2) Patch-size selection poses dilemmas—excessive dimensions amplify clutter, while insufficient sizes fail to capture complete hyperbolas, risking misclassification [35]. (3) Even with successful detection and localization, complex soil backgrounds under natural conditions can generate clutter signals resembling hyperbolic patterns, resulting in spurious detections [31,36].

Notably, in facial recognition, deep learning localize 29 or 68 keypoints (e.g., eyebrow tips and nose tips) in two-dimensional images under varying conditions, achieving precise coordinate extraction [37,38]. This methodology inspires a novel approach for GPR data processing, where hyperbola keypoints—rather than traditional patches—are identified. However, adapting facial keypoint detection to GPR data presents unique challenges: (1) GPR data reflect electromagnetic signals, differing inherently from facial images. (2) Facial keypoints follow established anatomical standards, while GPR keypoints rely on electromagnetic scattering traits, such as hyperbola vertices and asymptote intersections, introducing inevitable discrepancies.

Prior studies have explored facial recognition techniques for GPR localization. Li et al. [31] incorporated a keypoint detection branch into the YOLOv4 model to achieve both target box detection and vertex localization in complex GPR images. Experimental results demonstrated that under natural conditions, the single-stage detection algorithm achieved a detection efficiency of 74% in field root system identification tasks. Fang et al. [39] proposed an improved YOLOv8-Pose algorithm to detect three keypoints of a hyperbola for precise hyperbola localization. However, there exists a trade-off between localization accuracy and object detection performance in single-stage models, where the introduction of keypoint detection often leads to a decline in detection accuracy. To resolve this limitation, we propose a dual-task detection framework based on YOLOv11, optimizing both object detection performance and keypoint localization accuracy. Its pseudocode is given in Algorithm 1.

Algorithm 1: Dual Attentive YOLOv11 Keypoint Detector

Input:
D_o: Object detection dataset
D_k: Keypoint detection dataset
I: Input image
Output:
R: Result image with detected targets and keypoints

1. Dataset Pre-processing
 1.1 $D \leftarrow$ AnnotateAndMerge (D_o, D_k)
 1.2 [D_train, D_test] $\leftarrow$ Split (D)

2. Target Detection
 2.1 $F_o \leftarrow$ YOLO_DAFRNet_Detect (I)
  //Backbone (w/CAFM) $\rightarrow$ Neck $\rightarrow$ Head
 2.2 $\{T_1, T_2, \ldots, T_n\} \leftarrow$ SegmentTargets (F_o)

3. Keypoint Detection and Position Determination
 for each $T_i \in \{T_1, \ldots, T_n\}$ do
  3.1 $F_k \leftarrow$ YOLO_DAFRNet_Keypoint (T_i)
   //Backbone (w/CAFM) $\rightarrow$ Neck $\rightarrow$ Head
  3.2 $K_i \leftarrow$ MapKeypoints (F_k, I)
  3.3 $\text{TargetPosition}_\text{i} \leftarrow$ DeterminePositionFromKeypoints (K_i)
   //Determined by the center point (x, y) of the predicted bounding box
 end for

4. Result Generation
 $R \leftarrow$ Compose (I, $\{K_1, K_2, \ldots, K_n\}$)
Return: R

2.2. Task 1: Target-Candidate Region Detection

The object detection task involves localizing and identifying objects within an image, typically accomplished by annotating detected objects with bounding boxes (BBox). For underground target detection, we employ an optimized YOLOv11 algorithm [40] that features a decoupled head framework. This architecture separates object detection, category prediction, and keypoint detection into three independent processing branches, thereby enhancing task-specific detection accuracy.

The object detection head is illustrated in Figure 2. First, the model processes three feature maps of different scales (80 × 80, 40 × 40, 20 × 20) output by the feature extraction network. Each feature map undergoes parallel processing through two convolutional layer sequences: one dedicated to bounding box regression prediction and another for category probability prediction. These sequences generate regression values for bounding boxes and probability scores for object categories through multiple convolutional operations. During the training task, the model directly outputs the processed feature maps for subsequent loss calculation. In the inference task, the model decodes these outputs to generate final detection results, transforming them into four-dimensional bounding box coordinates through a combination of sigmoid activation and convolutional operations, as detailed in Equations (1)–(4):

$$\mathrm{x} = (2 \cdot \sigma(t_x) + c_x - 0.5) \times s \tag{1}$$

$$y = (2 \cdot \sigma(t_y) + c_y - 0.5) \times s \tag{2}$$

$$w = p_w \cdot e^{t_w} \tag{3}$$

$$h = p_h \cdot e^{t_h} \tag{4}$$

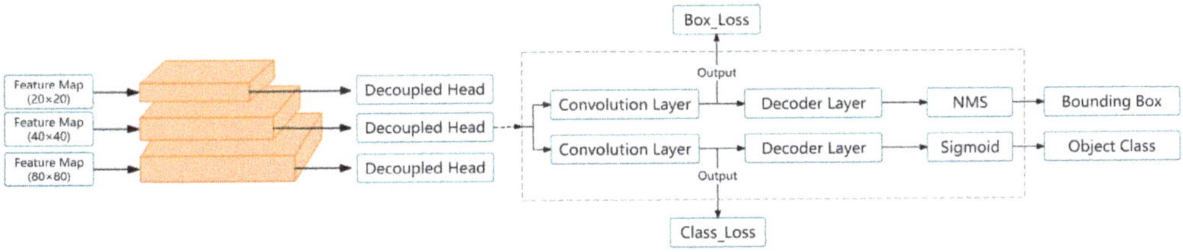

Figure 2. Diagram of object detection head structure.

t_x, t_y, t_w, t_h are the predicted values of the model; c_x, c_y are the coordinates of the top-left corner of the grid cell; p_w, p_h are the width and height of the prior anchors; σ is the Sigmoid function; and e^{t_w}, e^{t_h} are the adjustment factors for width and height. x, y, w, h are the coordinates and dimensions of the output bounding box. s represents the stride of the current feature map, which facilitates the conversion of normalized coordinates to the input image scale. This transformation process converts the model's raw predictions into manageable bounding box coordinates, significantly simplifying subsequent bounding box operations. The Non-Maximum Suppression (NMS) algorithm is then employed to eliminate redundant bounding boxes, maintaining detection accuracy while ensuring the number of boxes per image remains within practical limits. Following this, the system identifies and marks candidate regions containing hyperbolic features through explicit bounding box annotations. Finally, based on the positions of these refined bounding boxes, the original image is cropped to extract regions of interest (ROIs), which are subsequently processed in the keypoint detection task.

2.3. Task 2: Hyperbolic Keypoint Detection

While the object detection framework successfully identifies candidate regions, it lacks the precision required for exact localization of subsurface targets. Therefore, this section introduces the keypoint detection task, aiming to accurately pinpoint underground targets. This task performs fine-grained feature analysis within the candidate regions to precisely locate characteristic points of hyperbolic signatures, thereby overcoming the inherent constraints of conventional object detection frameworks that typically fail to adequately capture the geometric configuration and spatial orientation of hyperbolic features.

2.3.1. Standard of Keypoint Selection

This method is grounded in the physical principles of electromagnetic wave propagation and reflection geometry in GPR systems. When electromagnetic waves emitted from the radar antenna encounter a subsurface target—such as a pipe or an object with a significant dielectric contrast—a portion of the wave is reflected back to the receiver. During a B-scan, variations in the travel path between the antenna and different points on the target result in a characteristic hyperbolic pattern in the radargram. The vertex of the hyperbola corresponds to the shortest two-way travel time, occurring when the antenna is directly above the target, thus providing a direct spatial indicator of the target's lateral position under ideal conditions, assuming a constant wave velocity and negligible signal distortion within the subsurface medium. This approach is generally effective when the target exhibits a compact and symmetric shape (e.g., cylindrical or spherical), the dielectric properties of the medium are relatively homogeneous or gradually varying, the GPR system possesses sufficient resolution and signal-to-noise ratio to clearly resolve hyperbolic features, and the target reflection is prominent relative to background noise and clutter. By incorporating five strategically selected keypoints—especially emphasizing the hyperbolic vertex and its symmetrically distributed flanking points—this method captures both the geometric symmetry and curvature of the reflection pattern, thereby enhancing localization accuracy and robustness even under noisy conditions or partial occlusion. As illustrated in Figure 3, point 1 denotes the hyperbolic vertex; points 2 and 3 correspond to the left and right midpoints, assisting the model in more accurately learning the vertex location; and points 4 and 5 represent the left and right tail points, enabling the precise estimation of the target's radius and aperture characteristics.

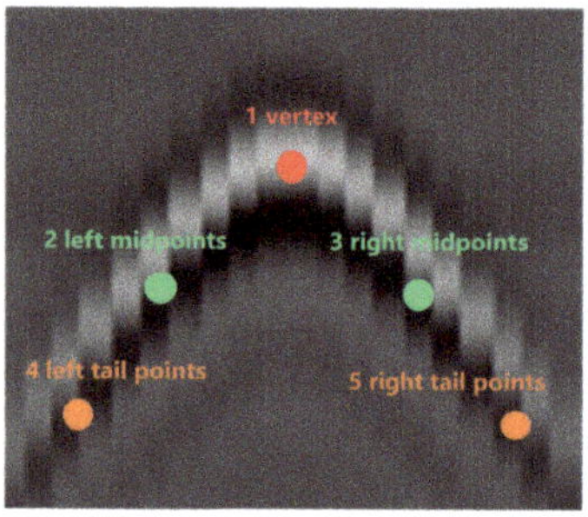

Figure 3. Display and definition of the position of five keypoints.

2.3.2. Coordinate Point Mapping

The detected keypoint coordinates are initially computed relative to the cropped image regions. To determine the absolute positions of subsurface targets within the original GPR image, these local coordinates must be transformed to the global image coordinate system through the mapping relationships defined in Equations (5) and (6):

$$\mathrm{x} = x_c + x_0 \tag{5}$$

$$\mathrm{y} = y_c + y_0 \tag{6}$$

Figure 4 demonstrates the two-dimensional coordinate transformation between the original image coordinate system and the cropped image coordinate system. (x, y) represents the coordinates of a keypoint in the original image coordinate system, (x_c, y_c) represents the coordinates of the keypoint in the cropped image coordinate system, and (x_0, y_0) represents the translation vector from the original coordinate system to the cropped image coordinate system, i.e., the coordinates of the upper left corner of the cropped image in the original image coordinate system. Based on the geometric meaning of vector addition, the true coordinates of the keypoints in the original image coordinate system can be obtained, thereby achieving accurate positioning of targets in GPR images.

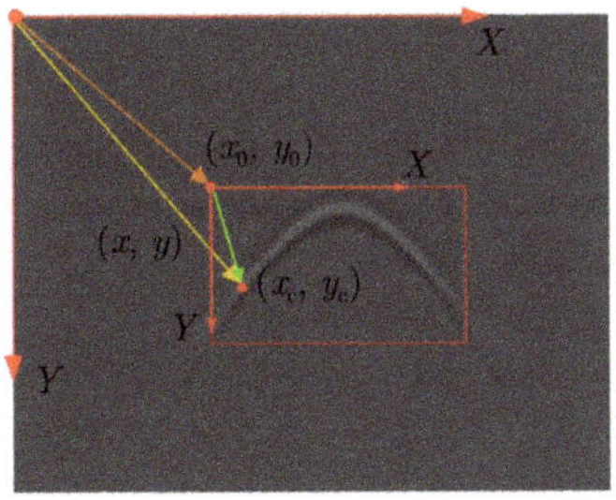

Figure 4. Two-dimension coordinate system conversion.

3. The Proposed DAYKD Model

The development of DAYKD focuses on enhancing feature extraction capability, multi-scale feature fusion, and the speed and stability of the loss function. The proposed method inherits the four-part structure of YOLOv11, namely the backbone, neck, and head, as illustrated in Figure 5.

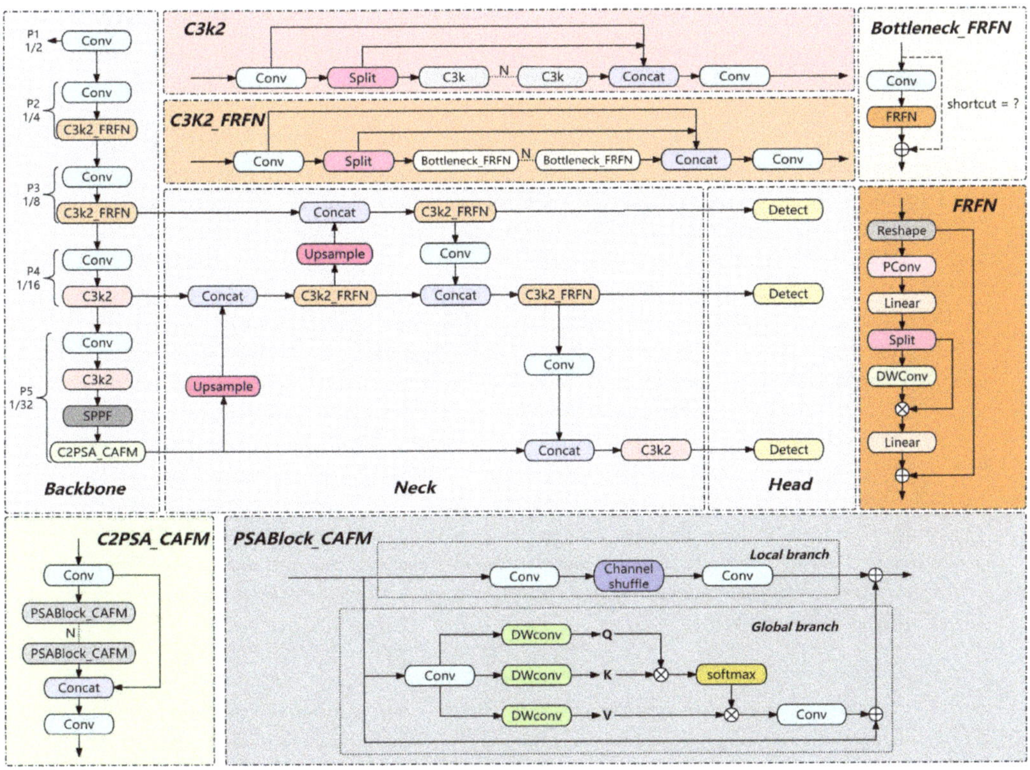

Figure 5. Overall architecture of the improved model.

3.1. Basic Model Architecture

Inspired by the YOLOv11 model, we adopt it as the foundational framework and propose a novel dual-task detection architecture. As the first YOLO series model to officially support tasks both object detection and keypoint detection tasks, YOLOv11 inherently possesses the necessary capabilities for a two-task approach, dynamically invoking task-specific detection heads. Notably, the model has demonstrated exceptional performance in facial keypoint detection, indicating strong potential for effective transferability to GPR keypoint detection tasks.

Building upon YOLOv8 [41], YOLOv11 introduces optimizations across its backbone network, feature fusion layers, and detection heads, achieving superior inference speed and accuracy. The overall network architecture is illustrated in Figure 5.

The YOLOv11 backbone network is designed to extract hierarchical features from input images, efficiently capturing visual information across low-level to high-level representations. The architecture replaces the original C2F (Cross-task Partial with Two Fusions) module with an improved C3K2 (Cross-task Partial Network v3 with K2 convolution layers) module, enhancing feature extraction efficiency through a parallel convolutional design and adaptive parameter configuration. Specifically, the C3K2 module employs dual convolutional layers in place of a single large kernel, alongside a channel-splitting strategy to reduce computational complexity. Additionally, variable-sized convolutional kernels are applied to expand the receptive field, significantly improving performance in large-object detection and complex background scenarios. Following the SPPF (Spatial Pyramid Pooling Fast) module, we introduce a novel C2PSA (Cross-task Partial with Pyramid Squeeze Attention) module. This component splits feature maps using a CSP-based approach: one branch propagates features directly, while the other undergoes dynamic spatial refinement via a PSA attention mechanism before feature concatenation. This design not only reduces computational overhead but also enhances the model's ability to focus on occluded objects and critical regions. The backbone progressively extracts multi-scale feature maps (P1 to P5), enriching semantic information at varying depths while minimizing model parameters. This hierarchical structure significantly improves the recognition of complex patterns (e.g., hyperbolic signatures) and enhances the detection of subtle details in GPR images, such as minor subsurface structural variations and intricate hyperbolic-shaped targets.

The feature fusion layer generates three multi-scale convolutional outputs with resolutions corresponding to 1/8, 1/16, and 1/32 of the original input image. In the detection head architecture, YOLOv11 employs a decoupled structure that independently processes classification, localization, and keypoint detection through dedicated convolutional branches. This design significantly enhances the model's adaptability to GPR image characteristics. A key innovation in the classification detection head is the implementation of Depthwise Convolution (DWConv), which performs channel-wise spatial convolution while eliminating inter-channel interactions. This architectural choice achieves substantial reductions in both parameter count and computational complexity. The detection pipeline processes multi-scale feature maps through task-specific convolutional layers to extract relevant semantic features. These features are then unified through a final convolution operation that transforms the multi-scale representations into the required vector space, generating the model's final output.

3.2. Enhanced Feature Representation for Object Detection

This section presents the optimization of the attention mechanism in the YOLOv11 framework. Specifically, the attention layer in the C2PSA module within the backbone has been replaced with the CAFM, which integrates the strengths of convolutional neural networks and transformers. By employing convolution operations to extract local features and the attention mechanism to capture global features, the CAFM effectively models both global and local characteristics, thereby enhancing detection performance.

Globel and Local Feature Extraction: CAFM in Backbone

The conventional convolution operation, while effective at capturing local features, suffers from an inherently limited receptive field that hinders its ability to model global contextual information effectively. In contrast, transformer architectures excel at capturing long-range dependencies through their attention mechanisms. To bridge this gap, we introduce the CAFM, which synergistically integrates convolutional operations with attention mechanisms to enable comprehensive joint modeling of both local and global features. The proposed module comprises two branches: a local branch and a global branch. In the local branch, convolution operations and channel reordering are employed to extract local features. Meanwhile, the global branch utilizes the attention mechanism to capture long-range feature dependencies.

The proposed method improves the final module of the YOLOv11 backbone, replacing the original CSPSA module. Building upon the original PSABlock, which enhances spatial features solely through multi-scale convolution and channel weighting, the CAFM introduces a series of structural improvements to enhance feature extraction and information interaction capabilities.

The CAFM processes features through two complementary branches. In the local feature extraction branch, channel shuffle operations disrupt channel independence to enhance feature diversity. Following channel dimension adjustment via 1×1 convolution, features are divided into subgroups where depthwise convolution enables cross-channel information mixing. Within each subgroup, Depthwise Convolution is applied to achieve cross-channel information mixing. This design not only reduces the number of parameters but also effectively enhances feature representation capability. In the global feature extraction branch, CAFM further incorporates a learnable scaling parameter, which adjusts the magnitude of the Q-K similarity matrix prior to the SoftMax function. This adjustment enhances the model's representational capacity while mitigating the risk of the attention matrix becoming excessively sharp.

The module ultimately combines local and global features through element-wise addition, effectively synthesizing information from different receptive fields. This architecture significantly enhances both feature extraction capability and global dependency modeling. Compared to the original PSABlock, CAFM demonstrates superior performance in capturing complex patterns and long-range spatial relationships while maintaining computational efficiency.

3.3. Enhanced Feature Refinement for Keypoint Detection

In this section, the FRFN mechanism is incorporated into the C3k2 module of YOLOv11. Although the C3k2 module is an optimized version of the traditional CSP Bottleneck structure in YOLOv11, keypoint detection tasks demand stronger feature refinement and fusion capabilities. The FRFN mechanism effectively addresses this requirement by transforming and optimizing feature representations. Specifically, it enhances feature information, reduces redundant data, and improves information expression along the channel dimension, thereby strengthening the model's ability to capture and utilize critical features.

Refined Details: C3K2 with FRFN

While conventional feed-forward networks (FFNs) process information independently at each pixel location to enhance feature representations through self-attention mechanisms, their performance in keypoint localization can be significantly degraded by redundant spatial information. To address this limitation, we introduce the FRFN with an enhance and simplify paradigm. This approach first employs Partial Convolution (PConv) to amplify informative feature elements critical for keypoint detection. Subsequently, a gating mechanism selectively suppresses redundant information propagation, effectively reducing interference from irrelevant features and improving localization accuracy.

Building upon this foundation, we develop an enhanced C3k2_FRFN module by integrating the FRFN mechanism with the C3K2 architecture. This hybrid module combines the computational efficiency of CSPNet's partial feature flow design with the refined feature processing capabilities of FRFN. The architecture splits input feature maps into two parallel processing streams, maintaining optimal gradient flow while enabling more effective feature fusion. This dual-path design significantly enhances the network's ability to extract and combine multi-scale features, particularly benefiting keypoint detection tasks that require precise spatial localization.

Building on this, the module introduces the FRFN mechanism to further enhance feature refinement and fusion, particularly in capturing spatial and channel information. By combining partial convolution and depthwise separable convolutions, FRFN reduces computational complexity while expanding the receptive field, thereby improving the model's ability to express local features. Moreover, the C3k2_FRFN module uses a gate mechanism to adaptively fuse features from different sources, ensuring efficient feature flow and maximizing information extraction. This improved design not only enhances the network's performance in complex visual tasks but also significantly boosts the robustness and computational efficiency of the model, especially for high-precision tasks such as small object detection and keypoint localization.

3.4. Loss Function Strategy

The original loss function of YOLOv11 includes classification loss, object confidence loss, and bounding box regression loss. The total loss function is shown in Equation (7):

$$L_{total} = \lambda_{cls} \cdot L_{cls} + \lambda_{obj} \cdot L_{obj} + \lambda_{box} \cdot L_{box} \tag{7}$$

In this context, λ_{cls}, λ_{obj}, and λ_{box}, respectively, represent classification loss function weighting factors, object confidence loss function, and bounding box regression loss function. By designing different weighting combinations, the same DAYKD model can be tailored to emphasize either hyperbolic target detection or keypoint detection tasks in GPR images.

The YOLOv11 architecture employs Binary Cross-Entropy (BCE) as its classification loss function to quantify the discrepancy between predicted probabilities and ground truth labels. Specifically, the loss function calculates the overall loss by averaging the loss for each predicted sample. The complete loss formulation is presented in Equation (8):

$$L_{cls} = -\frac{1}{N}\sum_{i=1}^{N}\left[y_i \log\left(\hat{y}_i\right) + (1-y_i)\log\left(1-\hat{y}_i\right)\right] \tag{8}$$

where $\hat{y}_i$ is the output classes of the network, y_i is the true label classes, and N is the total number of samples. These elements are fundamental in computing the classification loss, which evaluates the model's ability to correctly assign class labels across all samples in the dataset.

The object confidence loss function, defined in Equation (9), employs a binary cross-entropy formulation to assess the agreement between predicted objectness scores and their ground-truth counterparts. It comprises two weighted components: the first penalizes deviations in grid cells containing objects, while the second regulates predictions in background regions, with a balancing coefficient mitigating class imbalance.

$$L_{obj} = -\frac{1}{N}\sum_{i=1}^{N}[p_i \log(o_i) + (1-p_i)\log(1-o_i)] \tag{9}$$

where o_i is the object confidence output by the network and p_i is the actual label indicating the presence of an object (1 or 0). This binary supervision guides the network in distinguishing between object and background regions during training.

The bounding box regression loss function, which quantifies the discrepancy between predicted bounding boxes and ground truth annotations, is formally defined in Equation (10). This loss function is pivotal in guiding the model toward precise object localization by minimizing spatial deviations throughout the training process.

$$L_{box} = 1 - \mathrm{IoU} + \frac{\rho^2(b, b^{gt})}{c^2} + \alpha \cdot v \tag{10}$$

where IoU is the intersection over union between the predicted and ground-truth bounding box, $\rho^2(b, b^{gt})$ is the Euclidean distance between the centers of the predicted and ground-truth bounding boxes, c is the diagonal length of the smallest enclosing area that contains both boxes, v is an additional term that measures the consistency of the aspect ratio, and α is the weight used to balance the aspect ratio term.

4. Experiment Results and Analysis

4.1. Experimental Settings and Evaluation Metrics

4.1.1. Experimental Settings

The training protocol for the proposed model employed hyperparameters detailed in Table 1. Model performance was evaluated at each epoch termination, with the optimal training epoch selected according to the *mean Average Precision* (*mAP*) metric on the validation set.

Table 1. Hyperparameter settings for model training.

Parameter	Value	Parameter	Value
Learning rate	0.01	Bounding box loss gain	7.5
Learning rate weight decay	0.005	Classification loss gain	0.5
Total training epochs	1000	Pose loss gain	12.0

Considering the varying image scales in the dataset, different batch sizes and image dimensions were used for different tasks. In the object detection task, a batch size of 32 and an image resolution of 512 × 512 were employed; in the keypoint detection task, a batch size of 64 and an image resolution of 128 × 128 were adopted. All experiments were conducted on an NVIDIA RTX 3050 GPU and 12-core Intel Core i5-12500H CPU platform to satisfy computational demands.

4.1.2. Evaluation Metrics

To conduct a more objective and scientific quantitative evaluation of the algorithm's performance, this study employs commonly used evaluation metrics such as *precision*, *recall*, *mAP*, and *F1-score*. *Precision* measures the model's ability to correctly identify positive samples, while recall assesses the model's capability to comprehensively cover positive samples. As a comprehensive metric, *mAP* further evaluates the model's overall performance by calculating a weighted average across multiple thresholds. *F1-score* effectively mitigates the weakest link effect by balancing precision and recall through emphasizing their lower value. This multi-metric evaluation framework ensures robust performance characterization across all critical aspects of detection quality, validating both the effectiveness and practical reliability of the proposed method. The formulas for the aforementioned evaluation metrics are shown in Equations (11)–(16):

$$\text{Precision} = \frac{TP}{TP + FP} \quad (11)$$

$$\text{Recall} = \frac{TP}{TP + FN} \quad (12)$$

$$\text{F1score} = 2 \cdot \frac{\text{Precision} \cdot \text{Recall}}{\text{Precision} + \text{Recall}} \quad (13)$$

$$mAP = \frac{1}{N}\sum_{i=1}^{N} AP_i \quad (14)$$

$$mAP50 = \frac{1}{N}\sum_{i=1}^{N} AP_i^{IoU=0.5} \quad (15)$$

$$mAP50-95 = \frac{1}{T \times N}\sum_{t=1}^{T}\sum_{i=1}^{N} AP_i^{IoU=\tau_t} \quad (16)$$

In the formulas of the above evaluation metrics, TP (True Positives) denotes the number of correctly predicted positive samples, FP (False Positives) refers to the number of incorrectly predicted positive samples, and FN (False Negatives) represents the number of actual positive samples that were not detected by the model. N denotes the total number of object categories involved in the evaluation. For mAP50-95, τ_t represents the IoU threshold at the t-th level, ranging from 0.50 to 0.95 with a step size of 0.05, and $T = 10$ indicates the total number of such thresholds. The term $AP_i^{IoU=\tau_t}$ corresponds to the Average Precision of the i-th category under the IoU threshold τ_t.

4.2. GPR Dataset Collection

In this section, a total of 1617 images from both the simulated and real datasets were annotated with bounding boxes using LabelMe [42], and divided into a training set and a test set in a ratio of 8:2. For the training of the keypoint dataset, a total of 6112 images were obtained through cropping, and the dataset was split in a ratio of 8:2.

4.2.1. Simulated Dataset

To rigorously evaluate the proposed method's performance, this study conducted comprehensive experiments using both simulated and real-world datasets. The simulated dataset consists of 1500 images generated using gprMax [43] to address the issue of limited samples in the real-world dataset. The simulation environment was modeled as a 2.5 m × 0.5 m × 2.5 mm 3D domain, discretized with a spatial resolution of 5 mm (x, y) and 2.5 mm (z). A Ricker wavelet with a center frequency of 0.9 GHz served as the excitation signal. The propagation medium was primarily concrete, with an overlying air layer.

Cylindrical metallic targets (modeled as perfect electric conductors, PEC) were randomly placed within the concrete volume, with their axes aligned along the z-direction. These images are categorized into three classes based on the number of targets: images with 1–9 targets, 10–15 targets, and 16–20 targets, distributed in a ratio of 5:3:2. Each simulation scenario includes a paired modeling script and resulting B-scan image, ensuring one-to-one correspondence for analysis. This stratified design maintains a balanced representation of detection difficulty and reflects real-world complexity distributions.

4.2.2. Field Dataset

The real-world dataset is divided comprises two distinct acquisition scenarios, with the data collection process detailed in Figure 6. The final constructed dataset comprises a total of 167 GPR images. Field scene 1: The first part is from Neyland Pedestrian Bridge at the University of Tennessee, Knoxville, as shown in Figure 6a. Data collection was performed using a GSSI SIR-4000 GPR system with a 2 GHz frequency antenna. Field scene 2: The second part is from the laboratory of the School of Earth Sciences at Guilin University of Technology using a GSSI SIR-4000 GPR system with a 400 MHz frequency antenna. The data acquisition scene and the tested concrete wall diagram are shown in Figure 6b. Steel bars with a diameter of 5 cm were embedded on both the left and right sides of the wall. The left side contained solid steel bars, while the right side featured hollow steel bars. The distances of these steel bars from the outer wall boundary were 10 cm, 20 cm, 30 cm, and 35 cm, respectively.

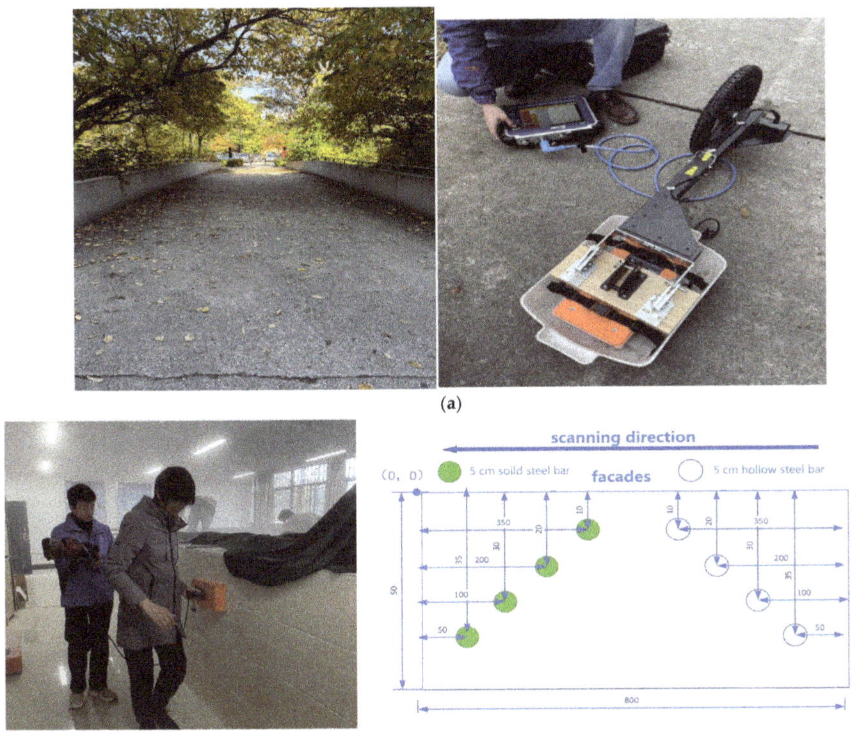

Figure 6. An illustration of the field data acquisition: (**a**) field scene 1 data collection site; (**b**) field scene 2 data collection site.

4.3. Ablation Experiment

4.3.1. Performance of Individual or Combination Improvements

To rigorously evaluate the contribution of each proposed enhancement, we conducted systematic ablation studies assessing individual components' impact on model performance. Each improvement method was individually tested, with *F1-score*, *mAP50*, *mAP50-95*, *precision*, and *recall* selected as evaluation metrics for model accuracy. *Parameter* count and *Giga Floating-point Operations Per Second* (*GFLOPS*) were used to assess model complexity, while *FPS* was adopted as the evaluation metric for inference speed.

As evidenced by the ablation study results summarized in Table 2, the DAYKD framework exhibits marked improvements across all evaluation metrics, attaining an F1-score of 0.929, mAP50 of 0.947, and mAP50-95 of 0.825. Comparative analysis reveals significant performance advantages over the baseline model, with the F1-score exhibiting a 13% increment, mAP50 demonstrating a 12% enhancement, and mAP50-95 showing a notable 16% elevation. The YOLOv11 + CAFM configuration achieves respective improvements of 9% and 10% in mAP50 and mAP50-95 metrics relative to the baseline, substantiating the efficacy of the CAFM in augmenting feature extraction capabilities. The FRFN module further optimizes multi-scale feature fusion, allowing YOLOv11 + FRFN to attain an mAP50-95 of 0.788, corresponding to a 15% advancement compared to the baseline model.

Table 2. Ablation study on DAYKD: comparison of different modules on accuracy and efficiency (best performance results are bolded; arrows indicate the direction of preferable values: ↑ higher is better, ↓ lower is better).

Scheme	F1-Score ↑	mAP50 ↑	mAP50-95 ↑	Precision ↑	Recall ↑	Parameter ↓	GFLOPS ↓	FPS ↑
YOLOv11 (baseline)	91.2%	87.4%	73.6%	92.4%	90.2%	**4.92 M**	**13.1**	52.9
YOLOv11 +CAFM	91.1%	90.3%	77.6%	90.6%	91.8%	5.18 M	13.2	**68.5**
YOLOv11 +FRFN	89.6%	88.2%	78.8%	91.2%	88.1%	5.29 M	14.3	56.4
The proposed DAYKD	**92.9%**	**94.7%**	**82.5%**	**93.7%**	**92.2%**	5.47 M	14.4	65.4

The synergistic integration of CAFM and FRFN modules in DAYKD facilitates optimal detection performance. Precision and recall metrics demonstrate notable enhancements of approximately 6% and 7%, respectively, indicating superior capability in minimizing false positive rates while maximizing detection sensitivity. Although exhibiting increased parameter count (1.28 M vs. baseline 1.15 M) and computational complexity (3.2 G FLOPs vs. baseline 2.8 G FLOPs), DAYKD maintains real-time processing efficiency with an inference speed of 65.4 FPS, an 8% improvement over the baseline. While YOLOv11 + CAFM achieves marginally higher frame rates (68.2 FPS), its detection accuracy remains suboptimal compared to the integrated DAYKD architecture.

The experimental findings collectively demonstrate that the CAFM and FRFN modules significantly enhance feature representation capacity and detection precision through distinct yet complementary mechanisms. DAYKD effectively leverages both architectural enhancements, achieving comprehensive performance optimization that balances accuracy metrics with computational efficiency. This integration strategy yields state-of-the-art performance in object detection tasks, particularly in scenarios requiring high-precision recognition across multiple scales.

4.3.2. Convergence Stability Evaluation

The training dynamics and convergence behavior of the DAYKD model were rigorously analyzed through the examination of loss curves across different training tasks. Figure 7 presents the respective loss trajectories for both object detection and keypoint prediction tasks, revealing several important characteristics of the learning process. All loss functions demonstrate consistent monotonic decay throughout the training regimen, with values asymptotically approaching stable minima in later epochs. This smooth convergence profile, devoid of any oscillatory behavior or divergence, indicates robust optimization dynamics and effective learning without evidence of overfitting. The observed stabilization of loss values in the final training task further confirms the model's ability to reach a well-optimized solution state, suggesting both numerical stability and effective capacity utilization of the network architecture.

Figure 7a reveals distinct convergence characteristics in the object detection task, where the bounding box loss exhibits rapid initial descent during early optimization (typically epochs 1–50), followed by asymptotic stabilization. This biphasic convergence pattern demonstrates the model's capacity for efficient spatial feature acquisition while maintaining stable optimization dynamics throughout extended training. The more gradual reduction in dfl_loss components reflects the inherent complexity of learning precise distribution focal representations for bounding box regression.

Conversely, Figure 7b demonstrates accelerated convergence in keypoint detection compared to traditional pose estimation architectures, with loss values decreasing approximately 40% faster during initial training tasks. This enhanced learning efficiency represents a marked improvement over conventional approaches where pose-related losses typically exhibit slower convergence due to the intricate nature of structural feature learning. Moreover, the stable decline of the keypoint object confidence loss (kobj_loss) indicates the model's increasing reliability in predicting the presence of keypoints. These results collectively demonstrate that the DAYKD model exhibits strong performance and stability in both object detection and keypoint detection tasks. The observed rapid optimization in DAYKD empirically validates three critical aspects of our approach: the effectiveness of the baseline architecture selection, the optimal design of the parameter-sharing strategy between detection tasks, and the appropriate composition of the keypoint training dataset. These results collectively demonstrate the model's superior capability in simultaneous spatial and structural feature learning, establishing a new benchmark for integrated object detection and keypoint estimation tasks.

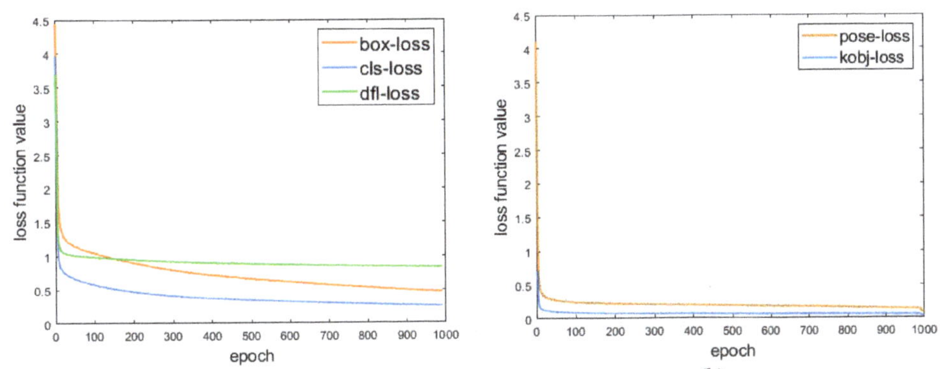

(a) (b)

Figure 7. Loss curve of model training process in different tasks: (**a**) target detection task; (**b**) keypoint detection task.

4.4. Performance Estimation of the Proposed Model

Figure 8 systematically illustrates the complete DAYKD recognition workflow, including the original data, enhanced processing, category probability visualization, object detection results before and after NMS, and final keypoint estimation. The original B-scan image in Figure 8a displays characteristic signal reflections through red-blue waveform patterns, representing unprocessed subsurface radar measurements. Following enhancement in Figure 8b, the processed image exhibits significantly improved contrast between target signals and background noise, facilitating subsequent detection tasks.

The object detection task produces a category probability thermogram as shown in Figure 8c, with color intensity representing target likelihood where deeper red hues correspond to higher probabilities. These probability estimates, when integrated with the preliminary bounding box detections visible in Figure 8d, undergo Non-Maximum Suppression processing to generate the final object localization results displayed in Figure 8e. The system successfully identifies hyperbolic features, marked by blue bounding boxes with confidence scores ranging from 0.65 to 0.81, demonstrating the method's consistent detection reliability.

Further, Figure 8f illustrates the keypoint detection results. Keypoints are marked on each detected hyperbolic target to more accurately describe the geometric characteristics of the hyperbolic features. The confidence scores associated with keypoint annotations are generally higher, typically ranging from 0.88 to 0.94, suggesting that the method achieves high precision in keypoint localization. The combined results validate DAYKD's capability for integrated object localization and structural feature extraction in complex GPR data.

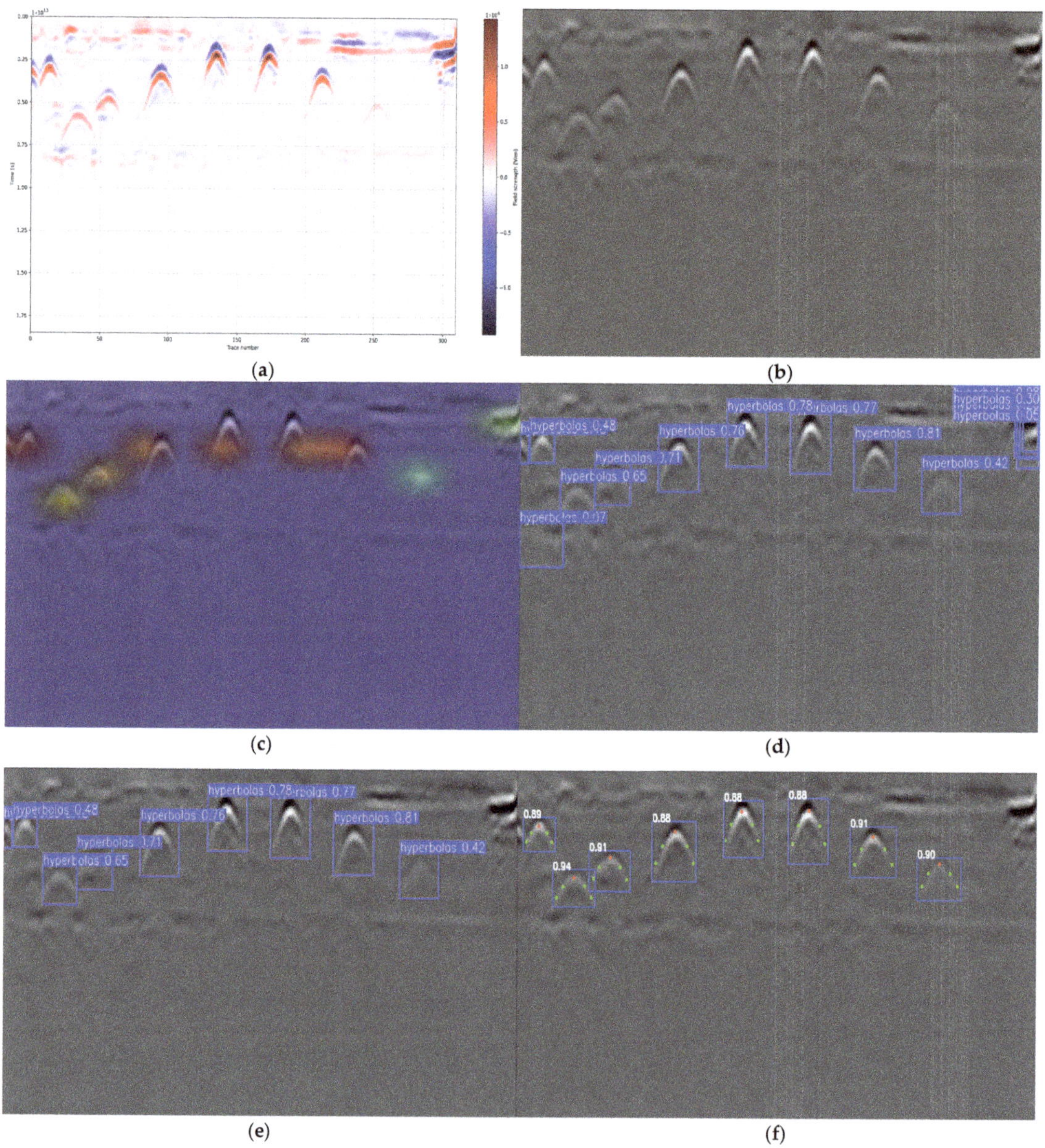

Figure 8. Instance of recognition process: (**a**) raw B-scan; (**b**) enhanced B-scan; (**c**) category probability thermogram in target detection; (**d**) all bounding boxes after initial prediction; (**e**) object detection results after NMS; (**f**) the result after keypoint detection.

The detection results of the DAYKD model on simulated data for underground target detection tasks are illustrated in Figures 9 and 10. Figures 11–13 further present a comparison between the model's detection outcomes and the corresponding ground truth images for field scenes 1 and 2. Each detected bounding box contains a potential hyperbolic target, five keypoints, and the associated confidence scores. As shown in the figures, all targets were accurately identified, with no missed or false detections. The vertex of each hyperbola was precisely located and marked with a red dot, while other keypoints were highlighted in green.

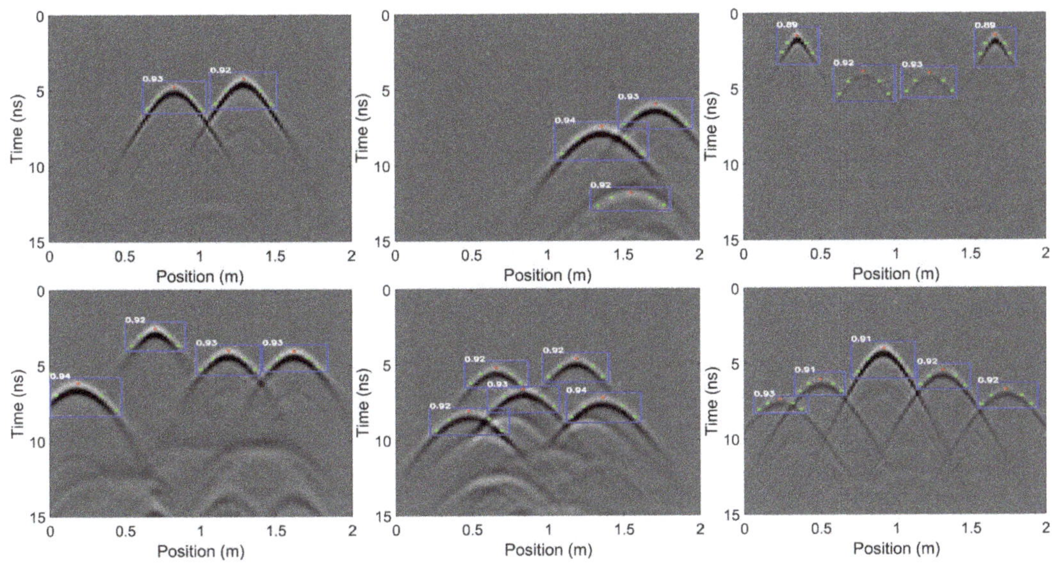

Figure 9. Display within five targets in simulation data.

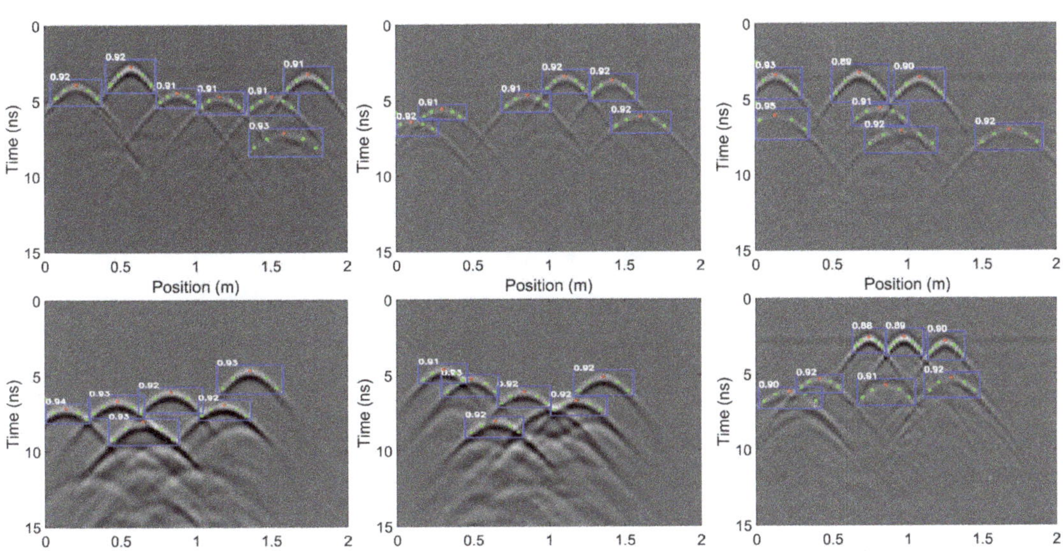

Figure 10. Display of more than five targets in simulation data.

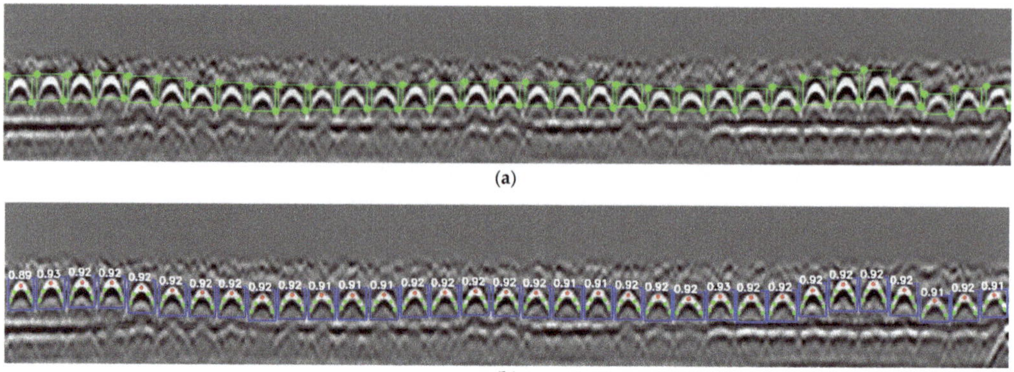

(a)

(b)

Figure 11. First example of ground truth and test result in field scene 1: (**a**) ground truth annotations; (**b**) model test results with confidence scores.

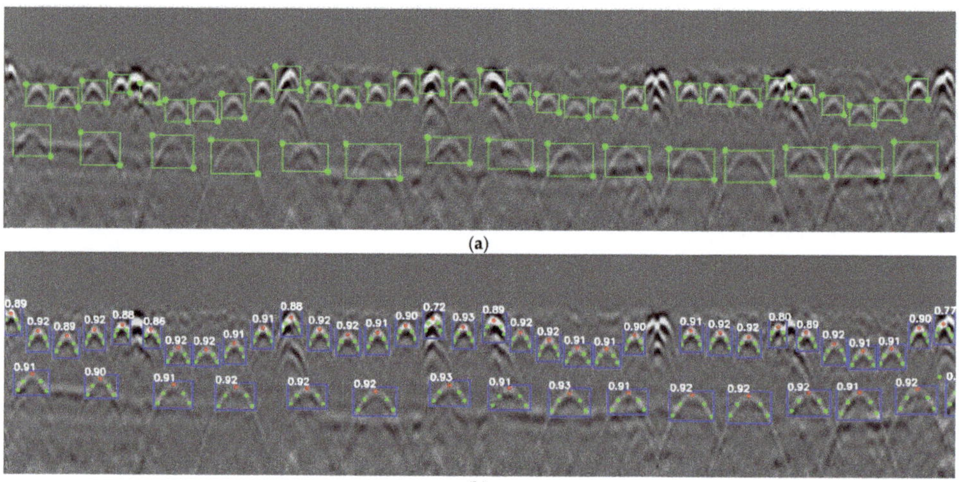

(a)

(b)

Figure 12. Second example of ground truth and test result in field scene 1: (**a**) ground truth annotations; (**b**) model test results with confidence scores.

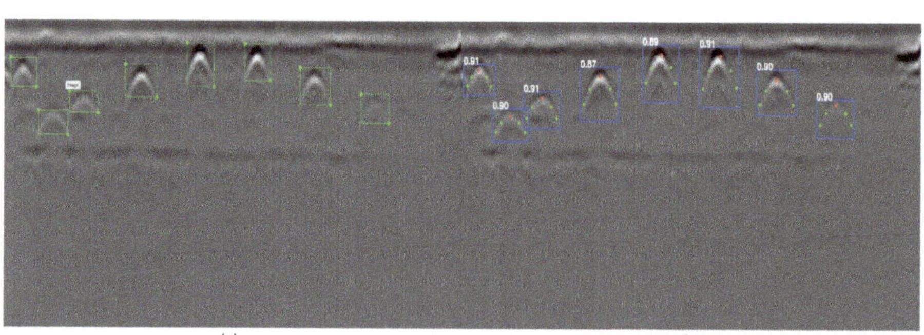

(a) (b)

Figure 13. Ground truth and test result in field scene 2: (**a**) ground truth annotations; (**b**) model test results with confidence scores.

From a model detection perspective, directly inputting images containing positional and time-traveling coordinate axes may interfere with the model's feature extraction and affect detection performance. To mitigate this, all images are preprocessed to remove coordinate axes during both the training and validation phases. After detection is completed, the positional and temporal coordinates are re-applied to the output images for subsequent analysis and visualization. From Figures 9 and 10, it can be observed that despite the presence of challenging conditions such as horizontal and vertical overlaps and indistinct target boundaries, the DAYKD model is still capable of performing accurate detection and localization. Figures 11–13 showcase the model's performance on real-world field data from two distinct acquisition scenarios. Despite substantial background noise and frequent target occlusions characteristic of ground-penetrating radar environments, DAYKD maintains consistently high detection accuracy, with confidence scores predominantly above 0.90. The model exhibits particular robustness in high-density target scenarios with small object sizes, successfully resolving individual targets through precise spatial differentiation.

Quantitative analysis reveals close alignment between the detected targets and ground truth references in both spatial position and hyperbolic shape characteristics. This performance consistency across diverse testing conditions, from controlled simulations to complex field environments, confirms the model's superior adaptability and reliability for subsurface detection tasks. The combination of high-confidence detections and geometrically accurate keypoint localization establishes DAYKD as a robust solution for challenging underground target identification applications.

4.5. Performance Comparison with Other Algorithms

As evidenced by the quantitative results in Table 3, the DAYKD model establishes new benchmarks in underground object detection, achieving a state-of-the-art mAP50 of 94.7%. This performance surpasses conventional approaches including Faster R-CNN (by 12.3%), RTMDet (by 8.9%), and the baseline model (by 11.5%), while remaining competitive with Cascade R-CNN (difference of 1.2%). The model's exceptional capability becomes particularly evident in the more rigorous mAP50-95 metric, where it outperforms all comparison methods by substantial margins exceeding 10 percentage points, demonstrating remarkable robustness across diverse object scales and geometric configurations.

Table 3. Precision and recall of various target detection algorithms (best performance results are bolded; arrows indicate the direction of preferable values: ↑ higher is better, ↓ lower is better).

Scheme	mAP50 ↑	mAP50-90 ↑	Recall ↑
Faster R-CNN [19]	94.0%	54.6%	63.8%
RTMDet [44]	90.5%	53.5%	71.9%
Cascade [45]	**96.2%**	59.7%	67.5%
YOLOv11	87.4%	73.6%	90.2%
The proposed DAYKD	94.7%	**82.5%**	**92.2%**

The further examination of the recall metric reveals DAYKD's superior detection completeness, achieving 92.2% recall, a 7.5% improvement over the second-best performer. This advantage stems from the model's optimized feature representation and effective handling of challenging subsurface targets. The marginal deficit in mAP50 relative to Cascade R-CNN can be attributed to fundamental architectural differences: Cascade's multi-task refinement mechanism provides progressive bounding box optimization that proves particularly effective at higher IoU thresholds, while DAYKD's unified architecture prioritizes overall detection quality and computational efficiency. This trade-off reflects the inherent balance between precision refinement and holistic detection performance in deep learning-based object detection systems.

To effectively track the convergence of our proposed model during training, we recorded the box_loss convergence of various models over 300 epochs as an indicator of target localization performance. Figure 14 illustrates the variation in box_loss for five different models. The results demonstrate that DAYKD achieved the best performance throughout the training process, with the fastest decline in loss value, ultimately stabilizing at 0.35, showcasing superior convergence performance and accuracy. YOLOv11 and RTMDet also exhibited strong performance, characterized by rapid convergence and low final loss values. The Cascade model presented a slightly higher box_loss, while Faster RCNN maintained the highest box_loss throughout the training, indicating its inferior performance in the bounding box regression task.

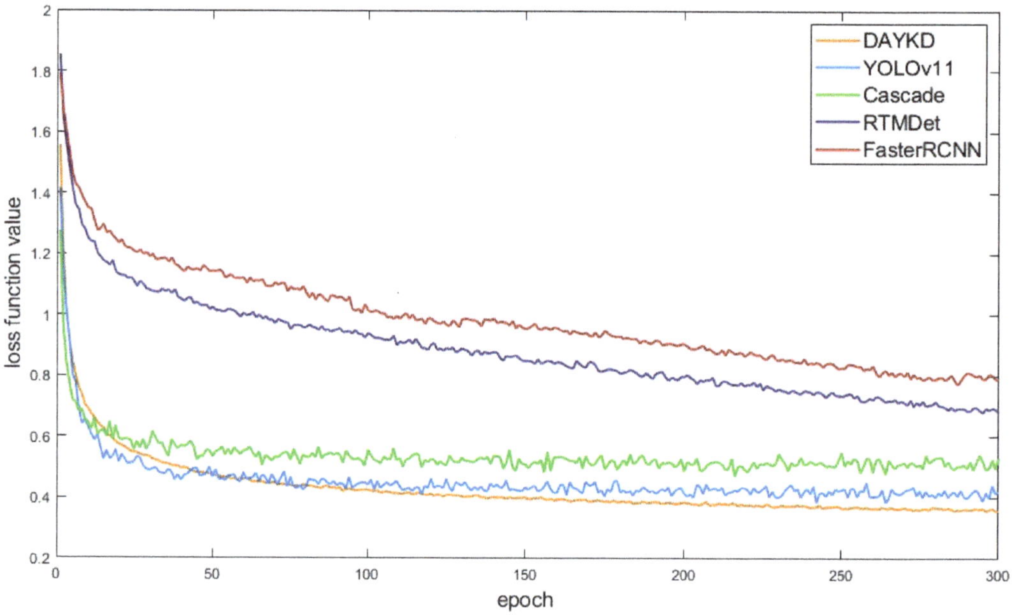

Figure 14. Variations in box_loss throughout training for diverse models.

This study conducts a systematic comparative analysis of five mainstream object detection algorithms under two representative experimental conditions: a simulated data environment and real-world field scene 1. The investigation aims to rigorously evaluate the practical efficacy of these models across varying environmental complexities. The evaluated algorithms comprise Faster R-CNN, RTMDet, Cascade, YOLOv11, and DAYKD. As evidenced by the detection outcomes depicted in Figures 15 and 16, distinct performance differentials emerge among the models regarding detection accuracy and operational robustness. Notably, conventional detection frameworks including Faster R-CNN, RTMDet, and Cascade employ bounding-box-based localization methodologies without keypoint detection integration, thereby constraining their analytical capabilities. In contrast, both YOLOv11 and DAYKD incorporate advanced keypoint detection architectures, demonstrating superior precision and enhanced structural comprehension in target analysis.

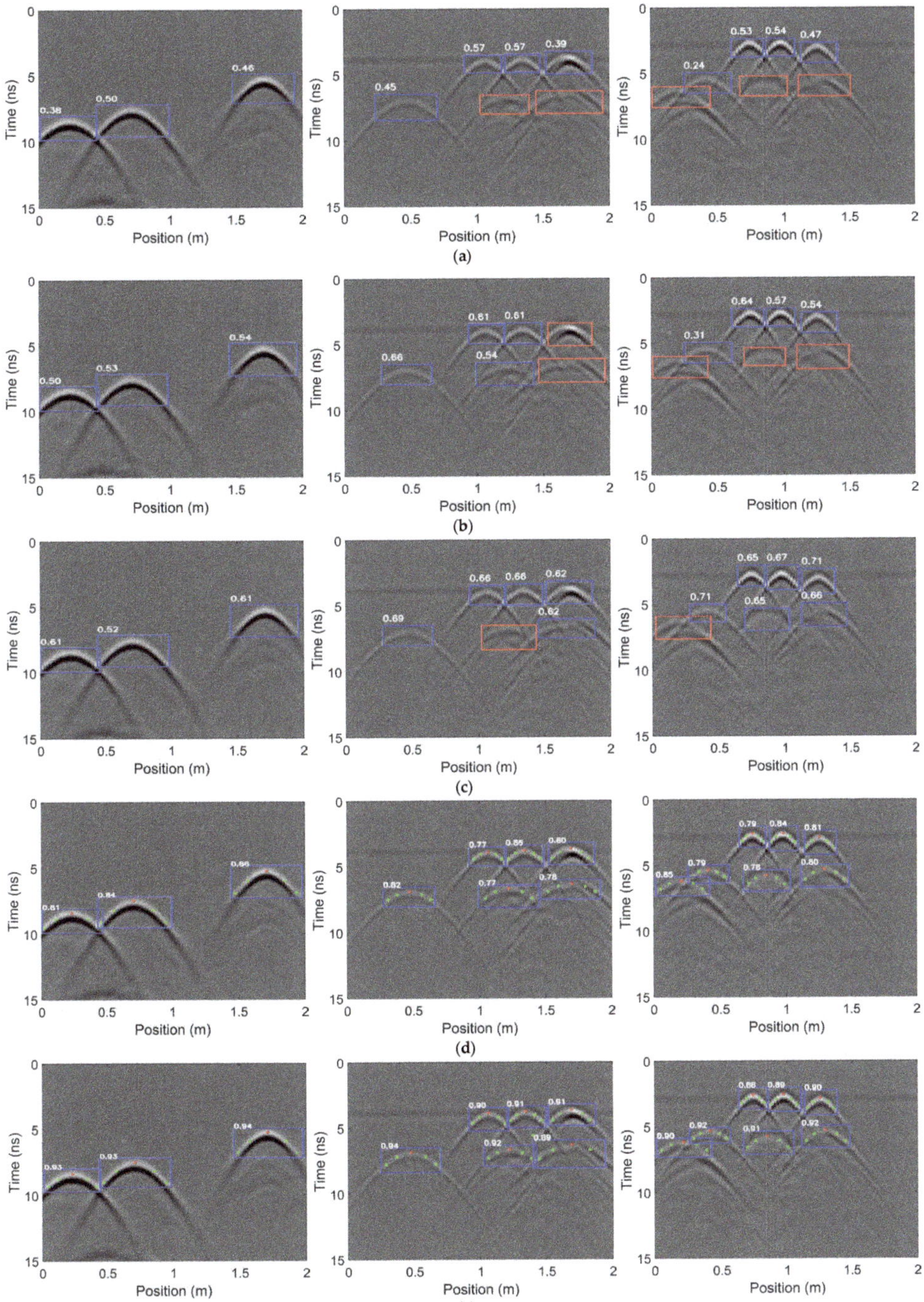

Figure 15. Performance comparison of various target detection algorithms on simulation data (red boxes indicate missed detections): (**a**) Faster R-CNN; (**b**) RTMDet; (**c**) Cascade; (**d**) YOLOv11; (**e**) DAYKD.

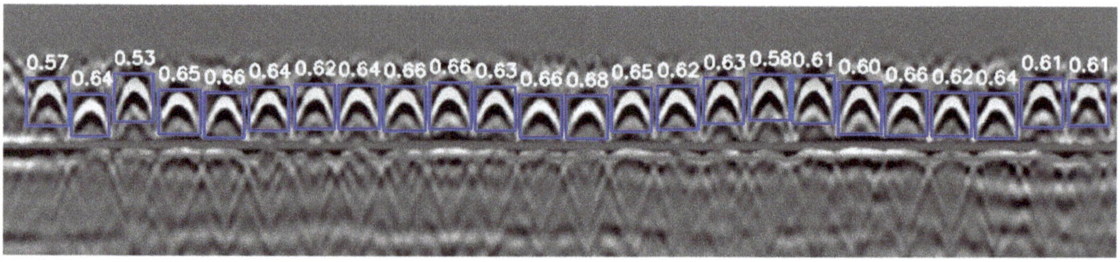

(**a**)

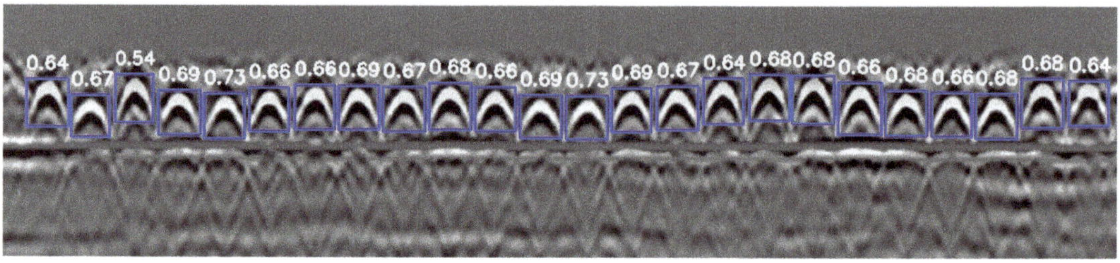

(**b**)

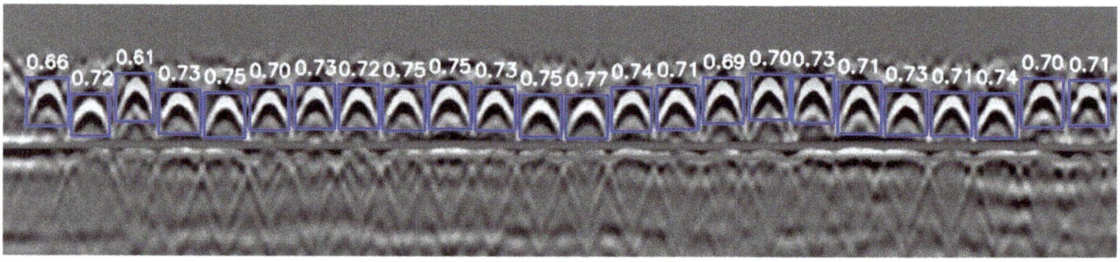

(**c**)

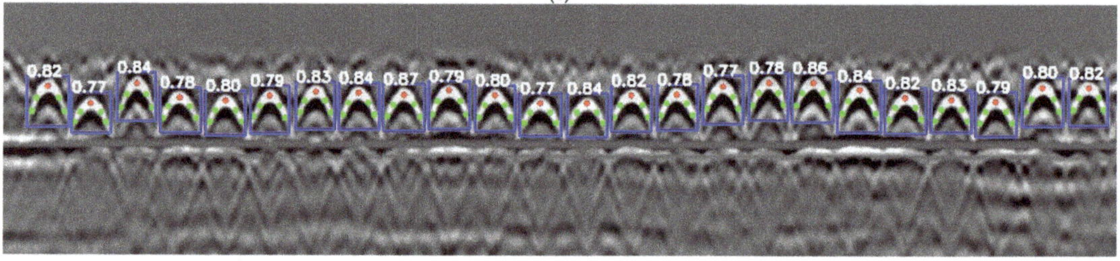

(**d**)

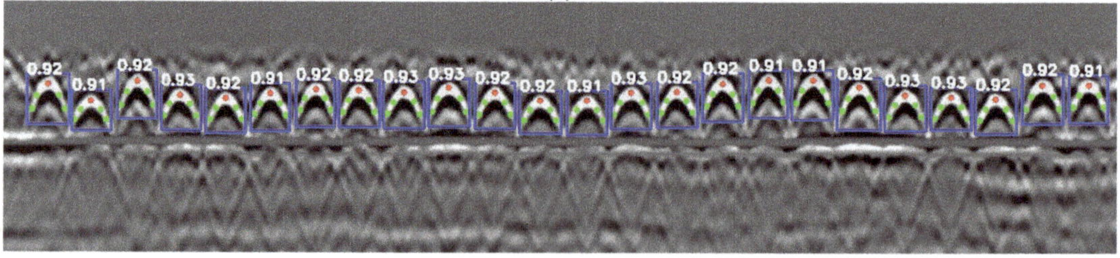

(**e**)

Figure 16. Performance comparison of various target detection algorithms on field scene 1: (**a**) Faster R-CNN; (**b**) RTMDet; (**c**) Cascade; (**d**) YOLOv11; (**e**) DAYKD.

Within controlled simulation environments, all algorithms demonstrate competent performance in basic recognition tasks, achieving target identification with negligible error margins. However, performance stratification becomes apparent with increasing target density and intersection scenarios. As quantitatively demonstrated in Figure 15, conventional models including Faster R-CNN, RTMDet, and Cascade manifest significant detection deficiencies in complex configurations, characterized by elevated rates of false negatives and suboptimal localization precision. These limitations indicate inherent constraints in high-density target processing capabilities. Comparative analysis further reveals that traditional architectures exhibit constrained feature representation capabilities and limited structural interpretability relative to their advanced counterparts.

In contrast, YOLOv11 demonstrates significant improvements in detection confidence through its integrated keypoint detection mechanism, thereby effectively mitigating false positive identifications and misdetections in regions devoid of distinct structural features. DAYKD exhibits the most balanced and robust performance metrics across all evaluation parameters, demonstrating exceptional boundary localization precision and superior small-object detection efficacy. The model achieves detection accuracy surpassing the 90% threshold, representing substantial performance enhancement over comparative architectures, which substantiates the effectiveness of its structural design and feature extraction methodology.

The evaluation on field scene 1, characterized by greater environmental complexity, further highlights the robustness of each model. Although Faster R-CNN, RTMDet, and Cascade can detect a majority of hyperbolic structures, their overall recognition performance is markedly inferior to that of YOLOv11 and DAYKD. In particular, DAYKD maintains high detection accuracy and boundary fitting in complex real-world conditions, with virtually no missed detections. This demonstrates its strong generalization ability and adaptability, making it the most robust and reliable model in this evaluation. In summary, DAYKD achieves the best detection performance across both tested environments, excelling in accuracy, robustness, and small-object recognition. It stands out as the most capable model among those evaluated. These results further validate the effectiveness of incorporating keypoint detection in enhancing both the precision and stability of object detection algorithms.

In another example from field scene 1, we conducted a detailed comparative analysis of the model combinations used in the ablation study. Unlike the relatively clear and scale-consistent detection targets in Figure 16, the objects in Figure 17 exhibit significant scale variations, along with complex backgrounds and occlusion effects, introducing substantial interference. These factors impose greater demands on the model, particularly in terms of global perception and multi-scale feature extraction capabilities.

As shown in Figure 17a–d, the performance differences among the models in handling this challenging scenario are evident. The baseline YOLOv11 model suffers from missed and false detections due to object scale variations and background noise. By integrating the CAFM during the object detection task, the model enhances its regional perception of targets, effectively mitigating background interference and significantly reducing missed detections. When the FRFN module is applied solely during the keypoint detection task, the model exhibits stronger feature fusion and scale modeling capabilities, leading to a notable improvement in keypoint confidence scores. Finally, the model incorporating the DAYKD strategy achieves the highest overall detection accuracy, demonstrating that this approach effectively enhances generalization and robustness.

However, it is worth noting that the proposed algorithm struggles to recognize incomplete hyperbolic targets, which may arise due to measurement line constraints or interference from adjacent objects. Although partial hyperbolic signatures often still indicate the presence of true subsurface targets, the model currently lacks the sensitivity to consistently interpret these incomplete patterns.

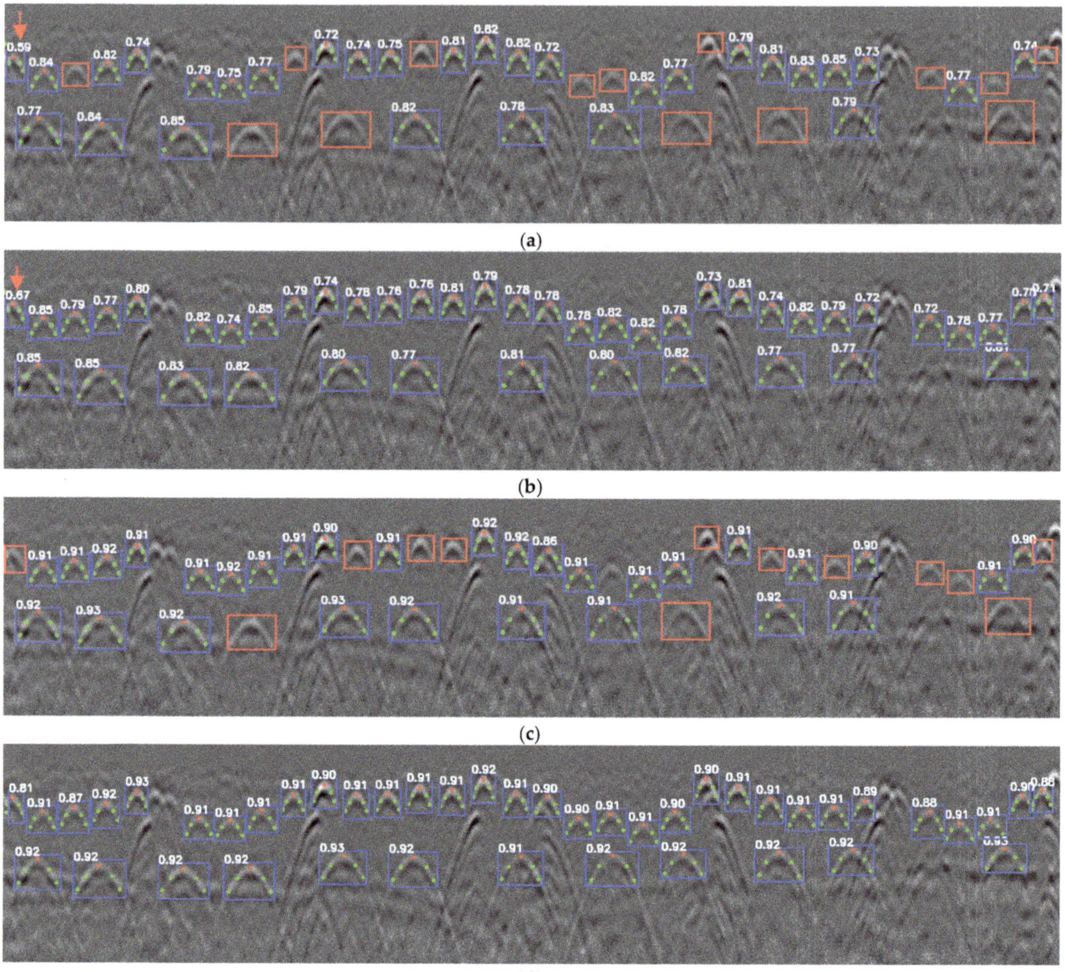

Figure 17. Performance comparison of ablation experiment on field scene 1 (red boxes indicate missed detections, and red arrows point to false positives): (**a**) YOLOv11, (**b**) YOLOv11 + CAFM, (**c**) YOLOv11 + FRFN, (**d**) DAYKD.

5. Conclusions

This paper proposes the DAYKD model, a dual-architecture framework based on YOLOv11, designed for efficient and precise underground target detection and keypoint localization in GPR images. DAYKD employs a two-task training strategy: the first task focuses on high-precision object detection, while the second task refines keypoint recognition accuracy. Ablation experiments demonstrate that DAYKD achieves a precision of 93.7% and an mAP@50 of 94.7% in object detection tasks. Compared to state-of-the-art models, such as YOLOv11, Cascade R-CNN, RTMDet, and Faster R-CNN, DAYKD exhibits superior performance, with an mAP@50-90 of 82.5% and a recall of 99.2%.

DAYKD demonstrates strong performance in GPR image interpretation, particularly in distinguishing overlapping subsurface targets and maintaining accuracy in cluttered, noisy environments. Its dual-task architecture and modular design enable efficient, robust inference across diverse imaging conditions and GPR systems. The model exhibits resilience to background noise and interference, with low computational overhead. However, its generalization is limited by the diversity of real-world datasets. DAYKD also struggles to detect incomplete hyperbolic signatures, especially near image boundaries, which may lead to missed detections under complex conditions.

Future work will aim to improve the generalization ability of the DAYKD model by expanding on-site data collection across a wider range of geological conditions and deployment scenarios. A more diverse and representative dataset is expected to enhance the model's robustness in real-world applications. In addition, we plan to explore the use of partially labeled samples, particularly those with incomplete hyperbolic features, to improve the model's ability to detect subtle or truncated subsurface targets under challenging conditions. To support deployment in resource-constrained environments, future research will also focus on improving computational efficiency. This includes investigating model compression techniques such as pruning, knowledge distillation, and low-bit quantization. These methods are expected to reduce inference time and memory consumption, enabling real-time performance while maintaining detection accuracy in practical engineering settings.

Author Contributions: Conceptualization, F.H.; Methodology, F.H.; Software, Y.Z.; Validation, Y.Z.; Formal analysis, F.H.; Resources, J.D.; Data curation, J.D.; Writing—original draft, Y.Z.; Writing—review & editing, F.H.; Supervision, J.F. All authors have read and agreed to the published version of the manuscript.

Funding: This research was funded by the National Natural Science Foundation of China (nos 62406346); the Natural Science Foundation of Hunan Province (nos 2025JJ60257); and the Degree and Graduate Education Reform Project of Central South University (nos 2025YJSKSO19).

Data Availability Statement: Data will be made available upon request.

Acknowledgments: The authors would also like to thank Wang Honghua from Guilin University of Technology and Wang Xiangxu from Guangzhou University for their valuable assistance in the joint effort of collecting the dataset.

Conflicts of Interest: The authors declare that they have no known competing financial interest or personal relationships that could have appeared to influence the work reported in this paper.

References

1. Jing, L.; Cai, L.; Zhaofa, Z.; Lingna, C. GPR Signal Denoising and Target Extraction with the CEEMD Method. *IEEE Geosci. Remote Sens. Lett.* **2015**, *12*, 1615–1619. [CrossRef]
2. Li, H.-L.; Lu, Q. Progress in application of ground penetrating radar in LNAPL contaminated soil detection of LNAPL. *Prog. Geophys.* **2020**, *35*, 1141–1148. [CrossRef]
3. Yin, D.; Ye, S.; Zhang, J.; Liu, J.; Liu, X. Study on the effect of highway structure and dielectric properties on GPR echo. *Electron. Meas. Technol.* **2018**, *41*, 51–56. [CrossRef]
4. Wang, Y.; Wang, H.; Li, Q.; Kang, W. Study on sedimentation regulation of silted soil in the front of reservoir dam based on GPR. *J. Soil Water Conserv.* **2023**, *34*, 102–110. [CrossRef]
5. Liu, H.; Yue, Y.; Lian, Y.; Meng, X.; Du, Y.; Cui, J. Reverse-time migration of GPR data for imaging cavities behind a reinforced shield tunnel. *Tunn. Undergr. Space Technol.* **2024**, *146*, 105649. [CrossRef]
6. Bai, D.-W.; Du, B.-R.; Zhang, P.-H. Weak signal processing technology of low frequency GPR based on Hilbert-Huang transform and its application to permafrost area gas hydrate exploration. *Geophys. Geochem. Explor.* **2017**, *41*, 1060–1067. [CrossRef]
7. Baek, J.; Yoon, J.S.; Lee, C.M.; Choi, Y. A Case Study on Detection of Subsurface Cavities of Urban Roads Using Ground-coupled GPR. In Proceedings of the 2018 17th Int. Conf. Ground Penetrating Radar (GPR), Rapperswil, Switzerland, 18–21 June 2018; pp. 1–4. [CrossRef]
8. Li, H.; Chou, C.; Fan, L.; Li, B.; Wang, D.; Song, D. Toward Automatic Subsurface Pipeline Mapping by Fusing a Ground-Penetrating Radar and a Camera. *IEEE Trans. Autom. Sci. Eng.* **2020**, *17*, 722–734. [CrossRef]
9. Liu, Y.; Wang, M.; Cai, Q. The target detection for GPR images based on curve fitting. In Proceedings of the 2010 3rd International Congress on Image and Signal Processing, Yantai, China, 16–18 October 2010; Volume 6, pp. 2876–2879. [CrossRef]

10. Li, W.; Cui, X.; Guo, L.; Chen, J.; Chen, X.; Cao, X. Tree Root Automatic Recognition in Ground Penetrating Radar Profiles Based on Randomized Hough Transform. *Remote Sens.* **2016**, *8*, 430. [CrossRef]
11. Maas, C.; Schmalzl, J. Using pattern recognition to automatically localize reflection hyperbolas in data from ground penetrating radar. *Comput. Geosci.* **2013**, *58*, 116–125. [CrossRef]
12. Sharafeldin, S.M.; Essa, K.S.; Youssef, M.A.S.; Karsli, H.; Diab, Z.E.; Sayil, N. Shallow geophysical techniques to investigate the groundwater table at the Great Pyramids of Giza, Egypt. *Geosci. Instrum. Method. Data Syst.* **2019**, *8*, 29–43. [CrossRef]
13. Sharafeldin, M.; Essa, K.S.; Sayıl, N.; Youssef, M.A.S.; Diab, Z.E.; Karslı, H. Geophysical Investigation of Ground Water Hazards in Giza Pyramids and Sphinx Using Electrical Resistivity Tomography and Ground Penetrating Radar: A Case Study. In Proceedings of the 9th Congress of the Balkan Geophysical Society, Antalya, Turkey, 5–9 November 2017; Volume 2017, pp. 1–5. [CrossRef]
14. Luo, X.; He, J.; Zhang, D.; Zhu, J.; Li, M.; Zhang, B.; Li, Q. Evaluating subsurface cavities detection using innovative laser dynamic deflectometer for efficient and large-scale urban road network inspections. *Tunn. Undergr. Space Technol.* **2025**, *159*, 106471. [CrossRef]
15. Dou, Q.; Wei, L.; Magee, D.R.; Cohn, A.G. Real-Time Hyperbola Recognition and Fitting in GPR Data. *IEEE Trans. Geosci. Remote Sens.* **2017**, *55*, 51–62. [CrossRef]
16. Zhang, X.; Xue, F.; Wang, Z.; Wen, J.; Guan, C.; Wang, F.; Han, L.; Ying, N. A Novel Method of Hyperbola Recognition in Ground Penetrating Radar (GPR) B-Scan Image for Tree Roots Detection. *Forests* **2021**, *12*, 1019. [CrossRef]
17. Xiang, D.; Pan, X.; Ding, H.; Cheng, J.; Sun, X. Two-Stage Registration of SAR Images With Large Distortion Based on Superpixel Segmentation. *IEEE Trans. Geosci. Remote Sens.* **2024**, *62*, 5211115. [CrossRef]
18. Pham, M.T.; Lefèvre, S. Buried Object Detection from B-Scan Ground Penetrating Radar Data Using Faster-RCNN. In Proceedings of the IGARSS 2018—2018 IEEE International Geoscience and Remote Sensing Symposium, Valencia, Spain, 22–27 July 2018; pp. 6804–6807. [CrossRef]
19. Ren, S.; He, K.; Girshick, R.; Sun, J. Faster R-CNN: Towards Real-Time Object Detection with Region Proposal Networks. *IEEE Trans. Pattern Anal. Mach. Intell.* **2017**, *39*, 1137–1149. [CrossRef]
20. Cui, F.; Ning, M.; Shen, J.; Shu, X. Automatic recognition and tracking of highway layer-interface using Faster R-CNN. *J. Appl. Geophys.* **2022**, *196*, 104477. [CrossRef]
21. Li, H.; Zhang, J.; Ye, M.; Wang, Q.; Wei, M.; Zhang, Z.; Guo, Y.; Tian, Y.; Zhang, Y.; Wang, C.; et al. High-accuracy intelligent detection of centimeter-level voids in cement pavement via ground-penetrating radar. *Measurement* **2025**, *256*, 118354. [CrossRef]
22. Redmon, J.; Divvala, S.; Girshick, R.; Farhadi, A. You Only Look Once: Unified, Real-Time Object Detection. In Proceedings of the 2016 IEEE Conference on Computer Vision and Pattern Recognition (CVPR), Las Vegas, NV, USA, 27–30 June 2016; pp. 779–788. [CrossRef]
23. Liu, H.; Lin, C.; Cui, J.; Fan, L.; Xie, X.; Spencer, B.F. Detection and localization of rebar in concrete by deep learning using ground penetrating radar. *Autom. Constr.* **2020**, *118*, 103279. [CrossRef]
24. Sun, X.; Qidao, Z.; Da, Y.; Youyuan, Z. Research on ground penetrating radar image target detection based on YOLOv5s. *World Transp.* **2023**, *7*, 3–6. [CrossRef]
25. Wang, Z.; Lan, T.; Qu, X.; Gao, S.; Yu, Z.; Yang, X. Improved SSD Framework for Automatic Subsurface Object Indentification for GPR Data Processing. In Proceedings of the 2021 CIE International Conference on Radar (Radar), Haikou, China, 15–19 December 2021; pp. 2078–2081. [CrossRef]
26. Qiu, Z.; Zhao, Z.; Chen, S.; Zeng, J.; Huang, Y.; Xiang, B. Application of an Improved YOLOv5 Algorithm in Real-Time Detection of Foreign Objects by Ground Penetrating Radar. *Remote Sens.* **2022**, *14*, 1895. [CrossRef]
27. Hu, H.; Fang, H.; Wang, N.; Ma, D.; Dong, J.; Li, B.; Di, D.; Zheng, H.; Wu, J. Defects identification and location of underground space for ground penetrating radar based on deep learning. *Tunn. Undergr. Space Technol.* **2023**, *140*, 105278. [CrossRef]
28. Wang, Y.; Qin, H.; Zhang, D.; Wu, T.; Pan, S. Domain Adaption YOLO Network to Enhance Target Detection in GPR Images. *IEEE Trans. Geosci. Remote Sens.* **2024**, *62*, 5936415. [CrossRef]
29. Lan, T.; Zhao, Y.; Guo, C.; Gong, J.; Yang, X. A State-Space-Model-Based Hyperbola Detection Method for Arbitrarily Long GPR B-Scan. *IEEE Geosci. Remote Sens. Lett.* **2025**, *22*, 7504805. [CrossRef]
30. Wang, N.; Zhang, Z.; Hu, H.; Li, B.; Lei, J. Underground Defects Detection Based on GPR by Fusing Simple Linear Iterative Clustering Phash (SLIC-Phash) and Convolutional Block Attention Module (CBAM)-YOLOv8. *IEEE Access* **2024**, *12*, 25888–25905. [CrossRef]
31. Li, S.; Cui, X.; Guo, L.; Zhang, L.; Chen, X.; Cao, X. Enhanced Automatic Root Recognition and Localization in GPR Images Through a YOLOv4-Based Deep Learning Approach. *IEEE Trans. Geosci. Remote Sens.* **2022**, *60*, 5114314. [CrossRef]
32. Hu, S.; Gao, F.; Zhou, X.; Dong, J.; Du, Q. Hybrid Convolutional and Attention Network for Hyperspectral Image Denoising. *IEEE Geosci. Remote Sens. Lett.* **2024**, *21*, 5504005. [CrossRef]
33. Zhou, S.; Chen, D.; Pan, J.; Shi, J.; Yang, J. Adapt or Perish: Adaptive Sparse Transformer with Attentive Feature Refinement for Image Restoration. In Proceedings of the 2024 IEEE/CVF Conference on Computer Vision and Pattern Recognition (CVPR), Seattle, WA, USA, 16–22 June 2024; pp. 2952–2963. [CrossRef]

34. Yang, G.; Yuan, D.; Xu, T.; Li, B. An Adaptive Clutter-Immune Method for Pipeline Detection with GPR. *IEEE Sens. J.* **2023**, *23*, 22984–22997. [CrossRef]
35. Terrasse, G.; Nicolas, J.M.; Trouvé, E.; Drouet, É. Automatic localization of gas pipes from GPR imagery. In Proceedings of the 2016 24th European Signal Processing Conference (EUSIPCO), Budapest, Hungary, 29 August–2 September 2016; pp. 2395–2399. [CrossRef]
36. Ma, Y.; Lei, W.; Pang, Z.; Zheng, Z.; Tan, X. Rebar Clutter Suppression and Road Defects Localization in GPR B-Scan Images Based on SuppRebar-GAN and EC-Yolov7 Networks. *IEEE Trans. Geosci. Remote Sens.* **2024**, *62*, 1–14. [CrossRef]
37. Kazemi, V.; Sullivan, J. One millisecond face alignment with an ensemble of regression trees. In Proceedings of the 2014 IEEE Conference on Computer Vision and Pattern Recognition, Columbus, OH, USA, 23–28 June 2014; pp. 1867–1874. [CrossRef]
38. Vilaça, L.; Viana, P.; Carvalho, P.; Andrade, M.T. Improving Efficiency in Facial Recognition Tasks Through a Dataset Optimization Approach. *IEEE Access* **2024**, *12*, 32532–32544. [CrossRef]
39. Fang, T.; Wang, C.; Wang, J.; Du, Y. Ground penetrating radar image pipeline location based on YOLO v8n. *Foreign Electron. Meas. Technol.* **2023**, *42*, 170–177. [CrossRef]
40. Jocher, G.; Qiu, J. Ultralytics YOLO11. Available online: https://github.com/ultralytics/ultralytics (accessed on 11 December 2024).
41. Jocher, G.; Chaurasia, A.; Qiu, J. Ultralytics YOLOv8. Available online: https://github.com/ultralytics/ultralytics (accessed on 10 December 2024).
42. Russell, B.C.; Torralba, A.; Murphy, K.P.; Freeman, W.T. LabelMe: A Database and Web-Based Tool for Image Annotation. *Int. J. Comput. Vis.* **2008**, *77*, 157–173. [CrossRef]
43. Warren, C.; Giannopoulos, A.; Giannakis, I. gprMax: Open source software to simulate electromagnetic wave propagation for Ground Penetrating Radar. *Comput. Phys. Commun.* **2016**, *209*, 163–170. [CrossRef]
44. Lyu, C.; Zhang, W.; Huang, H.; Zhou, Y.; Wang, Y.; Liu, Y.; Zhang, S.; Chen, K. RTMDet: An Empirical Study of Designing Real-Time Object Detectors. *arXiv* **2022**. [CrossRef]
45. Cai, Z.; Vasconcelos, N. Cascade R-CNN: High Quality Object Detection and Instance Segmentation. *IEEE Trans. Pattern Anal. Mach. Intell.* **2021**, *43*, 1483–1498. [CrossRef]

remote sensing

MDPI

Article

Synergistic Multi-Model Approach for GPR Data Interpretation: Forward Modeling and Robust Object Detection

Hang Zhang, Zhijie Ma, Xinyu Fan and Feifei Hou *

School of Automation, Central South University, Changsha 410083, China; zhang22@csu.edu.cn (H.Z.); mazhijie@csu.edu.cn (Z.M.); auxyfan@csu.edu.cn (X.F.)
* Correspondence: houfeifei@csu.edu.cn

Abstract

Ground penetrating radar (GPR) is widely used for subsurface object detection, but manual interpretation of hyperbolic features in B-scan images remains inefficient and error-prone. In addition, traditional forward modeling methods suffer from low computational efficiency and strong dependence on field measurements. To address these challenges, we propose an unsupervised data augmentation framework that utilizes CycleGAN-based model to generate diverse synthetic B-scan images by simulating varying geological parameters and scanning configurations. This approach achieves GPR data forward modeling and enhances the scenario coverage of training data. We then apply the EfficientDet architecture, which incorporates a bidirectional feature pyramid network (BiFPN) for multi-scale feature fusion, to enhance the detection capability of hyperbolic signatures in B-scan images under challenging conditions such as partial occlusions and background noise. The proposed method achieves a mean average precision (mAP) of 0.579 on synthetic datasets, outperforming YOLOv3 and RetinaNet by 16.0% and 23.5%, respectively, while maintaining robust multi-object detection in complex field conditions.

Keywords: ground penetrating radar (GPR); GPR data forward modeling; hyperbolic characteristic detection; data augmentation; deep learning

1. Introduction

As a non-invasive electromagnetic (EM) detection technology, Ground Penetrating Radar (GPR) is a cornerstone technique in modern non-destructive testing of civil engineering structures, owing to its superior operational efficiency and versatility across multiple scenarios. Its principal engineering applications encompass: 3D reconstruction of subsurface utility networks [1], quantitative analysis of internal defects in reinforced concrete components [2–4], interlayer damage evaluation in concrete bridge decks [5–7], void detection in pavement structural layers [8–10], the recognition of moisture damage in asphalt pavement [11], high-accuracy intelligent detection of centimeter-level voids in cement pavement [12] and assessment of construction process conformance [13]. Notably, GPR has also enabled groundbreaking planetary discoveries, exemplified by its critical role in identifying ancient Martian coastal deposits through rover-mounted radar systems [14]. The system operates by emitting nanosecond-scale EM pulses into layered media to capture reflected waveforms generated by dielectric discontinuities, which are subsequently synthesized into time-domain radargrams containing subsurface target signatures. Of particular significance is the observation that typical detection targets such as buried metallic conduits, concrete reinforcement, and geotechnical voids exhibit distinct hyperbolic diffraction patterns in B-scan images [15,16]. Crucially, the reliability of GPR data interpretation remains dependent upon precise extraction and physical characterization of target reflection signatures. Traditional manual interpretation methods face significant challenges when processing large-scale GPR scan data, including signal distortion due to complex noise interference, low processing efficiency, and susceptibility to subjective errors [8].

To address the inherent limitations of traditional manual interpretation, recent advancements in artificial intelligence have provided new pathways for enhancing GPR technology. Deep learning (DL), particularly convolutional neural networks (CNNs), has demonstrated exceptional performance in computer vision tasks such as image classification [17,18], semantic segmentation [19–22], and object detection [23–25], owing to its hierarchical feature extraction and spatial correlation capture mechanisms [26]. This data-driven pattern recognition paradigm aligns closely with the core requirements of GPR image interpretation. These requirements span feature extraction to target discrimination. The paradigm establishes end-to-end mapping relationships. These relationships directly associate EM reflection signatures with the spatial distributions of subsurface targets. This approach overcomes the limitations of empirical human interpretation.

In this context, DL is reshaping GPR data processing methodologies, offering significant potential for automating the critical task of object detection in B-scan images. However, the successful deployment of DL, particularly for supervised object detection, is highly contingent upon access to large-scale, high-quality training datasets. To mitigate the critical issue of data scarcity, forward modeling has been employed to generate synthetic training data. While theoretically effective, conventional numerical methods exhibit low computational efficiency and frequently fail to capture the full complexity and variability of real-world scenarios (e.g., heterogeneous soil layers, target interactions, complex noise) [27]. Notably, recent years have witnessed the emergence of DL approaches addressing the inverse problem in GPR, such as reconstructing permittivity distributions or target geometries from B-scan data [27–29]. These studies demonstrate DL's capacity to learn complex mappings governing EM wave phenomena.

Inspired by the success of DL in addressing GPR inverse problems and acknowledging the limitations of conventional forward modeling, we explore the under-explored application of DL specifically for efficient and flexible forward modeling. By leveraging the powerful pattern learning and generation capabilities of deep neural networks, particularly generative adversarial networks (GANs) [30], this approach offers the potential to overcome the computational bottlenecks and scenario constraints inherent in traditional methods. The methodology aims to rapidly generate diverse synthetic B-scans, tailored to enhance datasets for training robust object detection models.

This study proposes a synergistic framework integrating a CycleGAN [31]-based data augmentation module with an EfficientDet-based interpretation module. The CycleGAN component establishes an image-to-image mapping that directly translates permittivity distribution patterns into synthetic B-scan images while preserving hyperbolic signatures. This capability expands training data diversity without modifying existing scans, serving as a critical augmentation strategy.

The core interpretation task—detecting hyperbolic signatures in B-scans—is performed by the EfficientDet architecture. EfficientDet incorporates a Bidirectional Feature Pyramid Network (BiFPN) module [32], which optimizes multi-scale detection through bidirectional cross-scale feature fusion. By dynamically integrating multi-resolution features, the BiFPN enhances the detection of small targets (e.g., shallow voids) and hyperbolic signatures.

This integrated framework operates bidirectionally: Data augmentation diversifies training samples, while interpretation extracts semantic subsurface information. Leveraging CycleGAN-generated cross-domain data, the system learns robust representations under complex conditions, overcoming feature diversity limitations of conventional methods and addressing critical challenges in real-data scarcity and complex feature interpretation.

The principal contributions of this work are:

(i). Unsupervised GPR Data Augmentation Framework: A CycleGAN-based generation methodology is proposed, synthesizing diverse B-scan images through simulation of varying geological parameter and scanning configurations(This process is also known as GPR data forward modeling). This significantly enhances the geological rationality and generalization of synthetic data.

(ii). BiFPN-Driven B-scan Feature Recognition: A detection architecture incorporating BiFPN is engineered for multi-scale hyperbolic targets in GPR imagery. By utilizing cross-scale dynamically weighted fusion mechanisms, the framework optimizes feature representation for small targets and hyperbolae while suppressing time-varying noise, thereby striking a balance between geometric feature preservation and detection efficiency.

2. Methodology

GPR data interpretation confronts two critical challenges: First, well-annotated B-scan samples are scarce in practical engineering contexts, and the complex variability of subsurface medium EM parameters severely constrains data diversity. Second, targets in B-scan images (e.g., pipelines and voids) typically manifest time-varying hyperbolic signatures, whose vertex positions and geometric characteristics are susceptible to noise interference, rendering conventional detection algorithms inadequate in both accuracy and generalization for field applications.

To address these issues, this study proposes a hybrid data-driven and model-driven framework. As illustrated in Figure 1, the implemented solution establishes a three-stage methodological pipeline: (1) Multi-source dataset preparation combining open-access measured data and numerical simulation models as foundational inputs for CycleGAN training; (2) forward-augmented data generation via CycleGAN-based cross-domain translation, which enhances training set diversity while preserving consistency between hyperbolic signatures and subsurface EM properties; (3) multi-scale hyperbolic object detection employing the EfficientDet architecture with compound scaling, specifically optimized for time-varying geometric patterns in GPR imagery. This workflow constructs an integrated pipeline from raw data preparation to intelligent interpretation, directly resolving the dual challenges of data scarcity and hyperbolic feature instability.

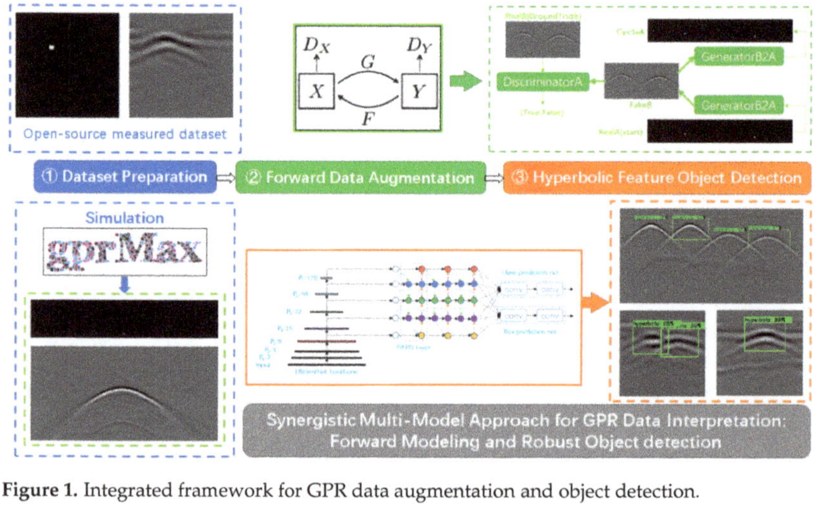

Figure 1. Integrated framework for GPR data augmentation and object detection.

2.1. Architecture of Data Augmentation Model

2.1.1. CycleGAN Model

DL models for GPR image interpretation depend critically on large-scale, diverse training datasets. However, acquiring well-annotated B-scan samples in practical engineering contexts presents significant challenges: First, the distribution of subsurface targets (e.g., pipelines and voids) is inherently complex, rendering measured data collection labor-intensive, while public datasets remain limited due to confidentiality constraints on critical infrastructure. Second, conventional simulation methods (e.g., the finite-difference time-domain (FDTD)-based gprMax platform) can generate high-fidelity synthetic data; however, their computational inefficiency and resource-intensive nature restrict feasibility for large-scale data generation.

GANs have demonstrated exceptional performance in image generation [33,34] and image editing [35,36], powered by their adversarial loss mechanism that aligns generated data with target domain distributions. This loss continuously evaluates the realism of synthetic samples through a discriminator network, achieving the long-pursued objective of photorealistic rendering in computer graphics. In GPR data augmentation tasks, this adversarial training framework is adapted to learn mappings from subsurface permittivity distributions to corresponding B-scan images. However, acquiring strictly paired training samples (e.g., permittivity maps spatially aligned with B-scan images) remains highly challenging in GPR applications. To overcome this limitation, we implement the CycleGAN framework as illustrated in Figure 2, which employs bidirectional generators ($G : X \rightarrow Y$ and $F : Y \rightarrow X$) to establish unpaired domain mappings, thereby accounting for spatial variability in subsurface medium parameters.

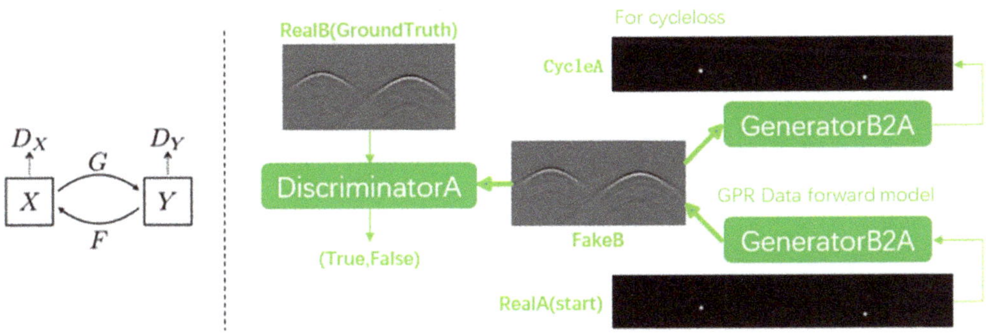

Figure 2. Schematic diagram of cycle-consistent adversarial architecture.

The generator G translates subsurface relative permittivity distributions (domain X, e.g., laboratory-standard permittivity maps) into B-scan radar images (domain Y, e.g., EM echoes under complex field conditions), while the generator F performs the inverse mapping from B-scan images back to permittivity distributions. Each generator employs an enhanced U-Net architecture with an encoder–decoder structure to extract and reconstruct input features. The discriminator network adopts a PatchGAN architecture, consisting of two independent branches (D_X and D_Y). The PatchGAN architecture divides input images into overlapping $N \times N$ pixel patches. Each discriminator independently evaluates patch authenticity through patch-level analysis, enabling precise capture of localized features in GPR signals.

During CycleGAN training, the generators G and F are iteratively optimized through adversarial training to produce realistic images that deceive the discriminators D_X and D_Y. Simultaneously, the discriminators strive to distinguish real samples from generated ones. This dynamic competition drives the generators to learn domain-specific features while preserving key properties of the input data, achieving high-fidelity cross-domain translation.

2.1.2. Loss Function

To establish effective cross-domain mapping, CycleGAN incorporates multiple loss functions that enforce realism in synthesized images and structural fidelity. The core loss components comprise adversarial loss, cycle-consistency loss, and identity consistency loss.

The adversarial loss is applied to the mapping functions G and F, along with their corresponding discriminators D_Y and D_X. Specifically, the adversarial loss for generator G is defined as:

$$\mathcal{L}_{GAN}(G, D_Y, X, Y) = \mathbb{E}_{Y \sim p_{data}(y)}[logD_Y(y)] + \mathbb{E}_{X \sim p_{data}(x)}[log(1 - D_X(G(x)))] \quad (1)$$

This loss enforces domain realism for images synthesized by generator G. Generator G seeks to maximize the discriminator D_Y's assessment confidence for its generated outputs $G(x)$, while D_Y learns to differentiate authentic samples y from synthesized counterparts $G(x)$. Through adversarial optimization, G assimilates target domain features. Similarly, generator F and discriminator D_X are governed by a symmetric adversarial loss formulation:

$$\mathcal{L}_{GAN}(F, D_X, Y, X) = \mathbb{E}_{x \sim p_{data}(x)}[logD_X(x)] + \mathbb{E}_{y \sim p_{data}(y)}[log(1 - D_X(G(y)))] \quad (2)$$

For a source domain sample x, generator G maps it to the target domain as $y' = G(x)$. Generator F subsequently reconstructs the sample to the source domain as $x' = F(y')$. The cycle-consistency loss minimizes the L1-norm distance between the original sample x and its reconstruction x':

$$\mathcal{L}_{cyc}(G, F) = \mathbb{E}_{x \sim p_{data}(x)}[\| F(G(x)) - x \|_1] + \mathbb{E}_{y \sim p_{data}(y)}[\| G(F(y)) - y \|_1] \quad (3)$$

This loss constitutes the fundamental innovation distinguishing CycleGAN from conventional unidirectional GANs. As depicted in Figure 2, the cyclic reconstruction mechanism enforces structural integrity and feature preservation during bidirectional domain transitions ($X \rightarrow Y \rightarrow X$ and $Y \rightarrow X \rightarrow Y$).

The identity consistency loss ensures that generators approximate identity transformations for samples already residing in the target domain. This loss is mathematically defined as:

$$\mathcal{L}_{identity}(G, F) = \mathbb{E}_{y \sim p_{data}(y)}[\| G(y) - y \|_1] + \mathbb{E}_{x \sim p_{data}(x)}[\| F(x) - x \|_1] \quad (4)$$

The total loss is a weighted sum of the above components:

$$\mathcal{L}(G, F, D_X, D_Y) = \mathcal{L}_{GAN}(G, D_Y, X, Y) + \mathcal{L}_{GAN}(F, D_X, Y, X) + \lambda_1 \mathcal{L}_{cyc}(G, F) + \lambda_2 \mathcal{L}_{identity}(G, F) \quad (5)$$

Here, λ_1 and λ_2 are hyperparameters that balance the contributions of the cycle-consistency and identity losses, respectively.

2.2. Architecture of Object Detection Model

2.2.1. EfficientDet Model

The EfficientDet architecture adopts a single-stage object detection framework. Its core design principle optimizes the trade-off between detection accuracy and computational efficiency through an efficient backbone network and multi-scale feature fusion mechanism. The modular architecture comprises three cascaded components: a backbone network, a BiFPN, and a prediction head, as illustrated in Figure 3.

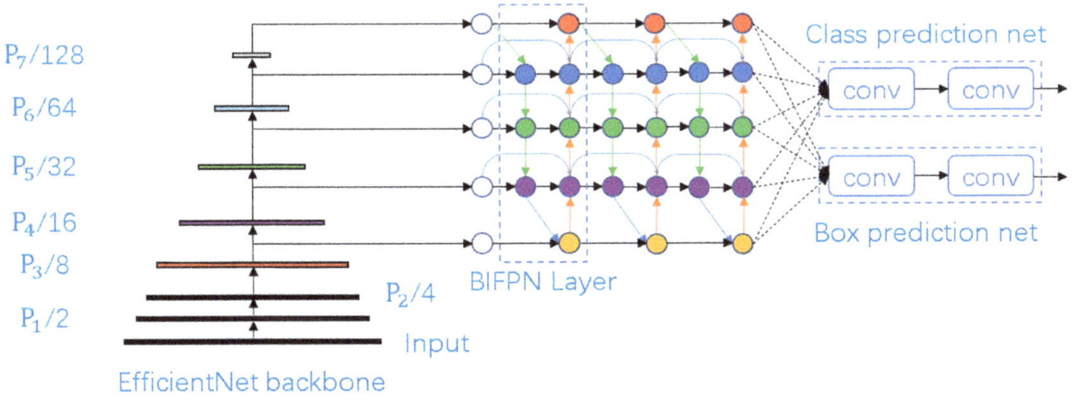

Figure 3. EfficientDet architecture—it employs EfficientNet as the backbone network, BiFPN as the feature network, and shared class/box prediction network.

The backbone utilizes EfficientNet, pre-trained on ImageNet. It employs compound scaling to optimize depth, width, and resolution. This approach maintains robust feature representations while minimizing computational complexity. The backbone extracts hierarchical feature maps that provide multi-scale spatial and semantic information for subsequent fusion.

BiFPN functions as the core feature fusion module, recursively stacking bidirectional pathways to establish cross-scale feature interactions. This architecture integrates top-down and bottom-up information flows through weighted fusion of multi-resolution features, enabling synergistic combination of high-level semantics and low-level spatial details. The recurrent bidirectional topology significantly enhances multi-scale object characterization.

The prediction heads comprise parallel branches for classification and bounding box regression, employing shared weights across all feature levels. This parameter-sharing strategy reduces model complexity while enhancing robustness via multi-level supervision. The entire network undergoes end-to-end training to jointly optimize detection accuracy and inference efficiency.

2.2.2. BiFPN: Bidirectional Feature Pyramid Network for Multi-Scale Fusion

The evolution of multi-scale feature fusion methodologies has been marked by significant architectural advancements in feature pyramid networks. The foundational feature pyramid network (FPN) architecture (Figure 4a) establishes a unidirectional top-down pathway that propagates high-level semantic features to lower-resolution layers, enabling elementary cross-scale feature interactions. However, this unidirectional information flow intrinsically limits effective integration of low-level spatial details with high-level semantic representations, particularly degrading detection performance for small-scale targets.

To overcome these limitations, BiFPN (Figure 4b) implements an optimized recurrent fusion mechanism. This architecture first streamlines the feature hierarchy by pruning single-input nodes exhibiting marginal contribution to detection outcomes, selectively retaining core layers supporting multi-directional connectivity. It subsequently constructs a bidirectional propagation framework:

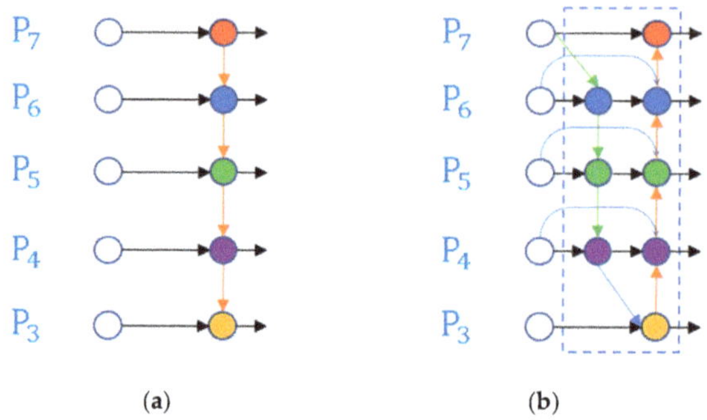

Figure 4. (**a**) FPN: Unidirectional top-down fusion; (**b**) BiFPN: Pruned cross-scale recurrent connections.

The top-down pathway, visually represented as the green part, integrates upsampled high-level semantic features with high-resolution low-level features, effectively conveying rich contextual information. Conversely, the bottom-up pathway, depicted as the orange part, facilitates hierarchical refinement by propagating low-level spatial features (conveying precise positional information) through downsampling operations to semantically enriched higher layers. Critically, each fusion node dynamically aggregates dual-source inputs: intrinsic features from the current resolution level and cross-scale features from adjacent layers accessed via skip connections. Furthermore, blue edges represent the additional connections added between input nodes residing on the same layer, enhancing intra-level feature exchange. This integrated design, combining the bidirectional flows and the novel same-layer connections, enables simultaneous feature reuse and information enhancement, achieving the balanced spatial precision and contextual awareness essential for hyperbolic target detection in GPR B-scan imagery.

3. GPR Dataset Construction

This study establishes an integrated multi-source GPR dataset for subsurface object detection, integrating three distinct data modalities: numerically simulated B-scan images, synthetic B-scan images, and field-measured data. Each sample is paired with corresponding EM property characterization through relative permittivity distribution maps. To mitigate insufficient subsurface target samples, 1000 high-fidelity simulated samples were generated using the gprMax3.0 [37] EM simulation platform. Augmented with 98 open-source field datasets [38], these simulated samples trained a CycleGAN model to synthesize 500 additional data pairs (synthetic data), expanding the original dataset. The standardized composite dataset comprises 1598 samples (1000 simulated + 500 synthetic + 98 measured). From this repository, 1469 B-scan images were annotated with pixel-level hyperbolic target delineations using LabelImg1.8.1, forming an object detection dataset with detailed composition in Table 1.

Table 1. Composition and characteristic parameters of multi-source GPR dataset.

Dataset	Source	Quantity	Train/Val	Description
Simulated data	gprMax3.0 [37]	1000 pairs	1386 pairs/212 pairs	Paired permittivity maps and B-scan images
Measured data	Open-source [38]	98 pairs		
Synthetic data	CycleGAN	500 pairs		
Object Detection data	LabelImg1.8.1	1469 images	1214/255 images	Annotated B-scans

3.1. Simulated Dataset

This study addresses the data augmentation requirements for GPR B-scan images by constructing a multi-domain fusion dataset based on a collaborative framework integrating EM simulation and DL techniques. To overcome challenges such as the scarcity of real GPR data and high annotation costs, the open-source software gprMax3.0 [37] was utilized to establish refined subsurface models. Large-scale simulated B-scan images were generated using the FDTD method as the foundational dataset.

Key modeling parameters were configured as follows: The computational domain comprised a 2.5 m × 0.5 m 3D space. Spatial discretization employed an isotropic grid ($\Delta x = \Delta y = \Delta z = 2.5$ mm), with a 15 ns time window configured to align with typical road inspection scenarios. Stratigraphic layers included (top-to-bottom): 10 cm air layer, 5 cm asphalt layer, and 10 cm concrete layer. Grid resolution was set at 0.0025 m × 0.0025 m, yielding 1000 × 100 computational cells. EM excitation utilized a z-polarized Hertzian dipole with a Gaussian pulse waveform (center frequency: 900 MHz). Transmitting and receiving antennas were positioned 2.5 mm above the concrete surface. During acquisition, the transceiver system traversed synchronously along the horizontal axis at 20 mm intervals, capturing 1000 A-scans per B-scan. To ensure diversity, tubular targets with randomized geometries and positions were embedded subsurface, ultimately yielding 2000 B-scan datasets through parametric scripting. Relative permittivity and conductivity values per medium are detailed in Table 2.

Table 2. Model medium parameter information.

Model Medium	Concrete	Asphalt	Free_Space (Cavity)
Relative permittivity	6.4	4	1
Conductivity	0.01	0.005	0

To suppress interference from direct waves in radar images, exponential gain compensation is applied to enhance the hyperbolic response characteristics in B-scan data, while median filtering is employed to effectively separate direct wave components. This processing method significantly improves the discernibility of target reflection signals in GPR imaging data. The final simulated B-scan data are shown in Figure 5a. To enable the model to focus on the geometric distribution of subsurface targets in relative permittivity maps, the input permittivity maps are uniformly processed by rendering solid layers (excluding air and cavities) in black and air/cavity layers in white. The processed relative permittivity maps are illustrated in Figure 5b.

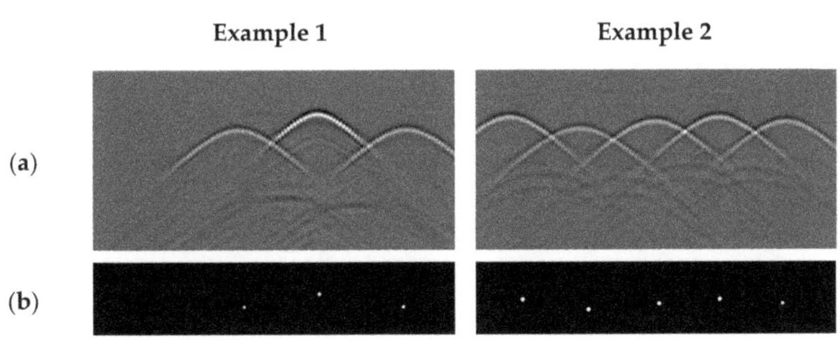

Figure 5. Acquired GPR simulation dataset. (**a**) B-scan time-domain waveform. (**b**) Spatial distribution of relative permittivity in subsurface media.

3.2. Measured Dataset

The measured GPR data were sourced from a publicly available open-source [38] dataset. The original measurements were collected by researchers using a commercial GSSI Utility Scan Pro GPR system in an outdoor sandy environment. The dataset includes multiple B-scan radar images and corresponding subsurface relative permittivity distribution information, making it suitable for algorithm validation and subsurface target reconstruction. During data acquisition, a 400 MHz antenna and control unit were employed to capture data along predefined scanning paths. Each scan trajectory spans 1 m, containing 88 A-scan signals with a sampling time window of 20 ns. The dataset underwent standardized preprocessing, including normalization, resizing, and mean subtraction, and provides denoised reference B-scan images and ground-truth relative permittivity maps. The final processed measured B-scan images are shown in Figure 6a. Similar to the simulated dataset, to enable the model to focus on the geometric distribution of subsurface targets in relative permittivity maps, the measured dataset's permittivity maps were converted to black-and-white representations, with solid layers rendered in black and air/cavity layers in white. The final processed relative permittivity maps are illustrated in Figure 6b.

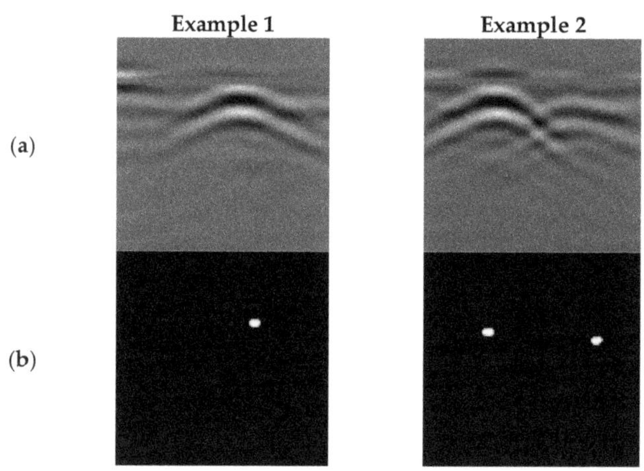

Figure 6. Acquired GPR measured dataset and permittivity map. (**a**) B-scan time-domain waveform. (**b**) Spatial distribution of relative permittivity in subsurface media.

3.3. GPR Data Augmentation Results by CycleGAN

The CycleGAN model was trained using the Adam optimizer (initial learning rate = 0.0002, $\beta_1 = 0.5$) for 100 epochs, with the first 50 epochs at a fixed learning rate and the subsequent 50 epochs linearly decaying to zero. The loss function integrated LSGAN (least squares GAN), cycle-consistency loss ($\lambda_1 = 10$), and identity consistency loss ($\lambda_2 = 0.5$) to jointly optimize the plausibility of generated images and geometric alignment in cross-domain mapping.

To objectively demonstrate the computational efficiency advantage of the proposed CycleGAN-based forward modeling approach over traditional numerical simulation (e.g., gprMax), a comprehensive time cost analysis was conducted. As shown in Table 3, experiments were performed on a workstation equipped with an NVIDIA RTX 4070 GPU and an Intel i5-13600KF CPU. The conventional gprMax (v3.0) platform, based on the FDTD method, was used to generate high-fidelity B-scan images. Under a typical road inspection scenario configuration (center frequency: 900 MHz antenna, spatial grid resolution $\Delta x = \Delta y = \Delta z = 2.5$ mm, time window: 15 ns, 1000 A-scans per B-scan), generating a high-fidelity simulated dataset comprising 1000 pairs of samples (B-scan images and corresponding relative permittivity maps) took approximately 10 days.

Table 3. Computational efficiency comparison of forward modeling approaches.

Method	Quantity	Time Cost	Resolution	Notes
gprMax (v3.0)	1000 pairs	20 days	3543 × 1772 pixels	Numerical simulation (B-scans + permittivity maps)
Proposed CycleGAN	-	10 h	-	Model training
	200 B-scans	2 min	1000 × 180 pixels	Image generation (inference)
Measured data	98 pairs	-	128 × 128 pixels	Open-source dataset; field acquisition excluded

In contrast, the proposed CycleGAN model, once trained (training time: 10 h), demonstrated extremely high efficiency in generating new synthetic B-scan images: generating 200 B-scan images with equivalent resolution (1000 × 100 pixels) and complexity to those from gprMax3.0 required only about 2 min. The measured data were sourced from an open-access public dataset; the original acquisition time involving fieldwork is not included in this computational efficiency comparison. Although the CycleGAN model training requires a significant computational investment (10 h), once trained, it exhibits orders of magnitude efficiency improvement (minutes vs. days) in generating new samples. This "train once, generate efficiently" paradigm is particularly advantageous for scenarios demanding large-scale, diverse synthetic data to support DL model training, effectively overcoming the primary bottlenecks of computational intensity and long time consumption inherent in traditional numerical simulation methods.

To objectively demonstrate the computational efficiency advantage of the proposed CycleGAN-based forward modeling approach over traditional numerical simulation (e.g., gprMax), a comprehensive time cost analysis was conducted under a typical road inspection scenario. Experiments were performed on a workstation equipped with an NVIDIA RTX 4070 GPU and an Intel i5-13600KF CPU (Table 3). The conventional gprMax platform, based on the FDTD method, was used to generate high-fidelity B-scan images. For a dataset comprising 1000 sample pairs (B-scan images at 3543 × 1772 pixels and corresponding permittivity maps), the generation process took approximately 20 days. In contrast, the proposed pre-trained CycleGAN model generated 200 synthetic B-scan images at a resolution of 1000 × 180 pixels in only about 2 min. It is important to note that while the gprMax simulations model the complex physical processes resulting in high-resolution, physically accurate outputs, the CycleGAN generates synthetic data primarily for efficient data augmentation to support DL model training. The measured validation data originated from an open-access public dataset; its field acquisition time is excluded from this computational comparison.

The key observation is the dramatic difference in sample generation throughput after the initial investment. Although training the CycleGAN model required a significant 10 h computational investment, the subsequent generation of new synthetic samples is orders of magnitude faster (minutes vs. over two weeks) compared to FDTD simulation. This approach, characterized by a substantial upfront training cost followed by highly efficient generation, fundamentally shifts the computational bottleneck. The one-time training investment enables the rapid, scalable production of synthetic data tailored for DL. This efficiency is particularly transformative for applications demanding large volumes of diverse training data, effectively overcoming the prohibitive computational cost and time constraints inherent in traditional physics-based simulation methods like gprMax for iterative data generation tasks.

To validate the effectiveness of CycleGAN in expanding GPR B-scan image datasets, this study systematically compared its performance with the traditional pixel-aligned model (pixel to pixel) [39]. Note that the ground-truth data in Figure 7 (second row) were generated using classical FDTD-based simulations via gprMax3.0, providing the benchmark for evaluating CycleGAN outputs.

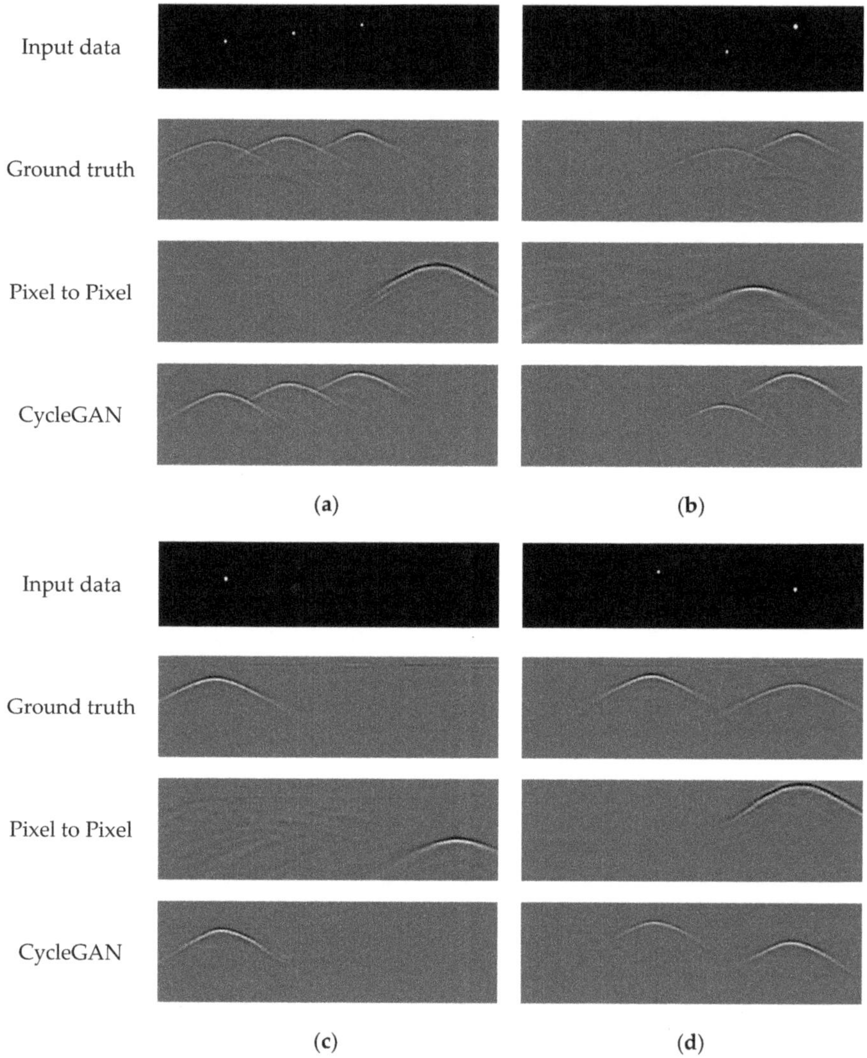

Figure 7. Data augmentation results based on simulated GPR B-scan images. (**a**) First example. (**b**) Second example. (**c**) Third example. (**d**) Fourth example.

As shown in Figure 7c, while pixel-to-pixel-generated B-scan images exhibited basic hyperbolic contours, the vertex positions deviated significantly from the spatial coordinates of targets in the input permittivity maps; in multi-target scenarios (Figure 7a), it failed to match the number of input targets, with exacerbated positional errors at vertices, indicating its inability to model the complex coupling effects of EM reflections among multiple targets. In contrast, CycleGAN, through the joint optimization of cycle-consistency loss and adversarial training, achieved superior geometric fidelity in cross-domain mapping.

As illustrated in Figure 7b,d, CycleGAN-generated B-scan images strictly preserved the geometric distribution of targets in the input permittivity maps and precisely replicated the spatiotemporal evolution of hyperbolic vertices; for instance, when the input contained two adjacent pipelines (Figure 7b), the generated image displayed independent hyperbolic reflections with vertex spacing fully consistent with the input's spatial distribution, confirming its capability to achieve high-fidelity visual mapping that preserves the geometric relationships implied by the permittivity distribution.

Further validation on field-measured scenarios (Figure 8) demonstrated CycleGAN's robust generation capability under strong clutter interference. Despite significantly higher background noise in measured B-scan images compared to synthetic data, CycleGAN-generated images accurately identified target reflection features while dynamically suppressing clutter, maintaining precise localization of hyperbolic vertices. In contrast, Pixel to Pixel exhibited consistent limitations in measured data: generated hyperbola vertices retained positional deviations and failed to reflect multi-target quantitative relationships, merely approximating hyperbolic shapes. This indicates that pixel to pixel struggles to adapt to the complex requirements of GPR B-scan data augmentation tasks.

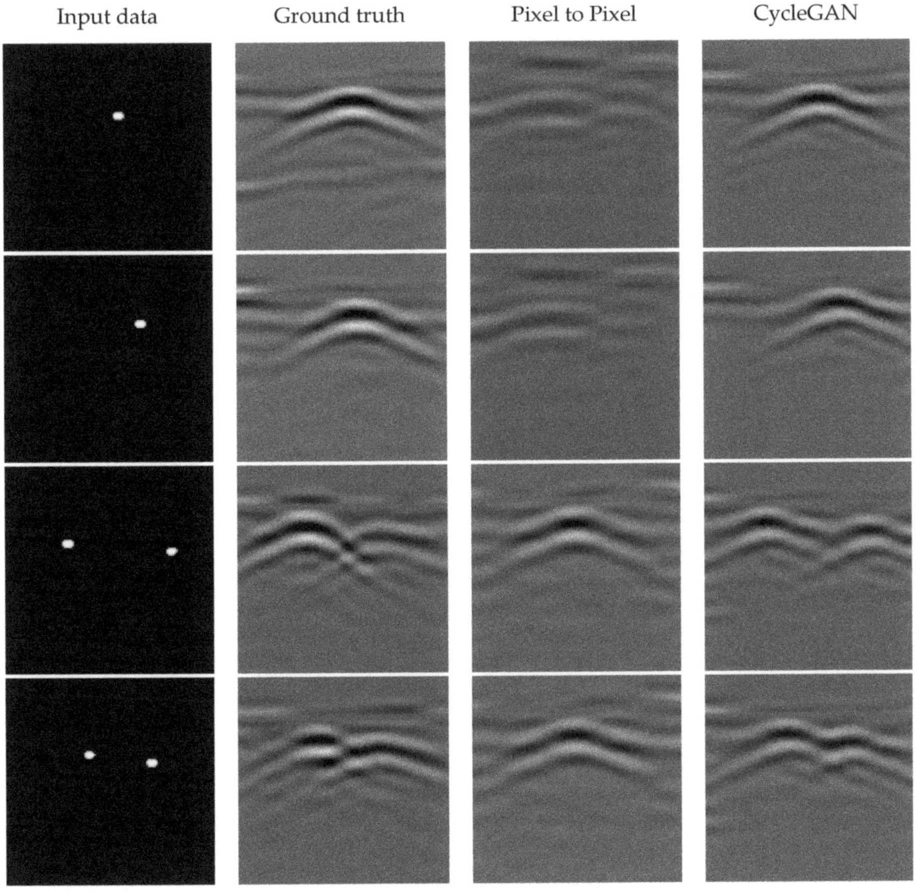

Figure 8. Data augmentation results in measured GPR B-scan images.

4. Results and Discussion

4.1. Experimental Setup and Evaluation Metrics

The experimental platform was developed on the PyTorch 2.4.0 framework with an Intel i5-13600KF CPU and NVIDIA RTX 4070 GPU hardware. Mixed-precision training acceleration was implemented via CUDA 12.1 to fully exploit GPU parallel computing capabilities.

The model was trained using the AdamW optimizer with an initial learning rate of 1×10^{-4}. A dynamic learning rate adjustment strategy (ReduceLROnPlateau) was implemented, automatically reducing the learning rate by 50% when validation loss plateaued for three consecutive epochs. Training spanned 100 epochs with a fixed batch size of 3. To ensure robustness, intermediate model weights were saved at every 202nd iteration, enabling recovery from interruptions and optimal checkpoint selection.

To comprehensively evaluate the performance of the object detection model, four key metrics were adopted, covering both precision and recall aspects across varying Intersection over Union (IoU) thresholds: mean average precision (mAP) comprehensively reflects the overall detection performance averaged over IoU thresholds from 0.5 to 0.95 (step size 0.05), mAP50 represents the average accuracy at an IoU threshold of 0.5, mAP75 corresponds to the detection accuracy at an IoU threshold of 0.75, and AR (Average Recall) quantifies the mean maximum recall rate of each target category within the IoU range of 0.5 to 0.95.

4.2. Performance Evaluation

The EfficientDet was trained and tested on the object detection dataset. Experimental results, as illustrated in Figure 9, indicate accurate target identification on both datasets. For simulation data, we restored the original coordinate system. The horizontal axis corresponds to 125 measurement positions along a 2.5 m scan line with 20 mm intervals, indexed from 1 to 125. The vertical axis represents the time range of 0–15 ns divided into three segments: 0–5 ns, 5–10 ns, and 10–15 ns. During CycleGAN training, we intentionally removed axis labels. This was done to help the model focus on learning the intrinsic features of the B-scans without being distracted by coordinate information. In the revised manuscript, we have reinstated full coordinate annotations to all synthetic B-scans.

Figure 10 further validates the model's adaptability in real-world environments. Measured data present challenges like background clutter and random noise interference. Despite this, the model accurately identifies low-contrast hyperbolic features. It achieves this without a miss or false alarm. The results demonstrate that EfficientDet delivers consistent multi-object detection performance across cross-domain data (simulated and measured), proving its effectiveness and stability in hyperbolic feature detection tasks for GPR B-scan images. It should be noted that the measured dataset is an open-source collection that inherently lacked corresponding ground truth bounding box coordinate information upon acquisition. Consequently, the prediction results for measured data in the final figures were not supplemented with this information. Regarding the simulation data, to prevent coordinate axis interference during CycleGAN's image-to-image translation, we intentionally removed axis labels during training. For the final result figures, we have added explicit time-scale annotations to all synthetic B-scan images. These annotations directly reflect the two-way travel time of EM waves, which correlates with target depth (i.e., greater depth corresponds to longer return time).

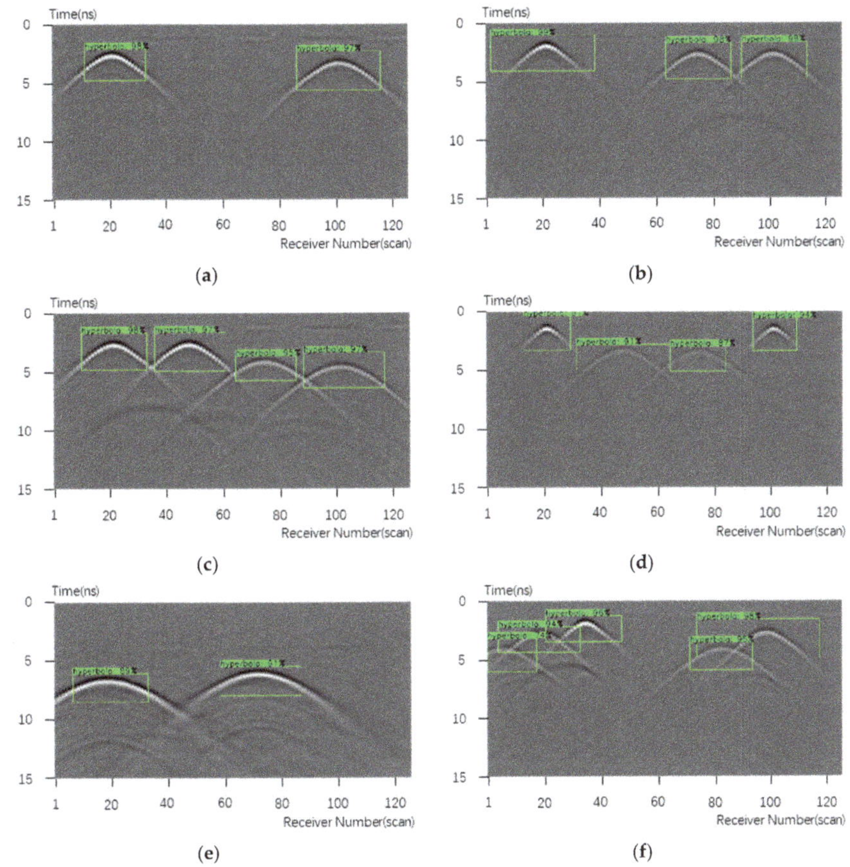

Figure 9. Detection results of simulated data. (**a**) First example. (**b**) Second example. (**c**) Third example. (**d**) Fourth example. (**e**) Fifth example. (**f**) Sixth example.

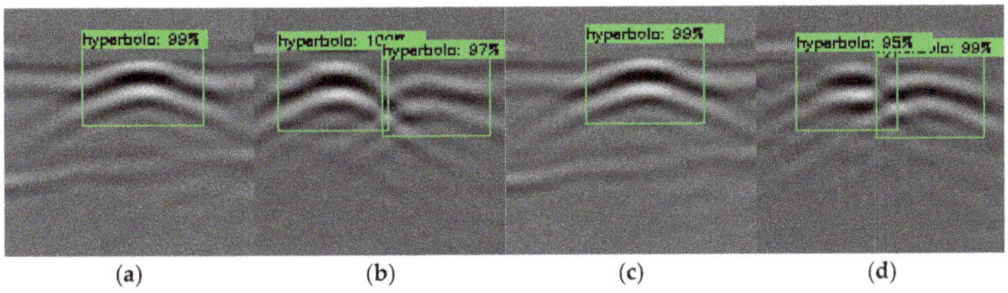

Figure 10. Detection results of measured GPR data. (**a**) First example. (**b**) Second example. (**c**) Third example. (**d**) Fourth example.

During training, both classification and regression losses exhibit stable convergence trends as iterations progress (Figure 11). As shown in Figure 11a, the classification loss decreases steadily to low values, indicating progressive optimization of target category recognition. As illustrated in Figure 11b, the regression loss initially fluctuates but eventually converges, reflecting effective learning of bounding box localization. Neither loss shows significant oscillations or secondary increases, suggesting that the training process remains unaffected by gradient anomalies or overfitting. The synchronous convergence of both losses underscores the model's balanced optimization of classification and localization tasks, validating the efficacy of the EfficientDet framework in detecting hyperbolic features in B-scan images.

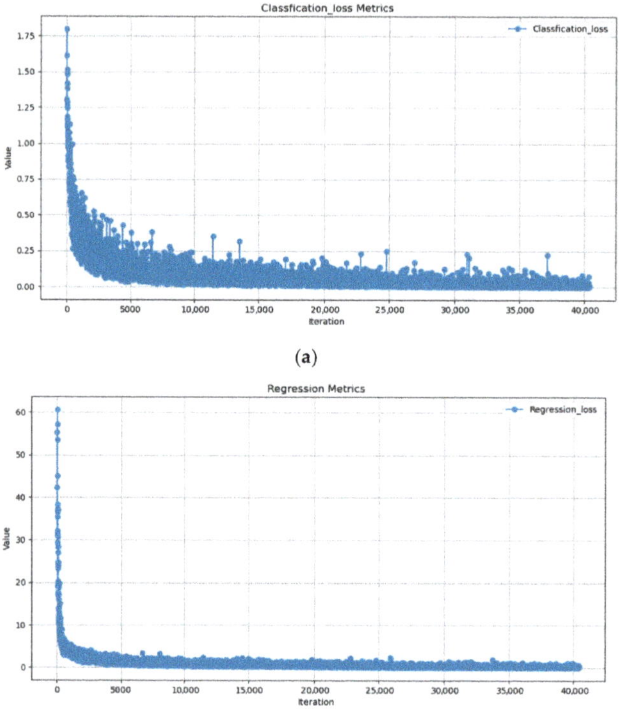

(b)

Figure 11. Training loss convergence: (**a**) classification loss and (**b**) regression loss.

4.3. Comparison with Different Object Detection Models

To prove our model works better, we compared it with two popular detection models—RetinaNet and YOLOv3. To keep things fair, all models were trained with exactly the same data for 100 cycles and tested under the exact same settings. Experimental results on the simulated dataset, as summarized in Table 4, demonstrate the comprehensive advantages of the EfficientDet in detection performance. Its overall detection capability (mAP = 0.579) surpasses YOLOv3 and RetinaNet by 16.0% and 23.5%, respectively, validating the effectiveness of the proposed architecture. Under lenient localization requirements (mAP50 = 0.986), the detection accuracy approaches theoretical limits, indicating extremely high confidence in target presence determination. However, in stringent localization scenarios (mAP75 = 0.578), the performance improvement narrows, which may be attributed to target edge ambiguity in complex environments. Notably, the model achieves a significantly higher average recall rate (AR = 0.667) compared to baseline models (YOLOv3: 0.551; RetinaNet: 0.528), proving its enhanced capability to suppress missed detections for multi-scale targets.

Table 4. Detection method evaluation results.

Model	mAP	mAP_{50}	mAP_{75}	AR
RetinaNet	0.469	0.752	0.516	0.528
YOLOv3	0.499	0.796	0.563	0.551
EfficientDet	0.579	0.986	0.578	0.667

Experimental results demonstrate that EfficientDet exhibits significant advantages in both simulated and real-world scenarios (Figures 12 and 13). Specifically, its detection confidence remains stably distributed within the 0.9–1.0 range, and the IoU values of predicted bounding boxes relative to hyperbolic vertices (mAP75 = 0.578) significantly outperform baseline models, validating the robustness of BiFPN multi-scale feature fusion against geometric distortions and noise interference. In contrast, RetinaNet shows an abnormal concentration of detection confidence in high-value intervals (Figure 12b), indicating an overfitting tendency that leads to a higher false negative rate (AR = 0.528) in low-contrast scenarios. Meanwhile, YOLOv3's positioning accuracy (mAP75 = 0.563) is constrained by the preset anchor box mechanism's insufficient adaptability to the dynamic morphology of hyperbolic features.

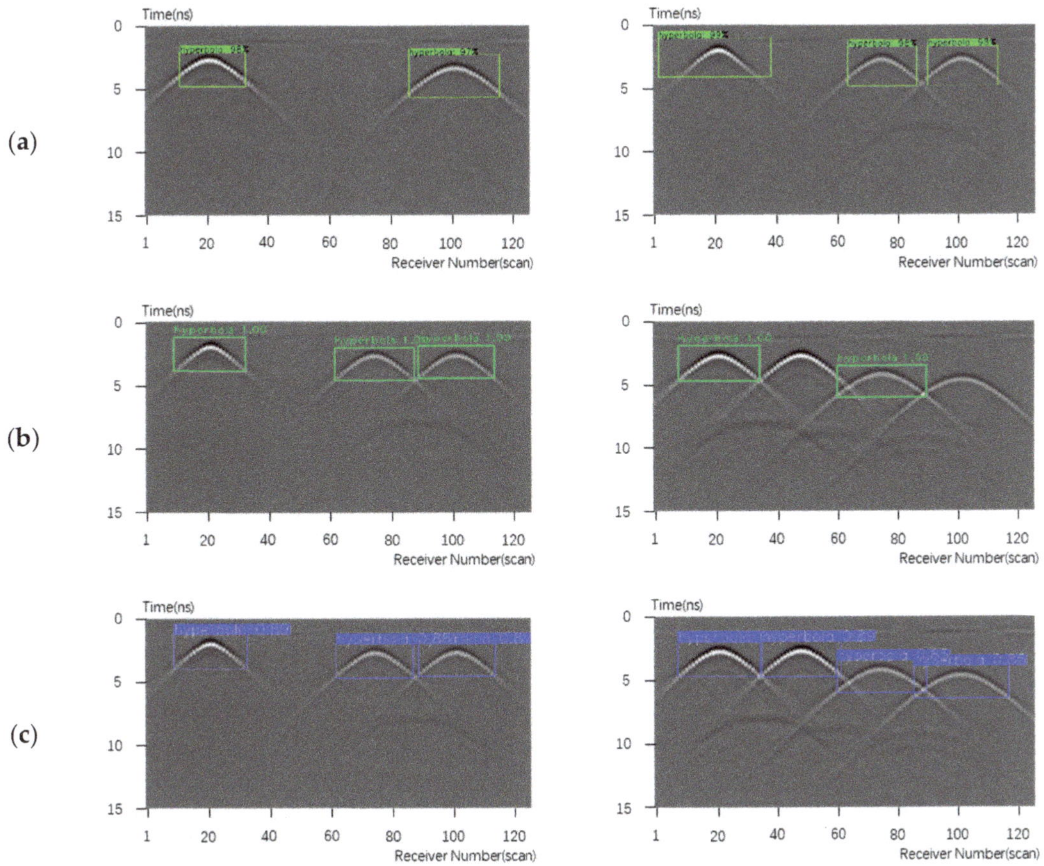

Figure 12. Comparative performance on simulated data: (**a**) EfficientDet; (**b**) RetinaNet; (**c**) YOLOv3.

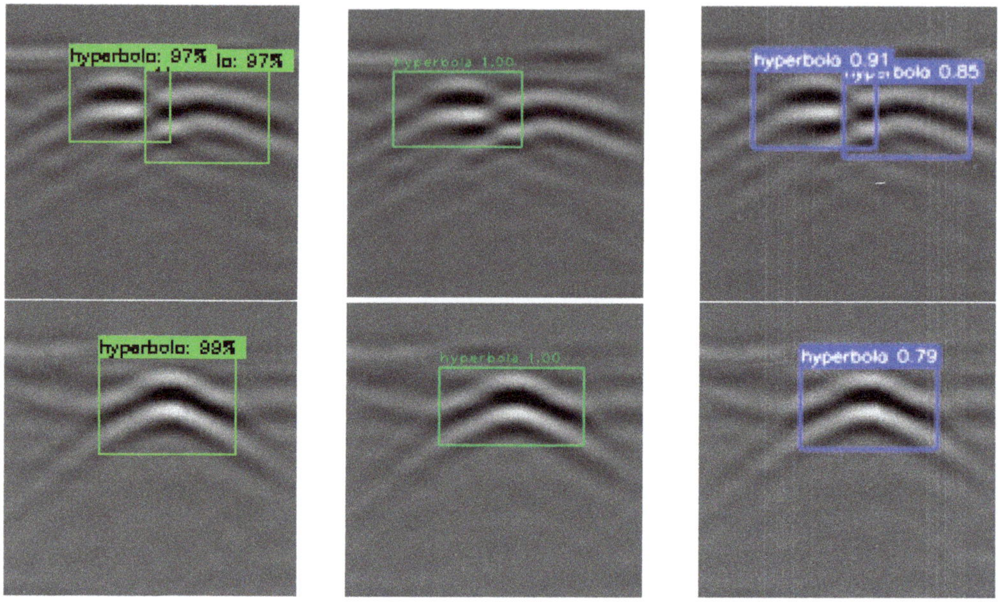

(**a**) (**b**) (**c**)

Figure 13. Comparative performance on measured GPR data: (**a**) EfficientDet; (**b**) RetinaNet; (**c**) YOLOv3.

In measured data (Figure 13), EfficientDet further validates its engineering applicability: it maintains high detection accuracy without missed or false alarms under strong background clutter and noise interference, while RetinaNet and YOLOv3 exhibit significantly weaker performance in complex scenarios. The consistency between simulated and measured results confirms that CycleGAN data augmentation and BiFPN dynamic fusion strategies effectively enhance the model's cross-domain generalization capability.

5. Conclusions

This study addresses the critical challenge of automating hyperbolic target detection in GPR B-scan images, hindered by scarce annotated real-world data and complex signatures under noise. We propose a novel framework combining CycleGAN for unsupervised data augmentation and EfficientDet for robust detection. CycleGAN generates diverse synthetic images by translating limited real data across geological domains. EfficientDet, trained on this augmented data, demonstrates substantial improvements in handling distortions and noise. Experimental validation confirms effectiveness, achieving a 0.579 mAP on synthetic datasets—surpassing YOLOv3 and RetinaNet by 16.0% and 23.5%, respectively. Analysis reveals persistent challenges:

(i). Recall-Precision Trade-off: While precision under lenient localization thresholds is exceptional (mAP50 = 0.986), the average recall (AR = 0.667), though superior to baselines, indicates potential missed detections, particularly for low-contrast targets or in high-clutter environments.

(ii). Localization Precision under Strict Criteria: The significant performance gap between mAP50 (0.986) and mAP75 (0.578) highlights difficulties in precisely localizing hyperbolic vertices under stringent IoU thresholds, likely due to residual sensitivity to geometric distortions and noise affecting vertex accuracy.

Future research will pursue several pivotal objectives, with these efforts targeting current methodological limitations to advance non-destructive testing technologies toward higher precision and broader applicability in complex environments:

(i) Replacing binary permittivity maps to enable the generation of B-scans depicting more complex subsurface scenarios, such as soil-concrete interfaces and gradual dielectric transitions, thereby increasing the visual complexity and diversity of the synthetic training data.
(ii) Developing detection methodologies for closely spaced or overlapping subsurface targets to address current limitations in resolving adjacent objects.
(iii) Establishing systematic C-scan processing workflows to shift focus beyond current B-scan hyperbolic signature detection toward exploring 3D subsurface interpretation.
(iv) Exploring technical pathways for integrating EM wave propagation laws (Maxwell's equations) into neural networks to enhance the physical plausibility of synthetic B-scans. This direction aims to combine physics-guided DL frameworks with generative models, improving the realism of reflection/attenuation behavior while maintaining data generation efficiency.

These efforts collectively target current methodological limitations. The goal is to enhance non-destructive testing technologies, achieving higher precision and practical utility in complex environments.

Author Contributions: Validation, Funding acquisition, H.Z.; Writing—original draft, Z.M.; Writing—review & editing, F.H.; Validation, X.F. All authors have read and agreed to the published version of the manuscript.

Funding: The study was supported by National Natural Science Foundation of China (nos. 62406346); Natural Science Foundation of Hunan Province, China (nos. 2025JJ60257). Degree and Graduate Education Reform Project of Central South University (nos. 2025YJSKS019).

Data Availability Statement: Data will be made available on request.

Conflicts of Interest: The authors declare that they have no known competing financial interests or personal relationships that could have appeared to influence the work reported in this paper.

References

1. Xiao, X.; Chen, S.; Xu, Y.; Ge, L.; Wu, J. Ground Penetrating Radar (GPR) Based Mapping of Underground Pipeline Network Distribution. *IEEE Sens. J.* **2024**, *24*, 19562–19576. [CrossRef]
2. Chang, C.-W.; Tsai, C.-A.; Shiau, Y.-C. Inspection of steel bars corrosion in reinforced concrete structures by nondestructive ground penetrating radar. *Appl. Sci.* **2022**, *12*, 5567. [CrossRef]
3. Wang, Y.; Qin, H.; Miao, F. A Multi-Path Encoder Network for GPR Data Inversion to Improve Defect Detection in Reinforced Concrete. *Remote Sens.* **2022**, *14*, 5871. [CrossRef]
4. Faris, N.; Zayed, T.; Fares, A.; Abdelkhalek, S.; Abdelkader, E.M. Automated rebar recognition and corrosion assessment of concrete bridge decks using ground penetrating radar. *Autom. Constr.* **2024**, *166*, 105631. [CrossRef]
5. Ghodoosi, F.; Bagchi, A.; Zayed, T.; Hosseini, M.R. Method for developing and updating deterioration models for concrete bridge decks using GPR data. *Autom. Constr.* **2018**, *91*, 133–141. [CrossRef]
6. Abouhamad, M.; Dawood, T.; Jabri, A.; Alsharqawi, M.; Zayed, T. Corrosiveness mapping of bridge decks using image-based analysis of GPR data. *Autom. Constr.* **2017**, *80*, 104–117. [CrossRef]
7. Asadi, P.; Gindy, M.; Alvarez, M.; Asadi, A. A computer vision based rebar detection chain for automatic processing of concrete bridge deck GPR data. *Autom. Constr.* **2020**, *112*, 103106. [CrossRef]
8. Rasol, M.; Pais, J.C.; Pérez-Gracia, V.; Solla, M.; Fernandes, F.M.; Fontul, S.; Ayala-Cabrera, D.; Schmidt, F.; Assadollahi, H. GPR monitoring for road transport infrastructure: A systematic review and machine learning insights. *Constr. Build. Mater.* **2022**, *324*, 126686. [CrossRef]
9. Liu, H.; Shi, Z.; Li, J.; Liu, C.; Meng, X.; Du, Y.; Chen, J. Detection of Cavities In Urban Cities by 3D Ground Penetrating Radar. *Geophysics* **2021**, *86*, A27–WB97. [CrossRef]

10. Luo, X.H.; Lai, W.L. GPR pattern recognition of shallow subsurface air voids. *Tunn. Undergr. Space Technol.* **2020**, *99*, 103355. [CrossRef]
11. Li, H.; Zhang, J.; Yang, X.; Ye, M.; Jiang, W.; Gong, J.; Tian, Y.; Zhao, L.; Wang, W.; Xu, Z. Bayesian optimization based extreme gradient boosting and GPR time-frequency features for the recognition of moisture damage in asphalt pavement. *Constr. Build. Mater.* **2024**, *434*, 136675. [CrossRef]
12. Li, H.; Zhang, J.; Ye, M.; Wang, Q.; Wei, M.; Zhang, Z.; Guo, Y.; Tian, Y.; Zhang, Y.; Wang, C.; et al. High-accuracy intelligent detection of centimeter-level voids in cement pavement via ground-penetrating radar. *Measurement* **2025**, *228*, 118354. [CrossRef]
13. Prego, F.J.; Solla, M.; Núñez-Nieto, X.; Arias, P. Assessing the applicability of ground-penetrating radar to quality control in tunneling construction. *J. Constr. Eng. Manag.* **2016**, *142*, 06015006. [CrossRef]
14. Li, J.; Liu, H.; Meng, X.; Duan, D.; Lu, H.; Zhang, J.; Zhang, F.; Elsworth, D.; Cardenas, B.T.; Manga, M.; et al. Ancient ocean coastal deposits imaged on Mars. *Proc. Natl. Acad. Sci. USA* **2025**, *122*, e2422213122. [CrossRef] [PubMed]
15. Hou, F.; Qiao, B.; Dong, J.; Ma, Z. S-cyclegan: A novel target signature segmentation method for gpr image interpretation. *IEEE Geosci. Remote Sens. Lett.* **2024**, *21*, 7502005. [CrossRef]
16. Xiao, J.; Gao, Q.; Ling, Y.; Yan, J.; Liu, B. Research on Hyperbola Detection and Fitting in GPR B-Scan Image. In Proceedings of the International Conference on Communications, Information System and Computer Engineering (CISCE), Beijing, China, 14–16 May 2021; IEEE: New York City, NY, USA, 2021; pp. 266–270.
17. Chen, R.; Wang, Z.; Zhang, K.Y.; Wu, S.; Sun, J.; Wang, S.; Yao, T.; Ding, S. Decoupled data augmentation for improving image classification. *arXiv* **2024**, arXiv:2411.02592.
18. Zhang, L.; Ma, R.; Zhang, Y. Application of improved LeNet—5 model in apple image recognition. *Comput. Eng. Des.* **2018**, *39*, 3570–3575.
19. Huang, Z.; Huang, L.; Gong, Y.; Huang, C.; Wang, X. Mask scoring r-cnn. In Proceedings of the IEEE/CVF Conference on Computer Vision and Pattern Recognition, Long Beach, CA, USA, 15–20 June 2019; pp. 6409–6418.
20. Kirillov, A.; Wu, Y.; He, K.; Girshick, R. Pointrend: Image segmentation as rendering. In Proceedings of the IEEE/CVF Conference on Computer Vision and Pattern Recognition, Seattle, WA, USA, 13–19 June 2020; pp. 9799–9808.
21. Chen, K.; Cao, Y.; Loy, C.C.; Lin, D. Feature pyramid grids. *arXiv* **2020**, arXiv:2004.03580. [CrossRef]
22. Hou, F.; Yin, X.; Fan, X. Refined Target Signature Segmentation from GPR B-scans based on the MaskScoreNet-E Model and RATS Method. *IEEE Trans. Geosci. Remote Sens.* **2024**, *62*, 5108614. [CrossRef]
23. Wang, A.; Chen, H.; Liu, L.; Chen, K.; Lin, Z.; Han, J. Yolov10, Real-time end-to-end object detection. *Adv. Neural Inf. Process. Syst.* **2024**, *37*, 107984–108011.
24. Wang, Y.; Qin, H.; Zhang, D.; Wu, T.; Pan, S. Domain Adaption YOLO Network to Enhance Target Detection in GPR Images. *IEEE Trans. Geosci. Remote Sens.* **2024**, *62*, 5936415. [CrossRef]
25. Qin, H.; Zhang, D.; Tang, Y.; Wang, Y. Automatic recognition of tunnel lining elements from GPR images using deep convolutional networks with data augmentation. *Autom. Constr.* **2021**, *130*, 103830. [CrossRef]
26. LeCun, Y.; Bottou, L.; Bengio, Y.; Haffner, P. Gradient-based learning applied to document recognition. *Proc. IEEE* **1998**, *86*, 2278–2324. [CrossRef]
27. Dai, Q.; Lee, Y.H.; Sun, H.H.; Qian, J.; Ow, G.; Yusof, M.L.; Yucel, A.C. A deep learning-based GPR forward solver for predicting B-scans of subsurface objects. *IEEE Geosci. Remote Sens. Lett.* **2022**, *19*, 4025805. [CrossRef]
28. Hou, F.; Qian, X.; Meng, Q.; Dong, J.; Lyu, F. Multi-scale GAN-driven GPR data inversion for monitoring urban road substructure. *Autom. Constr.* **2025**, *174*, 106140. [CrossRef]
29. Wang, X.; Yu, T.; Feng, D.; Li, B.; Ding, S. Spatiotemporal Optimization of GPR Full Waveform Inversion Based on Super-Resolution Technology. *IEEE Trans. Geosci. Remote Sens.* **2025**, *63*, 5908413. [CrossRef]
30. Goodfellow, I.; Pouget-Abadie, J.; Mirza, M.; Xu, B.; Warde-Farley, D.; Ozair, S.; Courville, A.; Bengio, Y. Generative adversarial nets. *Adv. Neural Inf. Process. Syst.* **2014**, *27*. [CrossRef]
31. Zhu, J.Y.; Park, T.; Isola, P.; Efros, A.A. Unpaired image-to-image translation using cycle-consistent adversarial networks. In Proceedings of the IEEE International Conference on Computer Vision, Venice, Italy, 22–29 October 2017; pp. 2223–2232.
32. Tan, M.; Pang, R.; Le, Q.V. Efficientdet: Scalable and efficient object detection. In Proceedings of the IEEE/CVF Conference on Computer Vision and Pattern Recognition, Seattle, WA, USA, 13–19 June 2020; pp. 10781–10790.
33. Denton, E.L.; Chintala, S.; Fergus, R. Deep generative image models using a laplacian pyramid of adversarial networks. *Adv. Neural. Inf. Process. Syst.* **2015**, *28*. [CrossRef]
34. Radford, A.; Metz, L.; Chintala, S. Unsupervised representation learning with deep convolutional generative adversarial networks. *arXiv* **2015**, arXiv:1511.06434.
35. Zhao, J.; Mathieu, M.; LeCun, Y. Energy-based generative adversarial network. *arXiv* **2016**, arXiv:1609.03126.
36. Zhu, J.Y.; Krähenbühl, P.; Shechtman, E.; Efros, A.A. Generative Visual Manipulation on the Natural Image Manifold. In Proceedings of the Computer Vision–ECCV 2016: 14th European Conference, Proceedings, Part v 14, Amsterdam, The Netherlands, 11–14 October, 2016; Springer International Publishing: Cham, Switzerland, 2016; pp. 597–613.

37. Warren, C.; Giannopoulos, A.; Giannakis, I. gprMax: Open source software to simulate electromagnetic wave propagation for Ground Penetrating Radar. *Comput. Phys. Commun.* **2016**, *209*, 163–170. [CrossRef]
38. Dai, Q.; Lee, Y.H.; Sun, H.H.; Ow, G.; Yusof, M.L.M.; Yucel, A.C. DMRF-UNet: A two-stage deep learning scheme for GPR data inversion under heterogeneous soil conditions. *IEEE Trans. Antennas Propag.* **2022**, *70*, 6313–6328. [CrossRef]
39. Isola, P.; Zhu, J.Y.; Zhou, T.; Efros, A.A. Image-to-image translation with conditional adversarial networks. In Proceedings of the IEEE Conference on Computer Vision and Pattern Recognition, Honolulu, HI, USA, 21–26 July 2017; pp. 1125–1134.

remote sensing

MDPI

Article

Estimating Subsurface Geostatistical Properties from GPR Reflection Data Using a Supervised Deep Learning Approach

Yu Liu [1,2], James Irving [2,*] and Klaus Holliger [2]

1 National Supercomputing Center in Zhengzhou, Zhengzhou University, Zhengzhou 450000, China; yuliu@zzu.edu.cn
2 Institute of Earth Sciences, University of Lausanne, CH-1015 Lausanne, Switzerland; klaus.holliger@unil.ch
* Correspondence: james.irving@unil.ch

Abstract

The quantitative characterization of near-surface heterogeneity using ground-penetrating radar (GPR) is an important but challenging task. The estimation of subsurface geostatistical parameters from surface-based common-offset GPR reflection data has so far relied upon a Monte-Carlo-type inversion approach. This allows for a comprehensive exploration of the parameter space and provides some measure of uncertainty with regard to the inferred results. However, the associated computational costs are inherently high. To alleviate this problem, we present an alternative deep-learning-based technique, that, once trained in a supervised context, allows us to perform the same task in a highly efficient manner. The proposed approach uses a convolutional neural network (CNN), which is trained on a vast database of autocorrelations obtained from synthetic GPR images for a comprehensive range of stochastic subsurface models. An important aspect of the training process is that the synthetic GPR data are generated using a computationally efficient approximate solution of the underlying physical problem. This strategy effectively addresses the notorious challenge of insufficient training data, which frequently impedes the application of deep-learning-based methods in applied geophysics. Tests on a wide range of realistic synthetic GPR data generated using a finite-difference time-domain (FDTD) solution of Maxwell's equations, as well as a comparison with the results of the traditional Monte Carlo approach on a pertinent field dataset, confirm the viability of the proposed method, even in the presence of significant levels of data noise. Our results also demonstrate that typical mismatches between the dominant frequencies of the analyzed and training data can be readily alleviated through simple spectral shifting.

Keywords: subsurface heterogeneity; GPR; geostatistical inversion; supervised learning; CNN

1. Introduction

The shallow subsurface consists of soil, glacial and alluvial deposits, as well as weathered bedrock. It is from this thin, surficial layer that we draw our drinking water and produce our food, and it is where we ultimately discharge much of our waste. In densely populated and highly industrialized regions, the sustainable use of this increasingly fragile surficial environment has become a societal and political issue of critical importance. Near-surface geophysics, in general, and ground-penetrating radar (GPR) in particular allow for the exploration and characterization of this surficial environment and thus have an essential role to play in this endeavor.

Surface-based GPR reflection profiling has the potential to provide images of shallow subsurface structure with unsurpassed spatial resolution. However, due to the inherently complex and heterogeneous nature of the surficial environment, these images are often better suited to stochastic than to deterministic interpretation approaches. This means that we aim to estimate parameters describing the geostatistical nature of heterogeneity. Specifically, the second-order statistical properties of GPR reflection data are expected to provide information about the correlation structure of the subsurface velocity field through which the high-frequency electromagnetic waves have traveled. Rea and Knight and Dafflon et al. observed a far-reaching qualitative agreement between the spatial correlation structure of cliff-face photographs and those of GPR images collected along the tops of the cliffs [1,2]. This led them to conclude that the lateral statistics of a GPR image and the underlying water content distribution are directly related. Knight et al. also found similarities between the lateral correlation statistics of GPR data and neutron probe measurements [3]. However, Knight et al. and Oldenborger et al. noticed that the lateral correlation structure of a GPR image is significantly affected by its vertical resolution, which is controlled by the frequency content of the emitted source signal [4,5]. These findings indicate that the assumption of a direct relation, let alone an equivalence, between the lateral statistics of water content and GPR reflection data is not valid and that the physics of the underlying wave propagation phenomena should be accounted for more appropriately. Irving et al. and Irving and Holliger followed up on these pioneering studies, notably by clarifying the underlying theoretical and methodological foundations [6,7]. They found that GPR reflection images allow us to infer the structural aspect ratio a_x/a_z of the subsurface heterogeneity, whereas the associated values of the horizontal and vertical correlation lengths a_x and a_z cannot be independently resolved. GPR data also seem to exhibit only a weak sensitivity to the decay of the associated power spectrum. These findings have been corroborated through a wide range of follow-up studies not only for GPR [8,9], but also for seismic reflection data [10,11]. In this context, it is important to note that knowledge regarding the structural aspect ratio a_x/a_z of the underlying heterogeneity is essential for the geostatistical characterization of the probed subsurface region, as this information cannot be easily inferred otherwise. The vertical correlation length a_z as well as the spectral decay of the power spectrum, in contrast, can often be estimated using logging data from nearby boreholes [9,12]. Geostatistical parameters such as correlation lengths and structural aspect ratio provide a description of small-scale subsurface heterogeneity. In GPR inversion, these parameters can serve as prior information to guide the selection of regularization parameters and reduce solution non-uniqueness [13]. In hydrogeological modeling, the spatial heterogeneity of subsurface media is typically represented as a random field, and these parameters act as key inputs to stochastic flow and transport models. They help to characterize the spatial structure of hydraulic properties and thereby influence predictions of groundwater flow and contaminant transport.

To date, the estimation of subsurface geostatistical parameters from reflection GPR data has been based on Monte-Carlo-type inversion approaches (e.g., Irving et al., 2009; Xu et al., 2020), whereby sets of model parameters drawn from prescribed prior distributions are either accepted or rejected depending on how well they allow for prediction of the observed autocorrelation structure of the GPR image. While this approach has the advantage of allowing for a comprehensive exploration of the model parameter space and providing a corresponding estimate of uncertainty, the associated computational costs are high, in particular for 3D datasets [6,8]. In this work, we seek to explore a potential alternative using a fundamentally different approach based on supervised deep learning. Deep learning techniques employ multi-layer artificial neural networks, especially convolutional neural networks (CNN), to unravel complex relationships between input and output [14,15]. In recent years, these methods have attracted increasing attention in the geophysical community as a novel data-driven technique compared to conventional model-driven approaches [16–18]. For wave-based geophysical exploration techniques, such as seismic and GPR reflection imaging, deep learning has demonstrated remarkable utility in many areas, including data processing [19–21], interpretation [22,23], and inversion [24–26].

Here, we explore the use of a standard CNN architecture to establish a relationship between the spatial autocorrelations of surface-based GPR reflection images and the geostatistical parameters characterizing the distribution of the dielectric permittivity in the probed subsurface region. An important aspect of our workflow is that the training of the CNN is based on synthetic GPR reflection data, which are generated using a computationally highly efficient convolution-based modeling technique. While the latter is clearly an approximation of the underlying physics of high-frequency electromagnetic wave propagation, it is generally considered adequate for small incidence angles and has proven to be highly effective in previous waveform-based GPR inversion studies [9,13,27]. In the following, we first present the methodological foundations of our proposed approach, followed by its validation for realistic synthetic GPR data generated using a finite-difference time-domain (FDTD) solution of Maxwell's equations. We then examine the impact of noise on the estimation process and how typical spectral mismatches between the training database and the analyzed data can be accounted for. Finally, we apply the proposed method to an observed GPR dataset from a typical surficial alluvial environment and compare the results with those obtained using the well-established Monte-Carlo-based inversion approach.

2. Methodology

The overall idea behind our proposed approach is as follows: (i) to generate a vast synthetic GPR database for a comprehensive range of stochastic subsurface models; (ii) to train a CNN in a supervised setting in order to build a link between the 2D autocorrelation images of the synthetic images and the targeted geostatistical parameters of the probed subsurface region; and (iii) to use the trained CNN to estimate the geostatistical parameters corresponding to a particular GPR dataset from its 2D autocorrelation image (Figure 1). A distinction of our approach compared to other deep learning applications in GPR research is that the input to the network is not the GPR reflection data themselves, but their 2D spatial autocorrelation. This transformation enables the CNN to more directly extract information related to subsurface geostatistical parameters.

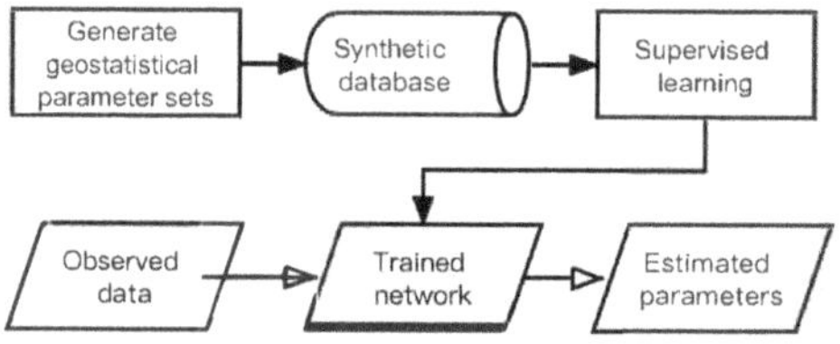

Figure 1. Flowchart illustrating the proposed supervised-learning-based approach for the estimation of subsurface geostatistical parameters from surface-based common-offset GPR reflection images.

2.1. Geostatistical Background

A fundamental assumption underlying our work is that the key characteristics of the prevailing subsurface heterogeneity can be adequately described by its second-order statistics. Our experience from previous Monte-Carlo-based inversions for near-surface geostatistical parameters, as well as evidence from the characterization of hydrocarbon reservoirs, indicate that, to a first approximation, this is generally the case as long as the analysis focuses on a given lithological and/or hydraulic unit [1,6–9]. To relate the second-order statistics of the electromagnetic velocity structure of the subsurface to the corresponding GPR reflection image, we briefly summarize the essential elements of the derivation presented in Irving et al. [9]. We begin with the assumption that the subsurface electromagnetic velocity distribution $v(x,z)$ can be characterized as the superposition of a slowly varying or constant background velocity model $v_0(x,z)$, upon which small-scale stochastic fluctuations $\Delta v(x,z)$ are superposed as follows:

$$v(x,z) = v_0(x,z) + \Delta v(x,z). \tag{1}$$

It is the stochastic component of the velocity structure $\Delta v(x,z)$ that gives rise to the reflected GPR wavefield recorded at the surface and whose second-order statistics we wish to estimate. To this end, the underlying subsurface GPR reflectivity distribution $r(x,z)$ can be approximated as the vertical derivative of the velocity perturbation field [28,29]

$$r(x,z) \approx \frac{\partial}{\partial z}\Delta v(x,z). \tag{2}$$

Assuming that the recorded GPR wavefield is largely non-dispersive and dominated by single scattering, a typical common-offset GPR reflection section, after processing and migration, can be approximated as the convolution of the subsurface reflectivity r with the source wavelet w as well as with a filter h that accounts for the limited lateral resolution of the GPR image [6,10].

$$\begin{aligned} p(x,z) &= w(z) \ast r(x,z) \ast h(x) \\ &= w(z) \ast \tfrac{\partial}{\partial z}\Delta v(x,z) \ast h(x) \\ &= \Delta v(x,z) \ast f(x,z), \end{aligned} \tag{3}$$

where $p(x,z)$ is the so-called primary reflection section and $f(x,z) = \frac{\partial}{\partial z}w(z) \ast h(x)$.

Equation (3) provides a simple convolutional relationship between the velocity perturbation field $\Delta v(x,z)$ and the GPR reflection image $p(x,z)$. Making use of the Fourier transform and the Wiener–Khinchin theorem linking the power spectrum and autocorrelation functions, a similar convolutional relationship is implied for the spatial autocorrelations of all quantities.

$$R_{pp}(\delta x,\delta z) = R_{vv}(\delta x,\delta z) \ast R_{ff}(\delta x,\delta z). \tag{4}$$

Equation (4) states that the 2D spatial autocorrelation of the GPR image $R_{pp}(\delta x,\delta z)$, where δx and δz refer to the horizontal and vertical lags, respectively, is related to the 2D spatial autocorrelation of the velocity perturbation field $R_{vv}(\delta x,\delta z)$ through 2D convolution with the filter autocorrelation $R_{ff}(\delta x,\delta z)$. This equation provides an effective link between the second-order statistics of the high-frequency electromagnetic wave velocity field with those of the corresponding GPR reflection section. Based on this link, we use 2D autocorrelation images of GPR reflection data R_{pp} to train our CNN to predict the geostatistical parameters governing R_{vv}.

To describe the expected range of near-surface subsurface velocity heterogeneity, we consider the highly versatile von Karman autocorrelation function [30], which is particularly well-established for this purpose [8,9,12,31] and given by the following:

$$R_{vv}(\delta x,\delta z) = \frac{r^v K_v(r)}{2^{v-1}\Gamma(v)}, \tag{5}$$

where $K_v(r)$ is the modified Bessel function of the second kind of order $0 \leq v \leq 1$, Γ is the gamma function, and

$$r = \sqrt{(x/a_x)^2 + (z/a_z)^2} \tag{6}$$

is the weighted radial autocorrelation lag with a_x and a_z denoting the horizontal and vertical correlation lengths, respectively. The v-value, also known as the Hurst number, determines the slope of the power spectrum $P_{vv} = \mathcal{F}\{R_{vv}\}$, with $\mathcal{F}$ denoting the Fourier transform, or, equivalently, the rate of decay of the autocorrelation R_{vv} at short lags [32]. Values of v close to zero characterize the seemingly ubiquitous pink or flicker noise. A value of $v = 0.5$ corresponds to the well-known exponential correlation function describing brown noise. Finally, for v-values close to 1, which characterize so-called black noise, the von Karman autocorrelation function resembles its Gaussian equivalent.

2.2. Generation of Training Database

Based on the von Karman autocorrelation function described in Equation (5), we generated 3000 stochastic subsurface GPR velocity models representing a wide range of combinations of a_x, a_z, and v. The models are 20 m wide and 10 m deep. We considered values of a_z ranging from 0.1 to 1 m, and we chose the aspect ratio a_x/a_z within the range of 1 to 20, which allowed us to achieve the corresponding values of a_x. The v-value varies between 0.1 and 0.9. Stochastically layered models of this kind are widely regarded as being characteristic of surficial alluvial environments [8,9,12,33]. For all 3000 realizations, the values of a_z, a_x/a_z, and v were chosen randomly, while the mean value of the velocity and its standard deviation are fixed at 0.1 m/ns and 0.01 m/ns, respectively. In addition, we generated 100 additional stochastic models using the same method for validation during network training.

From the subsurface velocity models, we generated corresponding synthetic GPR reflection images by convolving the vertical reflectivity series at all transmitter–receiver locations, spaced at 0.1 m, with a Ricker source wavelet having a central frequency of 100 MHz. This is followed by the application of a Gaussian filter $h(x)$ (Equation (3)) with a width of approximately one dominant wavelength to emulate the limited lateral resolution of a migrated GPR image [7]. It is important to note that we use this approximate convolution method to generate our synthetic database because it has a much lower computational cost than comprehensive approaches to solve Maxwell's equations, such as the FDTD method. The generation of the 3000 training datasets takes only tens of seconds on a 2018 MacBook Pro (Apple Inc., Cupertino, CA, USA). Although the convolution model does not capture the full physics of a GPR reflexion experiment, previous work has shown that it is capable of emulating a remarkable degree of realism present in GPR field data, even in complex scenarios [9,13]. The resulting synthetic GPR data were contaminated with zero–mean Gaussian random noise having a standard deviation of 2% with regard to the maximum absolute amplitude of the synthetic data.

2.3. Supervised Learning

We computed the 2D autocorrelations of the 3000 synthetic GPR reflection images, which we then used together with the corresponding values of a_x/a_z to perform the supervised training of the CNN. Figure 2 shows an example of a stochastic subsurface velocity model characterized by $a_x = 2$ m, $a_z = 0.2$ m, and $v = 0.5$ (Figure 2a), the corresponding synthetic convolution-based GPR image (Figure 2b), and its 2D autocorrelation function (Figure 2c). For the design and training of the CNN, we employed the MATLAB R2016 Deep Learning Toolbox (MathWorks, Natick, MA, USA). Our choice of a standard CNN architecture with six layers, comprising an input layer, a convolution layer, a rectified linear unit (ReLU) layer, a max pooling layer, a fully connected layer, and a regression layer, was motivated by its well-established efficacy and robustness with regard to image-based regression tasks [34,35]. The dataset used for training consisted of 3000 autocorrelation images with dimensions of 401 pixels in the horizontal direction and 201 pixels in the vertical direction, each labeled with the values of a_x/a_z corresponding to their underlying velocity models. For the training, we employed the stochastic gradient descent (SGD) with momentum algorithm, which demonstrated favorable characteristics with regard to speed of convergence and training efficiency in various deep learning applications [36]. The mini-batch size and epoch number are set to 128 and 30, respectively. These values strike a balance between computational efficiency and effective convergence during training [37]. For the learning rate schedule, we initiated training with a learning rate of 0.001 and then reduced it after 20 epochs using a drop factor of 0.1. This dynamic learning rate adjustment strategy helps to fine-tune the model as training progresses, preventing overshooting and improving generalization performance [38]. The entire training process was completed within approximately 50 min using a 2018 MacBook Pro (Apple Inc., Cupertino, CA, USA) with 16 GB of memory and a 2.2 GHz Intel i7-8750 processor.

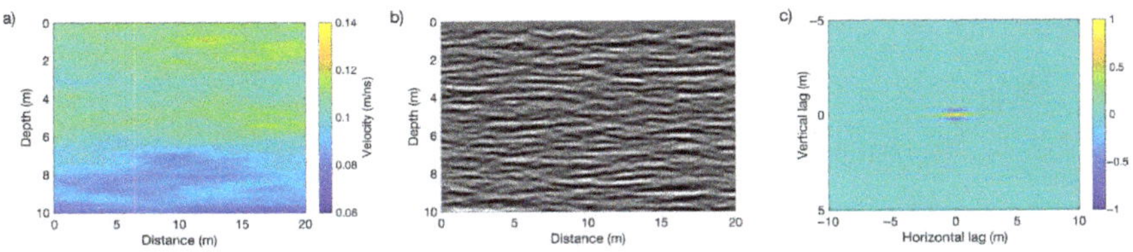

Figure 2. Example of (**a**) stochastic subsurface GPR velocity model, (**b**) corresponding convolution-based synthetic GPR image, and (**c**) 2D autocorrelation of the GPR image.

3. Results

3.1. FDTD-Based Synthetic Data

To assess whether and to what extent the trained CNN is capable of estimating the values of a_x/a_z from the 2D GPR autocorrelation images, we first performed tests based on synthetic GPR reflection data, which are generated using a FDTD solution of Maxwell's equations. It is important to note that data generated in this manner comprehensively account for the physics governing high-frequency electromagnetic wave propagation. As such, they represent the complete GPR wavefield, which, in addition to the primary reflections accounted for by the convolution-based approximation, may comprise multiple reflections and diffractions. To this end, we first generate 20 stochastic subsurface models (Figure 3) with randomly selected values of a_x, a_z, and v. We then simulated the corresponding GPR data using the FDTD algorithm of Irving and Knight [39]. For each simulation, the transmitter and receiver antennas were considered to be spaced 1 m apart and moved with a lateral increment of 0.1 m along the survey profile. The transmitter antenna current function was specified to be the first derivative of a Blackman–Harris window with a dominant frequency of 100 MHz. While the time–history and spectral characteristics of this function are similar to those of the Ricker wavelet used to create the training database, they are importantly not identical, which emulates corresponding discrepancies when applying the trained network to observed GPR data.

The resulting synthetic GPR data were contaminated with zero–mean Gaussian random noise having a standard deviation of 2% with regard to the maximum absolute signal amplitude below the direct wave, and were then subjected to a standard processing sequence involving the following: (i) elimination of the direct air and ground wave arrivals by removing the mean trace computed over a sliding window; (ii) 30–300 MHz band-pass filtering; (iii) amplitude scaling to compensate for the geometric spreading of energy along with scattering and transmission losses using a smooth gain function based on the mean absolute amplitude curve for the whole section; (iv) 2D F-K migration using a constant velocity of 0.1 m/ns, which was found to optimally focus diffraction events; and

(v) depth-conversion with a constant velocity of 0.1 m/ns. Figure 4 shows the processed FDTD-simulated GPR data.

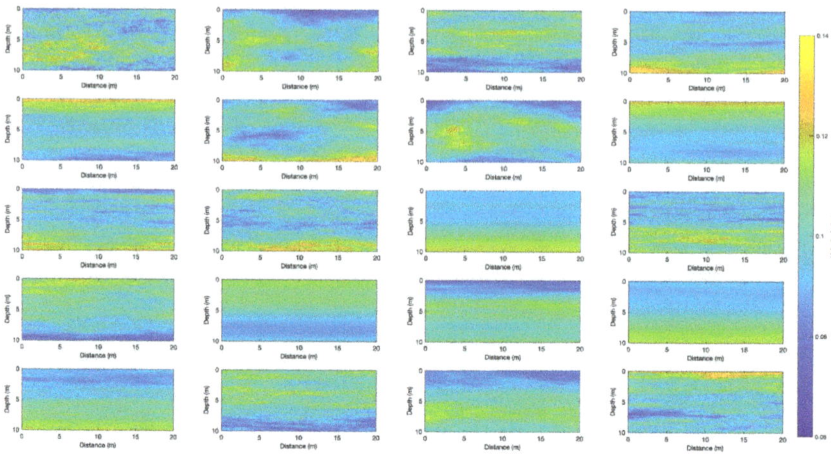

Figure 3. Velocity models used to generate FDTD-based synthetic GPR datasets.

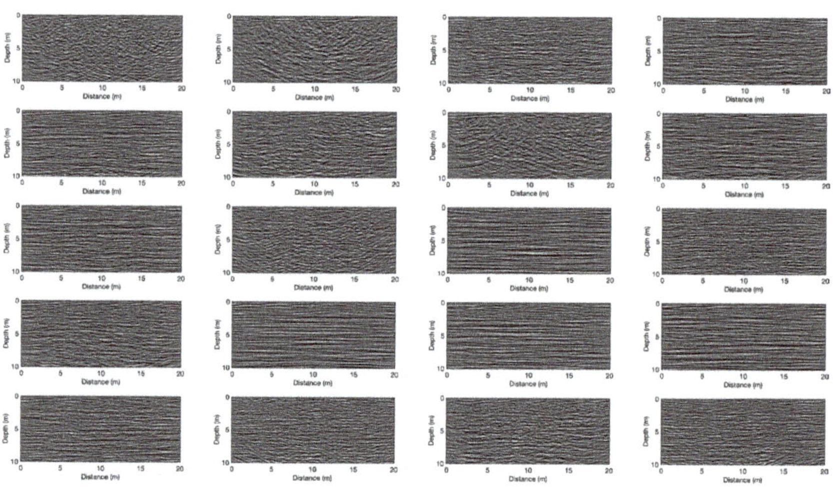

Figure 4. Processed 100 MHz FDTD-based synthetic GPR datasets corresponding to the velocity models shown in Figure 3.

We then compute the corresponding 20 autocorrelation images and submit them to the trained CNN to infer the geostatistical parameters of the underlying subsurface models. Figure 5 shows the estimated versus true values of the structural aspect ratio a_x/a_z. Overall, the estimations are remarkably accurate and closely follow the 45-degree trend line, which finds its expression in a rather high correlation coefficient of 0.93. The root mean square error (RMSE) is 1.93, further demonstrating the reliability of the predictions. This result is consistent with previous studies demonstrating that the autocorrelation of GPR reflection images is inherently sensitive to the aspect ratio a_x/a_z of the underlying heterogeneity [6,7]. Most importantly, however, it clearly underscores the potential of the novel estimation approach proposed in this study.

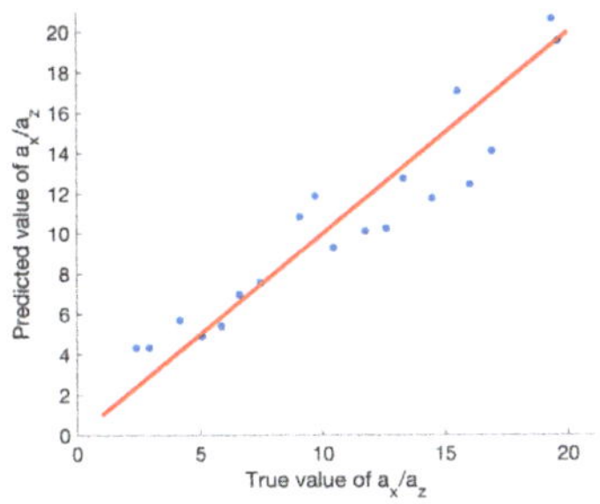

Figure 5. Scatter plot of estimated versus true aspect ratios a_x/a_z for the 100 MHz FDTD-based synthetic GPR datasets shown in Figure 4. The red line denotes perfect agreement.

3.2. Sensitivity to Noise

To test the sensitivity of the proposed method to the level of noise in the GPR data, we consider the stochastic model presented in Figure 2a, which was not included in the training database, and generate corresponding synthetic data using the FDTD method [39]. The synthetic data are then contaminated with gradually increasing amounts of zero-mean Gaussian random noise before being subjected to the same signal processing flow used in the previous test. In each case, the noise level is set relative to the maximum absolute signal amplitude below the direct arrivals.

Figure 6a,b show the resulting GPR images for noise levels of 2% and 50%, respectively. Comparing Figures 6a and 2a, we observe that the FDTD-based GPR image exhibits a remarkably strong resemblance to its convolution-based counterpart, which importantly corroborates the validity of using the latter type of approximate synthetic data for training the CNN. Although the GPR image shown in Figure 6b retains a clear similarity to its low-noise counterpart (Figure 6a), it is contaminated by a large number of "smiles", which typically arise when migrating pure random noise or very noisy data [40].

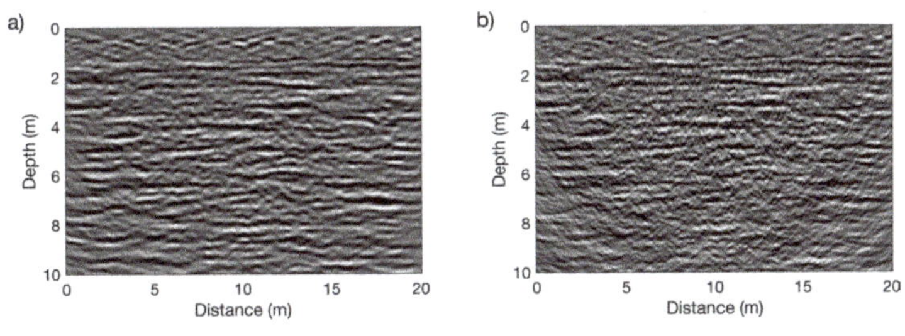

Figure 6. Processed 100 MHz FDTD-based synthetic GPR data corresponding to the velocity model shown in Figure 2a with noise levels of (**a**) 2% and (**b**) 50% with regard to the average amplitude.

Figure 7 shows the inferred values for a_x/a_z as a function of the noise level. Quite interestingly, these results demonstrate that the estimated values remain remarkably stable even in the presence of significant noise. We observed that the values of a_x/a_z gradually decreased with increasing noise, which is expected due to the decreasing lateral coherence of the GPR reflection images and the corresponding reduction in the lateral autocorrelation length. In this regard, it is important to note that the estimates of a_x/a_z progressively approach their target value with increasing noise content. This is due to the fact that, for the underlying data, our method overestimates the aspect ratio by ~10% at low noise levels (Figures 5 and 7).

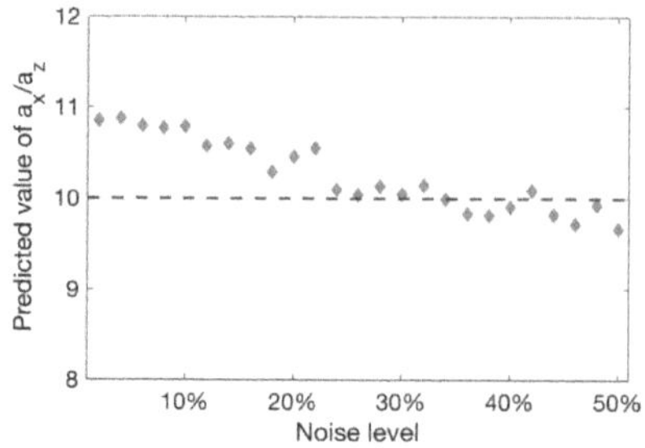

Figure 7. Aspect ratio a_x/a_z inferred from the 100 MHz FDTD-based synthetic GPR data shown in Figure 6 as a function of the level of added noise. The dashed line denotes the reference value.

3.3. *Differences in Spectral Characteristics of Analyzed Data and Training Database*

The synthetic GPR data used to train the CNN have a dominant frequency of 100 MHz and, hence, it is implicitly assumed that the spectral characteristics of the analyzed data are similar. For field GPR data, however, this is not guaranteed as, depending on near-surface electrical properties, the dominant frequencies often differ substantially from the corresponding nominal values of the transmitter antennae. To explore the impact of such differences, we again used 20 stochastic models shown in Figure 3 and simulated FDTD-based GPR data having a dominant frequency of 70 MHz. All other simulation parameters are kept unchanged. The resulting GPR reflection data were again contaminated with 2% Gaussian random noise and subjected to the same processing as the 100 MHz data (Figure 8). We then applied the following procedure to compensate for the differences in the spectral characteristics with regard to the training data: (i) calculate the Fourier transform of each trace; (ii) shift the amplitude spectrum upwards by 30 MHz by padding with the appropriate number of zeros; and (iii) transform the shifted spectrum back to the time domain retaining the original phase information. The original and shifted average amplitude spectra are shown in Figure 9.

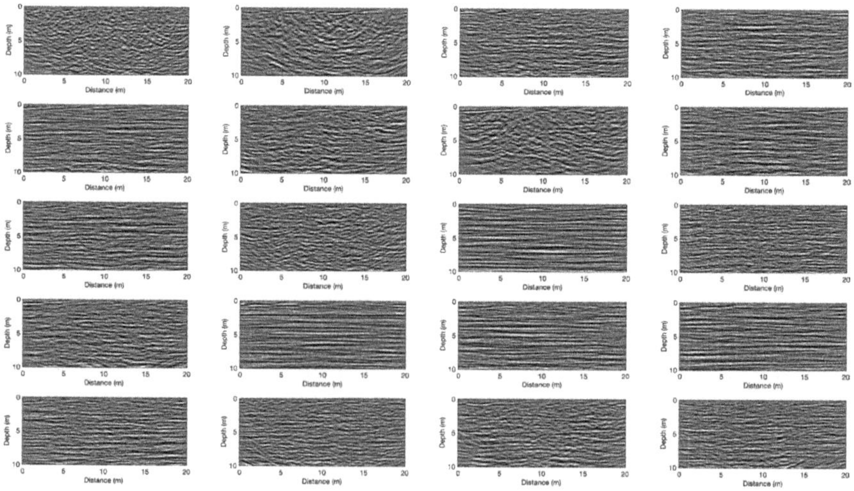

Figure 8. Processed 70 MHz FDTD-based synthetic GPR datasets corresponding to the velocity models shown in Figure 3.

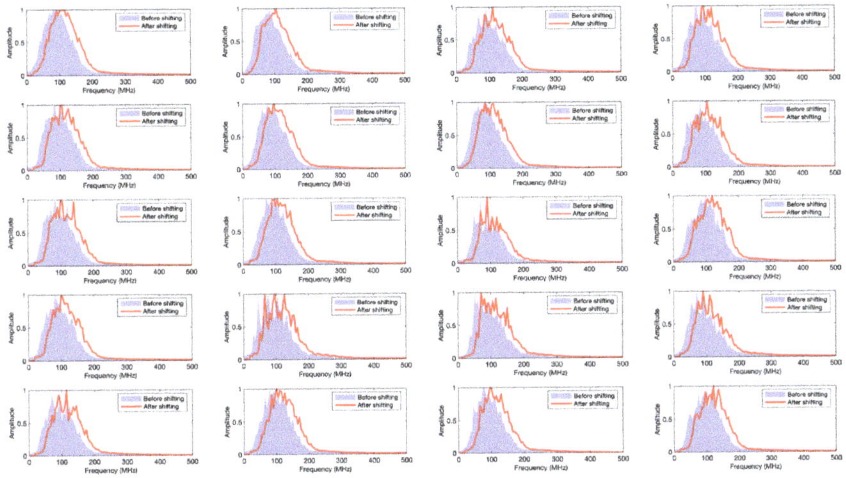

Figure 9. Average amplitude spectra corresponding to the 70 MHz FDTD-based synthetic GPR datasets shown in Figure 8 before and after spectral adjustment to a dominant frequency of ~100 MHz.

The 2D autocorrelation images of the GPR data before and after spectral shifting were submitted to the CNN trained on the original 100 MHz database. Figure 10 summarizes the resulting estimations of the structural aspect ratio a_x/a_z. The predicted values for the 70 MHz data show a systematic bias towards too low values (Figure 10a), whereas the estimates for the spectrally adjusted data (Figure 10b) agree as well with their target values as those for original 100 MHz FDTD-based synthetic GPR images (Figure 5). The correlation coefficient is 0.95, and the root mean squared error is 2.29. This indicates that the simple spectral adjustment procedure outlined above is a viable tool to compensate for differences in the spectral characteristics of the observed data with regard to the training database. In this context, it is important to note that most commercial GPR systems employ a limited number of "standard" antenna frequencies, for which the spectral bandwidth of the recorded data is quite uniformly in the range of 2 to 3 octaves. We therefore expect that the global training effort of the CNN can be restricted to these nominal values.

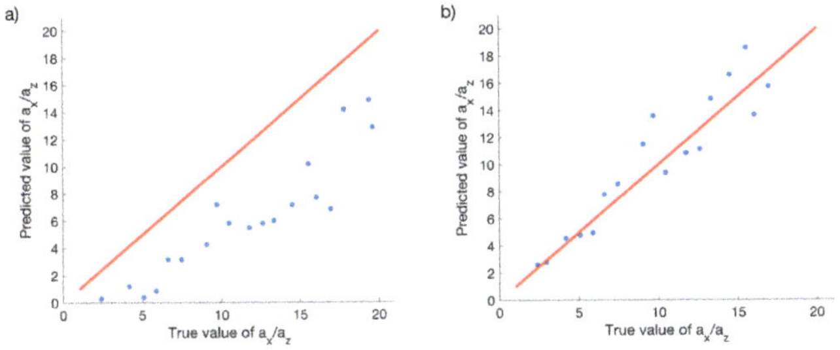

Figure 10. Scatter plots of estimated versus true aspect ratios a_x/a_z for the 70 MHz FDTD-based synthetic GPR datasets shown in Figure 8 (**a**) before and (**b**) after spectral adjustment. The red lines denote perfect agreement.

3.4. Application to Field Data

Following the preceding tests of our proposed methodology on FDTD-based synthetic data, we now consider its application to common-offset GPR reflection data acquired at the Boise Hydrogeophysical Research Site (BHRS), which is located on a gravel bar adjacent to the Boise River near Boise, Idaho, USA [41]. The surficial, unconfined alluvial aquifer at the BHRS is bound by a clay layer at approximately 20 m depth. The depth to the groundwater table varies seasonally between 0 and 3 m. Over the past two decades, a wide variety of geophysical and hydrogeological studies have been conducted at this site for characterizing heterogeneous alluvial aquifers [13,33,42–44]. The considered GPR reflection profile is part of a legacy 3D survey, which was performed in 1998 using a PulseEkko Pro 100 system (Sensors & Software, Inc., Mississauga, ON, Canada) with a nominal antenna center frequency of 100 MHz. The data were collected in common-offset mode using a transmitter–receiver antenna spacing of 1 m. Traces were recorded every 0.2 m along 18 m long profiles. The time sampling interval was 0.8 ns and 32 vertical stacks were performed for each recorded GPR trace to enhance the signal-to-noise ratio. As outlined earlier, second-order geostatistical characterizations should be confined to distinct lithologies and/or hydraulic units to be realistic and representative [9]. In the following, we therefore limit our analysis to the water-saturated zone and remove the top 40 ns comprising the vadose zone.

The considered GPR reflection data were subjected to a processing flow similar to their synthetic counterparts involving time-zero correction, dewow filtering, elimination of direct air and ground waves, 30–300 MHz bandpass filtering, smooth time-varying amplitude gain, and 2D F-K migration and depth conversion using a constant velocity of 0.08 m/ns. This velocity was found to optimally focus diffraction events below the water table and is consistent with the corresponding average value for the saturated zone at the BHRS as determined by Bradford et al. [43]. Figure 11a shows the processed GPR image. The somewhat jagged nature of the amplitude spectrum of the observed GPR data (Figure 11c) renders the estimation of the dominant frequency less straightforward than in the case of the synthetic data (Figure 9). Unlike in the synthetic case, where the center frequency is known, we must estimate the center frequency for the field data. One possibility we explored is to assess the average centroid frequency of the observed data and to use the corresponding difference with the training database to estimate the required spectral shift. Another possibility we considered is to estimate the dominant frequency of the observed data using a half-width-type approach, that is, to determine the mid-point of the amplitude spectrum at approximately half its maximum value. For the considered GPR data, both of these approaches yield an estimated central frequency of ~80 MHz. To make the spectrum of the observed data consistent with that of the training database, we therefore adjusted the former, as outlined above, by an upwards shift of 20 MHz. The comparison of the original and spectrally shifted data in Figure 11a,b is interesting and indicates that this procedure does not introduce any artifacts or otherwise alter the character of the original data apart from rendering them considerably crisper. Clearly, additional options for stabilizing the estimate of the dominant data frequency exist, such as, for example, spectral smoothing, which we have not explored given the simplicity and robustness of the considered approaches.

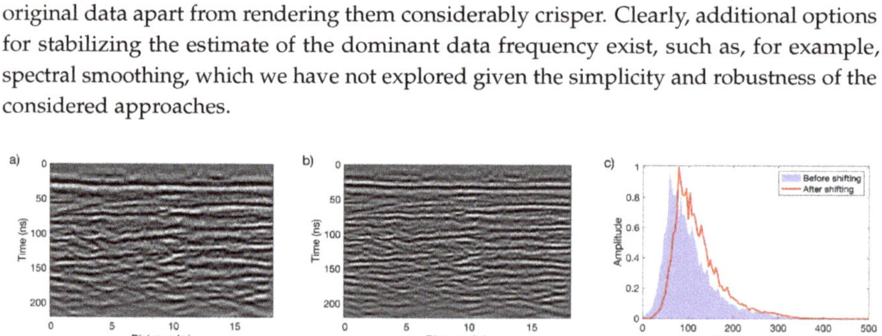

Figure 11. Processed GPR field data from the BHRS (**a**) before and (**b**) after spectral adjustment. (**c**) Corresponding average amplitude spectra before and after spectral adjustment to a dominant frequency of ~100 MHz.

The autocorrelation image of the spectrally adjusted section was then submitted to the trained CNN, which yields an estimated aspect ratio a_x/a_z of 9.25. For comparison, we also used the original Monte Carlo inversion approach to estimate the aspect ratio for the corresponding original, that is, spectrally not adjusted, processed GPR data [6,7]. The prior range of acceptable values for the horizontal and vertical correlation lengths was set between 0.1 and 3 m. The Monte Carlo estimation was run until 4000 accepted sets of model parameters were obtained. The resulting histogram for the horizontal-to-vertical aspect ratio of the underlying heterogeneity is shown in Figure 12. The corresponding values for the mean and standard deviation are 8.92 and 1.03, respectively, which implies that the value estimated with the newly proposed CNN-based technique is fully consistent with the range of estimates provided by the well-established Monte Carlo approach.

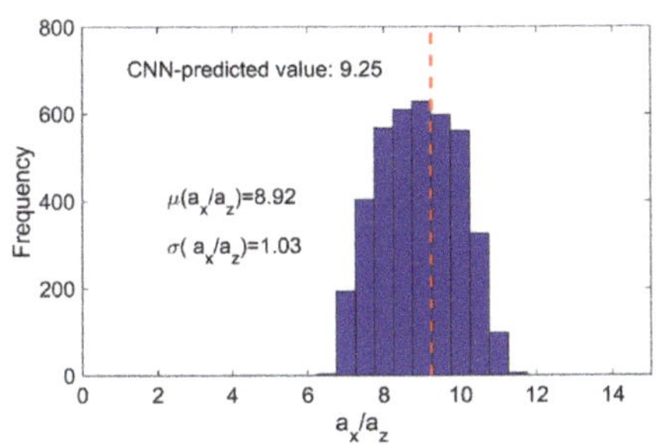

Figure 12. Comparison of results obtained through Monte-Carlo-type inversion with those obtained using the proposed CNN-based estimation. The blue histogram represents the distribution of acceptable aspect ratios a_x/a_z resulting from the Monte Carlo inversion with μ denoting the mean value and σ the standard deviation. The red dashed vertical line denotes the aspect ratio inferred using the CNN approach.

4. Discussion

Our tests on synthetic data illustrate that discrepancies between the dominant frequency of the training database and that of the analyzed GPR data may lead to biased estimates of the structural aspect ratio a_x/a_z. For the specific scenario we considered, where the dominant frequency of the analyzed GPR data was 30% lower than that used to generate the training database, our results indicate that a_x/a_z can be underestimated by up to 50%. Given that discrepancies of this magnitude between the nominal and actually recorded center frequencies are quite typical for commonly used GPR acquisition systems, a careful analysis of the average data amplitude spectrum followed, if necessary, by corresponding adjustments is essential. While our results indicate that the latter can be effectively achieved through a simple spectral shift, more sophisticated approaches such as shaping filters [45] are also conceivable. For larger discrepancies between the center frequencies of the analyzed and training data, it may be advisable to retrain the CNN on a correspondingly targeted database. As typical GPR antenna systems for geological applications tend to be available roughly in octave-type increments of the nominal center frequency (e.g., 50, 100, and 200 MHz), the generation of a widely applicable suite of trained CNNs is straightforward. Finally, the results of our method might also be biased, albeit probably to a lesser extent, by pronounced differences in the spectral bandwidth between the observed and training GPR data. We have not explored this possibility because the spectral bandwidth of approximately two to three octaves, which we used to create our training database, is indeed quite typical for the most commercially available GPR acquisition systems.

Biased estimates of the structural aspect ratio a_x/a_z could also arise when the processed and migrated GPR data are converted to depth with an incorrect average subsurface velocity. In the absence of complementary velocity measurements, such as common-mid-point (CMP) gathers, the most obvious way to estimate the average subsurface velocity from common-offset GPR reflection data is through migration focusing analysis [46]. The inferred velocities do, however, have a tendency to be too low, as artifacts resulting from the use of too high migration velocities are visually more obvious than imperfect focusing of diffracted and/or scattered wavefield components due to the use of too low migration velocities. Our experience indicates that the corresponding underestimation of the average velocity may amount up to 10 to 20%, which results in a bias of comparable magnitude in the estimated structural aspect ratio a_x/a_z. In this context, it is, however, important to note that corresponding estimations of a_x/a_z of based on the classical Monte Carlo approach would be subject to the same kind of bias.

Following the original work of Irving et al. [6], our proposed methodology is based on the analysis of the autocorrelation of the GPR image. This second-order approach provides a clear and simple relationship between the input and the output of the CNN and is generally regarded to be valid for distinct lithological and/or hydraulic units. However, it also results in an inherent loss of valuable information, such as, for example, the phase characteristics of the GPR data. Future work should therefore explore the use of additional inputs for training the CNN, such as the original GPR image as well as one or several versions enhanced by suitable attributes that are sensitive to the structural, lithological, and/or petrophysical characteristics [47]. Another inherent assumption of our approach is that the principal axes of the heterogeneity ellipsoid are aligned with the cardinal directions used to process and display the data. Although there is a prevalence of sub-horizontal stratification in surficial environments [48], this is clearly not ubiquitously adequate. As illustrated by Xu et al. [8], notable deviations from the assumption of horizontal stratification can be readily accounted for through a corresponding coordinate transformation.

As with all deep learning applications in geophysics, the fundamental question arises as to whether the corresponding results, whose inference is devoid of a solid theoretical foundation, are universally valid and trustworthy [17]. This question could at least be partly addressed by estimating the uncertainties of the inferred results, which, in our case, would also allow for a more direct and more comprehensive comparison with the corresponding results inferred through the traditional Monte-Carlo approach. Recent work of this kind, which we expect to explore in our future research efforts, indicates that this can be achieved by utilizing probability distribution as the output target [49].

While the acquisition of surface-based common-offset GPR reflection data along 2D profiles is common, there is a growing tendency towards acquiring surveys in 3D and, correspondingly, the geostatistical analysis of such 3D data will become increasingly important in the future [43]. While, as illustrated by Xu et al. [8], the traditional Monte-Carlo-type inversion approach can be readily adapted and applied to 3D GPR reflection volumes, the associated computational costs are extremely high. In this regard, we expect the proposed CNN-based technique to offer a decisive practical advantage. We anticipate that, for such an extension from 2D to 3D, the size of the training database will likely need to be increased by one to two orders of magnitude, which, in turn, is expected to increase the time required for training the CNN by one to two orders of magnitude or more. Conversely, the generation of the synthetic 3D GPR database for training the CNN remains computationally feasible due to the fact that it can be achieved using the highly efficient convolution approximation. As in the 2D case considered in this work, the application of the trained CNN to observed 3D GPR data is expected to come at virtually no extra computational cost. In view of this, we anticipate that the most important technical challenge for the extension of the proposed technique from 2D to 3D is related to the high computational cost of generating an adequate number of 3D FDTD-based synthetic GPR reflection surveys for testing and benchmarking purposes.

5. Conclusions

In this study, we have presented a novel method for inferring the structural aspect ratio a_x/a_z of subsurface heterogeneity from 2D surface-based GPR reflection images based on supervised deep learning. This approach provides a simple and efficient alternative to the current Monte-Carlo-based inversion of GPR reflection data for this pertinent geostatistical parameter. The proposed method utilizes a CNN trained on a large database of autocorrelations derived from synthetic GPR reflection images. This synthetic training database is generated for a wide range of stochastic subsurface models using a computationally efficient but approximate convolution-based modeling algorithm. The effectiveness of our approach is assessed and validated using realistic FDTD-based synthetic common-offset GPR reflection data as well as field data. Our results also demonstrate that the proposed method is robust to the presence of noise and that typical differences with regard to the dominant frequencies between the analyzed GPR data and the synthetic database used to train the CNN can be readily accommodated through simple spectral shifting operations. Compared to the traditional Monte Carlo inversion approach, the considered CNN-based technique offers significant enhancements in computational efficiency, requiring only minimal additional overhead after training. The latter is expected to become particularly advantageous for an eventual extension of this technique to 3D, which is conceptually straightforward and practically feasible due to the fact that the synthetic training database can be generated through a computationally highly efficient approximate modeling approach. Although this study focuses on GPR reflection data, previous related experience with the traditional Monte-Carlo-based inversion approach indicates that the newly proposed methodology should also be readily applicable to seismic reflection data.

Author Contributions: Methodology, Y.L. and J.I.; writing—original draft, Y.L.; writing—review and editing, Y.L., J.I. and K.H.; supervision, J.I. and K.H. All authors have read and agreed to the published version of the manuscript.

Funding: This research was funded by the China Scholarship Council, Grant Number 201806320359. Article processing charges were kindly paid for by the Institute of Earth Sciences, University of Lausanne.

Data Availability Statement: The synthetic data referenced in this study are available in the cited paper. However, the field data were collected by a third party, and we do not have the right to share them.

Acknowledgments: The considered GPR field data were collected by Michael Knoll and Warren Barrash as part of the initial BHRS characterization and were kindly provided to us by John Bradford. The authors are grateful to the National Supercomputing Center in Zhengzhou for its essential support with computational tasks.

Conflicts of Interest: The authors declare no conflicts of interest.

References

1. Rea, J.; Knight, R. Geostatistical analysis of ground-penetrating radar data: A means of describing spatial variation in the subsurface. *Water Resour. Res.* **1998**, *34*, 329–339. [CrossRef]
2. Dafflon, B.; Tronicke, J.; Holliger, K. Inferring the lateral subsurface correlation structure from georadar data: Methodological background and experimental evidence. In *Geostatistics for Environmental Applications*; Springer: Berlin/Heidelberg, Germany, 2005; pp. 467–478.

3. Knight, R.J.; Irving, J.D.; Tercier, P.; Freeman, G.J.; Murray, C.J.; Rockhold, M.L. A comparison of the use of radar images and neutron probe data to determine the horizontal correlation length of water content. *Subsurf. Hydrol. Data Integr. Prop. Process.* **2007**, *171*, 31–44.
4. Knight, R.; Tercier, P.; Irving, J. The effect of vertical measurement resolution on the correlation structure of a ground penetrating radar reflection image. *Geophys. Res. Lett.* **2004**, *31*, 2–5. [CrossRef]
5. Oldenborger, G.A.; Knoll, M.D.; Barrash, W. Effects of signal processing and antenna frequency on the geostatistical structure of ground-penetrating radar data. *J. Environ. Eng. Geophys.* **2004**, *9*, 201–212. [CrossRef]
6. Irving, J.; Knight, R.; Holliger, K. Estimation of the lateral correlation structure of subsurface water content from surface-based ground-penetrating radar reflection images. *Water Resour. Res.* **2009**, *45*, W12404. [CrossRef]
7. Irving, J.; Holliger, K. Geostatistical inversion of seismic and ground-penetrating radar reflection images: What can we actually resolve? *Geophys. Res. Lett.* **2010**, *37*, L21306. [CrossRef]
8. Xu, Z.; Irving, J.; Lindsay, K.; Bradford, J.; Zhu, P.; Holliger, K. Estimation of the 3D correlation structure of an alluvial aquifer from surface-based multi-frequency ground-penetrating radar reflection data. *Geophys. Prospect.* **2020**, *68*, 678–689. [CrossRef]
9. Xu, Z.; Irving, J.; Liu, Y.; Zhu, P.; Holliger, K. Conditional stochastic inversion of common-offset ground-penetrating radar reflection data. *Geophysics* **2021**, *86*, WB89–WB99. [CrossRef]
10. Scholer, M.; Irving, J.; Holliger, K. Estimation of the correlation structure of crustal velocity heterogeneity from seismic reflection data. *Geophys. J. Int.* **2010**, *183*, 1408–1428. [CrossRef]
11. Liu, Y.; Greenwood, A.; Hetényi, G.; Baron, L.; Holliger, K. High-resolution seismic reflection survey crossing the Insubric Line into the Ivrea-Verbano Zone: Novel approaches for interpreting the seismic response of steeply dipping structures. *Tectonophysics* **2021**, *816*, 229035. [CrossRef]
12. Dafflon, B.; Irving, J.; Holliger, K. Use of high-resolution geophysical data to characterize heterogeneous aquifers: Influence of data integration method on hydrological predictions. *Water Resour. Res.* **2009**, *45*, W09413. [CrossRef]
13. Liu, Y.; Irving, J.; Holliger, K. High-resolution velocity estimation from surface-based common-offset GPR reflection data. *Geophys. J. Int.* **2022**, *230*, 131–144. [CrossRef]
14. LeCun, Y.; Bengio, Y.; Hinton, G. Deep learning. *Nature* **2015**, *521*, 436–444. [CrossRef]
15. Goodfellow, I.; Bengio, Y.; Courville, A. *Deep Learning*; MIT Press: Cambridge, MA, USA, 2016.
16. Dramsch, J.S. 70 years of machine learning in geoscience in review. *Adv. Geophys.* **2020**, *61*, 1–55.
17. Yu, S.; Ma, J. Deep learning for geophysics: Current and future trends. *Rev. Geophys.* **2021**, *59*, e2021RG000742. [CrossRef]
18. Mousavi, S.M.; Beroza, G.C. Deep-learning seismology. *Science* **2022**, *377*, eabm4470. [CrossRef]
19. Zhu, W.; Mousavi, S.M.; Beroza, G.C. Seismic signal denoising and decomposition using deep neural networks. *IEEE Trans. Geosci. Remote Sens.* **2019**, *57*, 9476–9488. [CrossRef]
20. Kaur, H.; Pham, N.; Fomel, S. Improving the resolution of migrated images by approximating the inverse Hessian using deep learning. *Geophysics* **2020**, *85*, WA173–WA183. [CrossRef]
21. Yang, L.; Chen, W.; Wang, H.; Chen, Y. Deep learning seismic random noise attenuation via improved residual convolutional neural network. *IEEE Trans. Geosci. Remote Sens.* **2021**, *59*, 7968–7981. [CrossRef]
22. Wu, X.; Liang, L.; Shi, Y.; Fomel, S. FaultSeg3D: Using synthetic data sets to train an end-to-end convolutional neural network for 3D seismic fault segmentation. *Geophysics* **2019**, *84*, IM35–IM45. [CrossRef]
23. Song, S.; Mukerji, T.; Hou, J. Bridging the Gap between Geophysics and Geology with Generative Adversarial Networks. *IEEE Trans. Geosci. Remote Sens.* **2022**, *60*, 5902411. [CrossRef]
24. Yang, F.; Ma, J. Deep-learning inversion: A next-generation seismic velocity model building method. *Geophysics* **2019**, *84*, R583–R599. [CrossRef]
25. Di, H.; Abubakar, A. Estimating subsurface properties using a semi-supervised neural networks approach. *Geophysics* **2021**, *87*, IM1–IM10. [CrossRef]
26. Geng, Z.; Zhao, Z.; Shi, Y.; Wu, X.; Fomel, S.; Sen, M. Deep learning for velocity model building with common-image gather volumes. *Geophys. J. Int.* **2022**, *228*, 1054–1070. [CrossRef]
27. Schmelzbach, C.; Tronicke, J.; Dietrich, P. High-resolution water content estimation from surface-based ground-penetrating radar reflection data by impedance inversion. *Water Resour. Res.* **2012**, *48*, W08505. [CrossRef]
28. Pullammanappallil, S.; Levander, A.; Larkin, S.P. Estimation of crustal stochastic parameters from seismic exploration data. *J. Geophys. Res. Solid Earth* **1997**, *102*, 15269–15286. [CrossRef]
29. Poppeliers, C. Estimating vertical stochastic scale parameters from seismic reflection data: Deconvolution with non-white reflectivity. *Geophys. J. Int.* **2007**, *168*, 769–778. [CrossRef]
30. Goff, J.A.; Jordan, T.H. Stochastic modeling of seafloor morphology: Inversion of sea beam data for second-order statistics. *J. Geophys. Res. Solid Earth* **1988**, *93*, 13589–13608. [CrossRef]
31. Tronicke, J.; Holliger, K. Quantitative integration of hydrogeophysical data: Conditional geostatistical simulation for characterizing heterogeneous alluvial aquifers. *Geophysics* **2005**, *70*, H1–H10. [CrossRef]

32. Holliger, K.; Levander, A.R.; Goff, J.A. Stochastic modeling of the reflective lower crust: Petrophysical and geological evidence from the Ivera Zone (northern Italy). *J. Geophys. Res. Solid Earth* **1993**, *98*, 11967–11980. [CrossRef]
33. Tronicke, J.; Holliger, K.; Barrash, W.; Knoll, M.D. Multivariate analysis of cross-hole georadar velocity and attenuation tomograms for aquifer zonation. *Water Resour. Res.* **2004**, *40*, W01519. [CrossRef]
34. LeCun, Y.; Bottou, L.; Bengio, Y.; Haffner, P. Gradient-based learning applied to document recognition. *Proc. IEEE* **1998**, *86*, 2278–2324. [CrossRef]
35. Krizhevsky, A.; Sutskever, I.; Hinton, G.E. Imagenet classification with deep convolutional neural networks. *Adv. Neural Inf. Process. Syst.* **2012**, *25*, 1–9. [CrossRef]
36. Qian, N. On the momentum term in gradient descent learning algorithms. *Neural Netw.* **1999**, *12*, 145–151. [CrossRef]
37. Bottou, L.; Curtis, F.E.; Nocedal, J. Optimization methods for large-scale machine learning. *SIAM Rev.* **2018**, *60*, 223–311. [CrossRef]
38. Smith, L.N. Cyclical learning rates for training neural networks. In Proceedings of the 2017 IEEE Winter Conference on Applications of Computer Vision (WACV), Santa Rosa, CA, USA, 24–31 March 2017; pp. 464–472.
39. Irving, J.; Knight, R. Numerical modeling of ground-penetrating radar in 2-D using MATLAB. *Comput. Geosci.* **2006**, *32*, 1247–1258. [CrossRef]
40. Yilmaz, Ö. *Seismic Data Analysis: Processing, Inversion, and Interpretation of Seismic Data*; Society of Exploration Geophysicists: Tulsa, OK, USA, 2001.
41. Barrash, W.; Clemo, T. Hierarchical geostatistics and multifacies systems: Boise Hydrogeophysical Research Site, Boise, Idaho. *Water Resour. Res.* **2002**, *38*, 14-1–14-18. [CrossRef]
42. Dafflon, B.; Barrash, W. Three-dimensional stochastic estimation of porosity distribution: Benefits of using ground-penetrating radar velocity tomograms in simulated-annealing- based or Bayesian sequential simulation approaches. *Water Resour. Res.* **2012**, *48*, W05553. [CrossRef]
43. Bradford, J.H.; Clement, W.P.; Barrash, W. Estimating porosity with ground-penetrating radar reflection tomography: A controlled 3-D experiment at the Boise Hydrogeophysical Research Site. *Water Resour. Res.* **2009**, *45*, W00D26-1–W00D26-11. [CrossRef]
44. Pirot, G.; Linde, N.; Mariethoz, G.; Bradford, J.H. Probabilistic inversion with graph cuts: Application to the Boise Hydrogeophysical Research Site. *Water Resour. Res.* **2017**, *53*, 1231–1250. [CrossRef]
45. van der Baan, M. Time-varying wavelet estimation and deconvolution by kurtosis maximization. *Geophysics* **2008**, *73*, 11–18. [CrossRef]
46. Liu, Y.; Irving, J.; Holliger, K. Weighted Diffraction-Based Migration Velocity Analysis of Common-Offset GPR Reflection Data. *IEEE Trans. Geosci. Remote Sens.* **2023**, *61*, 5901509. [CrossRef]
47. Lowney, B.; Lokmer, I.; O'Brien, G.S. Multi-domain diffraction identification: A supervised deep learning technique for seismic diffraction classification. *Comput. Geosci.* **2021**, *155*, 104845. [CrossRef]
48. Gelhar, L.W. *Stochastic Subsurface Hydrology*; Prentice-Hall: Englewood Cliffs, NJ, USA, 1993.
49. Grana, D.; Azevedo, L.; Liu, M. A comparison of deep machine learning and Monte Carlo methods for facies classification from seismic data. *Geophysics* **2020**, *85*, WA41–WA52. [CrossRef]

remote sensing

MDPI

Article

GPR-Based Leakage Reconstruction of Shallow-Buried Water Supply Pipelines Using an Improved UNet++ Network

Qingqi Xu, Qinghua Liu * and Shan Ouyang

School of Information and Communication, Guilin University of Electronic Technology, Guilin 541004, China; xuqingqi@mails.guet.edu.cn (Q.X.); hmoysh@guet.edu.cn (S.O.)

* Correspondence: qhliu@guet.edu.cn

Abstract

Ground-penetrating radar (GPR) plays a critical role in detecting underground targets, particularly locating and characterizing leaks in buried pipelines. However, the complex nature of GPR images related to pipeline leaks, combined with the limitations of existing neural network-based inversion methods, such as insufficient feature extraction and low inversion accuracy, poses significant challenges for effective leakage reconstruction. To address these challenges, this paper proposes an enhanced UNet++-based model: the Multi-Scale Directional Network PlusPlus (MSDNet++). The network employs an encoder–decoder architecture, in which the encoder incorporates multi-scale directional convolutions with coordinate attention to extract and compress features across different scales effectively. The decoder fuses multi-level features through dense skip connections and further enhances the representation of critical information via coordinate attention, enabling the accurate inversion of dielectric constant images. Experimental results on both simulated and real-world data demonstrate that MSDNet++ can accurately invert the location and extent of buried pipeline leaks from GPR B-scan images.

Keywords: ground penetrating radar; pipeline leakage reconstruction; multi-scale directional convolution; coordinate attention; dense skip connections

1. Introduction

In recent years, the rapid pace of urbanization has exacerbated the aging of water supply and drainage pipelines in cities, making it an increasingly severe issue. Many cities began constructing their water supply pipelines in the 1970s and 1980s. After decades of use, these pipelines have entered a phase of aging, leading to frequent ruptures and leaks. Leakage in drainage pipelines not only leads to significant water loss but also compromises the integrity of surrounding foundations, creates underground voids, and poses serious public safety risks [1].

Ground-penetrating radar has emerged as a rapidly developing, high-resolution, and efficient non-destructive testing technology. Its key characteristics—non-destructive nature, portability, continuity, high efficiency, broad applicability, and ease of use—have led to its widespread application across various fields, including road detection [2], tunnel defect identification [3], and military target detection [4]. Among these applications, the two-dimensional reconstruction of buried pipeline leaks using GPR is particularly valuable because it enables the early detection and localization of underground water losses, which helps prevent soil erosion, structural damage, and urban road collapses. In urban infrastructure maintenance, the timely identification of leaks can significantly reduce repair costs and water loss, thus enhancing public safety and resource efficiency.

GPR inversion, in the broader context of geophysics, refers to the quantitative process of estimating the subsurface physical properties from the observed electromagnetic wave data [5]. Specifically for GPR, this involves recovering parameters such as the dielectric permittivity, electrical conductivity, and magnetic permeability of the subsurface materials, which govern the propagation and reflection of the radar waves. While magnetic permeability is often close to the free-space value for most common non-magnetic subsurface materials, permittivity and conductivity are the primary targets for inversion. By analyzing the inversion results, researchers can characterize subsurface structures or detect underground defects within the surveyed area. GPR inversion algorithms are generally categorized into traditional inversion methods and deep learning-based methods. The commonly used conventional inversion method is the Full Waveform Inversion (FWI) [6] method, which aims to estimate subsurface parameters by minimizing the difference between the observed and simulated waveforms. However, these conventional inversion methods are highly dependent on the initial model and tend to fall into local minima during the optimization process, which not only affects the convergence of the algorithm but also reduces the accuracy of the inversion.

In recent years, deep neural networks (DNNs) have demonstrated exceptional performance in various applications, including image classification [7], object detection [8], semantic segmentation [9], and image synthesis [10]. To enhance inversion efficiency and accuracy, many researchers have leveraged DNNs to explore the relationship between GPR B-scans and dielectric constant maps. In 2018, Alvarez et al. [11] employed three different network architectures—encoder–decoder, U-Net, and generative adversarial network (GAN)—for reconstructing the dielectric constant from ground-penetrating radar data, demonstrating the feasibility of using deep learning for GPR inversion. Xie et al. [12] introduced data compression and output units based on the U-Net framework and improved the network with instance normalization to invert both target locations and dielectric constants. Liu et al. [13] proposed a DNN-based GPR inversion network, GPRInvNet, which integrates a path-to-path encoder and a dielectric constant decoder. By incorporating neighborhood information-enhanced GPR A-scan features and a fully connected layer-based feature encoding method in the path-to-path encoder, the network addressed the spatial correspondence issue between B-scan data and dielectric constant maps, thereby facilitating tunnel lining reconstruction. Ji et al. [14] introduced an improved dielectric constant inversion network, PINet, which first compresses B-scan dimensions and extracts features using a dimension compressor. It then optimizes and aligns these features using a global feature encoder and finally upsamples them in the dielectric constant decoder to complete the reconstruction. Although PINet improved imaging results compared to GPRInvNet, its use of 128 fully connected layers as global feature encoders significantly increased model complexity and computational demands. Wu et al. [15] proposed a two-stage inversion network, MHInvNet, which uses multi-scale dilated convolution (MSDC) and hybrid attention gates (HAG). The network first denoises B-scans using MHInvNet1 and then jointly inputs the original data into MHInvNet2 for dielectric constant reconstruction. By combining the multi-scale feature extraction capability of MSDC with the key feature enhancement mechanism of HAG, the network significantly improved target localization accuracy and parameter inversion in complex scenarios. However, existing methods still face challenges in specific scenarios, especially in reconstructing buried pipeline leaks, which require inversion techniques capable of accurately capturing the size and precise location of the target, and there is little relevant research in pipeline leak reconstruction.

As a result, the reconstruction of pipeline leak locations and characteristics using deep learning has become a key research focus.

In response to these challenges, this paper proposes a network called MSDNet++, which is based on an enhanced version of UNet++, for the reconstruction of buried pipeline leaks. The network employs an encoder–decoder architecture to achieve dielectric constant image inversion through feature compression and reconstruction. Specifically, the network extracts multi-scale features in the encoder by designing multi-scale directional convolutional blocks (MSD), thereby enhancing its ability to perceive features across multiple scales and directions. The decoder then effectively fuses multi-level feature information through dense skip connections, enabling the network to learn both low-level detailed features and high-level abstract features simultaneously. This fusion helps mitigate information loss and enhances the network's ability to restore image details during upsampling. Following the fusion process, the network incorporates the coordinate attention (CA) mechanism and MSD modules to emphasize important spatial regions and channels, further optimizing the fused features and improving the representation of key features. Residual connections are incorporated to alleviate the vanishing gradient problem, accelerate training, and, in combination with multi-depth supervision outputs, strengthen multi-level feature learning. This approach significantly improves both the inversion accuracy and stability. The effectiveness and applicability of the proposed network are validated through data obtained from numerical simulations and real-world measurements.

2. Inversion Method Based on Improved UNet++

UNet++ [16] is an enhanced, fully convolutional neural network specifically designed to improve segmentation performance. Building upon the traditional U-Net architecture, UNet++ introduces dense skip connections through nested and redesigned decoder modules, progressively bridging the semantic gap between features at different levels. Specifically, UNet++ not only includes encoder and decoder components but also establishes dense connections between them via multiple intermediate nodes, thereby facilitating hierarchical feature fusion. In the encoder, the network progressively downsamples the input image to extract multi-scale features, while in the decoder, it refines and reconstructs these features through upsampling and skip connections. This nested structure enables a more effective fusion of multi-level features, alleviating the vanishing gradient problem and significantly improving the accuracy and robustness of both image segmentation and regression tasks.

Additionally, UNet++ incorporates a deep supervision mechanism that enhances both training efficiency and prediction performance. In the context of GPR image inversion, the primary task is to establish a non-linear mapping from GPR B-scan images to the corresponding dielectric constant images. Leveraging deep supervision in deep learning, this non-linear mapping is learned during training by minimizing the loss function via gradient descent and backpropagation, effectively overcoming many challenges associated with direct solution methods. Unlike the traditional U-Net structure, where the loss is calculated solely from the final decoder output, UNet++ adds supervision signals at multiple intermediate decoder outputs, allowing features from various network depths to contribute to the training process. This deep supervision mechanism accelerates convergence, enhances feature representation, and improves the inversion and segmentation performance for targets of different scales. Moreover, deep supervision mitigates the vanishing gradient problem, facilitates optimization, and ultimately improves accuracy and robustness for both regression and segmentation tasks.

2.1. Overall Framework and Improvements

MSDNet++, proposed in this paper, is an enhanced version of the UNet++ architecture, specifically optimized to address key challenges in GPR image inversion. Although traditional UNet++ enhances feature fusion via dense skip connections, it still has limitations in directional perception and localization accuracy when processing GPR images. To address these limitations, MSDNet++ integrates three key components: the multi-scale directional convolutional Block (MSD), the coordinate attention (CA) mechanism, and dense skip connections.

The MSD module enhances the model's capacity to capture anisotropic features by employing parallel horizontal and vertical convolutional kernels combined with a multi-scale feature extraction strategy. The CA mechanism applies 1D directional pooling and spatial weight generation to enhance the network's spatial sensitivity while maintaining computational efficiency. Dense skip connections optimize gradient propagation paths and enhance detail reconstruction by facilitating cross-level feature fusion and incorporating residual connections.

The overall architecture of MSDNet++ is illustrated in Figure 1. The cooperation of these three components enables MSDNet++ to significantly improve the localization and shape inversion accuracy of pipeline leak targets, providing a more powerful and robust solution for GPR image inversion.

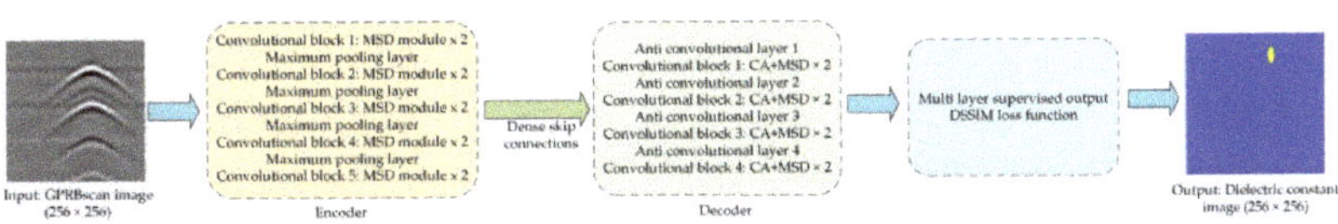

Figure 1. Network framework diagram.

2.2. Coordinate Attention Mechanism

The Coordinate attention mechanism decomposes channel attention into two one-dimensional encoding processes: one along the horizontal axis and the other along the vertical. This design facilitates the capture of long-range dependencies while preserving essential positional information. In contrast to the traditional 2D global pooling method used in channel attention, CA utilizes two parallel 1D global pooling operations to generate direction-specific feature maps, which are then encoded into spatial attention maps. This approach effectively enhances the perception of wide-area features in mobile neural networks while maintaining low computational overhead.

Figure 2 illustrates the structure of the CA mechanism. Given an input feature map $X \in R^{C\times H\times W}$, it first passes through a residual connection to preserve the original information. Then, global average pooling is applied along the height and width directions to obtain the horizontal feature $X_{avg}^{w} \in R^{C\times 1\times W}$ and vertical feature $X_{avg}^{h} \in R^{C\times H\times 1}$, respectively. The two directional features are concatenated and fused using a 1×1 convolution, followed by batch normalization and a non-linear activation function resulting in a fused representation $F \in R^{\frac{C}{r}\times(W+H)}$. A scaling factor r is introduced at this stage to reduce computational cost. F is then passed through two separate 1×1 convolutional layers with a Sigmoid activation function to generate the attention weights $A_w \in R^{C\times 1\times W}$ and $A_h \in R^{C\times H\times 1}$ for the height and width directions, respectively. Finally, these attention weights are applied to the corresponding spatial dimensions of the input feature map to produce the final weighted output:

$$Y = X \cdot A_h + X \cdot A_w \tag{1}$$

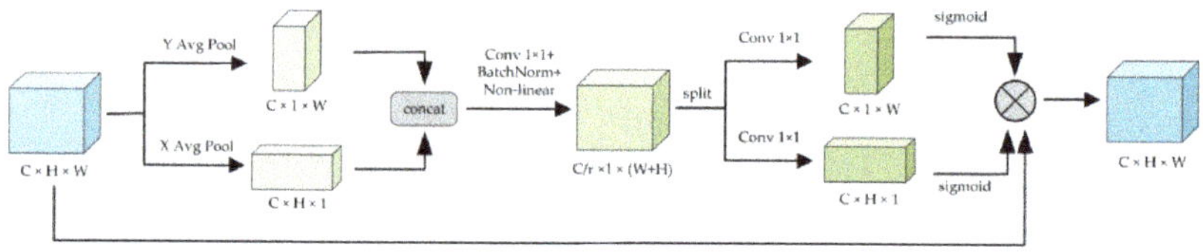

Figure 2. Coordinate attention mechanism.

This mechanism effectively captures long-range dependencies while retaining positional sensitivity, making it well suited for tasks requiring precise localization. Moreover, it incurs low computational overhead and can be seamlessly integrated into various network architectures.

2.3. *Multi-Scale Directional Convolution*

During the propagation of GPR signals, electromagnetic waves are scattered at the interfaces of differing dielectric properties in the subsurface medium, resulting in correlations among echoes across different traces. To enhance inversion accuracy, incorporating inter-trace correlation information is essential. Multi-scale convolution is a technique that uses parallel convolutional kernels of varying sizes to extract and fuse features across different scales. Traditional single-scale convolution can only capture features within a fixed receptive field and is limited in its ability to comprehensively represent multi-scale information. In contrast, multi-scale convolution combines global features from larger receptive fields with local features from smaller receptive fields, thereby enhancing the overall representational power and depth of feature extraction.

To address the complex feature extraction needs of GPR B-scan images, this paper proposes the multi-scale directional convolution module, illustrated in Figure 3. This module not only extracts features from multi-scale receptive fields but also integrates directional information, significantly enhancing the model's ability to identify anisotropic and complex structures. Specifically, the multi-scale directional convolution module includes the following key components:

(1) Multi-directional Convolution: The module incorporates horizontal and vertical convolutions to capture feature information from multiple directions within the image. These directional convolution layers use kernels of various scales, including four directional convolution layers with kernel sizes of 1×3, 3×1, 1×5, 5×1, ensuring the comprehensive extraction of features in each direction.

(2) Multi-scale Feature Extraction: The module combines small-scale kernels to capture fine-grained details and large-scale kernels to extract broader, global information, thereby enhancing the model's ability to capture a wide range of feature diversity.

(3) Deep Fusion: Each directional convolution is followed by batch normalization and an activation function (e.g., ReLU or LeakyReLU) to enhance feature expressiveness. The outputs from different directions are fused via channel concatenation, resulting in a high-dimensional representation that integrates both multi-scale and multi-directional information.

The concatenated feature map is then compressed through an additional convolution to generate a unified multi-scale, multi-directional representation. This design effectively captures information across diverse spatial resolutions and directions, thereby enriching the representational capacity of subsequent network layers.

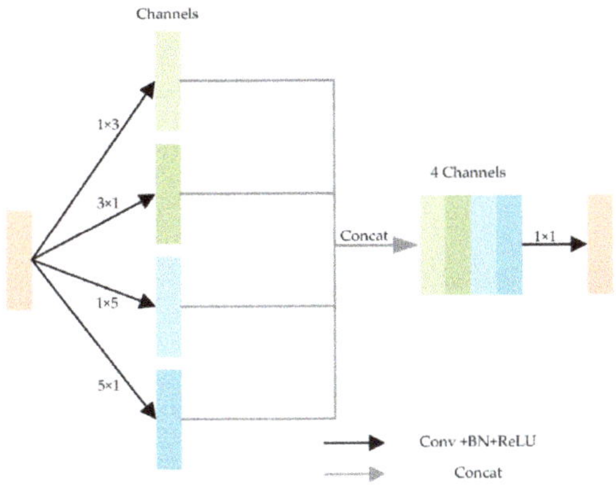

Figure 3. Multi-scale directional convolution module.

2.4. MSDNet++ Network Design

To establish a non-linear mapping between ground-penetrating radar (GPR) data and the corresponding dielectric constant images, this paper presents an enhanced GPR inversion network based on UNet++, termed MSDNet++. The overall network architecture is illustrated in Figure 4. MSDNet++ takes GPR B-scan images as input and uses the corresponding dielectric constant images as ground truth, performing inversion via feature compression and reconstruction.

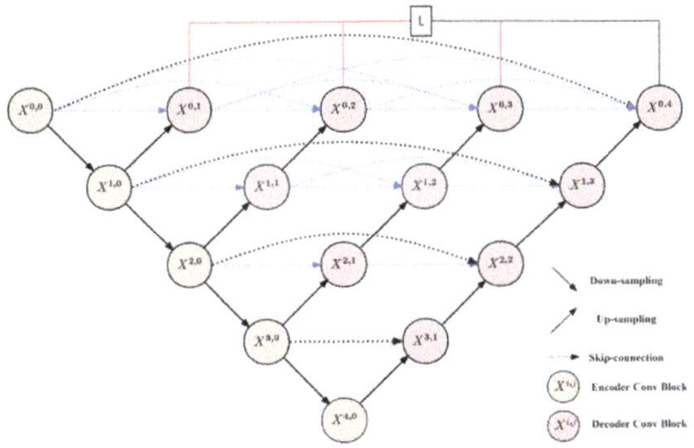

Figure 4. Network architecture of MSDNet++.

In Figure 4, the "L" marker represents the deep supervision mechanism, which is implemented by adding multiple output layers throughout the MSDNet++ network. The red solid lines indicate the connections between these output layers, and each output layer is supervised using a DSSIM loss function. These losses are combined into a total loss using a weighted average approach. This design enables the network to learn effectively at various levels by providing early-stage feedback during training. By supervising the network at different depths, the model learns features at multiple scales, thereby enhancing its ability to detect objects of varying sizes and shapes. This structure further enhances the fusion of low-level and high-level features, improving inversion accuracy in the presence of complex backgrounds and fine structural details.

The blue dashed lines in Figure 4 represent the dense skip connections, which are a feature unique to UNet++ compared to the standard UNet architecture which only has black solid skip connections. These dense skip connections allow each decoder layer to receive feature maps not only from its corresponding encoder layer but also from other encoder stages. This design enables a more effective fusion of multi-level features by learning both low-level details and high-level semantic representations. As a result, it helps reduce information loss and improves detail recovery during upsampling, thereby enhancing inversion accuracy.

Specifically, the feature extraction network comprises five encoder convolutional blocks and a max-pooling layer with a stride of 2, as illustrated in Figure 5. Each encoder convolutional block includes two MSD modules and a residual connection. The dielectric constant decoder consists of four decoder convolutional blocks and an upsampling layer, which together reconstruct the output image. Each decoder convolutional block contains a CA attention mechanism and two MSD modules, as shown in Figure 6. Residual connections are incorporated in both the encoder and decoder to mitigate the vanishing gradient problem and accelerate training. Additionally, the number of convolutional channels in the encoder and decoder layers is set to [32, 64, 128, 256, 512] and [256, 128, 64, 32], respectively.

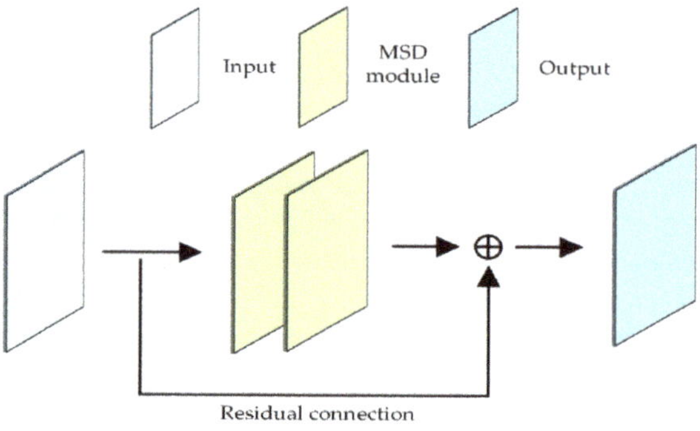

Figure 5. Encoder conv block.

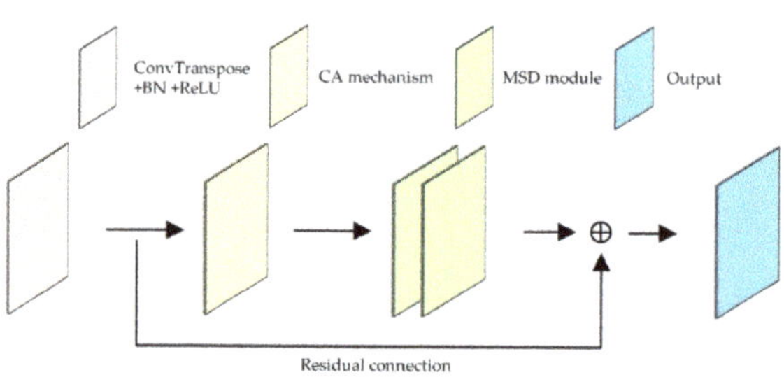

Figure 6. Decoder conv block.

2.5. Loss Function

In neural network models, the loss function plays a crucial role by quantifying the discrepancy between the model's predictions and the actual labels. Mean Squared Error (MSE) is one of the most commonly used loss functions. Minimizing the loss function via the backpropagation algorithm enables the effective adjustment of the network weights, thereby improving prediction accuracy. In this study, a variant of the Structural Similarity Index Measure (SSIM) [17] was adopted as the loss function for GPR image inversion, as defined below:

$$\mathrm{DSSIM}(x,y) = 1 - \mathrm{SSIM}(x,y) \tag{2}$$

SSIM (Structural Similarity Index Measure) evaluates the similarity between two images based on three key factors: luminance, contrast, and structure. Its specific formulation is as follows:

$$\mathrm{SSIM}(x,y) = \frac{(2\mu_x\mu_y + c_1)(2\sigma_{xy} + c_2)}{\left(\mu_x^2 + \mu_y^2 + c_1\right)\left(\sigma_x^2 + \sigma_y^2 + c_2\right)} \tag{3}$$

where μ_x and μ_y are the means of images x and y, respectively; σ_x^2 and σ_y^2 are the variances of images x and y, respectively; σ_{xy} is the covariance of images x and y; c_1 and c_2 are stabilizing constants.

3. Experimental Validation

3.1. Dataset and Evaluation Criteria

To evaluate the effectiveness of MSDNet++ for GPR dielectric constant inversion, this study constructed a dedicated inversion dataset for validation. GprMax, an open-source electromagnetic wave propagation simulator, was utilized to perform the forward modeling of GPR detection using the finite-difference time-domain (FDTD) method. Based on the characteristics of water supply pipeline leaks and their spatial distribution patterns, Shen Yupeng [18] used GeoStudio seepage field simulation software and found that water from a leak typically forms an approximately circular distribution around the pipeline, with the leak point at the center. Therefore, this study simulated buried pipeline leak scenarios with varying leakage radii and constructed a GPR inversion dataset using GprMax.

The simulation model had dimensions of 2 m × 1 m × 0.0025 m, with a uniform cell size of 0.0025 m in all directions and a perfectly matched layer (PML) thickness of 10 cells. A 25 ns time window was used with a Ricker wavelet centered at 400 MHz. The antenna transmit–receive distance was set to 0.1 m. In the common-offset mode, the GPR device moved horizontally from left to right with a measurement step of 0.02 m, resulting in 93 traces, each containing 4241 sampling points in the B-scan data. In the homogeneous half-space model, the background medium was dry sand, and the pipeline was made of PVC. The burial depth of the pipeline was randomly selected between 0.45 m and 0.75 m, with the leak point located directly beneath the pipeline. The radius of the leak area was randomly set between 0.05 m and 0.15 m, and the water content was set to either 30% or 40%. The electrical parameters of the predefined target media are listed in Table 1. In addition, in most natural media (e.g., air, soil, rock, etc.), for non-magnetic materials, the relative permeability is close to 1, i.e., these materials respond to the magnetic field similarly to a vacuum, and do not significantly enhance or diminish the magnetic field. Therefore, the magnetic permeability is set to 1 for all the materials simulated in this simulation.

Table 1. Dielectric constant and conductivity of the medium.

Media	Dielectric Constant	Conductivity (S/m)
Dry Sand	4	0.0068
PVC	3	0
Water	81	0
30% moisture content	17	0.0537
40% moisture content	25	0.1041

Based on the above settings, a total of 4000 data samples were generated, including 1000 non-leaking and 3000 leaking pipeline scenarios. All images were resized to 256 × 256 pixels and split into training and validation sets in a ratio of 8:2.

To further evaluate the performance of MSDNet++ under different time windows, an additional dataset with a 60 ns time window was generated while keeping all other parameters constant concerning the 25 ns dataset. This dataset also included 4000 samples, with the same distribution of 1000 non-leaking and 3000 leaking scenarios. A comparison of the inversion results from the two datasets was conducted to more comprehensively evaluate the adaptability and robustness of MSDNet++ under varying detection conditions.

To evaluate performance, standard image quality metrics were employed, including the Structural Similarity Index Measure (SSIM), Mean Absolute Error (MAE), and Mean Squared Error (MSE). SSIM quantifies the structural similarity between predicted and ground-truth images, where higher values indicate better perceptual similarity. MAE and MSE measure pixel-wise errors, with lower values indicating more accurate predictions.

3.2. Experimental Environment and Parameter Settings

The hardware and software environment during the experiment is shown in Table 2.

Table 2. Experimental software and hardware environment parameters.

Name	Particulars
System environment	Windows10 64bit
Graphics chip	NVIDIA TITAN RTX
Memory capacity	24 GB
Deep learning frameworks	TensorFlow

The MSDNet++ architecture was implemented in TensorFlow, with the DSSIM loss function employed for training. The initial learning rate was set to 0.0005, and the Adam optimizer was adopted with a batch size of four. The network was trained for 150 epochs to ensure convergence and achieve accurate GPR image inversion.

Regarding the model validation and preservation strategy, this study adopted an optimal weight-saving mechanism based on validation loss. Specifically, the validation loss was continuously monitored in real time through a callback function, which automatically saved model weights corresponding to the minimal validation loss. This approach ensures that the final model selected for evaluation represents the optimal weights demonstrating peak performance on the validation set throughout the entire training process, a design principle established to guarantee the reliability and generalization capability of model performance.

3.3. Simulation Data Experiments

3.3.1. The 25 ns Simulation Data Experiment

We compared our model with other deep learning-based inversion networks, such as UNet [19], MRF-UNet [20], and UNet++. Among them, MRF-UNet adopts a U-shaped architecture containing an encoder and decoder, which preserves spatial information by hopping connections. The Multi-Receptive Field (MRF) module in the network is capable of extracting features at different spatial scales, capturing multi-scale information in the image, and enhancing the network's ability to perceive image details and overall structure. Table 3 shows the comparison of the computational metrics MSE, MAE, and SSIM or the inversion of the different networks, as well as the number of parameters for each network and the average time required to invert each B-scan image. It can be observed from the table that MSDNet++ shows superior inversion accuracy, but there was a sacrifice in the number of parameters, and it takes longer to process each image.

Table 3. Comparison of inversion performance at 25 ns.

Method	MSE	MAE	SSIM	Parameter/M	Time/ms
UNet	0.00225	0.00506	0.97628	29.64	344.81
MRF-UNet	0.00037	0.00206	0.99297	59.75	476.33
UNet++	0.00285	0.01499	0.96491	34.55	519.51
MSDNet++	0.00018	0.00288	0.99667	75.13	1043.01

Figure 7a–c show the variation curves of MSE, MAE, and SSIM during the dataset training of different networks. It can be observed that MSDNet++ shows superior performance on the whole. Figure 8 illustrates the inversion results produced by the different networks. The color bar quantitatively represents the dielectric permittivity constant range of 0–81 for network inversion. Figure 8a–c are B-scan images of ground-penetrating radar. After the leakage of PVC water pipes, the electromagnetic wave will be reflected when it comes into contact with the wall of the pipe, the internal body of water, and the boundary of the leakage area, so that a complex oscillating hyperbolic signal is presented in the B-scan image. These oscillating waves show a specific shape and distribution pattern in the B-scan image. When the pipeline is not leaking, as shown in Figure 8a, the electromagnetic wave reflects at the interface of the pipeline wall and the homogeneous medium such as the water inside the pipeline, forming a relatively regular and symmetrical hyperbolic signal. After a leakage occurs in the pipeline, as illustrated in Figure 8b,c, the waveforms become more complex and irregular, and in addition to the reflected signals from the pipeline itself, the leakage region triggers additional scattering and reflections, forming complex oscillatory hyperbolic signals. In order to make full use of these oscillatory wave features, the MSDNet++ network is designed, which utilizes the MSD module to extract oscillatory wave features at different scales and directions. The small-scale convolution kernel captures the fine structure of the oscillatory waveform, while the large-scale convolution kernel helps to identify the overall waveform pattern and trend. Parallel horizontal and vertical convolution operations enhance the model's ability to perceive oscillatory waves in different directions. On top of this, dense skip connections incorporate multiple layers of feature information, enabling the decoder to learn both low-level detailed features and high-level abstract features. This helps to reduce information loss during the upsampling process and improves the recovery of image details. The CA mechanism is introduced to further optimize the fused features by emphasizing key spatial regions and channels. This mechanism highlights the most representative parts of the oscillating waveforms, resulting in a more accurate reconstruction of the leakage features.

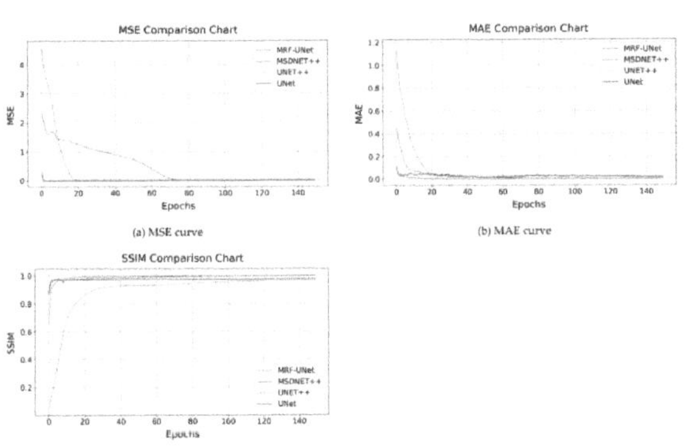

Figure 7. Comparison results of different network variation curves at 25 ns.

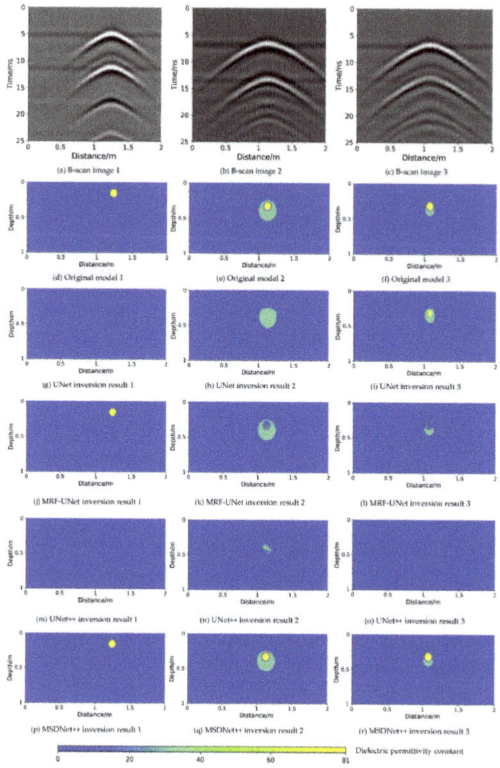

Figure 8. The inversion results of different networks at 25 ns.

Figure 8d–f are real dielectric constant images. As shown in Figure 8g–i, the pipeline contours generated by UNet are incomplete, and the leak areas are irregularly shaped, significantly deviating from the actual underground conditions. Figure 8j–l show that MRF-UNet, which incorporates a multi-receiver field module for multi-scale feature extraction, improves the inversion performance. However, the pipeline still exhibits unclear boundaries between the pipe wall and the internal water. Figure 8m–o indicate that the inversion results of UNet++ are blurry and suffer from ghosting artifacts. In contrast, Figure 8p–r show that MSDNet++, benefiting from the integration of multi-scale directional convolution, coordinate attention mechanisms, and dense skip connections, produces inversion results that closely match the ground truth model and outperform the other three networks.

3.3.2. The 25 ns Generalization Ability Test

In the previous section, we verified the superiority of MSDNet++ using simulation data, but in the actual detection process of ground-penetrating radar, the ground-penetrating radar data obtained is usually more complicated than the simulation data because of various noises and interferences. In order to verify the noise resistance of the model, we added Gaussian white noise with mean value 0 and standard deviation 1 to the ground-penetrating radar B-scan to simulate the background noise interference in the real detection environment, and at the same time, we used the peak signal-to-noise ratio, which is widely used in the field of denoising, as one of the indexes for evaluating the anti-noise performance of the network. The reconstruction test with background noise is obtained by fine-tuning 10 epochs on the trained model.

The results of the reconstruction test with background noise are shown in Table 4. It can be seen that compared with the other three networks, MSDNet++ still achieves better reconstruction performance.

Table 4. Comparison of noise data inversion index at 25 ns.

Method	MSE	MAE	PSNR (db)	Parameter/M	Time/ms
UNet	0.00435	0.02900	24.6693	29.64	399.24
MRF-UNet	0.00192	0.00897	27.5979	59.75	558.11
UNet++	0.05023	0.22237	23.8661	34.55	542.95
MSDNet++	0.00148	0.01106	29.6295	75.13	1082.50

Figure 9 illustrates the inversion results of the different networks for noisy data. Figure 9a–c show the B-scan images of the ground-penetrating radar with added Gaussian white noise, and Figure 9d–f are real dielectric constant images. As shown in Figure 9g–i, the UNet inversion of the target is not clear and is significantly inconsistent with the actual subsurface conditions. Figure 9j–l show that MRF-UNet is able to invert the pipeline target, but the leakage contour area is blurred. Figure 9m–o show that the UNet++ inversion results have artifacts and the inversion does not produce results. In contrast, Figure 9p–r show that the MSDNet++ inversion results are more consistent with the real model and superior to the other three networks, although the contours are somewhat blurred.

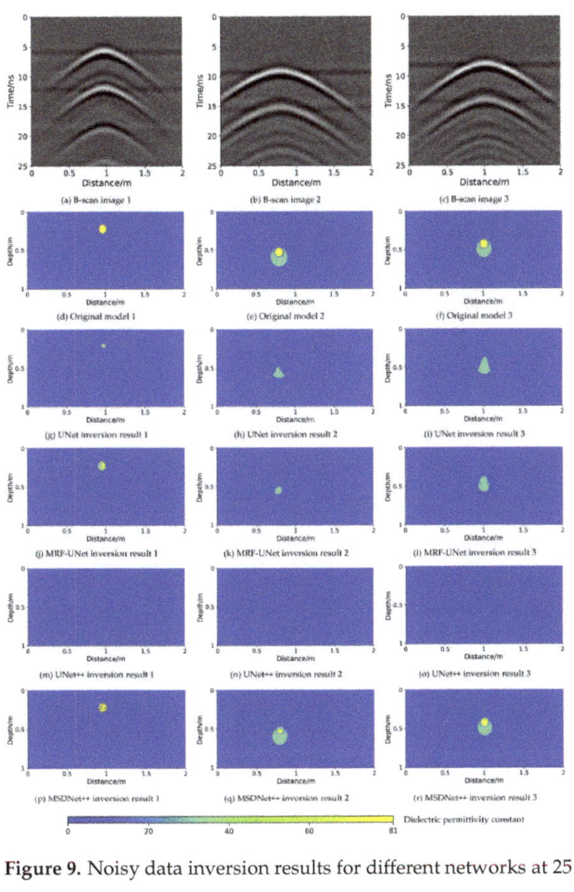

Figure 9. Noisy data inversion results for different networks at 25 ns.

3.3.3. The 60 ns Simulation Data Experiment

Similarly, a comparison was conducted with other inversion networks, including UNet, MRF-UNet, and UNet++, as shown in Table 5. The results indicate that MSDNet++ outperforms these networks in terms of overall performance.

Table 5. Comparison of inversion performance at 60 ns.

Method	MSE	MAE	SSIM	Parameter/M	Time/ms
UNet	0.00619	0.04033	0.96179	29.64	344.81
MRF-UNet	0.00148	0.02403	0.98175	59.75	476.33
UNet++	0.00241	0.00601	0.97471	34.55	519.51
MSDNet++	0.00058	0.00515	0.98838	75.13	1043.01

Figure 10a–c show the variation curves of MSE, MAE, and SSIM during the dataset training of different networks. It can be observed that MSDNet++ shows superior performance on the whole. Figure 11 displays the inversion results produced by each network. Figure 10a–c are B-scan images of ground-penetrating radar, and Figure 11d–f are real dielectric constant images. As shown in Figure 11g–i, UNet failed to generate effective inversion results. Figure 11j–l demonstrate that while MRF-UNet improved the inversion quality, the pipeline target remained blurry and the water–pipe interface was indistinct. Figure 11m–o indicate that UNet++ also produced blurry results, with poorly defined boundaries between the pipe wall and the internal water. In contrast, Figure 11p–r show that although the inversion results from MSDNet++ exhibited some ghosting artifacts, they were significantly more accurate and superior to those of the other three networks.

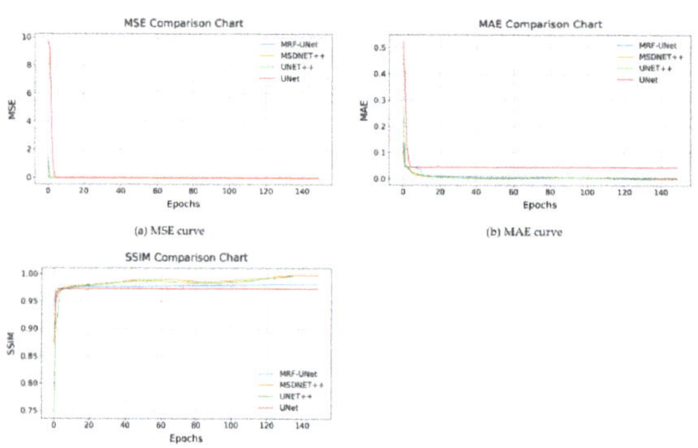

Figure 10. Comparison results of different network variation curves at 60 ns.

3.3.4. The 60 ns Generalization Ability Test

Similarly, to verify the model noise immunity at 60 ns, we also added Gaussian white noise with mean 0 and standard deviation 1 to the ground-penetrating radar B-scans. And the peak-to-noise ratio was used as a metric to assess the network's noise immunity performance. On the trained model micro-tuning for 10 epochs, we obtained the reconstruction test results with background noise. The results of the reconstruction test with background noise are shown in Table 6. It can be seen that compared with the other three networks, MSDNet++ still achieves better reconstruction performance.

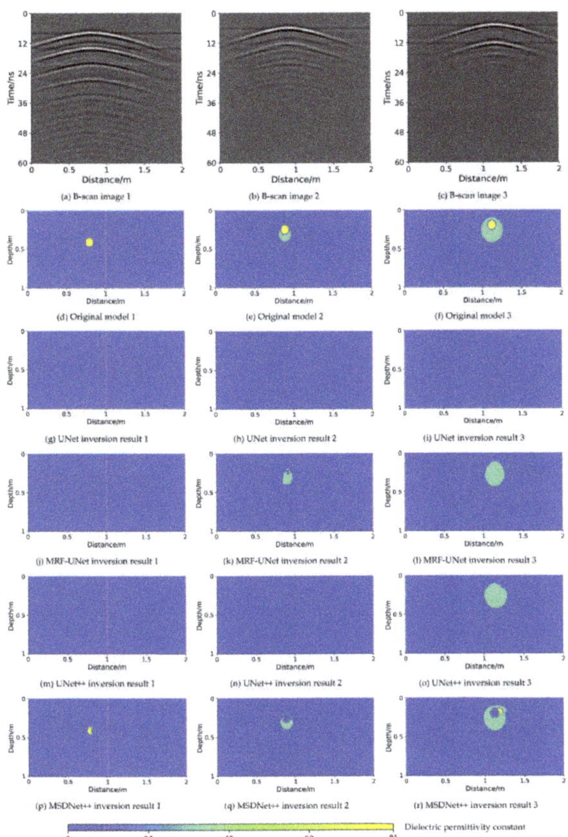

Figure 11. The inversion results of different networks at 60 ns.

Table 6. Comparison of noise data inversion index at 60 ns.

Method	MSE	MAE	PSNR (db)	Parameter/M	Time/ms
UNet	0.04086	0.20001	13.8894	29.64	399.24
MRF-UNet	0.02632	0.15819	15.9096	59.75	558.11
UNet++	0.02499	0.15415	16.0504	34.55	542.95
MSDNet++	0.00659	0.07954	21.8130	75.13	1082.50

Figure 12 illustrates the inversion results of different networks for noisy data. Figure 12a–c show the B-scan images of the ground-penetrating radar with added Gaussian white noise, and Figure 12d–f are real dielectric constant images. As shown in Figure 12g–i, the UNet inversion does not produce a target. Figure 12j–l show that the MRF-UNet inversion of the target has blurred contours, which are different from the real model. Figure 12m–o show that UNet++ is able to invert the pipeline target, but also suffers from blurred contours. In contrast, Figure 12p–r show that MSDNet++ inverts the inner water body of the pipeline indistinctly, but is more consistent with the real model, outperforming the other three networks.

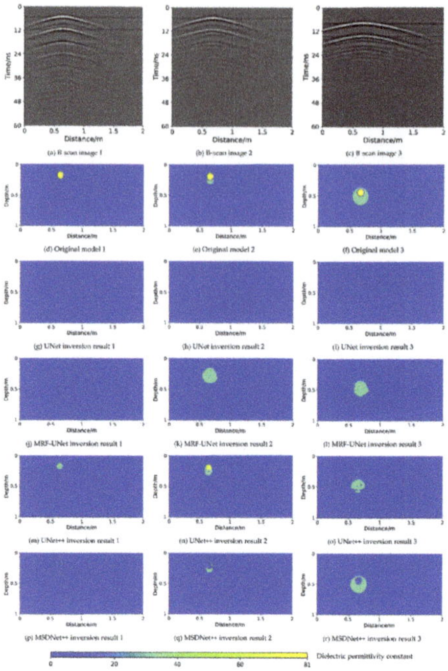

Figure 12. Noisy data inversion results for different networks at 60 ns.

3.4. Ablation Study

3.4.1. The 25 ns Ablation Experiment

To evaluate the effectiveness of the proposed MSD convolution module, the introduced CA mechanism, and the dense skip connections, ablation experiments were conducted. The inversion results on the test set were quantitatively analyzed using SSIM, MSE, and MAE as evaluation metrics, as presented in Table 7. The number of parameters per network and the average time required to invert each B-scan image can also be observed in Table 7.

Table 7. Test results under different modules at 25 ns.

Network Structure	MSE	MAE	SSIM	Parameter/M	Time/ms
① UNet++ only adds the CA mechanism	0.00112	0.00347	0.98408	36.60	647.27
② UNet++ only adds the MSD module	0.00122	0.00376	0.98538	74.74	913.06
③ UNet adds CA and MSD	0.00114	0.00455	0.97638	65.65	703.32
④ MSDNet++	0.00105	0.00341	0.99334	75.13	1043.01

The ablation strategy was based on the UNet++ architecture. First, only the CA attention mechanism was added, resulting in Network ①. Then, only the MSD module was added to form Network ②. These configurations were used to evaluate the individual contributions of each module. Subsequently, based on the UNet++ framework, both the CA mechanism and the MSD module were included, while the dense skip connections were removed to construct Network ③, aiming to assess the impact of dense skip connections. The comparative results of the ablation experiments are illustrated in Figure 13. Figure 13a–c are B-scan images of ground-penetrating radar, and Figure 13d–f are real dielectric constant images. As shown in the figure, Networks ① and ②—each incorporating only one of the modules—produced inversion results with blurred boundaries between the pipe wall and the internal water region. Similarly, Network ③, which lacked dense skip connections, exhibited comparable deficiencies. In contrast, the complete MSDNet++ model, which integrates all three components, achieved the most accurate inversion results, with well-defined contours and precise boundaries for the inverted targets.

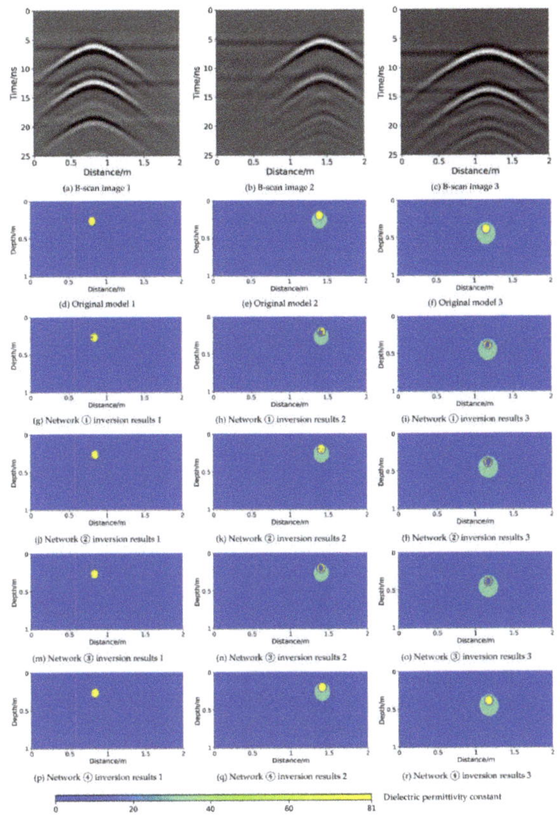

Figure 13. Comparison of ablation results at 25 ns.

3.4.2. The 60 ns Ablation Experiment

The ablation experiments were conducted using the same settings as the 25 ns time window experiments. SSIM, MSE, and MAE were again employed as evaluation metrics to quantitatively assess the inversion results on the test set, as presented in Table 8.

Table 8. Test results under different modules at 60 ns.

Network Structure	MSE	MAE	SSIM	Parameter/M	Time/ms
① UNet++ only adds the CA mechanism	0.00121	0.00802	0.98210	36.60	647.27
② UNet++ only adds the MSD module	0.00141	0.01647	0.98516	74.74	913.06
③ UNet adds CA and MSD	0.01854	0.13457	0.96331	65.65	703.32
④ MSDNet++	0.00111	0.00596	0.99121	75.13	1043.01

The comparative results of the ablation experiments are illustrated in Figure 14. Figure 14a–c are B-scan images of ground-penetrating radar, and Figure 14d–f are real dielectric constant images. As shown in the figure, Networks ① and ②, which incorporated only the CA attention mechanism or the MSD module, respectively, exhibited ghosting artifacts in the inversion results, with unclear boundaries between the pipe wall and the internal water region. Similarly, Network ③, in which the dense skip connections were removed, demonstrated comparable deficiencies. Although the MSDNet++ model—which integrates all three components—also exhibited some blurring at the boundaries between the pipe wall and the internal water region, it still achieved the best overall performance, producing inversion results with more apparent target contours. In conclusion, MSDNet++, which incorporates the CA mechanism, the MSD module, and dense skip connections, demonstrated superior performance compared to the other variants.

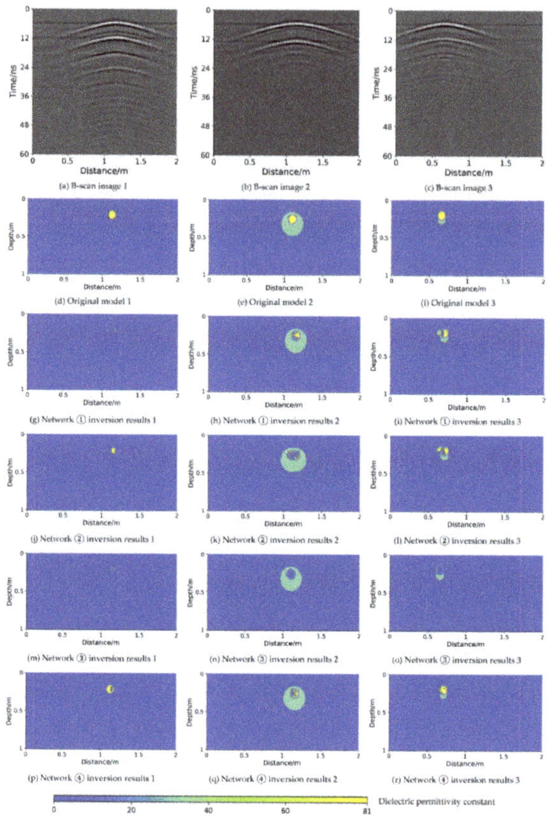

Figure 14. Comparison of ablation results at 60 ns.

By comparing the simulation experiments with 25 ns and 60 ns time windows, it was observed that the inversion results obtained from the 25 ns data were generally more accurate. This is primarily because the shorter 25 ns time window more effectively captures reflection signals from shallow layers, thereby highlighting the characteristics of pipeline leaks more clearly. And the 60 ns time window signals show multi-layer oscillations or multiple noise artifacts, which are more likely to occur in longer time windows, significantly affecting the clarity of the shallow reflection signals. Furthermore, the shorter time window reduces signal attenuation during propagation, preserving signal strength and clarity, which facilitates the extraction of useful features by the inversion network.

3.5. MSDNet++ Testing with Real Measurement Data

To validate the inversion results of MSDNet++ on real measurement data, an experimental setup was established, as shown in Figure 15. The overall experimental model consisted of a sandpit with dimensions of 1.8 m (length) × 0.5 m (width) × 0.7 m (height), as shown in Figure 15a. Inside the model, a PVC pipe with a diameter of 110 mm, wall thickness of 4.2 mm, and length of 0.5 m was buried at a depth of 0.25 m and positioned horizontally at 0.9 m. A leakage hole with a diameter of 5 mm was drilled at the midpoint of the pipe's bottom. One end of the pipe was connected to a one m long water pipe, as shown in the schematic diagram in Figure 15b. The small leakage holes were first plugged and the pipe was filled with water, and measurements were taken to obtain the results when the pipe was not leaking. Subsequently, the leakage hole was opened and a leakage experiment was conducted at a water pressure of 0.01 MPa, with a continuous filling of the pipe, which leaked 24 L of water after 45 min, followed by the excavation of sand for leakage assessment. The leaked water formed a circular pattern with a radius of approximately 10 cm, centered around the point of leakage. The experimental setting parameters are summarized in Table 9.

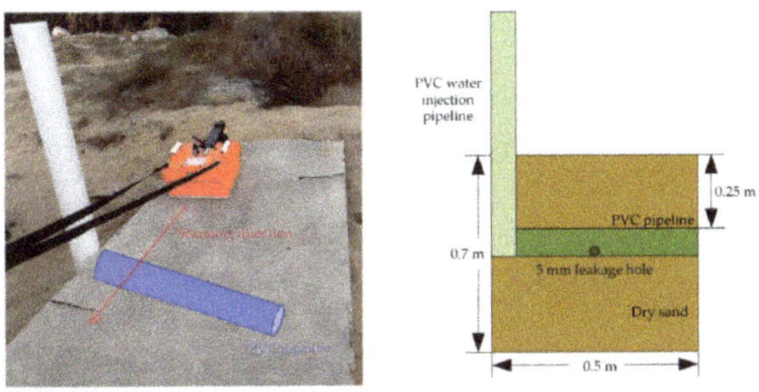

(a) Site diagram of test layout (b) Schematic diagram of the model

Figure 15. Experimental scenario.

Table 9. Summary of experimental setup parameters.

Name	Particulars
Dimensions	1.8 m × 0.5 m × 0.7 m
Pipeline diameter	110 mm
Depth of burial	0.25 m
Diameter of the leakage hole	5 mm
Capacity of leakage water	24 L
Leak time	45 min
Leakage radius	10 cm

In this experiment, a YL-GPR ground-penetrating radar from Y-Line Company, with a center frequency of 400 MHz, was used. During detection, the radar time window was set to 60 ns, the working mode was distance mode, and the sampling interval was 0.02 m. The collected data is preprocessed to include direct wave removal, filtering, gain, and finally cropping so that the processed data matches the size of the dataset data. Because the time window is long and the scanning distance is short, the hyperbolic curvature feature is not obvious when cropped to 256 × 256 size. Finally, the data was input into MSDNet++ for inversion testing, and the inversion results are shown in Figure 16.

Figure 16a,b are B-scan images of real scans from ground-penetrating radar, and Figure 16c,d show the original dielectric constant model images. As seen in Figure 16e,f, MSDNet++ successfully performed the inversion of real measurement data for both shallowly buried pipelines without leakage and those with leakage in the sand. However, deviations in the shape and size of the inverted targets were observed. This may be due to the presence of noise in the actual environment, which caused the B-scan image data to be less refined and significantly different from the simulated data, thereby affecting the network's prediction accuracy.

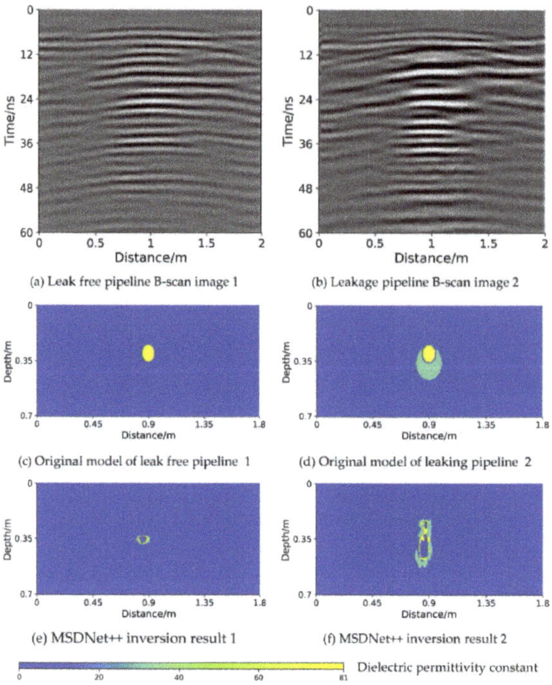

Figure 16. Results of inversion of measured data.

4. Discussion

This study proposes an improved UNet++-based neural network, MSDNet++, for reconstructing shallowly buried water supply pipeline leaks using ground-penetrating radar (GPR) data. By integrating multi-scale directional convolution (MSD), coordinate attention (CA) mechanisms, and dense skip connections, the network effectively addresses the challenges of feature extraction and spatial accuracy in GPR image inversion. Experimental validations were conducted on both simulated and real-world data. The key findings and implications are discussed as follows:

(1) The MSD Module: The MSD module combines multi-scale and directional convolutional kernels to capture anisotropic features in GPR images. Small-scale kernels extract fine-grained details of pipeline leaks, while large-scale kernels capture global structural information. Parallel horizontal and vertical convolutions enhance the model's ability to recognize features along different directions, which is crucial for accurately reconstructing irregular leak geometries. This design significantly improves the network's performance compared to traditional single-scale convolutions, as demonstrated by the clear leak boundaries in the inversion results (Figure 8p–r).

(2) The CA Mechanism: The CA mechanism decomposes channel attention into 1D feature encoding processes along the horizontal and vertical directions, preserving positional sensitivity while capturing long-range dependencies. This approach addresses the blurring and ghosting issues observed in UNet++ (Figure 8m–o) by emphasizing key spatial regions in the feature maps. The lightweight design of the CA mechanism

ensures computational efficiency, making it suitable for integration into complex network architectures.

(3) Dense Skip Connections: Building upon the UNet++ framework, MSDNet++ employs dense skip connections to fuse multi-level feature information. This design enables the decoder to simultaneously learn low-level, detailed features and high-level, abstract features, thereby minimizing information loss during upsampling. The residual connections further alleviate the vanishing gradient problem and accelerate the training process. The ablation experiments (Tables 7 and 8) confirm that the combination of these three components achieves the best inversion performance.

(4) Experimental Validation: The proposed network outperforms existing methods (UNet, MRF-UNet, and UNet++) in both 25 ns and 60 ns simulation experiments, achieving higher SSIM (0.99667) and lower MSE (0.00018) values (Tables 3 and 5). For real-world data, MSDNet++ successfully reconstructs pipeline leaks, though deviations in shape and size occur due to environmental noise (Figure 16). This highlights the need for noise-resistant training strategies or preprocessing techniques to bridge the gap between simulated and measured data performance.

(5) Limitations and Future Work: While MSDNet++ demonstrates promising results, its performance in harsh underground environments with strong interference signals requires further improvement. Future research should focus on the following:

   ① Enhancing noise robustness by incorporating advanced denoising modules or adversarial training.

   ② Expanding the training dataset with diverse leakage scenarios and soil conditions to improve generalization.

   ③ Optimizing the network for real-time deployment on portable GPR devices.

In conclusion, MSDNet++ provides an effective solution for GPR-based pipeline leak inversion by integrating multi-scale directional perception, attention mechanisms, and hierarchical feature fusion. Future efforts should enhance its adaptability to complex real-world conditions and extend its applications to other subsurface imaging tasks.

5. Conclusions

This paper presents an improved U-Net++ neural network architecture, MSDNet++, designed to efficiently convert ground-penetrating radar (GPR) B-scan images into dielectric constant distribution images of underground media. By integrating multi-scale directional convolution and coordinate attention mechanisms, the network enhances the extraction and representation of complex features in GPR images, enabling the more accurate inversion of the location and size of shallowly buried pipeline leaks. MSDNet++ combines multi-scale directional convolutional (MSD) blocks and coordinate attention mechanisms, performing feature extraction, compression, and reconstruction through an encoder–decoder architecture. The encoder utilizes MSD modules to extract multi-scale features, thereby enhancing the recognition of targets at various scales and directions. The decoder fuses multi-level feature information through dense skip connections, reducing information loss, and further optimizes feature representation with the coordinate attention mechanism to highlight key areas. Training and testing were conducted on a simulation dataset generated by GprMax, and the results demonstrate that MSDNet++ achieves higher inversion accuracy and robustness compared to traditional networks such as UNet, MRF-UNet, and UNet++. Testing with real measurement data further confirms the effectiveness of MSDNet++ in real-world GPR scenarios. However, due to environmental noise interference in real measurement data, discrepancies were observed between the inversion results and actual conditions. Future work will focus on developing more advanced GPR physical models, expanding the network's training dataset, and further optimizing the

MSDNet++ deep learning inversion method. By enhancing feature extraction and boosting the network's generalization capability, MSDNet++ will become even more applicable to real-world GPR signal inversion tasks, providing more accurate and reliable solutions for urban underground pipeline leak detection. However, the presence of ambient noise interference in the actual measurement data led to discrepancies between the inversion results and the actual situation. Future work will focus on developing more advanced physical models for GPR, expanding the size of the network training dataset, dealing with deeper buried pipeline leaks, and further investigating the inversion of underground pipeline targets and the inversion of specific values of dielectric constant images at the time of leakage for a more comprehensive and reliable 2D reconstruction of pipeline leaks. Meanwhile, the MSDNet++ deep learning inversion method will be further optimized to make MSDNet++ more suitable for practical GPR signal inversion tasks by enhancing feature extraction and improving the generalization ability of the network, so as to provide a more accurate and reliable solution for the reconstruction of urban underground pipeline leakage.

Author Contributions: Conceptualization, Q.X. and Q.L.; methodology, Q.X. and Q.L.; validation, Q.X. and Q.L.; writing—original draft preparation, Q.X.; writing—review and editing, Q.L. and S.O. All authors have read and agreed to the published version of the manuscript.

Funding: This work was supported by the National Natural Science Foundation of China under Grant 62361015 and the Guangxi Natural Science Foundation under Grant 2025GXNSFAA069204.

Data Availability Statement: The original contributions presented in this study are included in the article. Further inquiries can be directed to the corresponding author.

Conflicts of Interest: The authors declare no conflicts of interest.

References

1. Wang, S.C. Analysis and Study on Subsidence Mechanisms of Road Caused by Leakage of Urban Underground Pipeline. Master's Thesis, Zhengzhou University, Zhengzhou, China, 2017. (In Chinese)
2. Zhang, J.; Hu, Q.W.; Zhou, Y.M.; Zhao, P.; Duan, X. A multi-level robust positioning method for three-dimensional ground penetrating radar (3D GPR) road underground imaging in dense urban areas. *Remote Sens.* **2024**, *16*, 1559. [CrossRef]
3. Qin, H.; Xie, X.Y.; Tang, Y.; Wang, Z. Experimental study on GPR detection of voids inside and behind tunnel linings. *J. Environ. Eng. Geophys.* **2020**, *25*, 65–74. [CrossRef]
4. Daniels, D.J. A review of GPR for landmine detection. *Sens. Imaging* **2006**, *7*, 90–123. [CrossRef]
5. Abubakar, A.; Habashy, T.M.; Druskin, V.; Alumbaugh, D.; Knizhnerman, L. 2.5D forward and inversion modeling for interpreting low-frequency electromagnetic measurements. *Geophysics* **2008**, *73*, F165–F177. [CrossRef]
6. Lei, L.L.; Liu, S.X.; Fu, L.; Wu, J.J. Reverse time migration applied to GPR data based on full waveform inversion. *Chin. J. Geophys.* **2015**, *58*, 3346–3355. (In Chinese)
7. LeCun, Y.; Bengio, Y.; Hinton, G.E. Deep learning. *Nature* **2015**, *521*, 436–444. [CrossRef] [PubMed]
8. Ren, S.; He, K.; Girshick, R.; Sun, J. Faster R-CNN: Towards real-time object detection with region proposal networks. *IEEE Trans. Pattern Anal. Mach. Intell.* **2017**, *39*, 1137–1149. [CrossRef] [PubMed]
9. Miao, X.; Wang, J.; Wang, Z.; Sui, Q.; Gao, Y.; Jiang, P. Automatic recognition of highway tunnel defects based on an improved U-Net model. *IEEE Sens. J.* **2019**, *19*, 11413–11423. [CrossRef]
10. Goodfellow, I.; Pouget-Abadie, J.; Mirza, M.; Xu, B.; Warde-Farley, D.; Ozair, S.; Courville, A.; Bengio, A. Generative adversarial nets. *Proc. Adv. Neural Inf. Process. Syst.* **2014**, *3*, 2672–2680.
11. Alvarez, J.K.; Kodagoda, S. Application of deep learning image-to-image transformation networks to GPR radargrams for sub-surface imaging in infrastructure monitoring. In Proceedings of the 13th IEEE Conference on Industrial Electronics and Applications, Wuhan, China, 31 May–2 June 2018; pp. 611–616.
12. Xie, L.; Zhao, Q.; Ma, C.; Liao, B.; Huo, J. Ü-net: Deep-learning schemes for ground penetrating radar data inversion. *J. Environ. Eng. Geophys.* **2020**, *25*, 287–292. [CrossRef]
13. Liu, B.; Ren, Y.; Liu, H.; Xu, H.; Wang, Z.; Cohn, A.G.; Jiang, P. GPRInvNet: Deep learning-based ground-penetrating radar data inversion for tunnel linings. *IEEE Trans. Geosci. Remote Sens.* **2021**, *59*, 8305–8325. [CrossRef]
14. Ji, Y.; Zhang, F.; Wang, J.; Wang, Z.; Jiang, P.; Liu, H.; Sui, Q. Deep neural network-based permittivity inversions for ground penetrating radar data. *IEEE Sens. J.* **2021**, *21*, 8172–8183. [CrossRef]

15. Wu, M.; Liu, Q.; Ouyang, S. Two-stage GPR image inversion method based on multi-scale dilated convolution and hybrid attention gate. *Remote Sens.* **2025**, *17*, 322. [CrossRef]
16. Zhou, Z.; Siddiquee, M.M.R.; Tajbakhsh, N.; Liang, J. UNet++: Redesigning skip connections to exploit multiscale features in image segmentation. *IEEE Trans. Med. Imaging* **2020**, *39*, 1856–1867. [CrossRef] [PubMed]
17. Bakurov, I.; Buzzelli, M.; Schettini, R.; Castelli, M.; Vanneschi, L. Structural similarity index (SSIM) revisited: A data-driven approach. *Expert Syst. Appl.* **2022**, *189*, 116087. [CrossRef]
18. Shen, Y.P.; Dong, S.H.; Wang, Q.; Huang, L.Y. Seepage model analysis of urban water pipeline leakage degree and forward modeling of GPR signals. In Proceedings of the 2016 National Conference on Engineering Geology, Chengdu, China, 15–17 October 2016. (In Chinese)
19. Ronneberger, O.; Fischer, P.; Brox, T. U-Net: Convolutional networks for biomedical image segmentation. In Proceedings of the International Conference on Medical Image Computing and Computer-Assisted Intervention, Munich, Germany, 5–9 October 2015; pp. 234–241.
20. Dai, Q.; Lee, Y.H.; Sun, H.H.; Ow, G.; Yusof, M.L.M.; Yucel, A.C. DMRF-UNet: A two-stage deep learning scheme for GPR data inversion under heterogeneous soil conditions. *IEEE Trans. Antennas Propag.* **2022**, *70*, 6313–6328. [CrossRef]

remote sensing

MDPI

Article

Mapping Mountain Permafrost via GPR-Augmented Machine Learning in the Northeastern Qinghai–Tibet Plateau

Yao Xiao [1], Guangyue Liu [1,*], Guojie Hu [1], Defu Zou [1], Ren Li [1], Erji Du [1], Tonghua Wu [1], Xiaodong Wu [1], Guohui Zhao [2,3], Yonghua Zhao [1] and Lin Zhao [4]

[1] Cryosphere Research Station on the Qinghai-Tibet Plateau, State Key Laboratory of Cryospheric Science and Frozen Soil Engineering, Northwest Institute of Eco-Environment and Resources, Chinese Academy of Sciences, Lanzhou 730030, China; xiaoyao@lzb.ac.cn (Y.X.); huguojie123@lzb.ac.cn (G.H.); defuzou@lzb.ac.cn (D.Z.); liren@lzb.ac.cn (R.L.); thuawu@lzb.ac.cn (T.W.); wuxd@lzb.ac.cn (X.W.)

[2] Northwest Institute of Eco-Environment and Resources, Chinese Academy of Sciences, Lanzhou 730030, China

[3] National Cryosphere Desert Data Center, Lanzhou 730030, China

[4] School of Geographical Sciences, Nanjing University of Information Science & Technology (NUIST), Nanjing 210044, China; lzhao@nuist.edu.cn

* Correspondence: liuguangyue@lzb.ac.cn

Abstract: Accurate permafrost mapping in mountainous regions is hindered by sparse in situ observations and heterogeneous terrain. This study develops a GPR-augmented machine learning framework to map mountain permafrost in the northeastern Qinghai–Tibet Plateau. A total of 1037 presence–absence samples were compiled from boreholes, soil pits, 128 GPR transects collected in 2009, and 22 additional empirical points above 4700 m, covering diverse topographic and thermal conditions. Thirteen classification algorithms were evaluated using 5-fold cross-validation repeated 40 times, with LightGBM, CatBoost, XGBoost, and RF achieving top performance (F1 > 0.98). Elevation-based spatial comparisons revealed that LightGBM and CatBoost produced more terrain-adaptive predictions at high altitudes and slope transitions. Aspect-controlled permafrost boundaries were captured, with modeled lower elevation limits varying by >200 m across slope directions. SHAP analysis showed that climate and soil variables contributed nearly 80% to model outputs, with LST, FDD, BD, and TDD being dominant. Several predictors exhibited threshold or nonlinear responses, reinforcing their physical relevance. Additional experiments confirmed that integration of GPR and high-elevation constraint samples significantly improved model generalization, especially in underrepresented terrain zones. This study demonstrates that a GPR-augmented machine learning framework can support cost-effective, physically informed mapping of frozen ground in complex alpine environments.

Keywords: permafrost mapping; ground-penetrating radar (GPR); SHAP analysis; machine learning; Qinghai–Tibet Plateau; alpine permafrost

1. Introduction

Permafrost, defined as ground that remains at or below 0 °C for at least two consecutive years, plays a critical role in regional hydrology, carbon cycling, and infrastructure stability across high-latitude and high-altitude environments such as the Qinghai–Tibet Plateau (QTP), the Arctic, and sub-Arctic mountains [1–3]. With climate warming accelerating, permafrost degradation has emerged as a major concern due to its potential to amplify greenhouse gas emissions and disrupt surface stability [4–7]. Accurate mapping of permafrost extent is therefore essential for understanding environmental risk, managing hydrological systems, and guiding infrastructure design in cryospheric regions [8–11].

Traditional modeling approaches, including physics-based energy balance models [12–14] and empirical–statistical models [15–17], have provided foundational insights into permafrost dynamics. However, these methods are often constrained by high data requirements, strong assumptions, and limited capacity to model complex nonlinear interactions [16,18]. Machine learning (ML) models offer a promising alternative, capable of capturing nonlinear interactions across diverse environmental predictors [19,20]. Successful applications of ML for permafrost mapping have been demonstrated across Arctic and mountainous regions, including the Carpathians, northeastern China, and the Tibetan Plateau [21–25].

Nevertheless, ML-based permafrost studies face persistent challenges; observation data are typically sparse and spatially biased, especially in rugged alpine terrain, and model outputs often lack interpretability in terms of underlying physical mechanisms. Ground-penetrating radar (GPR) provides a cost-effective means of enhancing sample coverage, especially in inaccessible zones [26,27]. By integrating GPR-derived presence–absence information into machine learning frameworks, model training can better reflect terrain heterogeneity and subsurface variation. Furthermore, model interpretation techniques such as SHapley Additive exPlanations (SHAP) offer the ability to assess variable contributions and identify key environmental drivers in a statistically consistent and physically meaningful way [28,29]. Recent studies have begun exploring the role of aspect, elevation, and soil properties in shaping permafrost boundaries, highlighting the need for spatially adaptive modeling approaches [15,30].

To address these challenges, this study integrates GPR transect observations with machine learning to improve mountain permafrost mapping in the northeastern Qinghai–Tibet Plateau. A comprehensive presence–absence dataset was constructed from 1015 labeled samples derived from GPR and borehole surveys, combined with environmental variables representing climatic, topographic, and surface conditions. Thirteen machine learning classifiers were evaluated under repeated cross-validation, and spatial differences in predictive performance were analyzed across elevation bands and slope aspect. SHAP-based interpretation was further applied to identify key variables and their nonlinear relationships with permafrost occurrence. This approach aims to enhance spatial generalization in data-scarce, terrain-complex regions and provide physical insight into the environmental controls governing mountain permafrost distribution.

2. Materials and Methods

2.1. Study Area and Field Observations

The study was conducted in a mountainous catchment in the upper Yellow River basin, located on the northeastern Qinghai–Tibet Plateau (QTP), China (Figure 1). The region spans elevations from approximately 2000 m to 5000 m and encompasses a mosaic of continuous, discontinuous, sporadic, and seasonally frozen ground. This transitional permafrost zone is influenced by both the East Asian monsoon and midlatitude westerlies, resulting in steep hydrothermal gradients and highly heterogeneous terrain. These characteristics make it an ideal test site for evaluating permafrost mapping methods under complex environmental conditions.

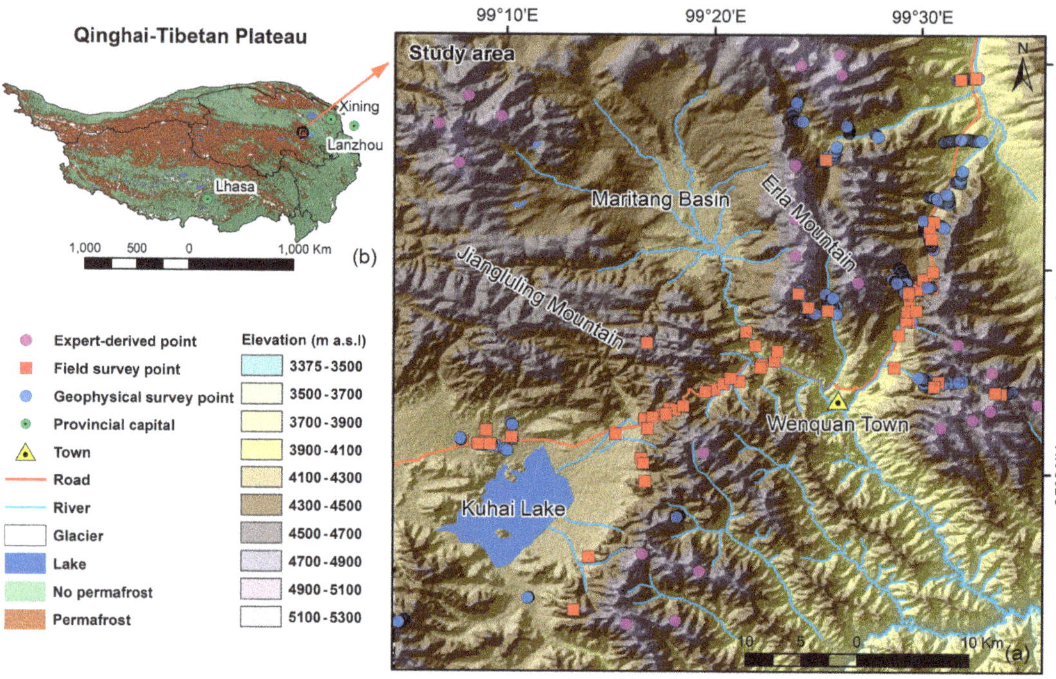

Figure 1. Study area and distribution of field and geophysical survey sites. (**a**) Elevation map of the study area showing field survey points (red squares), GPR-based geophysical survey points (blue circles), and high-elevation permafrost constraint points (purple triangles) manually selected to enhance upper-altitude representation. Major topographic features and infrastructure are also labeled. Permafrost distribution is based on Zou et al. [14], overlaid on a shaded relief map (DEM hillshade) to enhance topographic readability. (**b**) Regional location of the study area on the Qinghai–Tibet Plateau.

A comprehensive field campaign was conducted in September–October 2009 as part of a national permafrost baseline investigation. Multiple observational techniques were employed, including 73 soil pits, 21 shallow boreholes, and 128 ground-penetrating radar (GPR) transects [31]. The soil pits and boreholes provided point-scale evidence of permafrost presence or absence, while GPR surveys offered spatially continuous subsurface profiles across variable terrain.

GPR data were acquired using a 200 MHz antenna system along transects ranging from 500 m to 3 km in length, covering elevations between 3900 m and 4700 m. The 200 MHz antenna was chosen to balance resolution and penetration depth, with typical detection capabilities reaching 3–5 m under alpine soil conditions. This range is suitable for identifying the upper permafrost boundary, as demonstrated in prior studies [26,27]. Permafrost boundaries were interpreted from reflection horizons and verified against nearby ground-truth sites [26,32]. In total, 956 GPR-based samples were extracted, comprising 702 presence and 254 absence cases.

In addition to the 956 GPR-derived samples and 59 field-observed points from boreholes and pits, we manually selected 22 high-elevation presence samples above 4700 m as empirical constraints to enhance altitudinal representation, based on the established elevation dependence of permafrost and visual interpretation of terrain and landform characteristics. Altogether, 1037 labeled samples (753 presence, 284 absence) were compiled for model development and evaluation.

2.2. Environmental Variables and Data Preprocessing

To characterize the environmental factors influencing permafrost occurrence, we compiled 23 variables spanning three domains: climate, topography, and surface conditions (Table 1). These variables were selected based on their demonstrated relevance in permafrost and cold-region modeling over the Qinghai–Tibet Plateau [14,33–35]. All variables are derived from publicly available datasets and were chosen to balance predictive value with physical interpretability. To ensure consistency across heterogeneous spatial datasets, all environmental layers were first projected to a unified Albers equal-area conic coordinate system and then resampled to a 30 m spatial resolution, matching the native resolution of the ASTER GDEM. For continuous variables such as LST, TDD, FDD, and vegetation indices (NDVI, NDWI, NDMI), we applied bilinear interpolation to preserve spatial gradients and reduce smoothing artifacts. For categorical or index-type variables, including Aspect, TWI, and landform masks, we used nearest-neighbor interpolation to maintain class integrity. This preprocessing ensured spatial alignment and comparability across all input features in the modeling framework. Preprocessing steps included reprojection, spatial masking, and normalization. Multi-year means or climatological baselines were used to ensure temporal alignment with the 2009 field campaign.

Table 1. Summary of environmental variables used for permafrost distribution modeling.

Variable	Code	Description/Relevance	Data Source
Climatic factors			
Land surface temperature	LST	Proxy for near-surface energy input and surface heat budget	MODIS MOD11A1 (2003–2019)
Thawing degree days	TDD	Cumulative thermal energy during thaw season; key driver of permafrost degradation	Derived from MODIS
Freezing degree days	FDD	Accumulated freezing intensity; indicator of freeze duration	Derived from MODIS
Mean annual temperature	AT	Highly collinear with TDD; retained to enhance climatic context	Peng et al. [34]
Annual precipitation	Pre	Reflects hydrothermal conditions; potentially influences vegetation and insulation	Peng et al. [34]
Topographic factors			
Elevation	DEM	Controls surface temperature and soil water content via lapse rate	ASTER GDEM v2
Slope	Slope	Influences runoff and snow redistribution	Derived from DEM
Aspect	Aspect	Affects solar exposure and snowmelt timing	Derived from DEM
Topographic Wetness Index	TWI	Proxy for water accumulation and soil moisture	DEM-derived
Solar radiation	Sola	Controls energy input; relevant for melt and insulation processes	GIS-based solar model
Latitude	Lat	Proxy for regional climate gradient	ASTER
Longitude	Lon	Captures east–west climatic and vegetative differences	ASTER
Surface condition factors			
Normalized Difference Vegetation Index	NDVI	Proxy for vegetation cover; alters insulation and latent heat flux	Landsat 8 (2013–2015)
Normalized Difference Water Index	NDWI	Indicator of surface water content and wetness	Landsat 8 (2013–2015)
Normalized Difference Moisture Index	NDMI	Proxy for canopy and surface moisture	Landsat 8 (2013–2015)
Bulk density	BD	Key determinant of soil thermal conductivity	Liu et al. [35]
Sand content	Snd	Affects porosity, drainage, and freeze–thaw dynamics	
Clay content	Clay	Influences heat capacity and moisture retention	
Organic carbon density	Soc	Related to soil insulation and carbon storage; collinear with BD in some cases	
Gravel fraction	Cf	Impacts soil heat flux and infiltration pathways	

Climatic variables included land surface temperature (LST), thawing degree days (TDDs), and freezing degree days (FDDs), all derived from MODIS MOD11A1 daily records (2003–2019) at 1 km resolution. Mean annual temperature (AT) and annual precipitation (Pre) were obtained from a 1 km resolution monthly climate dataset developed by Peng et al. [34], which provides downscaled historical climate data over China. These five variables were jointly used to characterize near-surface thermal regimes and long-term hydroclimatic conditions relevant to permafrost occurrence. Topographic variables were extracted from ASTER GDEM v2 (30 m) and include elevation, slope, aspect, topographic wetness index (TWI), and potential solar radiation (Sola). Geographic coordinates (latitude and longitude) were retained to represent large-scale climatic gradients. These terrain-related variables influence energy balance, moisture redistribution, and snow cover retention.

Surface condition variables describe vegetation cover and soil physical properties. NDVI, NDMI, and NDWI were derived from Landsat 8 OLI imagery (30 m resolution, 2013–2015). Although these products postdate the field campaign, prior studies suggest minimal decadal change in alpine vegetation patterns, supporting their use in cold regions [15,30]. However, we note that localized, non-linear vegetation changes driven by warming and land use disturbances may have occurred, potentially introducing uncertainty in the derived indices. Soil variables—including bulk density (Bd), sand (Snd), clay (Clay), gravel content (Cf), and organic carbon density (Soc)—were obtained from a 250 m resolution soil texture dataset developed by Liu et al. [35], based on over 25,000 profile observations across China. These properties influence thermal conductivity and subsurface water dynamics.

2.3. Machine Learning Modeling and Evaluation

To predict permafrost presence or absence (PYN), we implemented thirteen machine learning (ML) algorithms representing four major categories: (1) tree-based models, (2) boosting-based models, (3) linear and kernel-based classifiers, and (4) other classical algorithms. The model types and their key chara cteristics are summarized in Table 2. All models were implemented in Python 3.8 using the Scikit-learn, XGBoost, LightGBM, and CatBoost libraries. Each model was embedded into a pipeline and optimized via grid search using algorithm-specific hyperparameter ranges. For example, Random Forest (RF) and Extremely Randomized Trees (ETs) were tuned for tree depth and the number of estimators; boosting models for learning rate and regularization; SVM for kernel and penalty settings; and neural networks for hidden layer structure and activation functions.

Table 2. Categories and key characteristics of the 13 machine learning models.

Category	Abbreviation	Model Name	Key Characteristics
Tree-Based Models	RF	Random Forest	Ensemble of decision trees based on bagging; robust to overfitting and handles high-dimensional data effectively.
	ET	Extremely Randomized Trees	Similar to RF but with more randomized splits; faster training and reduced variance.
	DT	Decision Tree	Single-tree structure; highly interpretable but prone to overfitting on complex datasets.

Table 2. *Cont.*

Category	Abbreviation	Model Name	Key Characteristics
Boosting-Based Models (GBDT)	XGBoost	Extreme Gradient Boosting	Highly efficient; incorporates advanced regularization techniques to prevent overfitting; excels in handling large datasets with complex patterns.
	LightGBM	Light Gradient Boosting Machine	Optimized for speed and memory efficiency; uses histogram-based learning; particularly suitable for large dataset.
	CatBoost	CatBoost	Specifically designed for categorical feature encoding and high performance; reduces the need for extensive preprocessing.
	AdaBoost	Adaptive Boosting	Using weighted weak classifiers to improve overall prediction accuracy.
Linear and Kernel-Based Models	LR	Logistic Regression	Linear model for binary classification; simple, interpretable, and works well with linearly separable data.
	SVM	Support Vector Machine	Kernel-based method for both linear and nonlinear classification; effective for small dataset with clear margins of separations.
	GP	Gaussian Process	Probabilistic non-parametric Bayesian model; provides uncertainty estimates but computationally expensive.
Other Models	BP-NN	BP Neural Network	Multi-layer perceptron; a classical neural network; suitable for capturing complex, nonlinear relationships in data.
	KNN	K-Nearest Neighbors	Non-parametric model relying on distance measures; sensitive to the choice of k and feature scaling.
	NB	Naive Bayes	Simple probabilistic classifier based on Bayes' theorem; assumes feature independence; included as a baseline for evaluating model performance in structured environmental data.

Model training and evaluation followed a stratified 5-fold cross-validation scheme repeated 40 times to enhance stability and robustness. This resulted in 200 train–validation iterations per model. Class stratification maintained the original class ratio (62.7% positive, 37.3% negative) within each fold. For tree-based models, out-of-bag (OOB) estimates were additionally used to assess generalization performance. Z-score normalization was applied to models sensitive to feature scale, such as SVM and logistic regression. Eight evaluation metrics were used to quantify classification performance: accuracy, precision, recall, F1 score, Cohen's kappa, Matthews correlation coefficient (MCC), area under the receiver operating characteristic curve (AUC-ROC), and area under the precision–recall curve (AUC-PR). The definitions and formulae of these metrics are listed in Table 3.

Table 3. Performance metrics and formulae for evaluating classification accuracy.

Performance Parameters	Formula	Description
Accuracy	$\text{Accuracy} = \frac{TP+TN}{TP+FP+TN+FN}$	The proportion of correctly classified samples among all samples, reflecting the overall predictive capability.
Precision	$\text{Precision} = \frac{TP}{TP+FP}$	The proportion of predicted positive samples that are true positives.
Recall	$\text{Recall} = \frac{TP}{TP+FN}$	The proportion of actual positive samples correctly identified.
F1 Score	$\text{F1} = 2 \cdot \frac{Precision \cdot Recall}{Precision+Recall}$	The harmonic mean of precision and recall, balancing both metrics.
Cohen's Kappa	$\kappa = \frac{P_o - P_e}{1-P_e}$	Measures agreement between predicted and observed classifications, considering chance agreement.
MCC	$\text{MCC} = \frac{TP \times TN - FP \times FN}{\sqrt{(\text{TP+FP})(\text{TP+FN})(\text{TN+FP})(\text{TN+FN})}}$	A balanced metric considering all four confusion matrix elements, suitable for imbalanced data.
AUC-ROC	$\text{AUC} = \int_0^1 TPR(FPR)dFPR$	Area under the ROC curve, measuring the ability to distinguish between positive and negative classes.
AUC-PR	$\text{AUC} - \text{PR} = \int_0^1 Precision(Recall)dRecall$	Area under the precision–recall curve, particularly useful for evaluating imbalanced data.

Notes: TP (true positives), TN (true negatives), FP (false positives), FN (false negatives), P_o (observed agreement), P_e (expected agreement by chance).

To interpret variable effects and improve model transparency, SHapley Additive exPlanations (SHAPs) were applied to selected tree-based models. SHAP values were computed using the TreeExplainer module (SHAP v0.47) to estimate the marginal contribution of each input feature. This method provides locally accurate and globally consistent interpretations of variable importance for ensemble decision trees [28]. While these components have been individually applied in previous studies, their combination in a reproducible and physically interpretable workflow offers a novel contribution to cold-region geospatial modeling.

3. Results

3.1. Model Performance Comparison of 13 Algorithms

The classification performance of thirteen machine learning models was evaluated using repeated cross-validation. Figure 2 summarizes their relative rankings across eight performance metrics. Ensemble tree-based models—particularly LightGBM, CatBoost, XGBoost, and RF—achieved consistently strong results, ranking among the top across nearly all metrics. These models demonstrated high F1 scores (0.98), MCC, and Cohen's kappa values exceeding 0.93, and an AUC-ROC above 0.96. Among them, LightGBM showed the most stable and balanced performance, ranking first in five of the eight metrics. CatBoost and XGBoost also performed robustly, with slight variations across individual metrics. RF displayed high classification accuracy but relatively lower AUC-ROC. In contrast, classical models such as SVM, logistic regression, and Gaussian process showed weaker performance, particularly in metrics sensitive to class imbalance (e.g., MCC and AUC-PR). Although these top models exhibited similar accuracy levels, their detailed metric profiles reveal nuanced differences in classification behavior. Therefore, we further compared the predicted permafrost distributions from the top four models to assess their spatial generalization capabilities.

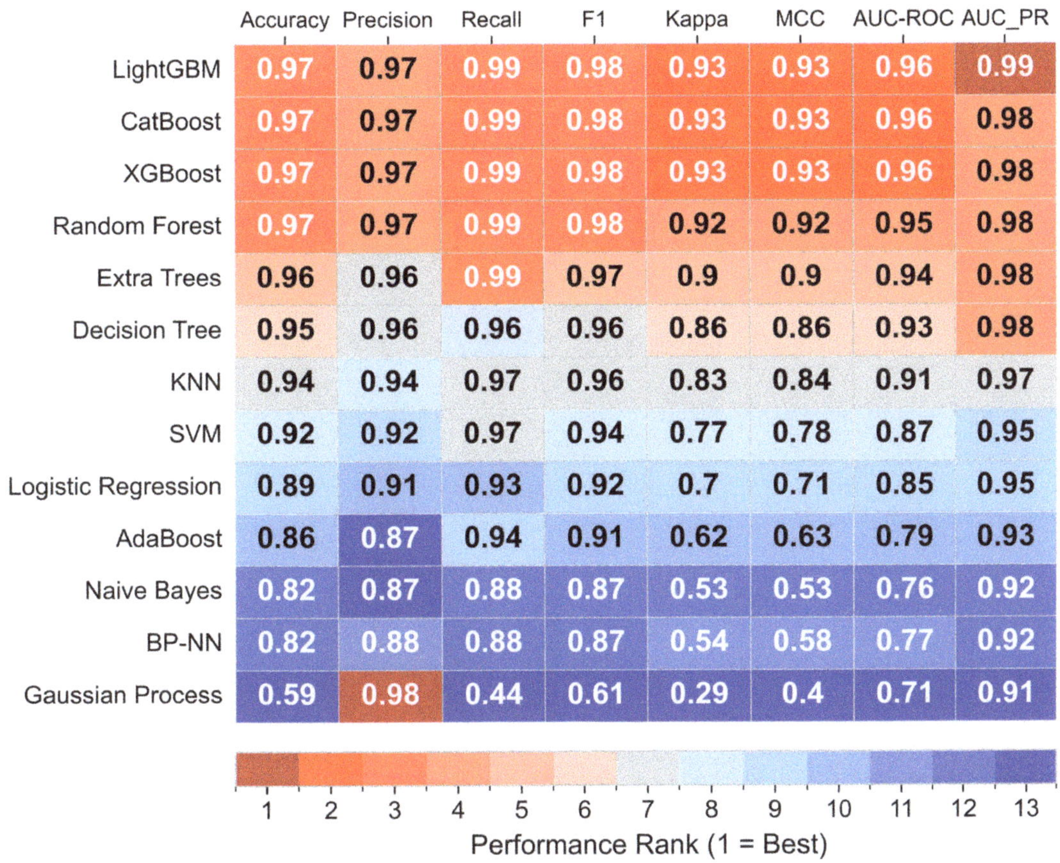

Figure 2. Cross-validated performance of 13 classification models evaluated using eight metrics. Values indicate the mean performance across 200 training–validation splits. Color shading denotes relative rank across models, with lower ranks indicating better performance (1 = best). Key metrics include Accuracy, F1 score, ROC-AUC, and MCC.

3.2. Spatial Prediction Comparison of the Top Models

To evaluate how model accuracy translates into spatial prediction, we compared the permafrost distribution maps generated by the four best-performing models (LightGBM, CatBoost, XGBoost, and RF) at 30 m resolution (Figure 3). All models captured the broad elevation dependence of mountain permafrost, with predicted presence largely concentrated between 4000 m and 5000 m. However, differences emerged in the continuity and boundary precision of predicted distributions. LightGBM, CatBoost, and XGBoost produced smoother and more spatially coherent patterns, whereas RF yielded more fragmented results, particularly in transitional zones and along south-facing slopes.

Notably, in the 5000–5500 m elevation band—where no original borehole or GPR training data were available—all models predicted nearly complete permafrost coverage (99.98%), as shown in Table 4. This consistency suggests that the inclusion of 22 manually selected high-elevation presence points effectively strengthened the upper bound extrapolation. In the 4500–5000 m zone, CatBoost predicted the highest permafrost proportion (97.58%), followed by Random Forest (96.63%), XGBoost (95.59%), and LightGBM (94.83%). Although differences among models were small in this band, LightGBM produced more spatially continuous and terrain-consistent outputs across the landscape. At mid-elevations

(4000–4500 m), which account for 63.6% of the terrain and contain the majority of observations, permafrost coverage ranged from 65.18% to 70.12% across models. In contrast, all models predicted less than 3% permafrost presence below 4000 m, aligning with the observed lower limit and supporting the modeled altitudinal sensitivity (Table 4).

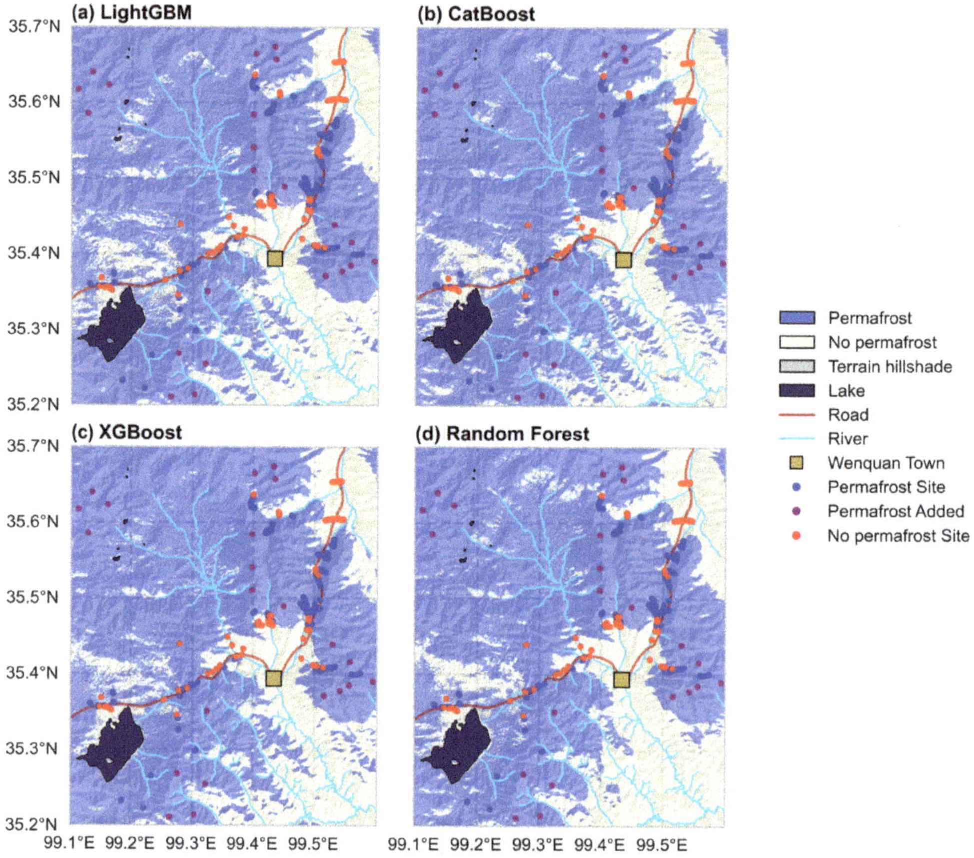

Figure 3. Spatial comparison of permafrost predictions from the top four performing models: (**a**) LightGBM, (**b**) CatBoost, (**c**) XGBoost, and (**d**) Random Forest. Each panel shows binary classification results (permafrost vs. non-permafrost) at 30 m resolution, overlaid on a terrain hillshade background. Colored points represent validation sites. Differences in boundary transitions and localized predictions reflect variations in model generalization.

Table 4. Elevation-based comparison of terrain area, observation site density, and predicted permafrost proportion across models.

Elevation Band	Area Proportion (%)	Observation Sites	Permafrost Proportion (%)			
			LightGBM	CatBoost	XGBoost	Random Forest
3000–3500 m	0.07	0	0	0	0	0
3500–4000 m	7.79	40	0.79	2.66	1.74	0
4000–4500 m	63.64	927	65.27	68.28	70.12	65.18
4500–5000 m	28.35	48	94.83	97.58	95.59	96.63
5000–5500 m	0.16	0	99.98	99.98	99.98	99.98

Note: "Permafrost Proportion (%)" refers to the percentage of grid cells within each elevation band predicted as permafrost.

3.3. Aspect-Dependent Lower Elevation Limits of Modeled Permafrost

In mountain permafrost environments, the lower elevation limit marks the climatic boundary below which permafrost can no longer persist. This threshold is sensitive to local microclimatic factors—particularly aspect—and serves as an indicator of terrain–climate interactions. To examine the role of slope orientation, we extracted the minimum elevation at which permafrost was predicted in each of eight aspect sectors, using the LightGBM model. Aspect classes were defined at 45° intervals, with the north-facing sector covering 337.5° to 22.5° (Figure 4). The results reveal clear directional asymmetries. On north-facing slopes (N, NE, NW), modeled permafrost did not appear below ~3730 m, while on south-facing slopes (S, SE, SW), the minimum elevation extended down to 3963 m. This pattern reflects the influence of differential solar radiation: south-facing slopes receive greater incoming energy, which raises the local thermal threshold for permafrost persistence. Although observational data were limited at higher elevations, the predicted aspect-dependent limits align with known microclimatic effects and suggest that the model effectively captures the influence of terrain controls on permafrost boundaries.

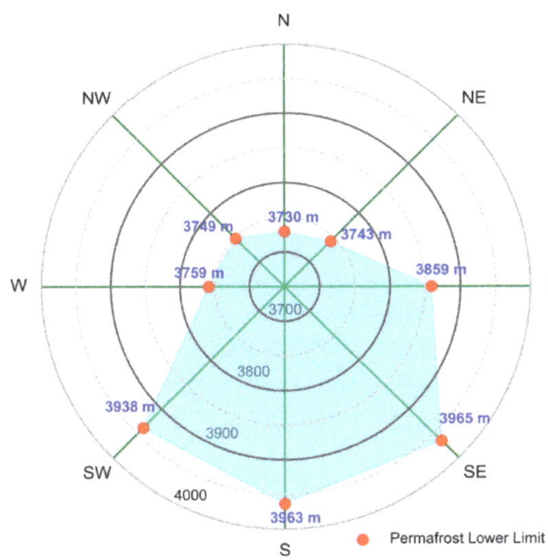

Figure 4. Aspect-dependent lower elevation limits of permafrost predicted by the LightGBM model. Each red dot represents the minimum elevation of predicted permafrost within a given aspect sector, defined at 45° intervals (e.g., N: 337.5°–22.5°, NE: 22.5°–67.5°, etc.). The shaded area connects these minimum values to illustrate the overall pattern. Solid and dashed rings indicate elevation intervals of 100 m and 50 m, respectively.

3.4. Feature Contributions Interpreted by SHAP in the LightGBM Model

To interpret the LightGBM classifier and assess variable importance, SHAP analysis was applied to quantify the marginal contribution of each predictor. As shown in Figure 5a, land surface temperature (LST), freezing degree days (FDDs), bulk density (BD), and thawing degree days (TDDs) were the four most important variables, contributing 15.4%, 14.5%, 13.3%, and 10.5% of the total SHAP values, respectively. These variables reflect key climatic and subsurface conditions relevant to permafrost occurrence. The grouped contribution plot (Figure 5b) further indicates that climate-related predictors accounted for 46.5% of the total contribution, followed by surface conditions (31.3%) and topographic factors (22.2%). These group-wise proportions suggest that temperature indices and substrate properties dominate over terrain geometry in determining model outputs.

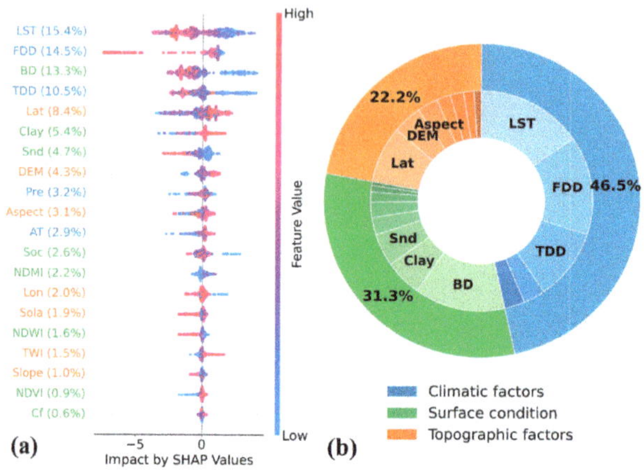

Figure 5. SHAP-based interpretation of feature contributions in the LightGBM model. (**a**) Summary plot of the top-ranked predictors showing their average SHAP values and effect directions. Colors indicate feature values. (**b**) Grouped contribution analysis based on SHAP values. The inner ring shows individual variable importance, and the outer ring summarizes the relative contributions of three factor categories: climatic (blue), surface condition (green), and topographic (orange).

Figure 6 presents the SHAP dependence relationships for four key variables. FDD showed a threshold-like effect, with SHAP values declining rapidly when values exceeded approximately −1250 °C·day (Figure 6a). BD exhibited a nonlinear negative relationship (Figure 6b), while elevation showed a relatively linear positive contribution within the 3900–4500 m range (Figure 6c). Clay content presented a segmented pattern, with SHAP values increasing more prominently above 15% (Figure 6d). BD also showed a notable change in SHAP response around 910 kg/m^3. These variable-specific patterns suggest that the model captured structured, non-linear responses along key environmental gradients.

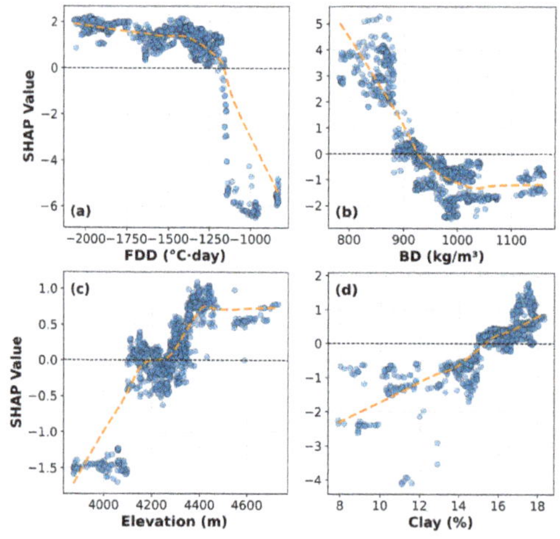

Figure 6. SHAP dependence plots for key predictors of permafrost in the LightGBM model. Each subplot shows how the SHAP value (y-axis) varies with the original variable value (x-axis), indicating the marginal effect of that variable on the predicted permafrost presence. Dashed lines indicate smoothed LOWESS trends. (**a**) Freezing degree days (FDDs), (**b**) bulk density (BD), (**c**) elevation (DEM), and (**d**) clay content demonstrate nonlinear and threshold responses, reflecting climatic and soil controls on permafrost presence.

4. Discussion

4.1. Comparison with Previous Studies

Permafrost distribution mapping on QTP has traditionally relied on empirical indicators such as elevation, air temperature, or vegetation proxies [13,14,36]. These models, while useful at continental scales, often assume fixed threshold relationships and lack the capacity to capture environmental heterogeneity in mountainous terrain. For instance, widely used zonation schemes based on mean annual ground temperature or TTOP models tend to underrepresent topographic and soil-related influences, limiting their applicability at landscape scale [37].

Our model framework, by contrast, incorporates multiple environmental domains—climate, soil, terrain, and vegetation—within a machine learning (ML) structure capable of capturing non-linear interactions. This multi-factor approach enables finer spatial representation and improved adaptability across elevation and slope gradients. In particular, ensemble-based algorithms such as LightGBM demonstrated more consistent performance in complex terrain, as supported by recent studies applying ML to permafrost prediction [38–40]. To further enhance performance in data-sparse high-elevation areas, we included 22 presence points above 4700 m as empirical constraints. This supplementation led to more realistic elevation gradients and aspect-dependent permafrost boundaries, underscoring the value of targeted data enhancement in cold-region ML applications.

Moreover, the integration of SHAP analysis provides a critical advancement beyond black-box ML applications. Instead of relying on model coefficients or assumed functional forms, SHAP allows interpretation of variable-specific effects and their interactions (Figure 6), which is especially important in permafrost systems influenced by coupled thermal–hydrological–geomorphic processes. The ability to detect spatially coherent patterns—such as aspect-modulated permafrost limits (Figure 5)—offers new potential for building transparent, decision-supportive models for cryosphere monitoring and infrastructure planning.

4.2. Added Value of GPR-Augmented Training Data

To improve training data representativeness across terrain gradients, we supplemented borehole records with GPR-derived thaw depth measurements (Figure 1). The additional GPR samples expanded coverage in high-elevation, shaded, and roadless areas, reducing sampling bias and enhancing the diversity of environmental conditions represented. The use of GPR in permafrost mapping has been validated in various alpine regions [26,27,31,32,41,42]. In our application, models trained with GPR-augmented data produced more terrain-adaptive and continuous predictions, particularly in transitional zones where borehole-only models tended to underrepresent permafrost (Figure 6). To quantify this improvement, we conducted a parallel experiment using only the 59 field survey samples (soil pits and boreholes). As shown in Figure 7, the classification performance of all top four models (LightGBM, CatBoost, XGBoost, and RF) dropped substantially relative to their GPR-augmented counterparts in Figure 2. Accuracy values fell to approximately 70%, accompanied by reductions across all major metrics, including F1 score, MCC, and AUC-PR. These results highlight the limitations of sparse training data in capturing spatial heterogeneity, particularly in mountainous terrain. The inclusion of GPR samples significantly improved the representativeness of training data, which proved critical for generalizing across heterogeneous terrain.

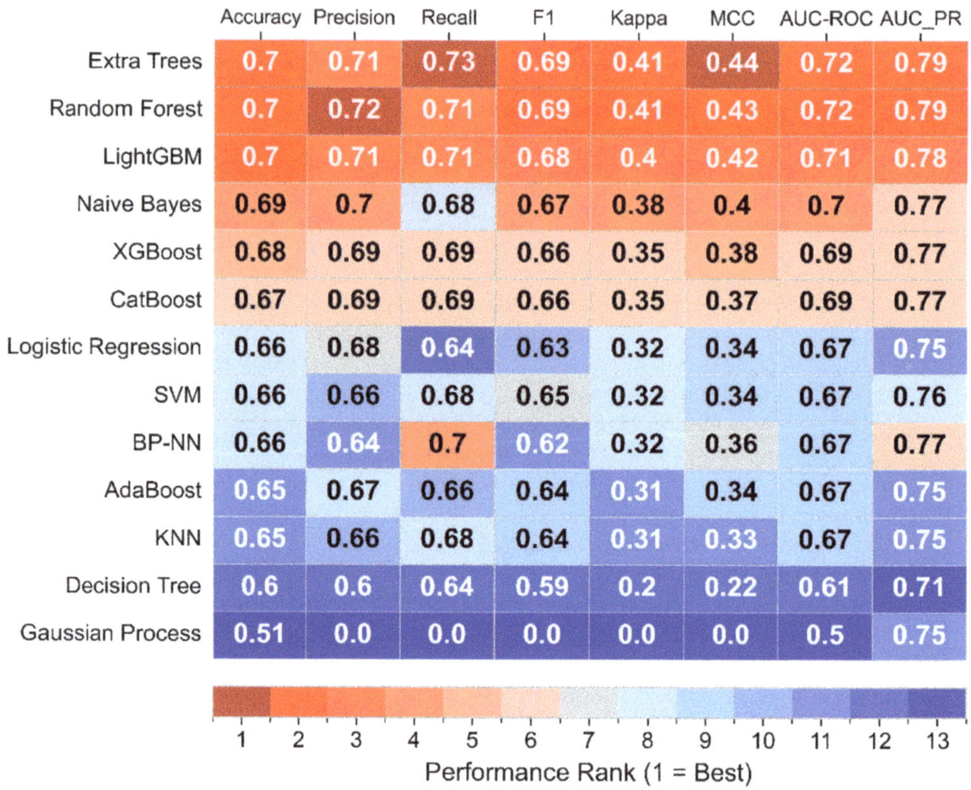

	Accuracy	Precision	Recall	F1	Kappa	MCC	AUC-ROC	AUC_PR
Extra Trees	0.7	0.71	0.73	0.69	0.41	0.44	0.72	0.79
Random Forest	0.7	0.72	0.71	0.69	0.41	0.43	0.72	0.79
LightGBM	0.7	0.71	0.71	0.68	0.4	0.42	0.71	0.78
Naive Bayes	0.69	0.7	0.68	0.67	0.38	0.4	0.7	0.77
XGBoost	0.68	0.69	0.69	0.66	0.35	0.38	0.69	0.77
CatBoost	0.67	0.69	0.69	0.66	0.35	0.37	0.69	0.77
Logistic Regression	0.66	0.68	0.64	0.63	0.32	0.34	0.67	0.75
SVM	0.66	0.66	0.68	0.65	0.32	0.34	0.67	0.76
BP-NN	0.66	0.64	0.7	0.62	0.32	0.36	0.67	0.77
AdaBoost	0.65	0.67	0.66	0.64	0.31	0.34	0.67	0.75
KNN	0.65	0.66	0.68	0.64	0.31	0.33	0.67	0.75
Decision Tree	0.6	0.6	0.64	0.59	0.2	0.22	0.61	0.71
Gaussian Process	0.51	0.0	0.0	0.0	0.0	0.0	0.5	0.75

Figure 7. Classification performance of 13 machine learning models trained without GPR-augmented training data. Same as Figure 2.

4.3. Feature Contributions and Environmental Controls

SHAP-based interpretation revealed that Climate variables dominated model responses, with LST, FDD, and TDD jointly accounting for nearly half of the total contribution. These variables represent cumulative thermal inputs and losses over seasonal timescales, reinforcing findings from previous pan-QTP studies [39,43]. However, beyond these known drivers, our results reveal additional layer-specific patterns.

Soil properties—particularly BD and clay content—exhibited threshold-like responses not captured by linear correlation alone. As shown in Figure 6b,d, BD showed a negative contribution shift near ~910 kg/m^3, while clay contribution increased more rapidly above 15%. These nonlinear effects reflect their role in regulating thermal conductivity and water retention, aligning with hydrothermal coupling processes previously reported [44,45]. Topographic influence, though secondary in total contribution, introduced spatial asymmetry. Aspect significantly modulated the lower elevation limit of predicted permafrost (Figure 4), consistent with solar radiation patterns [15,30,31]. While elevation displayed a quasi-linear SHAP trend across 3900–4500 m (Figure 6c), its predictive role was likely confounded by snow accumulation and radiation exposure.

In addition, Figure 6a shows a sharp decline in FDD's marginal contribution above ~−1250 °C·day, suggesting a possible climatic threshold below which permafrost becomes increasingly unlikely. Such thresholds provide valuable cues for future model-based extrapolation and for identifying tipping points under warming scenarios [46,47]. These results reinforce the need to jointly consider thermal, edaphic, and geomorphic controls when analyzing permafrost regimes.

4.4. Limitations and Outlook

This study relies on multi-source datasets that vary in spatial resolution and acquisition time. Some remote sensing variables, such as NDMI and NDVI, were derived from multi-year means (2013–2015), whereas other key predictors like LST and FDD were based on longer-term MODIS records (2003–2019). GPR observations were collected in 2009. Although this temporal offset was mitigated by focusing on stable land surface properties, minor inconsistencies cannot be fully excluded. In evaluating model performance, we observed that high cross-validation scores did not always translate into reliable spatial predictions, particularly in high-elevation or extrapolated zones. Recognizing this limitation, we incorporated the elevation dependence of permafrost into our assessment, which enabled the identification of models with stronger terrain sensitivity. This result supports recent findings that emphasize the need to complement statistical metrics with physical reasoning or empirical thresholds when selecting and interpreting machine learning models for geospatial applications [48–50].

Looking forward, future studies could benefit from integrating additional physical constraints, such as slope–aspect-adjusted thresholds or monotonic elevation trends, into both model evaluation and post-processing. Combining data-driven algorithms with empirical rules or process-based filters may enhance the robustness, realism, and transferability of permafrost predictions under changing climate conditions. Furthermore, while repeated cross-validation offers a robust internal estimate of model performance, our evaluation did not include an independent spatial or temporal test set. This limits the full assessment of the model's generalizability to unseen regions or conditions. Future work should incorporate spatially or temporally independent validation datasets—collected from different locations or time periods—to rigorously evaluate extrapolation capabilities and enhance confidence in large-scale deployment.

5. Conclusions

This study developed a high-resolution permafrost mapping framework by integrating GPR-augmented field observations with machine learning and SHAP-based interpretation. Using 1037 permafrost samples and 20 environmental predictors, we evaluated 13 classification algorithms and demonstrated that RF, LightGBM, CatBoost, and XGBoost achieved consistently high performance. In particular, cross-validation results showed that all top-performing models achieved average classification accuracies exceeding 97% within the training domain. Among them, LightGBM showed the strongest terrain adaptability and best preserved the altitudinal dependency of permafrost presence. The addition of empirically constrained high-altitude samples significantly improved model generalization, particularly in topographically complex and data-sparse regions.

SHAP analysis revealed that climate variables—especially LST, FDD, and TDD—were the most influential predictors, followed by soil-related factors such as BD and clay. Together, climate and soil variables contributed nearly 80% to model outputs. Several key predictors exhibited nonlinear or threshold-like responses, confirming their physical relevance. The modeled permafrost lower limit occurred near 4000 m, with over 94% predicted coverage above 4500 m and nearly full coverage in results above 5000 m. The elevation threshold varied by over 200 m across slope aspects, indicating strong topographic modulation. Additional comparative experiments showed that incorporating GPR-derived samples significantly improved model accuracy and generalization, particularly in high-altitude and data-sparse regions.

Despite existing limitations in temporal consistency, input resolution, and high-altitude sampling, the proposed model offers a robust, interpretable, and scalable approach to mountain permafrost mapping. This framework can be readily extended to other regions by incorporating localized observations and provides valuable support for environmental monitoring and infrastructure risk assessment in cryospheric environments.

Author Contributions: Conceptualization, Y.X., G.L. and L.Z.; methodology, Y.X. and G.H.; software, Y.X. and D.Z.; validation, R.L. and E.D.; formal analysis, Y.X.; resources, G.H.; data curation, G.L.; writing—original draft preparation, Y.X. and G.L.; writing—review and editing, T.W., X.W. and L.Z.; visualization, G.Z. and Y.Z.; funding acquisition, G.H. and L.Z. All authors have read and agreed to the published version of the manuscript.

Funding: This work was financially supported by the National Key Research and Development Program of China (2022YFF0711702), the National Natural Science Foundation of China (42322608, 42471168), the program of the Key Laboratory of Cryospheric Science and Frozen Soil Engineering, CAS (No. CSFSE-ZQ-2405, CSFSE-ZQ-2407 and CSFSE-ZZ-2408), and the Youth Innovation Promotion Association of the Chinese Academy of Sciences (2022430).

Data Availability Statement: All environmental variables used in this study were derived from publicly accessible datasets. Climatic data were obtained from the MODIS MOD11A1 product (https://doi.org/10.5067/MODIS/MOD11A1.061, accessed on 5 March 2025). Mean annual temperature and precipitation data were obtained from a 1 km gridded dataset developed by Peng et al. [34], accessible via the National Tibetan Plateau Data Center (http://data.tpdc.ac.cn, accessed on 5 March 2025). Topographic variables were derived from the ASTER GDEM v2 (https://asterweb.jpl.nasa.gov/gdem.asp, accessed on 5 March 2025), and vegetation indices (NDVI, NDWI, NDMI) were calculated from Landsat 8 OLI surface reflectance data (https://earthexplorer.usgs.gov/, accessed on 5 March 2025). Soil property data were derived from a 250 m resolution national dataset developed by Liu et al. [35], available at the National Tibetan Plateau Data Center (http://data.tpdc.ac.cn, accessed on 5 March 2025). All analyses were conducted in Python 3.9.21 using scikit-learn 1.6.1, LightGBM 4.5.0, XGBoost 2.1.4, CatBoost 1.2.7, and shap 0.46.0.

Acknowledgments: The authors appreciate all the data provided by each open database. The authors thank the anonymous reviewers and academic editors for their comments.

Conflicts of Interest: The authors declare no conflicts of interest.

References

1. Hugelius, G.; Loisel, J.; Chadburn, S.; Jackson, R.B.; Jones, M.; MacDonald, G.; Marushchak, M.; Olefeldt, D.; Packalen, M.; Siewert, M.B.; et al. Large Stocks of Peatland Carbon and Nitrogen Are Vulnerable to Permafrost Thaw. *Proc. Natl. Acad. Sci. USA* **2020**, *117*, 20438–20446. [CrossRef] [PubMed]
2. Vonk, J.E.; Tank, S.E.; Bowden, W.B.; Laurion, I.; Vincent, W.F.; Alekseychik, P.; Amyot, M.; Billet, M.F.; Canário, J.; Cory, R.M.; et al. Reviews and Syntheses: Effects of Permafrost Thaw on Arctic Aquatic Ecosystems. *Biogeosciences* **2015**, *12*, 7129–7167. [CrossRef]
3. Zhang, T. Influence of the Seasonal Snow Cover on the Ground Thermal Regime: An Overview. *Rev. Geophys.* **2005**, *43*, 2004RG000157. [CrossRef]
4. Miner, K.R.; Turetsky, M.R.; Malina, E.; Bartsch, A.; Tamminen, J.; McGuire, A.D.; Fix, A.; Sweeney, C.; Elder, C.D.; Miller, C.E. Permafrost Carbon Emissions in a Changing Arctic. *Nat. Rev. Earth Environ.* **2022**, *3*, 55–67. [CrossRef]
5. Turetsky, M.R.; Abbott, B.W.; Jones, M.C.; Anthony, K.W.; Olefeldt, D.; Schuur, E.A.G.; Grosse, G.; Kuhry, P.; Hugelius, G.; Koven, C.; et al. Carbon Release through Abrupt Permafrost Thaw. *Nat. Geosci.* **2020**, *13*, 138–143. [CrossRef]
6. Biskaborn, B.K.; Smith, S.L.; Noetzli, J.; Matthes, H.; Vieira, G.; Streletskiy, D.A.; Schoeneich, P.; Romanovsky, V.E.; Lewkowicz, A.G.; Abramov, A.; et al. Permafrost Is Warming at a Global Scale. *Nat. Commun.* **2019**, *10*, 264. [CrossRef]
7. Smith, S.L.; O'Neill, H.B.; Isaksen, K.; Noetzli, J.; Romanovsky, V.E. The Changing Thermal State of Permafrost. *Nat. Rev. Earth Environ.* **2022**, *3*, 10–23. [CrossRef]
8. Liljedahl, A.K.; Boike, J.; Daanen, R.P.; Fedorov, A.N.; Frost, G.V.; Grosse, G.; Hinzman, L.D.; Iijma, Y.; Jorgenson, J.C.; Matveyeva, N.; et al. Pan-Arctic Ice-Wedge Degradation in Warming Permafrost and Its Influence on Tundra Hydrology. *Nat. Geosci.* **2016**, *9*, 312–318. [CrossRef]

9. Mu, C.; Abbott, B.W.; Norris, A.J.; Mu, M.; Fan, C.; Chen, X.; Jia, L.; Yang, R.; Zhang, T.; Wang, K.; et al. The Status and Stability of Permafrost Carbon on the Tibetan Plateau. *Earth Sci. Rev.* **2020**, *211*, 103433. [CrossRef]
10. Peng, X.; Zhang, T.; Frauenfeld, O.W.; Mu, C.; Wang, K.; Wu, X.; Guo, D.; Luo, J.; Hjort, J.; Aalto, J.; et al. Active Layer Thickness and Permafrost Area Projections for the 21st Century. *Earth's Future* **2023**, *11*, e2023EF003573. [CrossRef]
11. Wang, T.; Yang, D.; Yang, Y.; Zheng, G.; Jin, H.; Li, X.; Yao, T.; Cheng, G. Pervasive Permafrost Thaw Exacerbates Future Risk of Water Shortage Across the Tibetan Plateau. *Earth's Future* **2023**, *11*, e2022EF003463. [CrossRef]
12. Nicolsky, D.J.; Romanovsky, V.E. Modeling Long-term Permafrost Degradation. *J. Geophys. Res. Earth Surf.* **2018**, *123*, 1756–1771. [CrossRef]
13. Obu, J.; Westermann, S.; Bartsch, A.; Berdnikov, N.; Christiansen, H.H.; Dashtseren, A.; Delaloye, R.; Elberling, B.; Etzelmüller, B.; Kholodov, A.; et al. Northern Hemisphere Permafrost Map Based on TTOP Modelling for 2000–2016 at 1 km^2 Scale. *Earth Sci. Rev.* **2019**, *193*, 299–316. [CrossRef]
14. Zou, D.; Zhao, L.; Sheng, Y.; Chen, J.; Hu, G.; Wu, T.; Wu, J.; Xie, C.; Wu, X.; Pang, Q.; et al. A New Map of Permafrost Distribution on the Tibetan Plateau. *Cryosphere* **2017**, *11*, 2527–2542. [CrossRef]
15. Boeckli, L.; Brenning, A.; Gruber, S.; Noetzli, J. Permafrost Distribution in the European Alps: Calculation and Evaluation of an Index Map and Summary Statistics. *Cryosphere* **2012**, *6*, 807–820. [CrossRef]
16. Deluigi, N.; Lambiel, C.; Kanevski, M. Data-Driven Mapping of the Potential Mountain Permafrost Distribution. *Sci. Total Environ.* **2017**, *590–591*, 370–380. [CrossRef]
17. Sun, W.; Cao, B.; Hao, J.; Wang, S.; Clow, G.D.; Sun, Y.; Fan, C.; Zhao, W.; Peng, X.; Yao, Y.; et al. Two-Dimensional Simulation of Island Permafrost Degradation in Northeastern Tibetan Plateau. *Geoderma* **2023**, *430*, 116330. [CrossRef]
18. Tahmasebi, P.; Kamrava, S.; Bai, T.; Sahimi, M. Machine Learning in Geo- and Environmental Sciences: From Small to Large Scale. *Adv. Water Resour.* **2020**, *142*, 103619. [CrossRef]
19. Ran, Y.; Li, X.; Cheng, G. Climate Warming over the Past Half Century Has Led to Thermal Degradation of Permafrost on the Qinghai–Tibet Plateau. *Cryosphere* **2018**, *12*, 595–608. [CrossRef]
20. Wang, X.; Ran, Y.; Pang, G.; Chen, D.; Su, B.; Chen, R.; Li, X.; Chen, H.W.; Yang, M.; Gou, X.; et al. Contrasting Characteristics, Changes, and Linkages of Permafrost between the Arctic and the Third Pole. *Earth-Sci. Rev.* **2022**, *230*, 104042. [CrossRef]
21. Baral, P.; Haq, M.A. Spatial Prediction of Permafrost Occurrence in Sikkim Himalayas Using Logistic Regression, Random Forests, Support Vector Machines and Neural Networks. *Geomorphology* **2020**, *371*, 107331. [CrossRef]
22. Li, X.; Ji, Y.; Zhou, G.; Zhou, L.; Li, X.; He, X.; Tian, Z. A New Method for Bare Permafrost Extraction on the Tibetan Plateau by Integrating Machine Learning and Multi-Source Information. *Remote Sens.* **2023**, *15*, 5328. [CrossRef]
23. Popescu, R.; Filhol, S.; Etzelmüller, B.; Vasile, M.; Pleşoianu, A.; Vîrghileanu, M.; Onaca, A.; Şandric, I.; Săvulescu, I.; Cruceru, N.; et al. Permafrost Distribution in the Southern Carpathians, Romania, Derived from Machine Learning Modeling. *Permafr. Periglac. Process.* **2024**, *35*, 243–261. [CrossRef]
24. Şerban, R.-D.; Şerban, M.; He, R.; Jin, H.; Li, Y.; Li, X.; Wang, X.; Li, G. 46-Year (1973–2019) Permafrost Landscape Changes in the Hola Basin, Northeast China Using Machine Learning and Object-Oriented Classification. *Remote Sens.* **2021**, *13*, 1910. [CrossRef]
25. Sheng, Q.; Zhang, Y.; Li, K.; Ling, X.; Li, J. Exploring the Seasonal Comparison of Land Surface Temperature Dominant Factors in the Tibetan Plateau. *ISPRS Ann. Photogramm. Remote Sens. Spat. Inf. Sci.* **2024**, *X-1-2024*, 197–203. [CrossRef]
26. Cao, B.; Gruber, S.; Zhang, T.; Li, L.; Peng, X.; Wang, K.; Zheng, L.; Shao, W.; Guo, H. Spatial Variability of Active Layer Thickness Detected by Ground-penetrating Radar in the Qilian Mountains, Western China. *J. Geophys. Res. Earth Surf.* **2017**, *122*, 574–591. [CrossRef]
27. Du, E.; Zhao, L.; Zou, D.; Li, R.; Wang, Z.; Wu, X.; Hu, G.; Zhao, Y.; Liu, G.; Sun, Z. Soil Moisture Calibration Equations for Active Layer GPR Detection—A Case Study Specially for the Qinghai–Tibet Plateau Permafrost Regions. *Remote Sens.* **2020**, *12*, 605. [CrossRef]
28. Lundberg, S.M.; Erion, G.; Chen, H.; DeGrave, A.; Prutkin, J.M.; Nair, B.; Katz, R.; Himmelfarb, J.; Bansal, N.; Lee, S.-I. From Local Explanations to Global Understanding with Explainable AI for Trees. *Nat. Mach. Intell.* **2020**, *2*, 56–67. [CrossRef]
29. Zhong, S.; Zhang, K.; Bagheri, M.; Burken, J.G.; Gu, A.; Li, B.; Ma, X.; Marrone, B.L.; Ren, Z.J.; Schrier, J.; et al. Machine Learning: New Ideas and Tools in Environmental Science and Engineering. *Environ. Sci. Technol.* **2021**, *55*, 12741–12754. [CrossRef]
30. Haberkorn, A.; Kenner, R.; Noetzli, J.; Phillips, M. Changes in Ground Temperature and Dynamics in Mountain Permafrost in the Swiss Alps. *Front. Earth Sci.* **2021**, *9*, 626686. [CrossRef]
31. Liu, G.; Zhao, L.; Xie, C.; Zou, D.; Wu, T.; Du, E.; Wang, L.; Sheng, Y.; Zhao, Y.; Xiao, Y.; et al. The Zonation of Mountain Frozen Ground under Aspect Adjustment Revealed by Ground-Penetrating Radar Survey—A Case Study of a Small Catchment in the Upper Reaches of the Yellow River, Northeastern Qinghai–Tibet Plateau. *Remote Sens.* **2022**, *14*, 2450. [CrossRef]
32. Angelopoulos, M.C.; Pollard, W.H.; Couture, N.J. The Application of CCR and GPR to Characterize Ground Ice Conditions at Parsons Lake, Northwest Territories. *Cold Reg. Sci. Technol.* **2013**, *85*, 22–33. [CrossRef]
33. Yang, J.; Dong, J.; Xiao, X.; Dai, J.; Wu, C.; Xia, J.; Zhao, G.; Zhao, M.; Li, Z.; Zhang, Y.; et al. Divergent Shifts in Peak Photosynthesis Timing of Temperate and Alpine Grasslands in China. *Remote Sens. Environ.* **2019**, *233*, 111395. [CrossRef]

34. Peng, S.; Ding, Y.; Liu, W.; Li, Z. 1 Km Monthly Temperature and Precipitation Dataset for China from 1901 to 2017. *Earth Syst. Sci. Data* **2019**, *11*, 1931–1946. [CrossRef]
35. Liu, F.; Zhang, G.-L.; Song, X.; Li, D.; Zhao, Y.; Yang, J.; Wu, H.; Yang, F. High-Resolution and Three-Dimensional Mapping of Soil Texture of China. *Geoderma* **2020**, *361*, 114061. [CrossRef]
36. Li, S.; Cheng, G. *Map of Frozen Ground on Qinghai-Xizang Plateau*; Gansu Culture Press: Lanzhou, China, 1996.
37. Li, W.; Weng, B.; Yan, D.; Lai, Y.; Li, M.; Wang, H. Underestimated Permafrost Degradation: Improving the TTOP Model Based on Soil Thermal Conductivity. *Sci. Total Environ.* **2023**, *854*, 158564. [CrossRef]
38. Liu, Y.; Ran, Y.; Li, X.; Che, T.; Wu, T. Multisite Evaluation of Physics-Informed Deep Learning for Permafrost Prediction in the Qinghai-Tibet Plateau. *Cold Reg. Sci. Technol.* **2023**, *216*, 104009. [CrossRef]
39. Ran, Y.; Li, X.; Cheng, G.; Nan, Z.; Che, J.; Sheng, Y.; Wu, Q.; Jin, H.; Luo, D.; Tang, Z.; et al. Mapping the Permafrost Stability on the Tibetan Plateau for 2005–2015. *Sci. China Earth Sci.* **2021**, *64*, 62–79. [CrossRef]
40. Zhang, Y.-Z.; Liang, S.-J.; Chen, J.-B.; Wang, M.; Jia, M.-T.; Jiang, Y.-T. Enhancing Artificial Permafrost Table Predictions Using Integrated Climate and Ground Temperature Data: A Case Study from the Qinghai-Xizang Highway. *Cold Reg. Sci. Technol.* **2025**, *229*, 104341. [CrossRef]
41. Shen, Y.; Zuo, R.; Liu, J.; Tian, Y.; Wang, Q. Characterization and Evaluation of Permafrost Thawing Using GPR Attributes in the Qinghai-Tibet Plateau. *Cold Reg. Sci. Technol.* **2018**, *151*, 302–313. [CrossRef]
42. Wu, T.; Wang, Q.; Watanabe, M.; Chen, J.; Battogtokh, D. Mapping Vertical Profile of Discontinuous Permafrost with Ground Penetrating Radar at Nalaikh Depression, Mongolia. *Environ. Geol.* **2009**, *56*, 1577–1583. [CrossRef]
43. Obu, J. How Much of the Earth's Surface Is Underlain by Permafrost? *J. Geophys. Res. Earth Surf.* **2021**, *126*, e2021JF006123. [CrossRef]
44. Hu, G.; Zhao, L.; Li, R.; Park, H.; Wu, X.; Su, Y.; Guggenberger, G.; Wu, T.; Zou, D.; Zhu, X.; et al. Water and Heat Coupling Processes and Its Simulation in Frozen Soils: Current Status and Future Research Directions. *Catena* **2023**, *222*, 106844. [CrossRef]
45. Gao, K.; Tang, Y.; Chen, D.; Wang, J.; Duan, A. Influence of Arctic Sea Ice and Interdecadal Pacific Oscillation on the Recent Increase of Winter Extreme Snowfall in Northeast China. *Atmos. Res.* **2023**, *295*, 107030. [CrossRef]
46. Li, R.; Ma, J.; Wu, T.; Wang, Q.; Wu, X.; Zhao, L.; Wang, S.; Hu, G.; Liu, W.; Jiao, Y.; et al. The Spatiotemporal Variations of Freezing Index and Its Relationship with Permafrost Degradation over the Qinghai–Tibet Plateau from 1977 to 2016. *Theor. Appl. Clim.* **2024**, *155*, 985–998. [CrossRef]
47. Martin, L.C.P.; Nitzbon, J.; Aas, K.S.; Etzelmüller, B.; Kristiansen, H.; Westermann, S. Stability Conditions of Peat Plateaus and Palsas in Northern Norway. *J. Geophys. Res. Earth Surf.* **2019**, *124*, 705–719. [CrossRef]
48. Reichstein, M.; Camps-Valls, G.; Stevens, B.; Jung, M.; Denzler, J.; Carvalhais, N. Prabhat Deep Learning and Process Understanding for Data-Driven Earth System Science. *Nature* **2019**, *566*, 195–204. [CrossRef]
49. Beddrich, J.; Gupta, S.; Wohlmuth, B.; Chiogna, G. The Importance of Topographic Gradients in Alpine Permafrost Modeling. *Adv. Water Resour.* **2022**, *170*, 104321. [CrossRef]
50. Bergen, K.J.; Johnson, P.A.; De Hoop, M.V.; Beroza, G.C. Machine Learning for Data-Driven Discovery in Solid Earth Geoscience. *Science* **2019**, *363*, eaau0323. [CrossRef]

remote sensing

MDPI

Article

Subsurface Cavity Imaging Based on UNET and Cross–Hole Radar Travel–Time Fingerprint Construction

Hui Cheng [1], Yonghui Zhao [1,*] and Kunwei Feng [2]

1 School of Ocean & Earth Science, Tongji University, Shanghai 200092, China; 2111322@tongji.edu.cn
2 Guizhou Transportation Planning Survey & Design Academe Co., Ltd., Guiyang 550081, China; kevin.feng91@foxmail.cn
* Correspondence: zhaoyh@tongji.edu.cn

Abstract: As a significant geological hazard in large–scale engineering construction, deep subsurface voids demand effective and precise detection methods. Cross–hole radar tomography overcomes depth limitations by transmitting/receiving electromagnetic (EM) waves between boreholes, enabling the accurate determination of the spatial distribution and EM properties of subsurface cavities. However, conventional inversion approaches, such as travel–time/attenuation tomography and full–waveform inversion, still face challenges in terms of their stability, accuracy, and computational efficiency. To address these limitations, this study proposes a deep learning–based imaging method that introduces the concept of travel–time fingerprints, which compress raw radar data into structured, low–dimensional inputs that retain key spatial features. A large synthetic dataset of irregular subsurface cavity models is used to pre–train a UNET model, enabling it to learn nonlinear mapping, from fingerprints to velocity structures. To enhance real–world applicability, transfer learning (TL) is employed to fine–tune the model using a small amount of field data. The refined model is then tested on cross–hole radar datasets collected from a highway construction site in Guizhou Province, China. The results demonstrate that the method can accurately recover the shape, location, and extent of underground cavities, outperforming traditional tomography in terms of clarity and interpretability. This approach offers a high–precision, computationally efficient solution for subsurface void detection, with strong engineering applicability in complex geological environments.

Keywords: cross–hole radar; cavity; UNET; travel–time fingerprint; transfer learning

1. Introduction

The rapid development of urban infrastructure, particularly large–scale transportation hubs and complex underground facilities, has driven the advancement of ultra–deep and clustered engineering. Accurately detecting underground cavities in such deep engineering environments remains a significant challenge in the field of geophysical exploration. Traditional ground–based geophysical methods are limited in resolution and detection depth, particularly when targeting features that are deeply buried or small–scale. In contrast, cross–hole radar tomography offers a promising alternative. It enables the accurate identification and characterization of cavity structures, such as their spatial distribution, shape, and fill materials, at significant depths, thereby providing an effective tool for detecting potential geological hazards in complex deep–earth environments.

Conventional cross–hole radar imaging techniques are primarily based on tomography, which relies on the travel time and amplitude of electromagnetic (EM) waves as influenced by the dielectric properties of the subsurface medium [1]. However, traditional ray–based tomography methods face significant resolution limitations in heterogeneous media because of wave scattering [2]. To improve imaging quality, full–waveform inversion (FWI) was introduced into ground–penetrating– radar (GPR) detection, offering more detailed subsurface reconstructions [3,4]. However, FWI demands a good initial model and substantial computational resources. Therefore, traditional methods face significant challenges in terms of stability, accuracy, and computational efficiency.

With recent advancements in computing power, deep learning (DL) has demonstrated considerable potential in geophysical imaging tasks, enabling the modeling of complex nonlinear relationships between inputs and outputs [5]. Several studies have applied DL to GPR–based detection [6,7]. For instance, Dinh et al. (2018) developed an algorithm that combines traditional image processing with deep convolutional neural networks (CNNs) to detect reinforcing bars in GPR images [8] automatically. Lei et al. (2019) integrated the Faster R–CNN framework with data augmentation to identify hyperbolic features in GPR data [9]. Similarly, DL approaches have also been applied to cross–hole radar. Zhang et al. (2023) developed an inversion framework based on a generative adversarial network (GAN) that converts radar images directly into two–dimensional subsurface structures [10]. Pongrac et al. (2023) introduced a supervised CNN–based method to estimate soil moisture using cross–hole radar records [11]. Liu et al. (2024) proposed a UNET–based approach for high–resolution slowness tomography from cross–hole radar data [12]. However, most of these studies rely on multi–channel radar profiles for the training of deep networks, which may lead to many problems and limitations. For example: (1) high computational cost—generating a large number of training datasets using finite-difference time-domain (FDTD) simulations is computationally intensive [13]; (2) huge data scale—due to the deep drilling depth (exceeding 50 m), a multi-channel GPR profile may contain upwards of tens of thousands of traces, containing excessive information, which will slow the training of the DL model [11]; (3) the difficulty of feature extraction—the waveform features in multi-channel radar profiles are highly susceptible to environmental interference, complicating feature extraction [10]; and (4) poor generalization capability—discrepancies between simulated and real-world data—due to differences in wavelets, instrumentation, and subsurface conditions—create a "simulation-to-reality" gap that reduces model performance in practical scenarios [14]. In summary, current DL-based methods for deep underground cavity imaging using cross-hole radar face substantial limitations in cost-effectiveness, feature robustness, and, most importantly, generalization to real–world conditions. To address these limitations, we conducted our research.

In 2015, Ronneberger et al. [15] proposed the "U" –shaped neural network to realize image feature segmentation and "end–to–end" learning. Based on the specific requirements of cross–hole radar travel-time imaging in this paper, the UNET architecture demonstrates three theoretical advantages: (1) Image–to–image mapping capability—UNET's encoder–decoder structure with symmetric feature extraction and reconstruction perfectly aligns with our need for "travel-time fingerprint to velocity model" conversion. (2) Learning potential—U—Net has demonstrated exceptional data efficiency in semantic segmentation tasks based on a small number of samples. This characteristic directly addresses our core challenge of limited real–world cross–hole radar data and scarce labeled samples. (3) Physical model integration—UNET's simple and straightforward architecture facilitates the integration of physical constraints, allowing for the straightforward incorporation of enhancement modules to improve physical consistency [16–18].

Based on the advantages of the UNET architecture, we propose a novel imaging method that utilizes a UNET model enhanced with transfer learning (TL). This approach adapts pre–trained models to new tasks using limited data [19]. The input features consist of travel–time fingerprints, a newly defined concept that compresses cross–hole radar records into structured 2D matrices. Initially, we proposed innovative construction methods for cavity models alongside their corresponding 2D travel–time fingerprints. Using this framework, datasets of cavity models paired with their travel–time fingerprints are generated. Subsequently, a UNET is pretrained to establish the mapping relationship between travel–time fingerprints and velocity models. To improve the generalization of real–world engineering applications, TL is employed to fine–tune the pre–trained network. Finally, numerical experiments validate the effectiveness and accuracy of the proposed method, while successful field tests conducted at a highway foundation site highlight its practical value for geophysical detection in complex environments.

2. Methodology

In DL applications, constructing a high–quality and diverse dataset is crucial for effective model training and robust generalization. Although DL is adept at learning intricate nonlinear relationships between inputs and outputs, its success is heavily reliant on the availability of large–scale, representative datasets. However, acquiring sufficient cross–hole radar data from real–world scenarios presents challenges due to limited site access and the considerable variability in geological conditions.

To address this issue, this section introduces methods for generating synthetic datasets, including the construction of cavity models and the generation of corresponding two–dimensional (2D) travel–time fingerprints. Subsequently, a modified UNET architecture is applied to map the travel–time fingerprints to subsurface velocity models, leveraging UNET's strong feature extraction and skip–connection capabilities.

2.1. Cavity Model Construction

2.1.1. Irregular Cavity Generation Method

To accurately simulate the complex and irregular cavity geometries commonly found in subsurface environments, we propose a novel cavity generation algorithm that employs dynamic random boundary evolution. This methodology constructs complex cavities by expanding the upper and lower boundaries using a controlled stochastic walk.

Step 1: Cavity Parameters Initialization

The number of cavities is randomly determined, and the computational domain $\Omega \in Z^2$ is defined with dimensions $N_x \times N_y$. Each time, random parameters x and y are generated, and represent the X–axis and Y–axis coordinates of the cavity's center point, respectively. The half–width of each cavity is also set randomly as L. Boundary conditions are applied to ensure that cavities do not extend beyond the limits of the model.

Step 2: Directional Sequences Generation

Each X–coordinate corresponds to a random direction sequence η_u (the value of the direction sequence is -1 or 1). A corresponding sequence η_l is obtained by flipping the signs of η_u. These antipodal sequences govern asymmetric boundary development through coordinated vertical displacement.

Step 3: Dynamic Boundary Expansion

Fix the position of an X–coordinate and calculate the variation in the upper and lower boundaries at each x–position with the width, adhering to the following formula:

$$\begin{cases} y_u(j) = y + \sum_{k=1}^{j} \eta_u(k), \ \textit{upper boundary} \\ y_l(j) = y + \sum_{k=1}^{j} \eta_l(k), \textit{lower boundary} \end{cases} \quad (1)$$

where y_u and y_l denote the upper and lower boundary positions at each vertical slice.

Step 4: Cavity Filling and Model Construction

At each vertical slice (x–position), the area from y_l to y_u in the y direction is filled with materials that have distinct dielectric properties, simulating the presence of cavities.

This process creates geometries with jagged or sawtooth–like features, effectively matching field cavities.

This method generates a 'stacked layers' effect, with each vertical slice exhibiting independently varying heights and irregular cavities, effectively enriching the variability and realism of the training datasets.

2.1.2. Generation of Cavity Model Datasets

Utilizing the methodology delineated in Section 2.1.1, a comprehensive set of 1500 synthetic half–space models, each with a size of 10 m × 10 m, was systematically generated. Within each model, the presence of one and three randomly situated irregular cavities was incorporated to enhance variability. A representative example of one such model is illustrated in Figure 1.

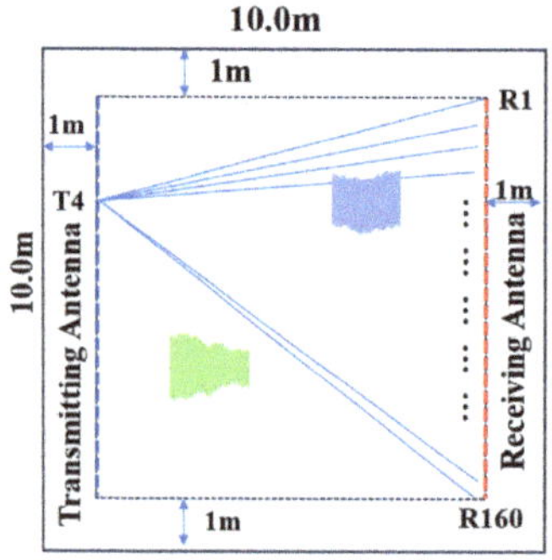

Figure 1. Schematic illustration of irregular cavity model construction.

To accurately simulate realistic geological conditions, the cavities were filled with typical materials such as air, water, and mud, each possessing distinctly different dielectric constants of 1, 81, and 5, respectively. The background medium was designated with a relative permittivity of 20. Table 1 provides a summary of the EM properties of all materials involved.

Table 1. EM properties of background and cavity–filling materials.

Material	Relative Dielectric Constant, ε_r	Electrical Conductivity, $\sigma(S/m)$
background	20.0	0.01
air	1.0	0
water	81.0	0.001
mud	5.0	0.0001

The corresponding EM wave propagation velocities are calculated using the relationship derived from Maxwell's equations [20]

$$v = \frac{c}{\sqrt{\mu_r \varepsilon_r}} \tag{2}$$

where c is the speed of light in vacuum (3×10^8 m/s), and ε_r is the material's relative permittivity. Under the assumption of weak conductivity and nonmagnetic materials ($\mu_r = 1$), Equation (2) provides a valid approximation of EM wave velocity in geological media.

By applying Equation (2), we convert the permittivity distributions into velocity models that reflect both the geometries of cavities and the properties of materials. These velocity models serve as output labels for training the DL network. Figure 2 illustrates six representative examples of these constructed velocity models.

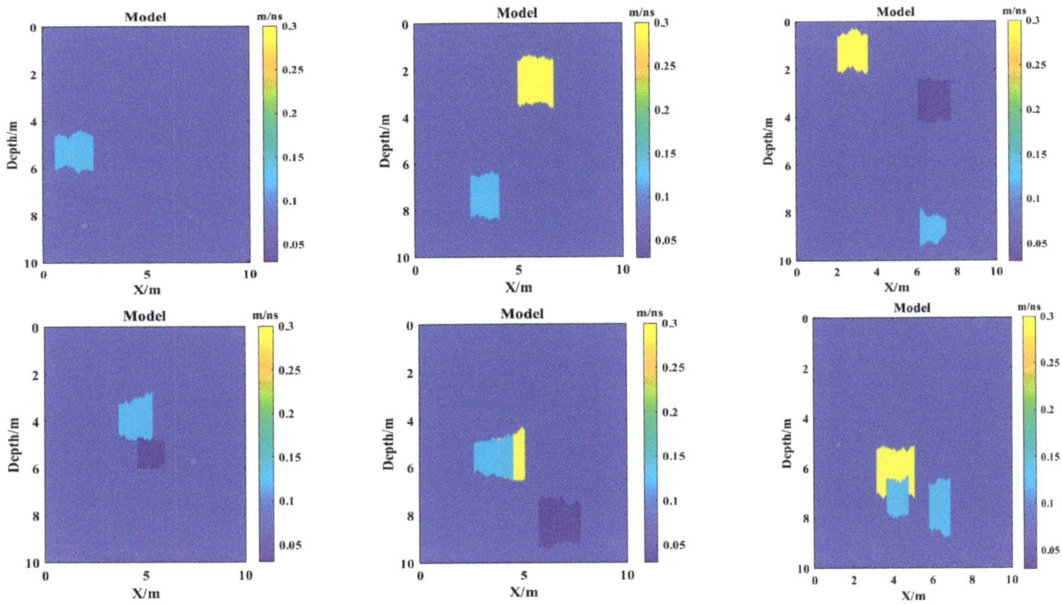

Figure 2. Six representative velocity models from the original domain dataset (yellow: air; dark blue: water; light blue: mud).

2.2. The Construction Method of Travel–Time Fingerprints

2.2.1. Construction Method of Travel–Time Fingerprints

The complexity of wave propagation in cross–hole radar surveys can be effectively reduced by using first–arrival travel times. We present a DL–based framework that leverages this physical compression by establishing travel–time fingerprints—unique identifiers that encode subsurface structural information through differential arrival patterns.

This study employs the fast sweeping method (FSM) to estimate radar wave travel times efficiently. The FSM, developed initially to solve first–order nonlinear hyperbolic partial differential equations, has been widely applied in seismic wave modeling [21]. It enables the fast computation of first–arrival travel times without solving the full wave equation. The travel–time field $T(x,y)$ satisfies the Eikonal equation

$$|\nabla T(x,y)| = s(x,y) \tag{3}$$

where $s(x,y) = \frac{1}{v(x,y)}$ denotes the slowness field, and $v(x,y)$ is the local wave velocity.

To systematically validate the FSM, a synthetic 2D velocity model is presented in Figure 3a. The transmitter is located at (0 m, 5.4 m) along the left domain boundary, and 80–element receiver points are deployed along x = 9.0 m with vertical sampling from z = 1.1 m to 9.0 m (Δz = 0.1 m). FSM is subsequently used to compute the travel–time curve, which is then overlaid on the radar waveform in Figure 3b. The curve aligns well with the actual radar record, validating the accuracy and reliability of the FSM.

Therefore, the FSM was used to calculate the travel time of every transmitter–receiver pair. All the computed travel–time data were organized into a matrix, where each row corresponds to a transmitter and each column corresponds to a receiver. The element at position (*i,j*) represents the travel time from the *i*th transmitter to the *j*th receiver. This original travel–time matrix was then normalized to the range [1,2], resulting in the final travel–time fingerprint.

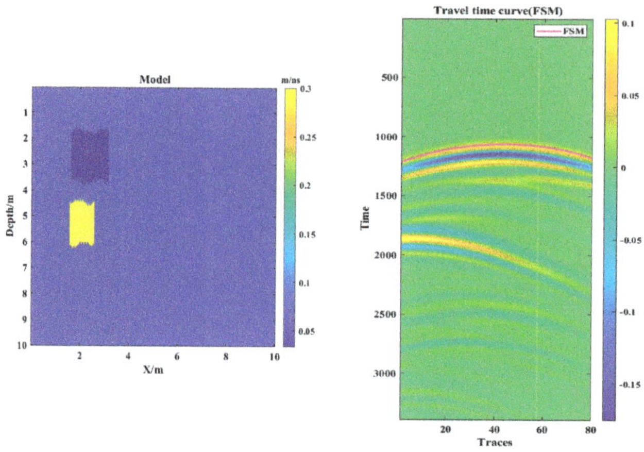

(a) (b)

Figure 3. The comparison of the GPR profile and the first–arrival travel–time calculate using FSM. (**a**) Sample velocity model; (**b**) FSM–calculated travel–time curve.

2.2.2. Generation of Travel–Time Fingerprint Datasets

Using the cavity models from Section 2.1.2 and the method described in Section 2.2.1, a total of 1500 travel–time fingerprints were generated. As illustrated in Figure 4a, the transmitting points are fixed on the left side of each model, with coordinates ranging from (1.0 m, 1.05 m) to (1.0 m, 9.0 m), spaced at intervals of 0.05 m. The receiving points are located on the right side of the model, spanning from (9.0 m, 1.05 m) to (9.0 m, 9.0 m), with a spacing of 0.05 m. This setup yields 160 transmitting points and 160 receiving points, forming 25,600 transmitter–receiver pairs for each model. The construction method of the two–dimensional travel–time fingerprint is illustrated in Figure 4b. As shown in Figure 4, taking the fourth transmitting point as an example, the first to the 160th receiving points receive signals in sequence and the corresponding travel–time data are calculated. Then, all the data are respectively filled into the fourth column of the matrix. The arrows in Figure 4 represent the data filling process.

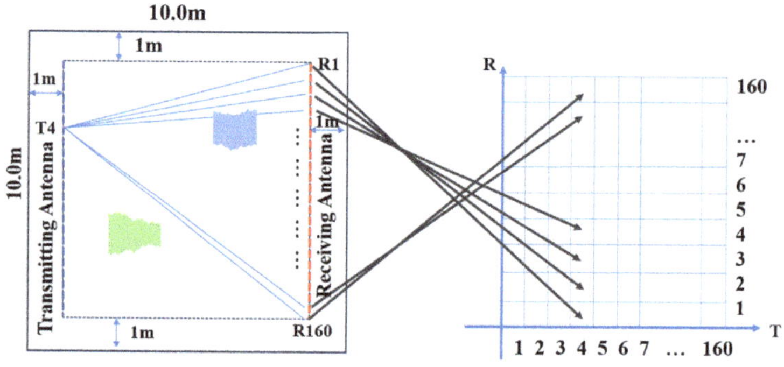

(a) (b)

Figure 4. The construction of the travel–time fingerprint dataset in the original domain. (**a**) An example of a model containing cavities and the setting of the receiving and transmitting antennas; (**b**) The construction of the corresponding travel–time fingerprint.

Figure 5 shows six representative travel–time fingerprints corresponding to the velocity models in Figure 2. The distinct spatial differences observed across these fingerprints reflect their sensitivity to the underlying cavity configurations, confirming their utility as input features for training DL models to identify and localize subsurface anomalies.

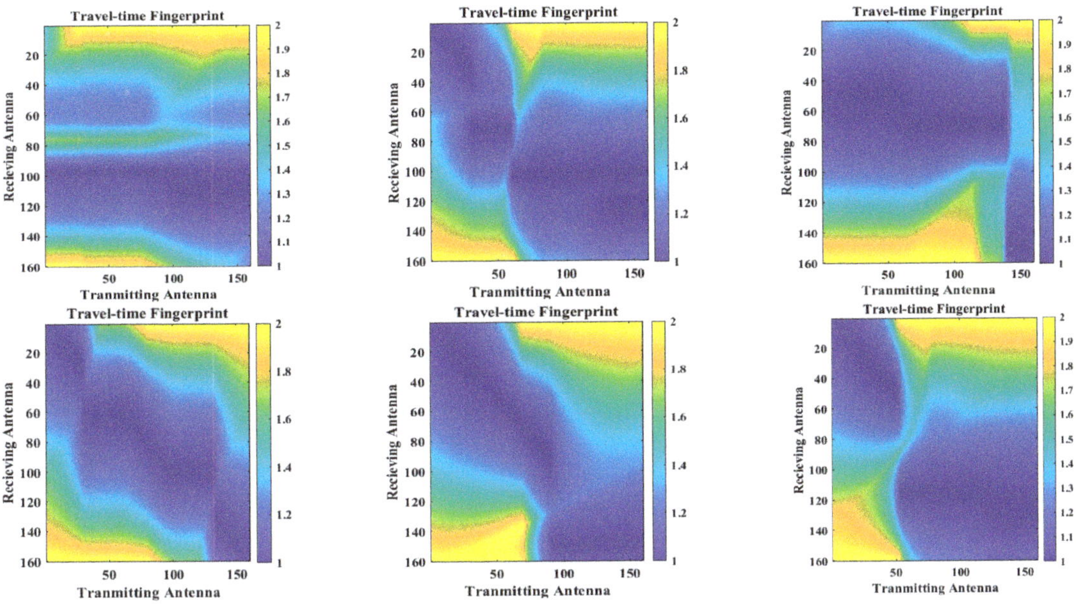

Figure 5. The corresponding travel–time fingerprints of the models.

2.3. Construction of the Pre–Trained UNET

To perform the inverse mapping from travel–time fingerprints to velocity models, a modified UNET architecture is employed, as illustrated in Figure 6. This architecture consists of an encoder, a decoder, and skip connections. The encoder includes five feature extraction blocks, with four blocks performing down–sampling operations and one utilizes a max pooling layer. A batch normalization layer was incorporated at each stage of the down–sampling process. In the decoder, four up–sampling blocks are followed by a 1×1 convolutional layer with a ReLU activation function. Skip connections were included to reduce information loss and contribute to the recovery of the cavity boundary during the encoding process. Lastly, the transformation of image size was achieved through linear interpolation. To mitigate the risk of overfitting, a dropout rate of 0.2 was set.

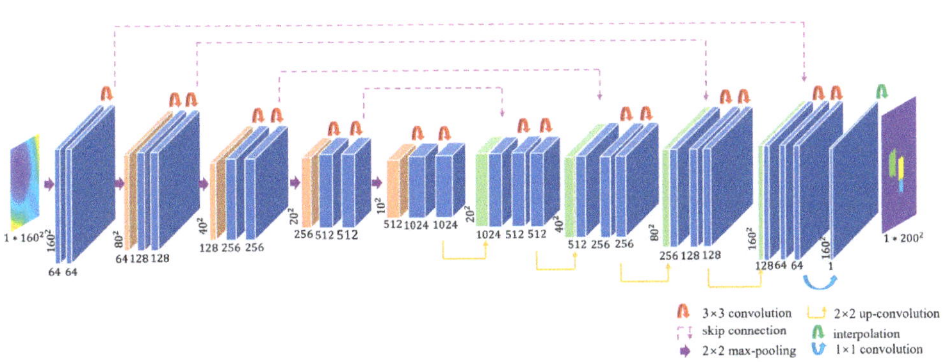

Figure 6. The constructed UNET structure of the pre–training.

The model was pre–trained using synthetic datasets derived from Sections 2.1.2 and 2.2.2. Of the entire dataset, 90% was allocated for training, while the remaining 10% was reserved for testing. The travel–time fingerprints serve as the input to the network, with the corresponding velocity models serving as the output. The primary training parameters for pre–training are outlined in Table 2.

Table 2. Parameter settings of the pre–trained UNET.

Task	Input Size	Learning Rate	Epochs	Batch	Gradient Descent Algorithm	Output Size
Pretraining	160 × 160	1.0×10^{-3}	600	16	Adam	200 × 200

Since the task is a regression problem, the mean squared error (*MSE*) loss function is employed to evaluate model performance:

$$MSE = \frac{1}{N}\sum_{i=1}^{N}(y_i - \hat{y}_i)^2 \quad (4)$$

where y_i is the ground truth value of sample i, $\hat{y}_i$ is the predicted value of sample i, and N is the total number of samples.

During the training process, the *MSE* loss demonstrates a clear downward trend as shown in Figure 7. Initially, there was rapid convergence in the early stages, followed by a gradual stabilization as the number of iterations increased. At the end of the training process, the training loss decreased significantly from 10.5208 to 0.1538, which is the minimal value.

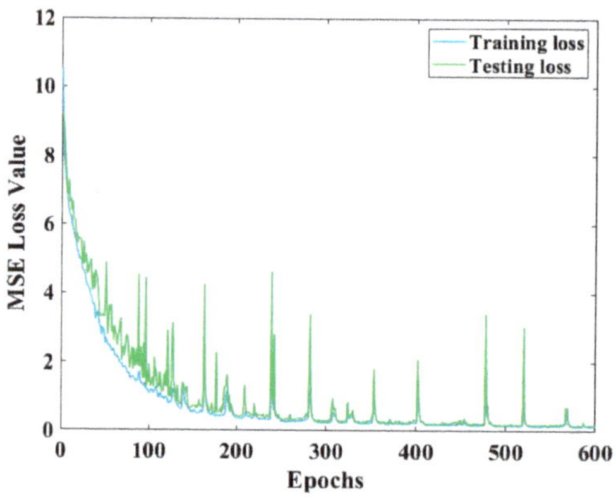

Figure 7. The MSE loss of the pre–training sequence.

To evaluate the model's performance, three random samples were selected from the test set. Each travel–time fingerprint was input into the pre–trained UNET, and the predicted velocity models derived from the UNET were compared to the ground truth. The results are shown in Figure 8.

In Figure 8, the predicted models (right column) exhibit strong consistency with the real models (middle column), effectively recovering the shape, location, and extent of the underground cavities, which verifies the effectiveness of the pre–trained UNET.

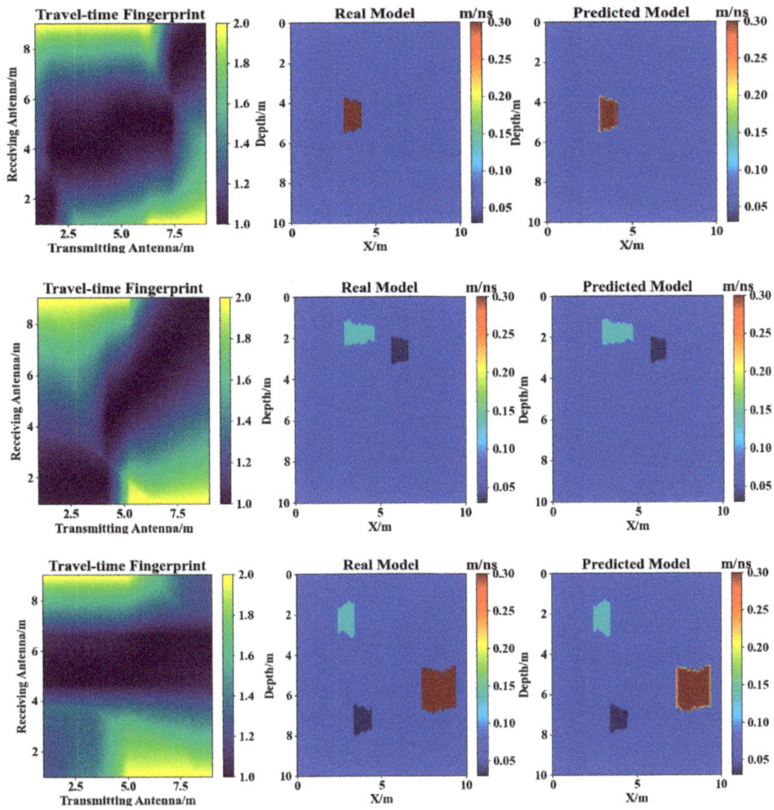

Figure 8. The imaging results of numerical models in the original domain. (**Left column**): input travel–time fingerprints; (**middle column**): true velocity models; (**right column**): predicted models).

3. Application and Discussion

Significant discrepancies frequently arose between synthetic training datasets and real–world field data, due to variations in wavelet characteristics, instrument responses, and complexities inherent in subsurface conditions. These divergences impede the direct application of models trained exclusively through simulation in practical engineering.

To improve the engineering applicability of the proposed method, a small set of field–inspired datasets was constructed to represent the target domain. A TL strategy was then employed, wherein the pre–trained UNET model was fine–tuned using the limited target domain data from the target domain, resulting in a retrained UNET. Finally, the effectiveness of this model was evaluated using actual field–scale synthetic datasets to validate its performance in more realistic geological conditions.

3.1. Transfer Learning and Retraining on Field–Scale Data

To validate the engineering applicability of TL, we developed an extensive velocity model dataset (62.25 m × 12.7 m, Figure 9) within the target domain. This model retains the irregular cavity generation methodology detailed in Section 2.1.1 and effectively simulates field–scale subsurface spatial structures under realistic geological conditions. This was achieved through strategic parameterization, with a relative dielectric constant of 9 assigned to the background medium.

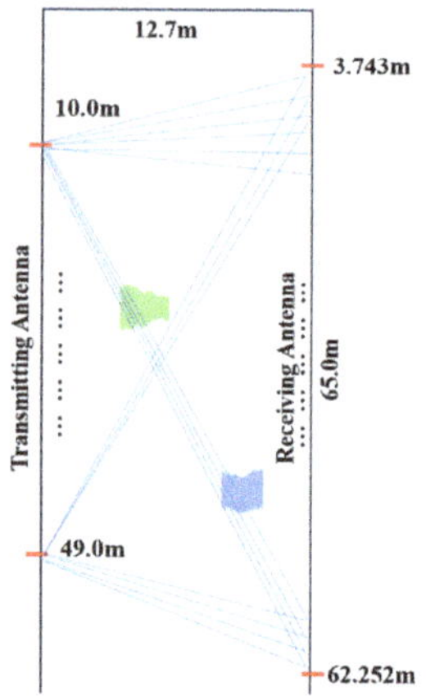

Figure 9. The transmitting and receiving antenna arrangement in the velocity model of the target domain.

The corresponding velocity distributions were derived using Equation (2) from Section 2.1.2, producing a set of field–scale velocity models. As illustrated in Figure 9, the antenna layout within the target domain is specified as follows: Transmitting antennas are positioned along the left boundary and are distributed between depths of 10 m to 49 m at 1 m intervals, yielding a total of 40 transmitters. Meanwhile, receiving antennas are fixed along the right boundary, spanning from 3.743 m to 62.252 m, with a spacing of 0.197 m, which corresponds to 298 receivers. Based on this antenna configuration, travel–time fingerprints were generated using the methodology outlined in Section 2.2.1, adapted for the scale of the new model.

Based on the target domain dataset, a TL strategy was applied to fine–tune the previously pre–trained UNET model, thereby enhancing the engineering applicability of the network. Initially, the pre–trained UNET model was loaded. Following that, specific layers were frozen, while the remaining layers were updated to fine–tune during TL process.

The training parameters are detailed in Table 3. The evaluation of the training process utilizes the MSE loss function. As illustrated in Figure 10, both training and testing losses demonstrate a clear downward trend, indicating successful adaptation. Notably, the training loss decreased from 6.941 to 0.4879, highlighting effective learning, even with a limited amount of new data.

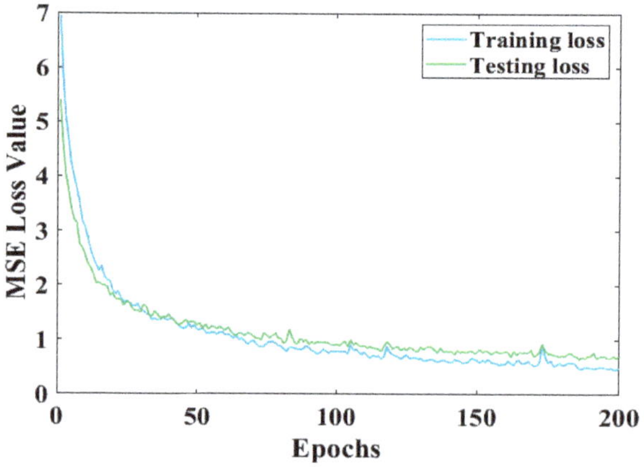

Figure 10. The *MSE* loss of the new UNET training in the target domain.

To assess the performance of the fine-tuned UNET in the target domain, three fingerprint–velocity model pairs were randomly selected from the test set. These fingerprints were fed into the retrained model, and the resulting predictions are shown in Figure 11.

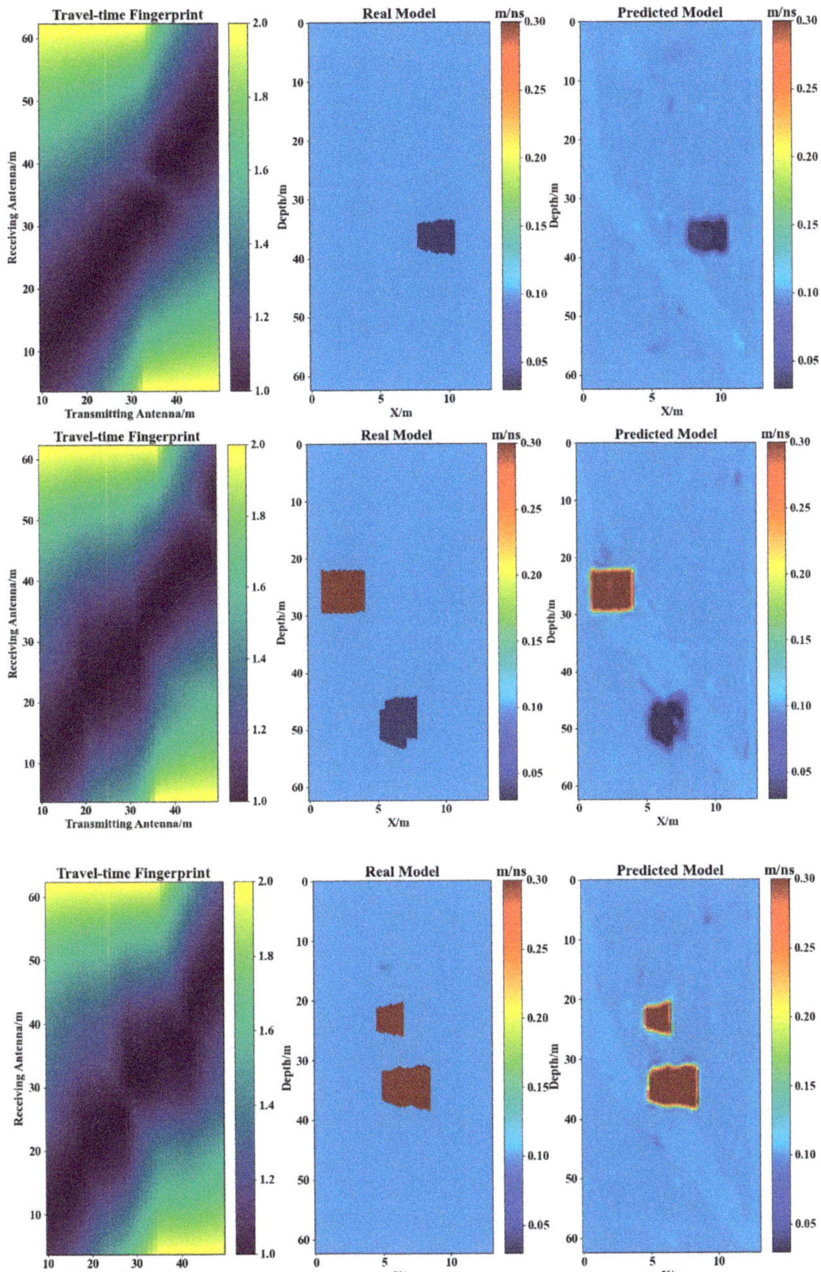

Figure 11. The imaging results of the numerical models in the target domain. (**Left column**): input travel–time fingerprints; (**middle column**): true velocity models; (**right column**): predicted models).

Table 3. The parameter settings of the new UNET in the target domain.

Task	Input Size	Learning Rate	Epochs	Batch	Gradient Descent Algorithm	Output Size
Retraining	298 × 40	1.0×10^{-3}	200	16	Adam	317 × 66

In Figure 11, the predicted models successfully reconstruct the shape, location, and spatial extent of underground cavities, showing strong agreement with the true models. This confirms that the fine–tuned UNET has effectively generalized to the target domain.

3.2. Engineering Application

To further validate the engineering applicability and practical effectiveness of the proposed method, it was applied to a real–world expressway construction project in Guizhou Province, China. The project area featured complex geological conditions, and multiple sets of cross–hole radar data had already been acquired.

This section selects a representative borehole dataset from the project for demonstration. The proposed deep learning–based travel–time imaging method was applied to the dataset to identify and visualize subsurface anomalies.

3.2.1. Project Background

The study area is situated near the Pingtang Special Bridge, a part of the Yuqing–Anlong Expressway (specifically, the Pingtang–Luodian section). Cross–hole radar data were collected using a RAMAC radar system with 100 MHz borehole antennas, manufactured by MALA Geoscience (Solna, Sweden).

Field–acquired radar data underwent standard preprocessing steps, including DC drift removal, static correction, AGC gain adjustment, and bandpass filtering. The calculation steps for processing the field data are shown in Figure 12.

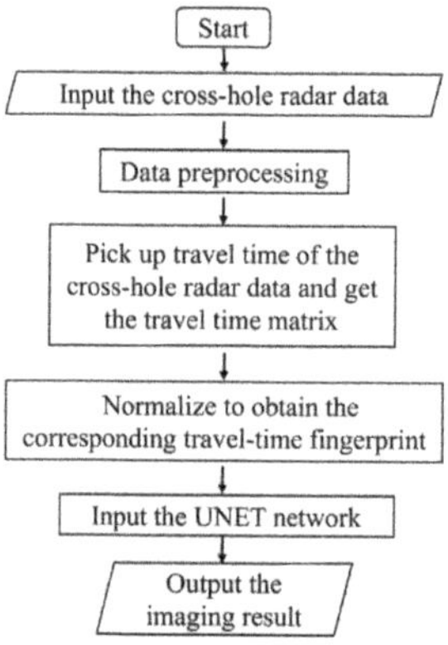

Figure 12. The data processing flow of the field data.

3.2.2. Case Study: Borehole Pair ZK113–ZK99

A total of 3830 ray paths were generated between borehole ZK113 (acting as the transmitter) and borehole ZK99 (serving as the receiver), with a borehole spacing of 12.7 m. The layout reflects the settings of the synthetic target domain settings: transmitting points comprise 40 positions, ranging from 10 m to 59 m in 1 m intervals in ZK113; the receiving points encompass 298 positions, spanning from 3.743 m to 62.252 m at 0.197 m intervals in ZK99.

Figure 13 shows the raw radar record for ZK113–ZK99, which has undergone preprocessing according to the flowchart illustrated in Figure 12. As shown in Figure 14a, the corresponding travel–time fingerprints for these field data were obtained using the data processing flow outlined in Figure 12. This travel–time fingerprint, shown in Figure 14a, is then input into the fine–tuned UNET model to produce the predicted model shown in Figure 14b. In Figure 14b, the anomalies highlighted in the red dotted areas represent results obtained through the proposed method, while the regions marked in yellow indicate cavity areas identified through geological interpretation via traditional methods. Two notable low–velocity anomalies—likely indicative of water–filled or clay–filled cavities—are clearly visible and emphasized in the red–dotted line areas: (1) spanning depth from 24 m to 39 m; horizontal extent from 2.5 m to 12.7 m from ZK113; (2) ranging from 42 m to 47 m in depth with a horizontal extent from 0 m to 8 m from ZK113.

Figure 13. The cross–hole radar record of ZK 113–ZK99.

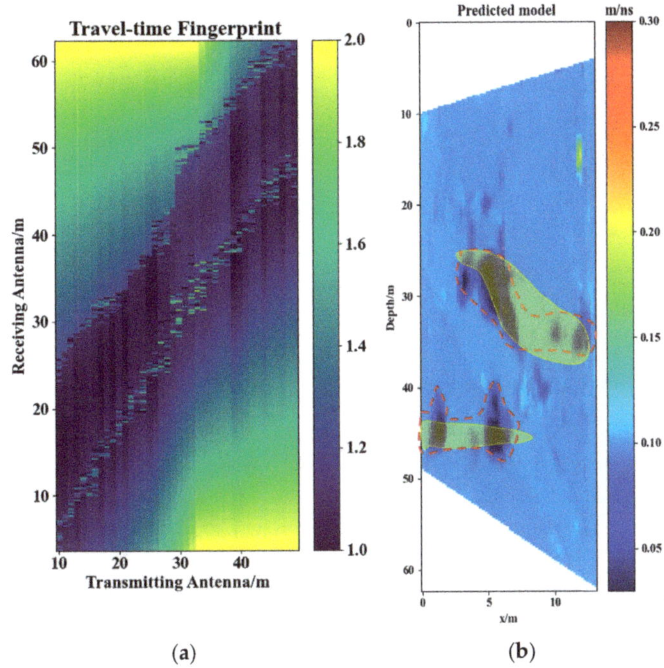

Figure 14. The prediction result of ZK113–ZK99 obtained by the proposed method. (**a**) The normalized travel–time fingerprint of ZK113–ZK99; (**b**) the model prediction result of the proposed method (red–dotted line areas: the result of the fine–tuned UNET; yellow areas: the result of the traditional tomography method).

Figure 15a presents the tomography imaging result obtained through the traditional tomography method. It reveals a weak velocity contrast and low spatial resolution, which make it difficult to clearly define the boundaries of cavities. In Figure 15b, the geological interpretation derived from the tomography results is displayed. The abnormal cavities indicated in this figure are also highlighted in yellow in Figure 14b, enabling the comparison of the results from both methods. According to the geophysical report of the CT result of this project: (1) In the section of the hole ZK113 ranging from a depth of 26.3 m to 39.5 m, the EM waves at a horizontal distance area of 3.2 m to 12.7 m from ZK113 to ZK99 exhibit characteristics of high absorption and low wave velocity. This phenomenon is speculated to indicate a semi–filled cavity, which contains water and a small amount of crushed stone soil at the bottom. (2) In the section of hole ZK113 ranging from a depth of 47.3 m to 46.9 m, the EM waves within a horizontal distance of 0.0 m to 8.0 m from ZK113 to ZK99 show high absorption and low wave velocity [22]. The results and comparative analysis of Figures 14 and 15 indicate a strong consistency in the cavity positions identified by both methods.

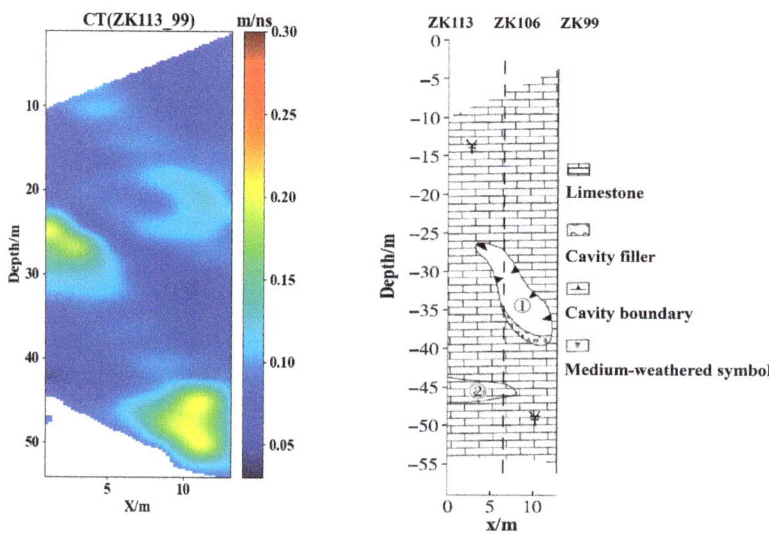

(a) (b)

Figure 15. The tomography result and geological interpretation of drilling at ZK113–ZK99. (**a**) The tomography result; (**b**) geological interpretation.

Figure 15b presents the geological interpretation and data from a verification borehole ZK106, situated 8 m along the alignment from ZK113 toward ZK99. The verification process revealed the presence of cavities at depths ranging from 24.5 m to 32.6 m and from 41.8 m to 43.6 m, which aligns well with the abnormal locations shown in Figure 14b. The results from borehole ZK106 confirm the existence of two underground cavities between borehole pairs ZK113 and ZK99, substantiating the effectiveness of the proposed method.

The cavity prediction result in Figure 14b was compared and analyzed with the tomography imaging result in Figure 15. In Figure 14b, the distribution and shape of underground low–speed cavities are clearly visible. Although the tomography result in Figure 15a also identify the anomaly zones, they lack the intuitive visual clarity found in Figure 14b, making geological interpretation challenging. In contrast, the proposed method offers enhanced visualization of the cavity shapes and extents, thereby facilitating a more profound understanding and informed decision–making in engineering.

In conclusion, the field–scale application of the proposed method demonstrates several key advantages: (1) accuracy—the imaging results closely align with traditional interpretations and physical borehole validation, verifying the method's reliability; (2) visual clarity—the predicted models deliver high–resolution, interpretable outputs, simplifying the detection and delineation of subsurface anomalies for engineers; and (3) efficiency—following the completion of transfer learning, the inference speed for processing field data is significantly increased, providing considerable benefits to practical engineering workflows.

4. Conclusions

This study introduces an innovative DL–based subsurface cavity imaging method using cross–hole radar. The approach features a unique framework comprising travel–time fingerprints and a TL–enhanced UNET architecture. The travel–time fingerprint construction method converts cross–hole radar measurements into structured inputs that capture essential spatial propagation features, establishing the foundation for accurate underground imaging.

A pre–trained UNET model has been developed to establish the relationship between travel–time fingerprints and subsurface velocity structures. To address domain shifts between simulated and field data, transfer learning was employed to fine–tune the model using limited field–scale travel–time fingerprints, significantly improving its adaptability and accuracy in complex geological environments.

Numerical simulations and field applications, including field–scale case studies from a highway construction site in Guizhou Province, demonstrate the method's precision, robustness, and practical feasibility. In comparison to conventional inversion techniques, the proposed approach exhibits enhanced interpretability and computational efficiency, facilitating the effective detection and delineation of deep subsurface cavities.

The methodology presented in this paper has undergone preliminary verification through numerical experiments and engineering applications; however, its practical implementation requires careful consideration. Furthermore, its applicability in various engineering scenarios requires additional testing and optimization. In practical engineering surveys, subsurface structures frequently demonstrate significant complexity with heterogeneous media distributions, which places heightened demands on the adaptability and stability of detection methods. Although the proposed approach has demonstrated impressive imaging performance under typical geological conditions, systematic validation across multiple practical engineering projects remains essential for broader applications. By accumulating field data from diverse geological settings (e.g., karst development zones, fracture zones, and multi–layer aquifers), we can optimize algorithm parameters further and refine network architecture design. These enhancements will improve the method's adaptability and stability in various complex detection scenarios, ultimately providing more reliable technical support for engineering geological investigations.

Author Contributions: Conceptualization, H.C. and Y.Z.; methodology, H.C.; validation, H.C.; data curation, H.C. and K.F.; writing—original draft preparation, H.C.; writing—review and editing, H.C. and Y.Z.; funding acquisition, Y.Z. All authors have read and agreed to the published version of the manuscript.

Funding: This work was supported by the National Key R&D Program of China under Grant 2023YFC3009300.

Data Availability Statement: Data will be made available on request.

Conflicts of Interest: Author Kunwei Feng was employed by the company Guizhou Transportation Planning Survey & Design Academe Co., Ltd. The remaining authors declare that the research was conducted in the absence of any commercial or financial relationships that could be construed as a potential conflict of interest.

References

1. Hanafy, S.; al Hagrey, S.A. Ground-penetrating radar tomography for soil-moisture heterogeneity. *Geophysics* **2006**, *71*, K9–K18. [CrossRef]
2. Williamson, P.R. A guide to the limits of resolution imposed by scattering in ray tomography. *Geophysics* **1991**, *56*, 202–207. [CrossRef]
3. Ernst, J.R.; Green, A.G.; Maurer, H.; Holliger, K. Application of a new 2D time-domain full-waveform inversion scheme to crosshole radar data. *Geophysics* **2007**, *72*, J53–J64. [CrossRef]
4. Kuroda, S.; Takeuchi, M.; Kim, H.J. Full-waveform inversion algorithm for interpreting crosshole radar data: A theoretical approach. *Geosci. J.* **2007**, *11*, 211–217. [CrossRef]
5. Zhu, J.-Y.; Park, T.; Isola, P.; Efros, A.A. Unpaired Image-to-Image Translation Using Cycle-Consistent Adversarial Networks. In Proceedings of the 2017 IEEE International Conference on Computer Vision (ICCV), Venice, Italy, 22–29 October 2017; pp. 2242–2251.
6. Liu, B.; Ren, Y.; Liu, H.; Xu, H.; Wang, Z.; Cohn, A.G.; Jiang, P. GPRInvNet: Deep Learning-Based Ground Penetrating Radar Data Inversion for Tunnel Lining. *IEEE Trans. Geosci. Remote Sens.* **2021**, *59*, 8305–8325. [CrossRef]
7. Ji, Y.; Zhang, F.; Wang, J.; Wang, Z.; Jiang, P.; Liu, H.; Sui, Q. Deep Neural Network-Based Permittivity Inversions for Ground Penetrating Radar Data. *IEEE Sens. J.* **2021**, *21*, 8172–8183. [CrossRef]
8. Dinh, K.; Gucunski, N.; Duong, T.H. An algorithm for automatic localization and detection of rebars from GPR data of concrete bridge decks. *Autom. Constr.* **2018**, *89*, 292–298. [CrossRef]
9. Lei, W.; Hou, F.; Xi, J.; Tan, Q.; Xu, M.; Jiang, X.; Liu, G.; Gu, Q. Automatic hyperbola detection and fitting in GPR B-scan image. *Autom. Constr.* **2019**, *106*, 102839. [CrossRef]
10. Zhang, D.; Wang, Z.; Qin, H.; Geng, T.; Pan, S. GAN-Based Inversion of Crosshole GPR Data to Characterize Subsurface Structures. *Remote Sens.* **2023**, *15*, 3650. [CrossRef]
11. Pongrac, B.; Gleich, D.; Malajner, M.; Sarjaš, A. Cross-Hole GPR for Soil Moisture Estimation Using Deep Learning. *Remote Sens.* **2023**, *15*, 2397. [CrossRef]
12. Liu, X.; Liu, S.; Tian, S.; Zhao, Q.; Lu, Q.; Wang, K. Slowness High-Resolution Tomography of Cross-Hole Radar Based on Deep Learning. *IEEE Geosci. Remote Sens. Lett.* **2024**, *21*, 3000505. [CrossRef]
13. Warren, C.; Giannopoulos, A.; Giannakis, I. gprMax: Open source software to simulate electromagnetic wave propagation for Ground Penetrating Radar. *Comput. Phys. Commun.* **2016**, *209*, 163–170. [CrossRef]
14. Tong, Z.; Gao, J.; Yuan, D. Advances of deep learning applications in ground-penetrating radar: A survey. *Constr. Build. Mater.* **2020**, *258*, 120371. [CrossRef]
15. Ronneberger, O.; Fischer, P.; Brox, T. U-Net: Convolutional Networks for Biomedical Image Segmentation. In *Medical Image Computing and Computer-Assisted Intervention—MICCAI 2015, Proceedings of the Medical Image Computing and Computer-Assisted Intervention, Munich, Germany, 5–9 October 2015*; Springer: Cham, Switzerland, 2015.
16. Berezsky, O.; Pitsun, O.; Derysh, B.; Pazdriy, I.; Melnyk, G.; Batko, Y. Automatic Segmentation of Immunohistochemical Images Based on U-net Architecture. In Proceedings of the 2021 IEEE 16th International Conference on Computer Sciences and Information Technologies (CSIT), Lviv, Ukraine, 22–25 September 2021; pp. 29–32.
17. Shang, K.; Zhang, F.; Song, A.; Ling, J.; Xiao, J.; Zhang, Z.; Qian, R. Fast Segmentation and Dynamic Monitoring of Time-Lapse 3D GPR Data Based on U-Net. *Remote Sens.* **2022**, *14*, 4190. [CrossRef]
18. Wang, S.; Gong, X.; Han, L. Monte Carlo Full-Waveform Inversion of Cross-Hole Ground-Penetrating Radar Data Based on Improved Residual Network. *Remote Sens.* **2024**, *16*, 243. [CrossRef]
19. Pan, S.J.; Yang, Q. A Survey on Transfer Learning. *IEEE Trans. Knowl. Data Eng.* **2010**, *22*, 1345–1359. [CrossRef]
20. Hunt, B.J. Maxwell, Measurement, and the Modes of Electromagnetic Theory. *Hist. Stud. Nat. Sci.* **2015**, *45*, 303–339. [CrossRef]
21. Zhao, H. A fast sweeping method for eikonal equations. *Math. Comput.* **2004**, *74*, 603–627. [CrossRef]
22. Liao, S.; Feng, K.; Li, F. Application of Cross-hole Electromagnetic Wave Tomography Detection Technology in Karst Exploration of Bridge Main Pier Foundation. *Technol. Highw. Transp.* **2020**, *36*, 49–54.

remote sensing

MDPI

Article

OEM-HWNet: A Prior Knowledge-Guided Network for Pavement Interlayer Distress Detection Based on Computer Vision Using GPR

Congde Lu [1], Senguo Cao [1,*], Xiao Wang [1], Guanglai Jin [2], Siqi Wang [3] and Wenlong Cai [2]

1 School of Mechanical and Electrical Engineering, Chengdu University of Technology, Chengdu 610059, China; cdlu@cdut.edu.cn (C.L.); 2023021194@stu.cdut.edu.cn (X.W.)
2 Jiangsu Sinoroad Engineering Technology Research Institute Co., Ltd., Nanjing 211800, China; jgl@sinoroad.com (G.J.); cwl@sinoroad.com (W.C.)
3 School of Transportation, Southeast University, Nanjing 211189, China; siqiwang@seu.edu.cn
* Correspondence: 2023051285@stu.cdut.edu.cn; Tel.: +86-028-84076139

Abstract: Accurate detection of interlayer distress based on ground-penetrating radar has been widely adopted for in-service asphalt pavement condition assessment to improve maintenance efficiency and reduce costs. However, accurate interlayer distress locating is challenging with limited adaptability to their large-scale variations, which significantly weakens the detection performance. This study proposed a novel automatic detection network based on YOLOv5s to detect interlayer distresses in asphalt pavement named OEM-HWNet. Firstly, an object enhancement module based on prior knowledge was designed to locate the regions of interlayer distress and enhance their characteristics. Then, wavelet convolution was added to increase the receptive field of the network and enhance the ability to capture low-frequency information. Finally, an additional detection head was added to improve the detection capability of interlayer distress with different sizes. Experiments demonstrated that the proposed network achieves a mean average precision (mAP) of 89.6%, outperforming other advanced models, such as YOLOv5s, YOLOv8s, YOLOv11s, and Faster R-CNN. Incorporating prior knowledge into deep learning networks could provide an effective solution to detect interlayer distress of asphalt pavement.

Keywords: interlayer distress; ground penetrating radar; object detection; prior knowledge; Canny operator

1. Introduction

Due to the influence of load, temperature, water, and other factors, various hidden distresses will occur during the operation of roads, which seriously affect their service life, such as interlayer distresses, concealed cracks, water-rich voids, loosening, etc. [1]. Therefore, efficient and accurate detection of hidden distress has become one of the focuses of road maintenance. Since ground penetrating radar (GPR) has advantages of high accuracy, high efficiency, low cost, and non-destructive testing, it has been widely used in hidden distress detection of roads [2,3]. Although GPR data can reflect the internal structure distribution of roads, it requires complex processing and analysis. At present, manual processing and interpretation are mainly used in practical projects, which are highly subjective and severely limit the efficiency of distress detection. Thus, to achieve efficient detection of road hidden distresses, the automation of GPR data analysis is crucial.

Many automated detection methods of hidden distress were presented from single-channel waveform (A-scan) [4–6], two-dimensional profile (B-scan) [7–9], and three-dimensional volume (C-scan) [10–14], respectively, according to the type of GPR data. A-scan signals represent only local information and are susceptible to noise, while C-scan includes a huge amount of data, requiring a powerful computing platform for analysis. In addition, the annotation of three-dimensional data is particularly difficult. As a main means, the automatic analysis methods based on B-scan images have been extensively applied in detecting hidden distress of roads [15], such as interlayer distresses, cracks, voids, and loosening. These hidden distresses may be classified into hyperbolic [16–20] (e.g., cracks and voids) and non-hyperbolic [21–23] (e.g., interlayer distresses and loosening). The above work has achieved good results, but most of them focus on hyperbolic distresses in the GPR B-Scan images, while the research on interlayer distresses is scarce.

In fact, as one of the early hidden distresses of pavements [24], interlayer distresses have been a priority for pavement maintenance [25]. In particular, most of the hidden distresses detected in Jiangsu province of China are interlayer distresses [26]. Therefore, it is necessary to detect interlayer distresses in a timely manner to prevent further development of the distresses. In addition, there are also differences in interlayer distress, and this difference will affect pavement maintenance options and costs. So, it needs not only to accurately detect interlayer distresses but also to further classify interlayer distresses as interlayer debonding and interlayer water seepage. For interlayer debonding, the maintenance program only needs to consider grouting, while both underground grouting and drainage system repair are carried out for interlayer water seepage as a more serious interlayer distress. Recently, there have been some works on classifying interlayer distresses in asphalt pavement [26,27], which is insufficient for practical engineering. Detection of interlayer distress requires not only classification but also precise localization. However, accurately identifying and locating different types of interlayer distress often faces the following difficulties. Firstly, as shown in Figure 1, the shape and texture of interlayer debonding and interlayer water seepage are very similar in GPR images, and these two types of interlayer distresses may also be similar to other distresses. Secondly, the size of interlayer distresses varies greatly, which demands that the network adapts to the various sizes, and the existing object detection methods are insufficient to deal with this problem. Thirdly, many object detection methods are limited in global feature extraction and low-frequency information capture.

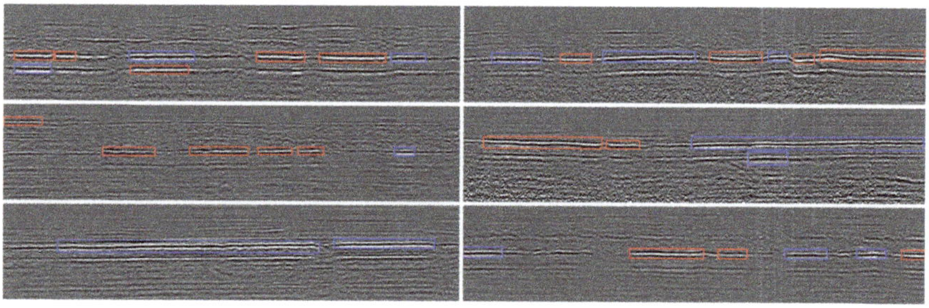

Figure 1. Examples of interlayer distress, where the red box indicates interlayer debonding and the blue indicates interlayer water seepage. Most of the two distresses are elongated, and the shapes and textures are similar.

Existing methods for identifying subsurface road distress based on GPR B-scan images can be broadly categorized into the rule-based recognition algorithm, machine learning algorithm, and deep learning algorithm [15]. For rule-based recognition algorithms, Bugarinovic et al. introduced the Canny edge detection operator to remove unnecessary edge pixels and horizontal reflections of the road surface and soil layer to improve the detection speed and positioning robustness of the hyperbola [28]. Similarly, Harkat and Benani used a modified Hough transform method to detect hyperbolic features caused by buried cavities [29]. Although rule-based algorithms employ expert knowledge to extract features from GPR images, these empirically derived features often fail to represent the intrinsic attributes for identifying interlayer distress. Building on rule-based algorithms, machine learning algorithms establish the relationship between features extracted and hidden distresses by training classifiers such as support vector machines (SVM) [30] and K-nearest Neighbors (KNN) [31]. Also, Sun et al. [32] utilized histograms of oriented gradients (HOG) to extract crack features, followed by an SVM classifier to identify cracks. Although machine learning algorithms offer multi-dimensional feature representation to significantly improve detection accuracy and applicability compared to rule-based models [33], they fail to recognize the interlayer distress with different sizes.

In recent years, the rapid advancement of deep learning has led to its extensive application in the field of object detection [34]. These object detection methods based on deep learning may be divided into two main categories. One is two-stage methods, which rely on candidate bounding boxes, such as Faster R-CNN [35] and Cascade R-CNN [36]. The other is one-stage methods based on regression, including the YOLO series [37,38], SSD [39], DETR [40], and RetinaNet [41]. While two-stage methods often achieve higher accuracy, one-stage methods are favored for their real-time detection capabilities. Furtherly, the YOLO series, as a mainstream one-stage model, has been extensively applied in detecting various distresses in GPR B-scan images [21–23]. However, these methods often cannot provide accurate location information for interlayer distress and demonstrate limited adaptability to their large-scale variations.

In general, the structure of the network needs to be modified for specific scenarios. For example, many researchers have modified the structure of the YOLO series network to improve the detection accuracy of their own datasets [42–46]. These improvement strategies have yielded positive results. However, they all ignore the benefits of prior knowledge. In fact, the introduction of prior knowledge in the design of deep networks or learning strategies can improve detection accuracy [47], as they render a knowledge-infused algorithm more suitable to the specific scenarios, lowering the training dataset's requirement. Moreover, it also enhances the interpretability of the network. Until now, some scholars have still been applying prior knowledge to GPR image recognition [48,49]. In this paper, shape and texture features as prior knowledge are of importance to identify interlayer distresses of asphalt pavement (see Figure 1), but the effective incorporation of specific prior knowledge to guide network architecture design remains a non-trivial challenge.

In this study, an improved YOLOv5s model was proposed to achieve interlayer distress detection of asphalt pavement with GPR images, named OEM-HWNet. In the method, shape and texture features of interlayer distress are used as prior knowledge to enhance object regions during the network learning. Firstly, by considering the requirements of real-time detection with few model parameters and high accuracy, YOLOv5s was selected as the baseline model [50]. Secondly, an object enhancement module (OEM) based on the Canny operator and morphological dilation was designed to enhance the representational ability of interlayer distresses. Thirdly, WTConv (the convolution in the wavelet domain) [51] was introduced to increase the capturing ability of low-frequency information and the receptive field. Finally, an additional detection head is incorporated to improve the detection performance of large objects. The OEM-HWNet effectively merges prior knowledge and automatic features. Figure 2 illustrates the workflow of interlayer distress detection. The primary contributions of this paper are as follows:

(1) This paper proposes a prior knowledge-guided network for interlayer distress detection based on GPR images, named OEM-HWNet, and demonstrates its effectiveness in comparison with the state-of-the-art algorithms such as Faster R-CNN, SSD, Reti-

naNet, RT-DETR [52], YOLOv3 [38], YOLOv5s, YOLOv7 [37], YOLOv8s [53], and YOLOv11s [54];

(2) The OEM, based on prior knowledge, is designed and integrated into the YOLOv5s model, which fully leverages edge position information to improve localization and feature representation abilities of the model. This module not only improves the performance of the model but also increases its interpretability of the model;

(3) WTConv is introduced into the YOLOv5s model. This approach effectively mitigates the limitations of a convolutional neural network (CNN) in global feature extraction and enhances the model's ability to capture low-frequency features;

(4) The number of detection heads for YOLOv5s was extended from three (small, medium, and large) to four (small, medium, large, and huge), which effectively addresses the issue of big variations in sizes of the interlayer distresses and enhances the detection ability of large objects.

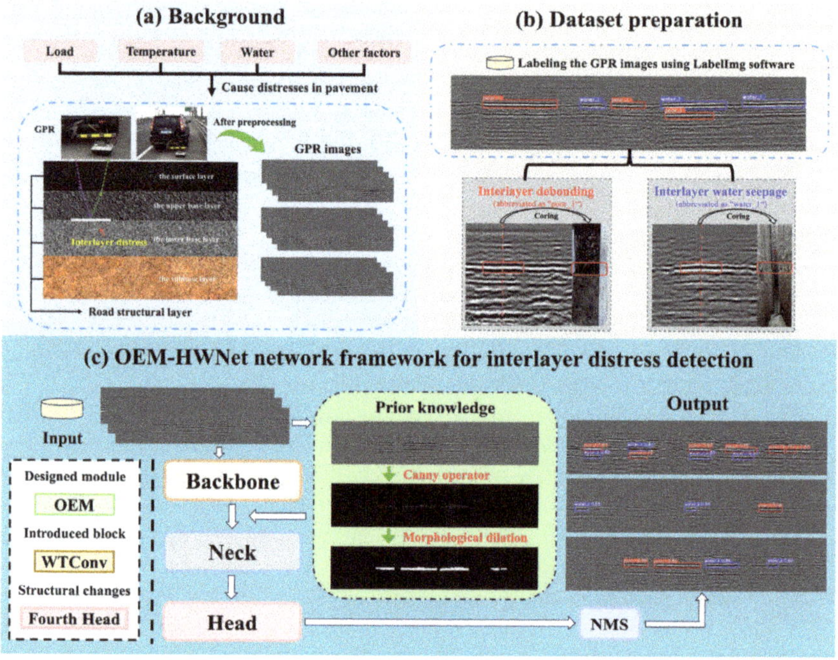

Figure 2. The workflow for interlayer distress detection of asphalt pavement. Includes three parts: (**a**) the causes of distress and the acquisition of GPR images; (**b**) GPR images annotation (including interlayer debonding and interlayer water seepages); (**c**) a prior knowledge-guided OEM-HWNet network for interlayer distress detection, where the white box is the improved methods and the green box is the prior knowledge generated by the Canny operator and morphological dilation.

The remainder of this paper is described as follows. Section 2 introduces the proposed OEM-HWNet model and the details of module design. Section 3 describes dataset construction and the experimental details. Section 4 analyzes the experimental results. Section 5 summarizes the conclusions of the study.

2. Methodology

To clearly describe the detection model developed in this paper, this section firstly introduces the OEM-HWNet model and its main components. Then, the design of the

OEM modules based on prior knowledge is described in detail. Finally, the C3WC module (WTConv embedded in the C3 module) is elaborated.

2.1. Overview of the OEM-HWNet Architecture

YOLOv5 is selected as our baseline model since it has fewer model parameters compared with the latest YOLOv11 [54] and high accuracy. As shown in Figure 3, OEM-HWNet is primarily divided into three parts: the backbone, neck, and head parts.

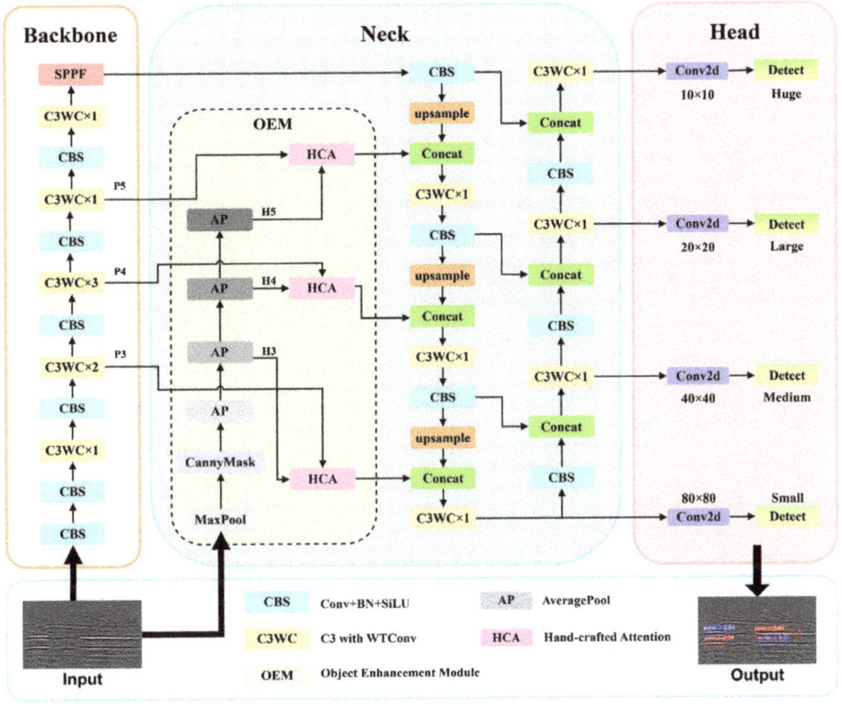

Figure 3. Overall framework of OEM-HWNet. The framework comprises three main parts: the backbone, neck, and head, for feature extraction, fusion, and detection, respectively.

First, the backbone part comprises the CBS module, C3WC module, and Spatial Pyramid Pooling Fast (SPPF) module. The CBS module includes a standard convolutional layer (Conv2d), a batch normalization (BN) layer, and the SiLU activation function. The original C3 module (three convolutional layers and several Bottleneck modules) is replaced by the C3WC module, integrating the WTConv into the Bottleneck to obtain a larger receptive field and improve the ability to capture low-frequency features. The SPPF module is an optimized version of spatial pyramid pooling (SPP) [55]. Second, the neck part combines Feature Pyramid Networks (FPN) [56] and Path Aggregation Network (PANet) [57], which realize the fusion of semantic information from low level to high level. Notably, a specially designed module called OEM is inserted between the neck and the backbone part. This module focuses on learning of the interlayer distresses and inhibits interference, which significantly improves the learning ability and robustness of the model. Finally, the original YOLOv5s model includes three detection heads with sizes of 20×20, 40×40, and 80×80, respectively. However, these detection heads cannot fully meet the needs of distress detection in the case of interlayer distress with huge scale changes in size. To correct this, we added an additional detection head with a size of 10×10 to improve the model's ability to detect large objects.

Loss function of the OEM-HWNet model consists of classification loss ($Loss_{cls}$), object confidence loss ($Loss_{obj}$), and box localization loss ($Loss_{box}$), which are expressed as follows:

$$Loss = Loss_{box} + Loss_{obj} + Loss_{cls} \tag{1}$$

The $Loss_{box}$ denotes the deviation between the truth box and predicted box. Here, the complete-IoU (CIoU) loss [58] is adopted in the localization of the bounding box, which is an improved object detection loss function. So, the $Loss_{box}$ is expressed as follows:

$$Loss_{box} = Loss_{CIOU} = 1 - IOU + \frac{\rho^2(b, b^{gt})}{c^2} + \alpha v \tag{2}$$

$$IOU = \left| \frac{B \cap B^{gt}}{B \cup B^{gt}} \right| \tag{3}$$

where B and B^{gt} denote the predicted and ground-truth box, respectively; b and b^{gt} denote the central points of B and B^{gt}, $\rho(\cdot)$ is the Euclidean distance; and c is the diagonal length of the smallest enclosing box covering the two boxes. The α is a positive trade-off parameter and v measures the consistency of the aspect ratio, as follow:

$$\alpha = \frac{v}{(1 - IOU) + v} \tag{4}$$

$$v = \frac{4}{\pi^2}(arctan\frac{w^{gt}}{h^{gt}} - arctan\frac{w}{h})^2 \tag{5}$$

where w and w^{gt} represent the widths of predicted box and ground-truth box, respectively, while h and h^{gt} correspond to their heights. Binary cross-entropy loss [41] is employed to calculate the object confidence ($Loss_{obj}$) and classification loss ($Loss_{cls}$). The former considers both positive and negative samples, while the latter only uses positive ones. The $Loss_{obj}$ and $Loss_{cls}$ can be expressed as follows:

$$\begin{aligned} Loss_{obj} &= \lambda_{obj} \sum_{i=0}^{N^2} \sum_{j=0}^{M} I_{ij}^{obj}[-\hat{C}_i lnC_i - (1 - \hat{C}_i)ln(1 - C_i)] \\ &+ \lambda_{nobj} \sum_{i=0}^{N^2} \sum_{j=0}^{M} I_{ij}^{nobj}[-\hat{C}_i lnC_i - (1 - \hat{C}_i)ln(1 - C_i)] \end{aligned} \tag{6}$$

$$Loss_{cls} = \sum_{i=0}^{N^2} \sum_{j=0}^{M} \sum_{c \subset cls} I_{ij}^{obj}[-\hat{p}_i(c)ln(p_i(c)) - (1 - \hat{p}_i(c))ln(1 - p_i(c))] \tag{7}$$

where N^2 stands for the number of grid cells; M represents the number of predicted bounding boxes per grid cell; λ_{obj} and λ_{nobj} represent the weights of the grid cell with or without the objects, respectively; I_{ij}^{obj} and I_{ij}^{nobj} denote whether there is an object to be predicted in the j-th anchor box of the i-th grid cell, respectively; $\hat{C}_i$ and C_i are the confidence of the predicted and truth values, respectively; and $\hat{p}_i(c)$, $p_i(c)$ are the predicted and truth probabilities of detecting an object belonging to interlayer distress type c in the i-th grid cell, respectively.

2.2. Prior Knowledge

2.2.1. Physical Basis

Figure 4a presents a schematic diagram of the GPR operation, where the antenna pair transmits an electromagnetic wave and receives the echo from the medium. In fact, the GPR obeying the wave equation [59] as

$$\nabla^2 E + \mu\sigma\frac{\partial E}{\partial t} + \mu\varepsilon\frac{\partial^2 E}{\partial t^2} = 0 \tag{8}$$

where E is electric field strength vector, μ is magnetic permeability, σ is conductivity, ε is permittivity, and t is time.

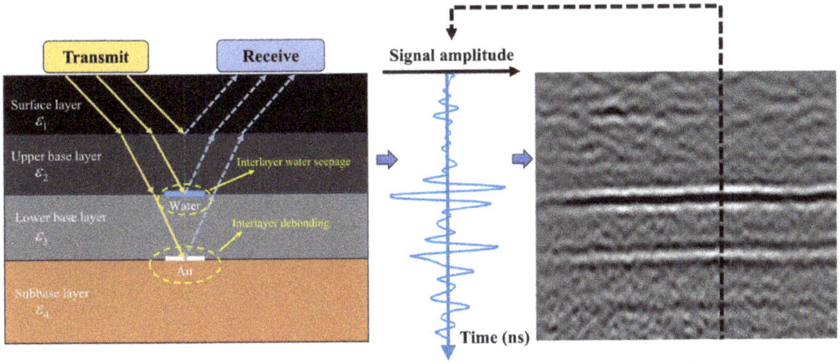

Figure 4. The physical background of GPR. (**a**) The working principle of GPR, where the ε indicates permittivity of the medium. (**b**) A-scan signals recorded by the receiving antenna. (**c**) The collection of A-scan signals along the driving direction forms the GPR B-scan images.

Strong reflections occur when the electromagnetic wave propagates to the interlayer distresses (Figure 4b). In general, the reflection intensity is determined by the difference in the permittivity of the two mediums, namely

$$R = \frac{\sqrt{\varepsilon_2} - \sqrt{\varepsilon_1}}{\sqrt{\varepsilon_2} + \sqrt{\varepsilon_1}} \tag{9}$$

where ε_1 and ε_2 are the permittivity of mediums in two adjacent layers. The greater the difference in permittivity, the stronger the reflection. When interlayer distress occurs, the permittivity of the distress is obviously different from that of the surrounding medium, so there is a strong reflected signal, as GPR B-scan images shown in Figure 4c. This principle is critical for non-destructive testing of interlayer distresses in asphalt pavement [60]. For example, interlayer debonding typically manifests as air filling between pavement layers, and the permittivity of air is 1, which is significantly different from the surrounding medium. This will result in a strong reflection, which is manifested in GPR images as the evidence for interlayer distress detection. In addition, higher-frequency antennas provide higher vertical resolution but limited penetration depth, whereas lower frequencies achieve deeper penetration at the cost of reduced resolution [59]. High-frequency antennas are essential to identify subtle changes in permittivity, while lower frequencies may fail to detect early-stage debonding. For the detection of interlayer distresses, the antenna frequency of GPR is typically set within the range of 700 MHz to 1000 MHz.

2.2.2. Object Region Segmentation

The shape and texture features are of importance to identify interlayer distresses in asphalt pavement, so these features extracted from GPR images as prior knowledge are used to segment interlayer distresses from the background. The Canny edge detection algorithm [61] utilizes dual thresholds (high and low thresholds) for object edge extraction and if any part of a contour is above the high threshold, those points are immediately output, as is the entire connected segment of contour, which contains the points and which lies above the low threshold. It has been proved to be effective in extracting the object edges [28]; however, these edges are one continuous line and cannot represent the object region. To separate the object region from the background, morphological dilation [62] is introduced as a complement to the Canny operator to connect neighboring lines to form the object region.

Specifically, the close distance between the upper and lower edges of the object region provides a clue for separating the object area from the background. The CannyMask module first uses the Canny operator to generate texture information and then uses morphological dilation to fill the gap between edges as

$$\text{CannyMask} = \text{Dilation}_{(5,1)}(\text{Canny}_{(90,225)}(X)) \tag{10}$$

where $\text{Dilation}_{(5,1)}$ denotes the dilation operation with a kernel size of 5 × 5 and iterations of 1 to fill the gap, $\text{Canny}_{(90,225)}$ represents the Canny operator with a low threshold of 90 and a high threshold of 225, X is the input image, and CannyMask is the mask tensor representing object area. For the mask tensor, the pixel value of the object region is set to 255, and the background region is set to 0.

As shown in Figure 5, CannyMask can separate the object region from the background, but there are still limitations. First of all, the edge detection method based on prior knowledge cannot adapt to various complex backgrounds, which refer to non-target signals including measurement noise, normal pavement structure, and other distresses. For example, the crack shown in the yellow circle of sample 2 in Figure 5 belongs to other distresses, but this area with continuous strong gray-scale variations and hyperbolic characteristics may be prone to be confused with interlayer distresses. The regions identified by CannyMask encompass the interlayer distresses and other types of distresses, such as cracks. However, not only can CannyMask suppress most of the background interference but it also enhances the region features identified by CannyMask, which significantly contributes to improving the recognition performance of deep learning-based networks. In addition, although CannyMask can locate the object region well, it cannot further classify the object region as interlayer debonding or interlayer water seepage. It is worth noting that deep learning-based networks can solve the above classification problems, but inaccurate localization often occurs in object detection. Therefore, as an a priori feature module, CannyMask may be embedded into the deep learning network to improve the location ability of detection network.

2.3. Object Enhancement Module (OEM) Based on Prior Knowledge

An OEM based on the prior knowledge is designed as shown in Figure 3. The OEM has two inputs: the original image and the input features (the output of the previous module). Firstly, the object area of the original image is extracted through the MaxPool and subsequent CannyMask module (Figure 6a). Secondly, in order to ensure the same scale of input features, the CannyMask module is followed by the AP (AveragePool) module. Thirdly, the output of the AveragePool module is normalized, and the weight matrix is generated by the softmax function shown in Figure 6b. Finally, the output features are obtained by multiplying the weight matrix with the input features shown in Figure 6c. This module not only enhances the object region but also adds no additional parameters. The mathematical expression of this process is as follows:

$$H_n = \text{AP}^{n-1}(\text{CM}[\text{MP}(X)]) \tag{11}$$

$$\text{Ouput} = P_n \times Softmax(Normalization(H_n)) \tag{12}$$

where H_n represents the hand-crafted features based on the Canny operator and morphological dilation for all stages and X is the original image. AP(·), CM(·), and MP(·) perform

AveragePool, CannyMask, and MaxPool, respectively. In Formula (12), P_n represents the input features for all stages and n is the stage number.

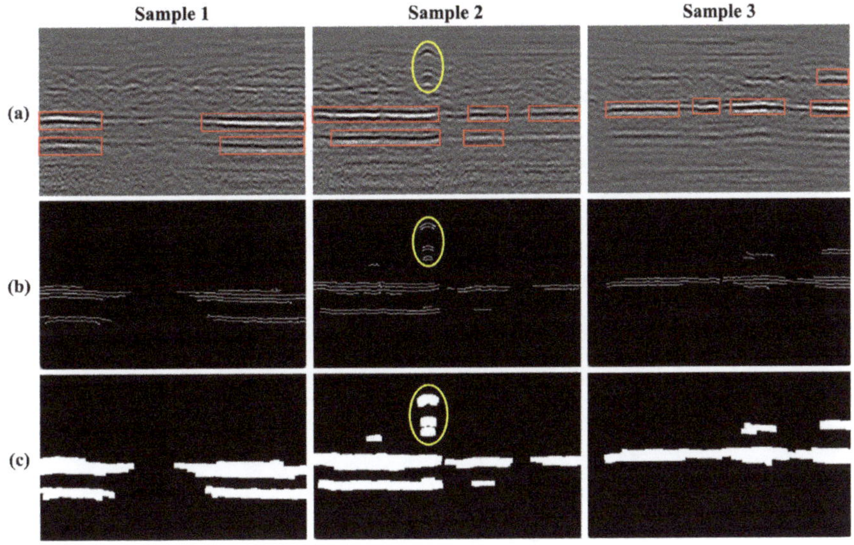

Figure 5. Examples of the CannyMask operation based on prior knowledge. (**a**) Original images, where the red box indicates the real object region. (**b**) The shape and texture of the interlayer distress obtained by the Canny operator as prior knowledge. (**c**) The results of morphological dilation, where the white regions indicate the object regions separated from the background.

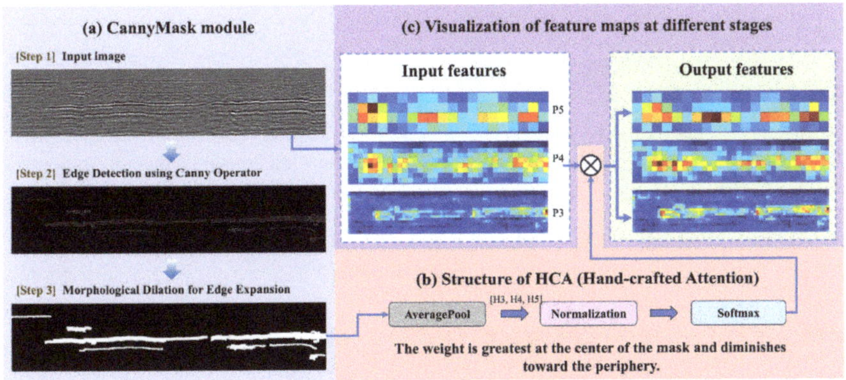

Figure 6. Schematic of the OEM processing flow. (**a**) The shape and texture of interlayer distress are used as prior knowledge to locate the object region. (**b**) The weight matrix is obtained by the softmax. (**c**) Visualization of feature maps generated by OEM, where the object regions are enhanced.

The OEM provides a new solution to some problems of the detection model, such as poor object positioning and unfocused learning. By incorporating prior knowledge, the model performance is significantly enhanced. A key advantage of this module is its parameter-free nature. Furthermore, the OEM helps to quickly locate the object region so that the learning of the model can be focused on that region, thus accelerating convergence.

2.4. C3WC Module

The size of the convolution kernel limits the global modeling capabilities of CNN. So, many researchers have attempted to increase the kernel size of CNN to mimic the global receptive field found in self-attention mechanisms. However, this method quickly hit an upper bound and saturated way before achieving a global receptive field. To address this challenge, Finder et al. proposed the WTConv based on wavelet transform, which excels at extracting features with different scales [51]. In this paper, the WTConv is integrated into the C3 module, named C3WC (Figure 7), to obtain a larger receptive field. As shown in the grid of Figure 8, a 1 × 1 convolution, which is performed on the low-frequency band of the second-level wavelet domain $X_{LL}^{(2)}$, is a response to lower frequencies of a 4 × 4 receptive field in the input X. The process is given by

$$Y = IWT(Conv(W, WT(X))) \quad (13)$$

where X is the input tensor, and W is the weight tensor of a k × k depth-wise kernel with four times as many input channels as X. In this study, we employ 5 × 5 kernel sizes for the convolutions.

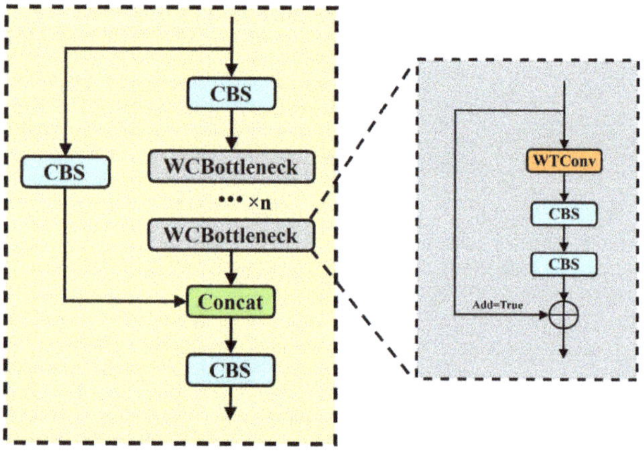

Figure 7. Structure of the C3WC module. The module is used for multiscale feature extraction, where the core WTConv in WCBottleneck enhances its ability to capture low frequency features and obtains a larger receptive field.

Figure 8 illustrates the WTConv process with a 2-level wavelet transform. The initial step performs decomposition of the input signal into multi-scale wavelets. Subsequently, different frequency bands are processed using depth-wise convolution [63] with small convolution kernels, followed by image reconstruction through an inverse wavelet transform (IWT) to obtain the final output. The decomposition is given by

$$(X_{LL}^{(i)}, X_{HL}^{(i)}, X_{LH}^{(i)}, X_{HH}^{(i)}) = \mathrm{WT}(X_{LL}^{(i-1)}\Big) \quad (14)$$

where $X_{LL}^{(0)} = X$ and i is the current level. $X_{LL}^{(i)}$ is the low-frequency component of X, while $X_{LH}^{(i)}$, $X_{HL}^{(i)}$, and $X_{HH}^{(i)}$ are its horizontal, vertical, and diagonal high-frequency components. The Formulas (15) and (16) describe convolution and inverse wavelet transform (IWT), respectively.

$$(Y_{LL}^{(i)}, Y_{HL}^{(i)}, Y_{LH}^{(i)}, Y_{HH}^{(i)}) = \mathrm{Conv}(W^{(i)}, (X_{LL}^{(i)}, X_{HL}^{(i)}, X_{LH}^{(i)}, X_{HH}^{(i)})) \quad (15)$$

$$Z^{(i)} = \mathrm{IWT}(Y_{LL}^{(i)} + Z^{(i+1)}, Y_{HL}^{(i)}, Y_{LH}^{(i)}, Y_{HH}^{(i)}) \quad (16)$$

where $Z^{(i)}$ is the aggregated outputs from level i onward. As observed in Figure 8, WTConv performs repetitive wavelet decomposition on the low-frequency band of the input image, continuously enhancing low-frequency features. WTConv has a superior capacity for capturing low-frequency information compared to standard convolution. Moreover, each level of WTConv increases the receptive field size while only marginally increasing the number of trainable parameters. This characteristic is a significant factor in integrating WTConv into the C3 module. In this study, we employ a 3-level wavelet decomposition to make a trade-off between accuracy and the number of parameters.

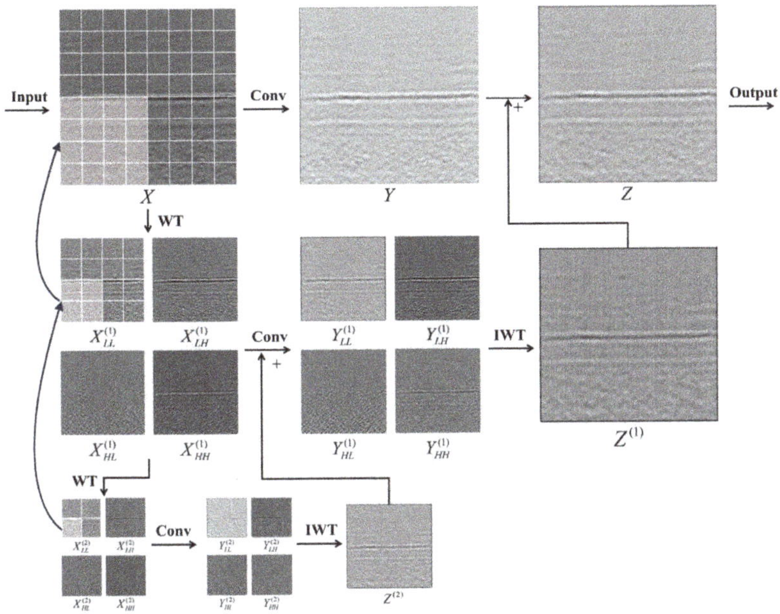

Figure 8. An example of the WTConv operation using a 2-level wavelet decomposition and 1×1 kernel sizes for the convolutions, where WT refers to the wavelet transform and IWT refers to the inverse wavelet transform.

3. Validation Using Field Tests

3.1. Dataset

3.1.1. Data Acquisition

In interlayer distress detection of asphalt pavement within GPR images, no public dataset currently exists. To evaluate the performance of the proposed model, this study collected thousands of GPR images from several highways, including the Lianyungang-Xuzhou Highway, Huai'an-Xuzhou Highway, Yancheng–Jingjiang Highway, and Shanghai-Nanjing Highway, in Jiangsu province of China, as shown in Figure 9a.

Figure 9. Data acquisition. (**a**) Several highways of GPR data have been collected. (**b**) Acquisition site. (**c**) GPR B-scan images.

Field data collection was conducted using the fourth-generation high-dynamic GPR, the MALA GX750 series (Figure 9b), produced by the Swedish company GuidelineGEO in Stockholm, Sweden. Its technical specifications are as follows: antenna frequency of 750 MHz, scanning speed of 1290 (traces/s), detection depth ranging from 0 to 1.5 m, air-coupling antenna mode, weight of 3.6 kg, and detection vehicle speed of 80 (km/h). After pre-processing, such as dewow, adjustment of start time, compensation for energy decay, background removal, deconvolution, and bandpass filtering, we obtained a total of 2105 GPR images (Figure 9c). Each image is with a width of 1005 pixels and a height of 210 pixels, corresponding to the length of 20 m and the time window of 20 ns, respectively.

3.1.2. Dataset Construction

Through the analysis of the above GPR images, we find that the main hidden distresses of asphalt pavement are interlayer distresses, so interlayer distresses become the research object of this paper. Further analysis of interlayer distresses shows that although the shape characteristics of interlayer distresses at different positions are similar, there are differences in vertical texture characteristics, as shown in Figure 10. By comparing and analyzing the image and the corresponding drilled sample core, combined with the experience of experts in the field, the interlayer distresses can be classified as interlayer debonding (Figure 10a) and interlayer water seepage (Figure 10b). This accurate classification of interlayer distresses is of importance to improve maintenance efficiency and reduce maintenance costs. For interlayer debonding, the pressure grouting is adopted to fill the gap between the layers of pavement [64], while interlayer water seepage is a more serious issue, requiring both pressure grouting and drainage system repair.

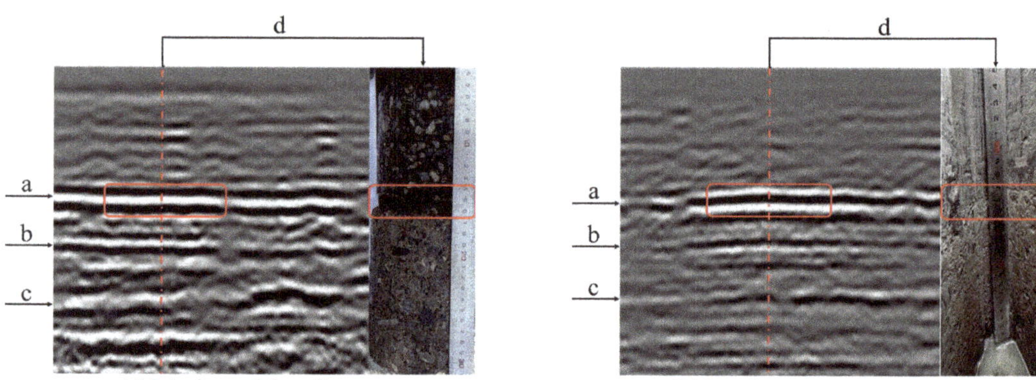

(**a**) Interlayer debonding. (**b**) Interlayer water seepage.

Figure 10. The images and sample cores of the two categories of distresses. a. The interface of the surface layer and the upper base layer; b. The interface of the upper base layer and the lower base layer; c. The interface of the the lower base layer and the subbase layer; d. Coring position; The red box indicates the location of the distress.

Figure 10 shows the GPR images and core samples of two categories of interlayer distress. In Figure 10, the red square represents the location of the distress, "a" indicates the interface of the surface layer and the upper base layer, "b" denotes the interface of the upper base layer and the lower base layer, "c" indicates the interface of the lower base layer and the subbase layer, and "d" denotes the coring position.

The characteristics of interlayer debonding and interlayer water seepage are described as follows:

- Interlayer debonding: temperature, load, or other factors may cause a continuous separation between the layers of pavement, and interlayer delamination occurs, named interlayer debonding. There is a strong reflection in the GPR image due to the difference in permittivity of the air contained between the layers and the surrounding medium. Specifically, the image features of interlayer debonding appear as black, white, and black in vertical order (Figure 10a). Interlayer debonding is abbreviated as "poor_1";
- Interlayer water seepage: when interlayer debonding has occurred without timely maintenance, interlayer water seepage distress between the layers occurs as rainwater seeps in. Unlike other air-containing distresses, this category of distress contains water. The permittivity of water and air are different, typically 81 for water and 1 for air. As a result, Figure 10b shows a clear polarity reversal for interlayer water seepage compared to interlayer debonding. Specifically, the image features of interlayer water seepage appear as white, black, and white in vertical order. Interlayer water seepage is abbreviated as "water_1".

We used LabelImg 1.8.6 software to annotate the 2105 GPR images. An annotated example is shown in Figure 11, where the red box indicates interlayer debonding and blue indicates interlayer water seepage. Table 1 summarizes the basic situation of the interlayer distresses, including the total number of samples and scale statistics of interlayer debonding, as well as interlayer water seepage. The average size indicates that the bounding boxes for interlayer debonding and interlayer water seepage are similar in size. In the field of object detection, the annotated areas less than 10% of an image belong to small objects [65]. However, the average area of interlayer distresses accounted for only 8% of the GPR images, which poses a challenge for the identification of interlayer debonding and interlayer water seepage. The maximum and minimum width values reveal that the size variability of the interlayer distresses predominantly occurs in width, with a difference of approximately 25 times. The big variation in size further increases the difficulty of detection. For all experiments conducted in this study, the dataset was partitioned into training, validation, and test sets in a ratio of 8:1:1.

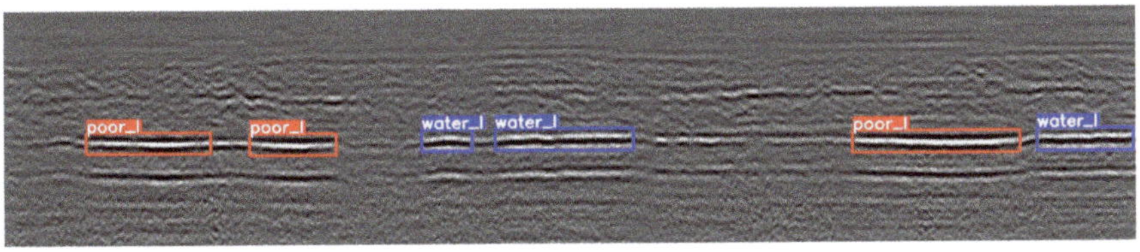

Figure 11. An example of sample annotation of GPR images, where the red box indicates interlayer debonding and the blue box indicates interlayer water seepage.

Table 1. Basic situation of various interlayer distresses.

Distress Category	Total	Width (Pixel)			Height (Pixel)		
		Max	Min	Average	Max	Min	Average
poor_l	2751	648	22	80.0	40	14	19.3
water_l	2862	574	24	82.5	27	19	20.3

3.2. Evaluation Metrics and Experimental Configuration

To accurately and fairly evaluate the performance of the OEM-HWNet, we employed several metrics, including precision (P), recall (R), and $F1$ score. The computations of these indicators are presented below, as follows:

$$P = \frac{TP}{TP + FP} \tag{17}$$

$$R = \frac{TP}{TP + FN} \tag{18}$$

$$F1 = \frac{2 * P * R}{P + R} \tag{19}$$

where TP is the true positive, FP is the false positive, FN is the false negative, and TN is the true negative. Mean average precision (mAP) is typically employed for the evaluation of multi-distress detection systems. In this context, it represents the average precision for interlayer distress detection. The precision–recall (P–R) curve is illustrated with recall on the horizontal axis and precision on the vertical axis. Average Precision (AP) denotes the area under the P–R curve, calculated at various thresholds for a specific distress category. The metrics $mAP@0.5$ can be expressed as follows:

$$mAP@0.5 = \sum_{i=0}^{n} \int_{0}^{1} PRdR(IOU_{thresh} = 0.5) \tag{20}$$

In this study, the proposed interlayer distress detection model of asphalt pavement was performed on a server with an Intel(R) Xeon(R) Gold 6133 CPU and a single NVIDIA GeForce RTX 3090 GPU. The software environment configuration of the model is shown in Table 2. The batch size and epoch for all experiments were set to 16 and 200, respectively. The initial learning rate was 0.01, and the learning rate decay followed a cosine schedule. The SGD [66] algorithm was selected as the optimizer with the momentum of 0.937 and decay weight of 0.0005.

Table 2. Software environment.

Name	Version
Pytorch	2.1.0
CUDA	12.1
Python	3.8.18
Numpy	1.23.5

4. Experimental Results

4.1. Preliminary Analysis of OEM-HWNet

4.1.1. Detection Results of OEM-HWNet

Confusion matrix is a common form of presenting and analyzing test results. The training and testing were carried out on the established GPR dataset, and the test results obtained were shown in Figure 12. The values in each cell represent the ratio of the predicted number and the true number of samples for each category. The "background" in the horizontal axis represents the situation in which the background is misjudged as the distress, whereas the "background" in the vertical axis represents the situation in which the distress is misjudged as the background. The analysis of Figure 12 is as follows. First, the main diagonal of the confusion matrix indicates that the OEM-HWNet model can identify the interlayer distresses with high accuracy, in which the scores of interlayer debonding and interlayer water seepage are 0.89 and 0.94, respectively. Then, the scores of interlayer debonding and interlayer water seepage from the "background" in the vertical axis are 0.08 and 0.05, respectively; it means that a small number of distresses are not recognized. In addition, the scores of interlayer debonding and interlayer water seepage from the "background" in the horizontal axis not only show that the interlayer distress and other distress in the background have similar characteristics but also the characteristics of interlayer debonding and interlayer water seepage are also very similar. Although interlayer disease detection is a difficult task, the OEM-HWNet still demonstrates excellent performance. To further demonstrate the detection performance of the proposed network, Figure 13 shows a comparison between the detection results and the ground-truth labels. From Figure 13, we can see that the detection results are almost consistent with the ground-truth label. This indicates that the OEM-HWNet performed well in accurately locating and identifying interlayer distresses.

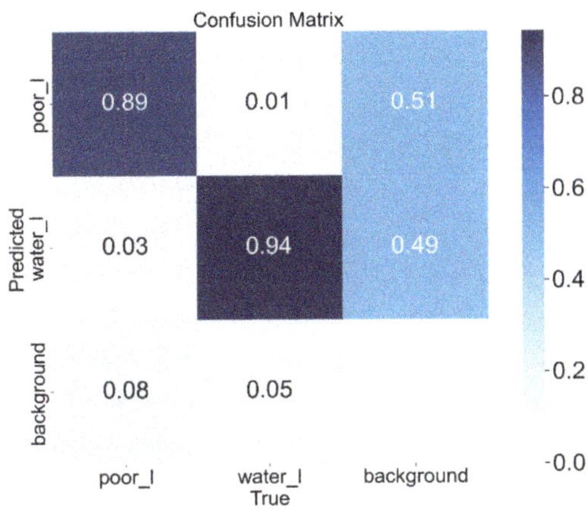

Figure 12. Confusion matrix obtained on test set using OEM-HWNet.

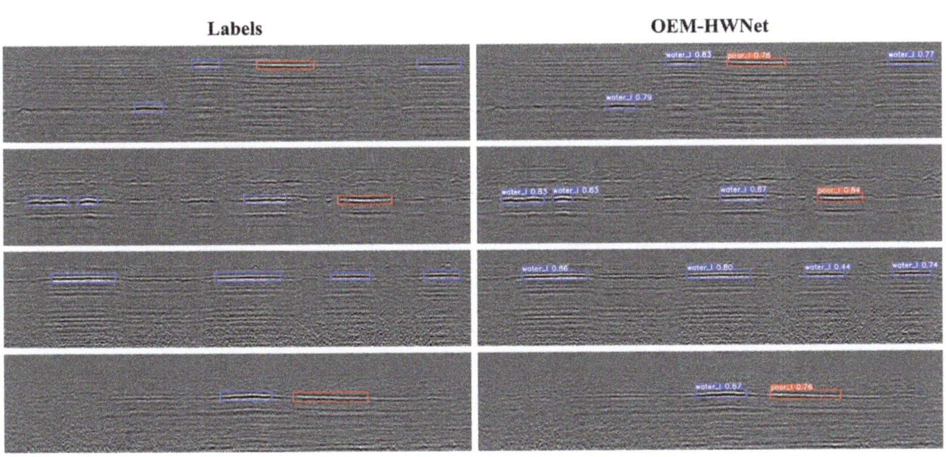

Figure 13. The detection results of the OEM-HWNet network.

4.1.2. Visualization of Attention Maps

To clearly understand the decision-making process of the OEM-HWNet network in recognizing interlayer distress, Grad-CAM [67] is employed to visualize the attention maps within the network architecture. Figure 14 presents the visualized attention maps from the output of the final layer. Notably, red indicates the areas that the network focuses on during the decision-making process. In contrast, blue indicates the less important areas. It can be observed that the red areas and the ground-truth labels are almost identical in position. This indicates that the prior knowledge-guided network can accurately locate and focus learning on interlayer distress. Although there are a few red or yellow areas appearing in the background, a clear distinction can be seen between interlayer distress and the background.

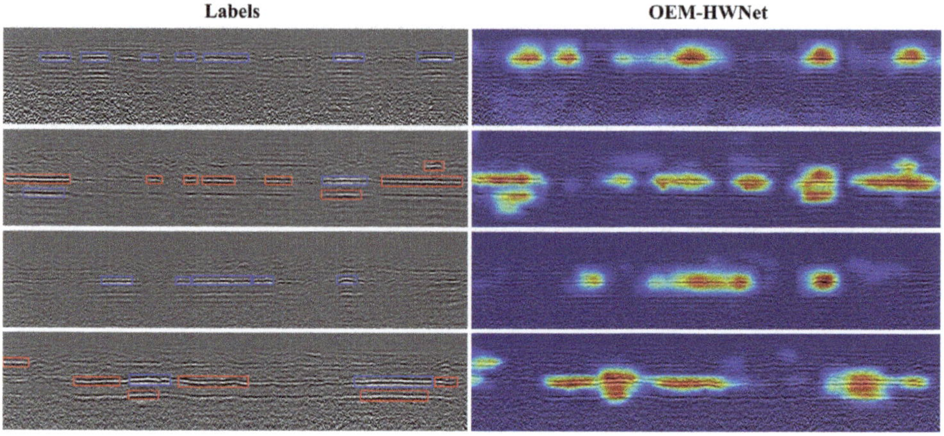

Figure 14. Visualization of attention maps produced by Grad-CAM tools.

4.2. Ablation Experiments

Relative to the YOLOv5s model, the improved parts of the OEM-HWNet model include the OEM, the fourth head, and the C3WC module. To demonstrate the effectiveness of each module and combination of modules, ablation experiments were performed. The evaluation indices used in ablation experiments include average precision (AP), mean average precision (mAP) at a threshold of 0.5, parameter count (parameters), precision (P), recall (R), and F1.

The experimental results are presented in Table 3. The analysis of Table 3 is described as follows. Firstly, adding a detection head (Fourth Head), introducing the OEM module, and incorporating the C3WC module obtains the improvements of 0.7%, 1.0%, and 1.5% in mAP, respectively. Secondly, the combination of the OEM and Fourth Head resulted in a 2.2% increase in mAP, while the joint use of the OEM and C3WC brought a mAP improvement of 2.5%. More importantly, the introduction of the OEM module does not add additional training load because of its parameter-free nature. Finally, the coupling of the three blocks resulted in a 3.3% increase in mAP. The experimental results verify that the proposed OEM-HWNet is a very competitive solution for the accurate detection of interlayer distresses.

Table 3. Ablation studies for different modules on the test set.

Model	AP		mAP@0.5	Parameters (M)	P	R	F1
	Poor_l	Water_l					
YOLOv5s (baseline)	84.0%	88.5%	86.3%	7.24	77.1%	82.3%	0.7962
+Fourth Head	84.9%	89.2%	87.0%	12.63	80.2%	79.8%	0.8000
+OEM	85.0%	89.5%	87.3%	7.24	81.3%	79.4%	0.8034
+C3WC	85.8%	89.7%	87.8%	7.70	76.2%	84.4%	0.8009
+Fourth Head + OEM	86.7%	90.2%	88.5%	12.63	81.3%	79.5%	0.8039
+Fourth Head + C3WC	85.5%	90.6%	88.1%	13.29	79.5%	81.8%	0.8063
+OEM + C3WC	87.6%	89.9%	88.8%	7.70	80.5%	80.5%	0.8050
+Fourth Head + OEM+ C3WC	87.8%	91.5%	89.6%	13.29	80.7%	82.4%	0.8154

The *P–R* curves provide an intuitive representation of model performance. Figure 15 compares the *P–R* curves of the different models within the dataset of this study. The larger the area under the P–R curve, the better the performance of the model. Compared with YOLOv5s, the area under the P–R curve of all recommended improved models has been increased, among which the OEM-HWNet model has the largest area.

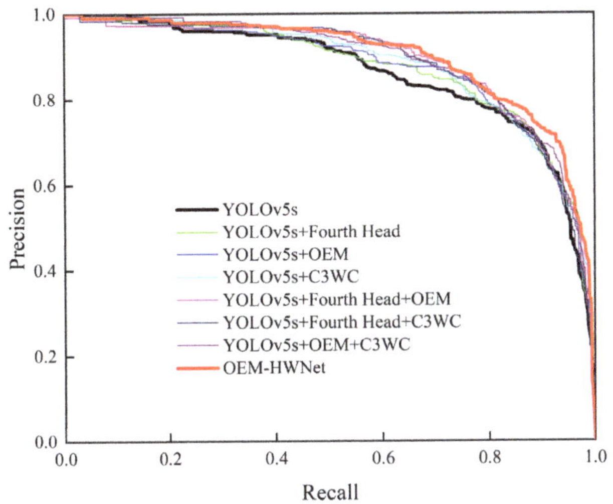

Figure 15. Comparison of P–R curve results for different models.

4.3. Comparison with the State-of-the-Art Methods

In order to verify the effectiveness of the proposed OEM-HWNet model, the current state-of-the-art object detection models, such as Faster R-CNN, RetinaNet, SSD, RT-DETR, YOLOv3, YOLOv5, YOLOv7, YOLOv8, and YOLOv11, are chosen to compare with our model. To ensure a fair comparison, a consistent operating environment and datasets are used.

Table 4 presents a quantitative comparison in Average Precision (AP), mean average precision (mAP), model size, and Frames Per Second (FPS) for the aforementioned models evaluated on the same test set. Firstly, the Faster R-CNN, RT-DETR, and YOLO series are similar in mAP, but YOLOv5s has the smallest size and the fastest detection speed. Therefore, we chose YOLOv5 as the baseline model. Secondly, compared with YOLOv5s, although the incorporation of the Fourth Head and C3WC modules increases the number of model parameters and leads to a decrease from 160 to 84 in FPS, the proposed OEM-HWNet model can still meet the requirement of real-time detection of interlayer distresses. Finally, the proposed OEM-HWNet model presents the highest AP and mAP for both categories of interlayer distresses, although YOLOv5s, YOLOv8s, and YOLOv11s have smaller model size and faster inference speed. In addition, for all models, the detection accuracy of interlayer water seepage is higher than that of interlayer debonding because the GPR image of interlayer water seepage has polarity reversal in the vertical direction, which makes its features more discriminative.

The qualitative assessment is crucial in understanding how these methods perform in real-time detection. Among the state-of-the-art methods, RT-DETR, YOLOv3, YOLOv5s, YOLOv7, and YOLOv8s have a higher mAP; therefore, these methods are selected to analyze furtherly. Figure 16 shows some detection results of the above methods and OEM-HWNet. As the position indicated by the yellow arrow in case 2 in Figure 16, the OEM-HWNet can detect the large-scale interlayer distress with accurate location and high confidence but YOLOv5s and YOLOv7 can not. Although RT-DETR, YOLOv3, and YOLOv8s may detect the large-scale interlayer distress, they may not detect it accurately.

The results indicate that the network requires a larger receptive field for detecting large-scale distress accurately. Moreover, as shown in case 1 of Figure 16, some distresses have not been detected in the detection results of YOLOv5s, while OEM-HWNet can detect all interlayer distresses. The comparison results indicate that OEM, based on prior knowledge, can help detect model precision positioning and focused learning. In summary, OEM-HWNet, including Fourth Head, OEM, and C3WC modules, demonstrates excellent performance in interlayer distress detection.

Table 4. Quantitative comparison with the state-of-the-art object detection methods.

Model	Average Precision		mAP@0.5	Size (MB)	FPS
	Poor_l	Water_l			
Faster R-CNN [35]	84.4%	87.9%	86.1%	330.35	13
RetinaNet [41]	79.6%	84.2%	81.9%	257.26	15
SSD [39]	77.6%	85.1%	81.4%	106.11	42
RT-DETR [52]	84.9%	88.5%	86.7%	82.08	80
YOLOv3 [38]	85.0%	88.4%	86.7%	117.69	71
YOLOv5s [50]	84.0%	88.5%	86.3%	13.63	160
YOLOv7 [37]	83.7%	90.0%	86.9%	71.30	110
YOLOv8s [53]	83.5%	89.9%	86.7%	21.45	145
YOLOv11s [54]	82.9%	88.3%	85.6%	18.27	149
OEM-HWNet (ours)	87.8%	91.5%	89.6%	25.31	84

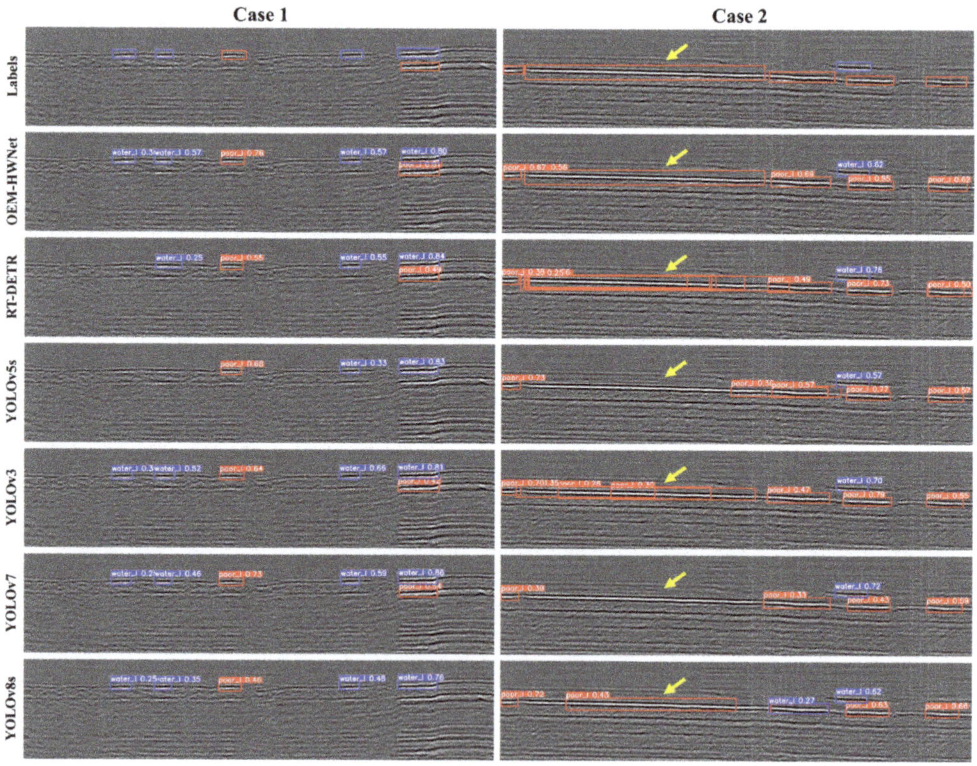

Figure 16. Qualitative analysis of different models for interlayer distress detection.

5. Conclusions

In this study, a prior knowledge-guided network called OEM-HWNet was designed to detect the interlayer distress of asphalt pavement with GPR images. Specifically, the OEM, based on prior knowledge, significantly improves network performance without imposing additional parameters. The C3WC module was introduced to obtain a larger receptive field and enhance the low-frequency information. An additional detection head was added to enhance the detection capability of large objects. The conclusions are drawn as follows:

(1) The existing methods may cause inaccurate locations for interlayer distress due to the interference of similar background features. Based on prior knowledge, the OEM can accurately locate the object region to allow the focusing of the learning of the model on that region without imposing additional parameters;

(2) The proposed OEM-HWNet model achieves an average precision (AP) of 87.8% for interlayer debonding and 91.5% for interlayer water seepage, resulting in a mAP of 89.6%. Meanwhile, the comparative results with the original YOLOv5s model represent a 3.3% increase in mAP. Moreover, the results also indicate that the OEM-HWNet model surpasses other advanced models in detection accuracy;

(3) The proposed method may be used for automatic and real-time interlayer distress detection of asphalt pavement using GPR. An extensive GPR dataset from four highways was constructed to evaluate the detection model rigorously. Future research may validate the proposed method with more testing datasets from asphalt pavement. Incorporating more interpretable prior knowledge to guide the design of the detection network is suggested.

Author Contributions: Conceptualization, C.L.; methodology, C.L. and S.C.; software, S.C.; validation, S.C. and X.W.; formal analysis, C.L.; investigation, C.L.; resources, C.L.; data curation, G.J.; writing—original draft preparation, C.L. and S.C.; writing—review and editing, C.L., S.C. and S.W.; visualization, W.C.; supervision, C.L.; project administration, C.L.; funding acquisition, C.L. All authors have read and agreed to the published version of the manuscript.

Funding: This research was funded by the National Key Research and Development Program of China (2022YFC3003202).

Data Availability Statement: Data will be made available on request.

Conflicts of Interest: Authors Guanglai Jin and Wenlong Cai were employed by the company Jiangsu Sinoroad Engineering Technology Research Institute Co., Ltd. The remaining authors declare that the research was conducted in the absence of any commercial or financial relationships that could be construed as a potential conflict of interest.

References

1. Solla, M.; Pérez-Gracia, V.; Fontul, S. A Review of GPR Application on Transport Infrastructures: Troubleshooting and Best Practices. *Remote Sens.* **2021**, *13*, 672. [CrossRef]
2. Wang, S.; Zhao, S.; Al-Qadi, I.L. Real-Time Density and Thickness Estimation of Thin Asphalt Pavement Overlay During Compaction Using Ground Penetrating Radar Data. *Surv. Geophys.* **2020**, *41*, 431–445. [CrossRef]
3. Xu, X.; Peng, S.; Xia, Y.; Ji, W. The Development of a Multi-Channel GPR System for Roadbed Damage Detection. *Microelectron. J.* **2014**, *45*, 1542–1555. [CrossRef]
4. Todkar, S.S.; Le Bastard, C.; Baltazart, V.; Ihamouten, A.; Derobort, X. Comparative Study of Classification Algorithms to Detect Interlayer Debondings within Pavement Structures from Step-Frequency Radar Data. In Proceedings of the IGARSS 2018—2018 IEEE International Geoscience and Remote Sensing Symposium, Valencia, Spain, 22–27 July 2018; IEEE: New York, NY, USA, 2018; pp. 6820–6823.
5. Guo, S.; Xu, Z.; Li, X.; Zhu, P. Detection and Characterization of Cracks in Highway Pavement with the Amplitude Variation of GPR Diffracted Waves: Insights from Forward Modeling and Field Data. *Remote Sens.* **2022**, *14*, 976. [CrossRef]

6. Xu, J.; Zhang, J.; Sun, W. Recognition of the Typical Distress in Concrete Pavement Based on GPR and 1D-CNN. *Remote Sens.* **2021**, *13*, 2375. [CrossRef]
7. Zhang, J.; Yang, X.; Li, W.; Zhang, S.; Jia, Y. Automatic Detection of Moisture Damages in Asphalt Pavements from GPR Data with Deep CNN and IRS Method. *Autom. Constr.* **2020**, *113*, 103119. [CrossRef]
8. Liang, X.; Yu, X.; Chen, C.; Jin, Y.; Huang, J. Automatic Classification of Pavement Distress Using 3D Ground-Penetrating Radar and Deep Convolutional Neural Network. *IEEE Trans. Intell. Transp. Syst.* **2022**, *23*, 22269–22277. [CrossRef]
9. Hou, F.; Lei, W.; Li, S.; Xi, J. Deep Learning-Based Subsurface Target Detection From GPR Scans. *IEEE Sens. J.* **2021**, *21*, 8161–8171. [CrossRef]
10. Yamaguchi, T.; Mizutani, T.; Meguro, K.; Hirano, T. Detecting Subsurface Voids From GPR Images by 3-D Convolutional Neural Network Using 2-D Finite Difference Time Domain Method. *IEEE J. Sel. Top. Appl. Earth Obs. Remote Sens.* **2022**, *15*, 3061–3073. [CrossRef]
11. Li, N.; Wu, R.; Li, H.; Wang, H.; Gui, Z.; Song, D. MV-GPRNet: Multi-View Subsurface Defect Detection Network for Airport Runway Inspection Based on GPR. *Remote Sens.* **2022**, *14*, 4472. [CrossRef]
12. Li, H.; Li, N.; Wu, R.; Wang, H.; Gui, Z.; Song, D. GPR-RCNN: An Algorithm of Subsurface Defect Detection for Airport Runway Based on GPR. *IEEE Robot. Autom. Lett.* **2021**, *6*, 3001–3008. [CrossRef]
13. Kim, N.; Kim, S.; An, Y.-K.; Lee, J.-J. Triplanar Imaging of 3-D GPR Data for Deep-Learning-Based Underground Object Detection. *IEEE J. Sel. Top. Appl. Earth Obs. Remote Sens.* **2019**, *12*, 4446–4456. [CrossRef]
14. Yang, J.; Ruan, K.; Gao, J.; Yang, S.; Zhang, L. Pavement Distress Detection Using Three-Dimension Ground Penetrating Radar and Deep Learning. *Appl. Sci.* **2022**, *12*, 5738. [CrossRef]
15. Liu, C.; Du, Y.; Yue, G.; Li, Y.; Wu, D.; Li, F. Advances in Automatic Identification of Road Subsurface Distress Using Ground Penetrating Radar: State of the Art and Future Trends. *Autom. Constr.* **2024**, *158*, 105185. [CrossRef]
16. Li, Y.; Liu, C.; Yue, G.; Gao, Q.; Du, Y. Deep Learning-Based Pavement Subsurface Distress Detection via Ground Penetrating Radar Data. *Autom. Constr.* **2022**, *142*, 104516. [CrossRef]
17. Li, S.; Gu, X.; Xu, X.; Xu, D.; Zhang, T.; Liu, Z.; Dong, Q. Detection of Concealed Cracks from Ground Penetrating Radar Images Based on Deep Learning Algorithm. *Constr. Build. Mater.* **2021**, *273*, 121949. [CrossRef]
18. Liu, Z.; Wu, W.; Gu, X.; Li, S.; Wang, L.; Zhang, T. Application of Combining YOLO Models and 3D GPR Images in Road Detection and Maintenance. *Remote Sens.* **2021**, *13*, 1081. [CrossRef]
19. Xiong, X.; Meng, A.; Lu, J.; Tan, Y.; Chen, B.; Tang, J.; Zhang, C.; Xiao, S.; Hu, J. Automatic Detection and Location of Pavement Internal Distresses from Ground Penetrating Radar Images Based on Deep Learning. *Constr. Build. Mater.* **2024**, *411*, 134483. [CrossRef]
20. Zhang, J.; Lu, Y.; Yang, Z.; Zhu, X.; Zheng, T.; Liu, X.; Tian, Y.; Li, W. Recognition of Void Defects in Airport Runways Using Ground-Penetrating Radar and Shallow CNN. *Autom. Constr.* **2022**, *138*, 104260. [CrossRef]
21. Liu, Q.; Yan, S. Measurement and Assessement of Road Poor Interlayer Bonding Assessment Using Ground Penetrating Radar. In Proceedings of the 2023 5th International Conference on Intelligent Control, Measurement and Signal Processing (ICMSP), Chengdu, China, 19–21 May 2023; pp. 671–674. [CrossRef]
22. Xue, B.; Gao, J.; Hu, S.; Li, Y.; Chen, J.; Pang, R. Ground Penetrating Radar Image Recognition for Earth Dam Disease Based on You Only Look Once V5s Algorithm. *Water* **2023**, *15*, 3506. [CrossRef]
23. Liu, C.; Yao, Y.; Li, J.; Qian, J.; Liu, L. Research on Lightweight GPR Road Surface Disease Image Recognition and Data Expansion Algorithm Based on YOLO and GAN. *Case Stud. Constr. Mater.* **2024**, *20*, e02779. [CrossRef]
24. Xiong, X.; Tan, Y.; Hu, J.; Hong, X.; Tang, J. Evaluation of Asphalt Pavement Internal Distresses Using Three-Dimensional Ground-Penetrating Radar. *Int. J. Pavement Res. Technol.* **2024**, 1–12. [CrossRef]
25. Jiang, B.; Xu, L.; Cao, Z.; Yang, Y.; Sun, Z.; Xiao, F. Interlayer Distress Characteristics and Evaluations of Semi-Rigid Base Asphalt Pavements: A Review. *Constr. Build. Mater.* **2024**, *431*, 136441. [CrossRef]
26. Jin, G.; Liu, Q.; Cai, W.; Li, M.; Lu, C. Performance Evaluation of Convolutional Neural Network Models for Classification of Highway Hidden Distresses with GPR B-Scan Images. *Appl. Sci.* **2024**, *14*, 4226. [CrossRef]
27. Cai, W.; Li, M.; Jin, G.; Liu, Q.; Lu, C. Comparison of Residual Network and Other Classical Models for Classification of Interlayer Distresses in Pavement. *Appl. Sci.* **2024**, *14*, 6568. [CrossRef]
28. Bugarinović, Ž.; Pajewski, L.; Ristić, A.; Vrtunski, M.; Govedarica, M.; Borisov, M. On the Introduction of Canny Operator in an Advanced Imaging Algorithm for Real-Time Detection of Hyperbolas in Ground-Penetrating Radar Data. *Electronics* **2020**, *9*, 541. [CrossRef]
29. Harkat, H.; Dosse Bennani, S. Ground Penetrating Radar Imaging for Buried Cavities in a Dispersive Medium: Profile Reconstruction Using a Modified Hough Transform Approach and a Time-Frequency Analysis. *Int. J. Commun. Antenna Propag. IRECAP* **2015**, *5*, 78. [CrossRef]

30. Todkar, S.S.; Le Bastard, C.; Baltazart, V.; Ihamouten, A.; Dérobert, X. Performance Assessment of SVM-Based Classification Techniques for the Detection of Artificial Debondings within Pavement Structures from Stepped-Frequency A-Scan Radar Data. *NDT E Int.* **2019**, *107*, 102128. [CrossRef]
31. Frigui, H.; Gader, P. Detection and Discrimination of Land Mines in Ground-Penetrating Radar Based on Edge Histogram Descriptors and a Possibilistic *K*-Nearest Neighbor Classifier. *IEEE Trans. Fuzzy Syst.* **2009**, *17*, 185–199. [CrossRef]
32. Sun, Z.; Caetano, E.; Pereira, S.; Moutinho, C. Employing Histogram of Oriented Gradient to Enhance Concrete Crack Detection Performance with Classification Algorithm and Bayesian Optimization. *Eng. Fail. Anal.* **2023**, *150*, 107351. [CrossRef]
33. Xie, X.; Li, P.; Qin, H.; Liu, L.; Nobes, D.C. GPR Identification of Voids inside Concrete Based on the Support Vector Machine Algorithm. *J. Geophys. Eng.* **2013**, *10*, 034002. [CrossRef]
34. LeCun, Y.; Bengio, Y.; Hinton, G. Deep Learning. *Nature* **2015**, *521*, 436–444. [CrossRef] [PubMed]
35. Ren, S.; He, K.; Girshick, R.; Sun, J. Faster R-CNN: Towards Real-Time Object Detection with Region Proposal Networks. *IEEE Trans. Pattern Anal. Mach. Intell.* **2017**, *39*, 1137–1149. [CrossRef] [PubMed]
36. Cai, Z.; Vasconcelos, N. Cascade R-CNN: Delving Into High Quality Object Detection. In Proceedings of the 2018 IEEE/CVF Conference on Computer Vision and Pattern Recognition, Salt Lake City, UT, USA, 18–23 June 2018; IEEE: New York, NY, USA, 2018; pp. 6154–6162.
37. Wang, C.-Y.; Bochkovskiy, A.; Liao, H.-Y.M. YOLOv7: Trainable Bag-of-Freebies Sets New State-of-the-Art for Real-Time Object Detectors. In Proceedings of the 2023 IEEE/CVF Conference on Computer Vision and Pattern Recognition (CVPR), Vancouver, BC, Canada, 18–22 June 2023; IEEE: New York, NY, USA; pp. 7464–7475.
38. Redmon, J.; Farhadi, A. YOLOv3: An Incremental Improvement. *arXiv* **2018**, arXiv:1804.02767.
39. Liu, W.; Anguelov, D.; Erhan, D.; Szegedy, C.; Reed, S.; Fu, C.-Y.; Berg, A.C. SSD: Single Shot MultiBox Detector. In *Computer Vision—ECCV 2016*; Leibe, B., Matas, J., Sebe, N., Welling, M., Eds.; Lecture Notes in Computer Science; Springer International Publishing: Cham, Germany, 2016; Volume 9905, pp. 21–37. ISBN 978-3-319-46447-3.
40. Carion, N.; Massa, F.; Synnaeve, G.; Usunier, N.; Kirillov, A.; Zagoruyko, S. End-to-End Object Detection with Transformers. In *Computer Vision—ECCV 2020*; Vedaldi, A., Bischof, H., Brox, T., Frahm, J.-M., Eds.; Lecture Notes in Computer Science; Springer International Publishing: Cham, Germany, 2020; Volume 12346, pp. 213–229. ISBN 978-3-030-58451-1.
41. Lin, T.-Y.; Goyal, P.; Girshick, R.; He, K.; Dollar, P. Focal Loss for Dense Object Detection. *IEEE Trans. Pattern Anal. Mach. Intell.* **2020**, *42*, 318–327. [CrossRef] [PubMed]
42. Luo, T.X.; Zhou, Y.; Zheng, Q.; Hou, F.; Lin, C. Lightweight Deep Learning Model for Identifying Tunnel Lining Defects Based on GPR Data. *Autom. Constr.* **2024**, *165*, 105506. [CrossRef]
43. Zheng, X.; Fang, S.; Chen, H.; Peng, L.; Ye, Z. Internal Detection of Ground-Penetrating Radar Images Using YOLOX-s with Modified Backbone. *Electronics* **2023**, *12*, 3520. [CrossRef]
44. Hu, H.; Fang, H.; Wang, N.; Ma, D.; Dong, J.; Li, B.; Di, D.; Zheng, H.; Wu, J. Defects Identification and Location of Underground Space for Ground Penetrating Radar Based on Deep Learning. *Tunn. Undergr. Space Technol.* **2023**, *140*, 105278. [CrossRef]
45. Zhou, Z.; Zhou, S.; Li, S.; Li, H.; Yang, H. Tunnel Lining Quality Detection Based on the YOLO-LD Algorithm. *Constr. Build. Mater.* **2024**, *449*, 138240. [CrossRef]
46. Liu, Z.; Gu, X.; Yang, H.; Wang, L.; Chen, Y.; Wang, D. Novel YOLOv3 Model With Structure and Hyperparameter Optimization for Detection of Pavement Concealed Cracks in GPR Images. *IEEE Trans. Intell. Transp. Syst.* **2022**, *23*, 22258–22268. [CrossRef]
47. Ma, Y.; Song, X.; Li, Z.; Li, H.; Qu, Z. A Prior Knowledge-Guided Semi-Supervised Deep Learning Method for Improving Buried Pipe Detection on GPR Data. *IEEE Trans. Geosci. Remote Sens.* **2024**, *62*, 1–15. [CrossRef]
48. Chen, H.; Yang, X.; Gong, J.; Lan, T. Multidirectional Enhancement Model Based on SIFT for GPR Underground Pipeline Recognition. *IEEE Trans. Geosci. Remote Sens.* **2024**, *62*, 5928614. [CrossRef]
49. Tag, A.; Shouman, O.; Heggy, E.; Khattab, T. Automatic Groundwater Detection from GPR Data Using YOLOv8. In Proceedings of the IGARSS 2024—2024 IEEE International Geoscience and Remote Sensing Symposium, Athens, Greece, 7–12 July 2024; IEEE: New York, NY, USA, 2024; pp. 8326–8329.
50. Jocher, G.; Chaurasia, A.; Stoken, A.; Borovec, J.; Kwon, Y.; Michael, K.; Fang, J.; Zeng, Y.; Wang, C.; Abhiram, V.; et al. Ultralytics/Yolov5: V7.0—YOLOv5 SOTA Realtime Instance Segmentation. Available online: https://zenodo.org/records/7347926 (accessed on 13 October 2024).
51. Finder, S.E.; Amoyal, R.; Treister, E.; Freifeld, O. Wavelet Convolutions for Large Receptive Fields. In *Computer Vision—ECCV 2024*; Springer: Cham, Germany, 2024.
52. Zhao, Y.; Lv, W.; Xu, S.; Wei, J.; Wang, G.; Dang, Q.; Liu, Y.; Chen, J. DETRs Beat YOLOs on Real-Time Object Detection. In Proceedings of the 2024 IEEE/CVF Conference on Computer Vision and Pattern Recognition (CVPR), Seattle, WA, USA, 16–22 June 2024; IEEE: New York, NY, USA; pp. 16965–16974.
53. Jocher, G. YOLO by Ultralytics. 2023. Available online: https://github.com/ultralytics/ultralytics (accessed on 10 January 2023).
54. Jocher, G. YOLO11. Available online: https://github.com/ultralytics/ultralytics/blob/main/docs/en/models/yolo11.md (accessed on 30 September 2024).

55. He, K.; Zhang, X.; Ren, S.; Sun, J. Spatial Pyramid Pooling in Deep Convolutional Networks for Visual Recognition. *IEEE Trans. Pattern Anal. Mach. Intell.* **2015**, *37*, 1904–1916. [CrossRef] [PubMed]
56. Lin, T.-Y.; Dollár, P.; Girshick, R.; He, K.; Hariharan, B.; Belongie, S. Feature Pyramid Networks for Object Detection. In Proceedings of the 2017 IEEE Conference on Computer Vision and Pattern Recognition (CVPR), Honolulu, HI, USA, 21–26 July 2017; pp. 936–944.
57. Liu, S.; Qi, L.; Qin, H.; Shi, J.; Jia, J. Path Aggregation Network for Instance Segmentation. In Proceedings of the 2018 IEEE/CVF Conference on Computer Vision and Pattern Recognition, Salt Lake City, UT, USA, 18–23 June 2018; pp. 8759–8768.
58. Zheng, Z.; Wang, P.; Liu, W.; Li, J.; Ye, R.; Ren, D. Distance-IoU Loss: Faster and Better Learning for Bounding Box Regression. In Proceedings of the AAAI Conference on Artificial Intelligence, New York, NY, USA, 7–12 February 2020; Volume 34, pp. 12993–13000. [CrossRef]
59. Jol, H.M. *Ground Penetrating Radar: Theory and Applications*, 1st ed.; Elsevier Science: Amsterdam, The Netherlands, 2009; ISBN 978-0-444-53348-7.
60. Daniels, D.J. *Ground Penetrating Radar*; Institution of Engineering and Technology: London, UK, 2004; ISBN 978-0-86341-360-5.
61. Canny, J. A Computational Approach to Edge Detection. *IEEE Trans. Pattern Anal. Mach. Intell.* **1986**, *PAMI-8*, 679–698. [CrossRef]
62. Haralick, R.M.; Sternberg, S.R.; Zhuang, X. Image Analysis Using Mathematical Morphology. *IEEE Trans. Pattern Anal. Mach. Intell.* **1987**, *PAMI-9*, 532–550. [CrossRef]
63. Chollet, F. Xception: Deep Learning with Depthwise Separable Convolutions. In Proceedings of the 2017 IEEE Conference on Computer Vision and Pattern Recognition (CVPR), Honolulu, HI, USA, 21–26 July 2017; IEEE: New York, NY, USA; pp. 1800–1807.
64. Raghuram, A.S.S.; Dasaka, S.M. Forensic Analysis of a Distressed RE Wall and Rigid Pavement in a Newly Constructed Highway Approach. *Int. J. Geosynth. Ground Eng.* **2022**, *8*, 38. [CrossRef]
65. Kisantal, M.; Wojna, Z.; Murawski, J.; Naruniec, J.; Cho, K. Augmentation for Small Object Detection. In Proceedings of the 9th International Conference on Advances in Computing and Information Technology (ACITY 2019), Sydney, Australia, 21–22 December 2019; Aircc Publishing Corporation: Chennai, India, 2019; pp. 119–133.
66. Ketkar, N. Stochastic Gradient Descent. In *Deep Learning with Python*; Apress: Berkeley, CA, USA, 2017; pp. 113–132, ISBN 978-1-4842-2765-7.
67. Selvaraju, R.R.; Cogswell, M.; Das, A.; Vedantam, R.; Parikh, D.; Batra, D. Grad-CAM: Visual Explanations from Deep Networks via Gradient-Based Localization. *Int. J. Comput. Vis.* **2020**, *128*, 336–359. [CrossRef]

remote sensing

MDPI

Article

Harris Hawks Optimization for Soil Water Content Estimation in Ground-Penetrating Radar Waveform Inversion

Hanqing Qiao [1], Minghe Zhang [2,*] and Maksim Bano [2]

[1] Institute of Geophysical and Geochemical Exploration, Chinese Academy of Geological Sciences, Langfang 065000, China; qiaohanqing@mail.cgs.gov.cn

[2] ITES (UMR7063), EOST, University of Strasbourg, F-67000 Strasbourg, France; maksim.bano@unistra.fr

* Correspondence: minghe.zhang@etu.unistra.fr

Abstract: Ground-penetrating radar (GPR) has emerged as a promising technology for estimating the soil water content (SWC) in the vadose zone. However, most current studies focus on partial GPR data, such as travel-time or amplitude, to achieve SWC estimation. Full waveform inversion (FWI) can produce more accurate results than inversion based solely on travel-time. However, it is subject to local minima when using a local optimization algorithm. In this paper, we propose a novel and powerful GPR waveform inversion scheme based on Harris hawks optimization (HHO) algorithm. The proposed strategy is tested on synthetic data, as well as on field experimental data. To further validate our approach, the results of the HHO algorithm are also compared with those of partial swarm optimization (PSO) and grey wolf optimizer (GWO). The inversion results from both synthetic and real experimental data demonstrate that the proposed inversion scheme can efficiently invert both SWC and layer thicknesses, thus achieving very fast convergence. These findings further confirm that the HHO algorithm can be effectively applied for the quantitative interpretation of GPR data.

Keywords: ground-penetrating radar; soil water content; waveform inversion; Harris hawks optimization; vadose zone

1. Introduction

Accurately characterizing the soil water content (SWC) in the vadose zone is essential crucial for various disciplines, including hydrology [1,2], agriculture [3,4], soil science, and environmental science. Traditional methods, like time-domain reflectometry (TDR) [5], exhibit significant limitations, such as being labor-intensive, time-consuming, and causing soil damage. Consequently, ground-penetrating radar (GPR) has emerged as a promising geophysical technique for SWC estimation due to its high sensitivity to changes in SWC [6]. It offers a non-destructive and efficient alternative to traditional methods.

Several review articles have been published to scrutinize the evolution of GPR applications in SWC estimation [6–9]. When estimating SWC, three types of GPR systems are utilized: surface GPR, borehole GPR, and off-ground GPR. Surface GPR is the most frequently employed technique, encompassing a single-offset mode [10–19] and multi-offset mode [17,18,20–24]. The multi-offset is an alternative when the air wave and the ground wave cannot be separated. However, the multi-offset measurement modes, including common mid-point and wide-angle reflection and refraction, are typically time-consuming and require significant measurement effort. The advancement of multi-channel GPR array antennas has helped to mitigate these drawbacks [25,26]. Surface GPR has a limited detection depth. To gain deeper information, borehole GPR is another measurement system that includes three types of measurement modes: zero-offset profile [27–31], multi-offset profile [32,33], and vertical radar profile [34]. Among these three modes, the vertical radar profile is less invasive and more cost-effective because it requires only one borehole. The third system is off-ground GPR, which has been developed because it is less invasive than surface and borehole GPR. Off-ground GPR is usually mounted on a vehicle or a drone, making it more suitable for rapid surface SWC estimation [35–43]. However, most current studies rely on GPR travel-time data to estimate the SWC, which offers limited resolution to characterize the subsurface because only some parts of the information contained in the GPR data are considered.

Full waveform inversion (FWI), which utilizes the entire waveform information, was initially proposed for seismic reflection exploration [44]. Compared with travel-time data-based inversion, FWI can provide higher resolution to characterize the subsurface. FWI has been rapidly developed for estimating subsurface properties using GPR data, as both seismic and GPR methods are founded on solving the wave equation [45–47]. In the context of SWC estimation, several studies have demonstrated the efficacy of GPR FWI [37,40,48]. These studies constructed an objective function by comparing Green's function between the calculated data and observed data, subsequently estimating the SWC based on off-ground reflection data [49].

In the field of FWI, the inversion techniques are mainly classified into two main groups: those based on local optimization and those based on global optimization algorithms. Many researchers have conducted studies on GPR FWI using local optimization algorithms [50–52]. However, the FWI inherently poses as a highly nonlinear and multi-parameter inversion challenge. As such, the problem space is characterized by the presence of numerous extreme points. A notable issue with local optimization algorithms is their tendency to become trapped in local minima. To overcome this obstacle, one of the strategies is to obtain a reasonably accurate model derived from ray-traced inversion, which will serve as the initial model for FWI. However, consistently obtaining an initial model that is proximate to the global minima within the problem space is a challenging endeavor. To mitigate the risk of falling into local minima, the inversion based on global optimization is a superior option. For the global optimization algorithm, a variety of models are initially defined in the problem space. These models are then updated together, and they gradually move towards the global minima. Global optimization methods are often inspired by the social or biological behavior of species. The literature presents a multitude of successful implementations of global optimization. By considering a wide range of initial models and employing sophisticated search strategies, the global optimization methods can significantly reduce the chances of being trapped in local minima and enhance the robustness of GPR FWI. In global optimization algorithms, the grey wolf optimizer (GWO) was used to estimate SWC from GPR waveform data [53]. However, the GWO has limitations when dealing with high-dimensional and multi-modal functions.

The Harris hawks optimization (HHO) algorithm simulates the hunting behavior of Harris's hawks, and it is an emerging global optimization algorithm proposed to solve optimization problems [54]. As a newcomer of population-based swarm intelligence optimization algorithms, the HHO has strong global search capability, which can efficiently explore complex search space and reduce the risk of falling into local minima. Additionally, the HHO shows a faster convergence rate on many test problems and allows for faster identification of the best solutions. It also can better maintain the diversity of the population, thus improving the comprehensiveness of the search by simulating Harris's hawks' hunting and strategy changes. Therefore, the HHO algorithm has begun to be applied in many areas [55–57].

In this paper, we demonstrate the performance of a HHO application on the inversion of GPR waveform data for the estimation of layer thicknesses and SWC in the unsaturated zone. We test the proposed scheme on data with and without noise. We also test it on field experimental data. Furthermore, in order to verify the performance of HHO, we compare its results with those of the PSO and GWO algorithms. The inversion results derived from both synthetic and experimental field data demonstrate that HHO exhibits high capacity and very fast convergence to invert SWC and layer thicknesses.

2. Materials and Methods

2.1. Harris Hawks Optimization (HHO)

The HHO algorithm was developed by Heidari et al. in 2019, which is inspired by the cooperative behavior and hunting process of Harris's hawks in nature [54]. The evolution of the population within this algorithm is achieved by mathematical modeling of various predation strategies. There are mainly three stages in the HHO algorithm: exploration, the transition from exploration to exploitation, and exploitation. In the subsequent sections, we will mathematically model the exploratory and exploitative phases of the HHO algorithm according to the Harris's hawks' techniques of exploring a prey, executing a surprise pounce, and employing diverse attacking strategies. Figure 1 displays all the phases for HHO.

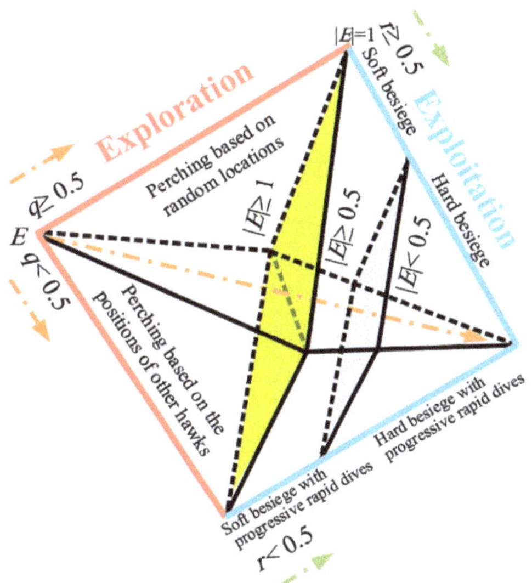

Figure 1. The description for different phases of HHO [54].

2.1.1. Exploration Phase

In this part, the exploration mechanism is first modeled mathematically. Harris's hawks track and detect the prey with their sharp vision, yet at times prey detection proves challenging. Therefore, there are two strategies for Harris's hawks to await prey detection. In HHO, the Harris's hawks are regarded as the candidate solutions and the optimal candidate solution at each stage is considered as the targeted prey. Considering both strategies, and assuming each perching strategy has an equal likelihood, the perch based on the locations of other family members and the prey can be modeled as follows:

$$\vec{X}(t+1) = \begin{cases} \vec{X}_{rand}(t) - r_1\left|\vec{X}_{rand}(t) - 2r_2\vec{X}(t)\right|, q \geq 0.5 \\ (\vec{X}_{prey}(t) - \vec{X}_m(t)) - r_3(LB + r_4(UB - LB)), q < 0.5 \end{cases} \quad (1)$$

where $\vec{X}(t+1)$ represents the position vector of hawks in the next iteration t, $\vec{X}_{rand}(t)$ is a randomly selected hawk from the current population, $\vec{X}(t)$ represents the current position vector of hawks, $\vec{X}_{prey}(t)$ stands for the position of rabbit, r_1, r_2, r_3, and r_4 are random numbers between 0 and 1, q is also random number in range of 0 and 1, which is used to decide which strategy to adopt and is updated in each iteration, and $\vec{X}_m(t)$ is the average position of the current population of hawks, which is calculated as follows:

$$\vec{X}_m(t) = \frac{1}{N}\sum_{i=1}^{N} X_i(t), \quad (2)$$

where $X_i(t)$ represents the position of each Harris's hawk in iteration t and t indicates the total number of hawks.

The first formula in Equation (1) updates the positions of individuals based on random position $\vec{X}_{rand}(t)$ in the population, considering the impact of random individuals in the population on Harris's hawks' search for prey. The second formula in Equation (1) uses the difference between the optimal position $\vec{X}_{prey}(t)$ of the current population and the average position $\vec{X}_m(t)$ of the population, augmented by a random scaling component of a variable range $[LB, UB]$, to update the position of the individual. The r_3 is a scaling coefficient. In this rule, r_3 and r_4 provide additional site alternatives for population individuals, aiding Harris's hawks in exploring different areas of the search area.

2.1.2. Transition Phase

The HHO algorithm can shift between the exploration and the exploitation according to the escaping energy of the prey. The energy of a prey decreases dynamically during the escaping process. The following equation is used to model the escaping energy of the prey:

$$E = 2E_0(1 - \frac{t}{T}), \quad (3)$$

where E indicates the escaping energy for the prey, E_0 represents the initial state of the escaping energy of the prey, and T is the maximum number of iterations.

In each iteration, when the E_0 decreases from 0 to -1, it means that the prey is physically weakening, whereas when the E_0 increases from 0 to 1, it indicates that the prey is strengthening. During the iterations, the dynamic escaping energy E presents a declining trend. When the escaping energy $|E|$ is greater than or equal to 1, the HHO is in the exploration phase, where Harris's hawks search varied areas to explore the location of prey. Conversely, when $|E|$ is less than 1, the algorithm tries to exploit the neighboring area of the current solution to locate the best solution, implying that the HHO is in the exploitation phase.

2.1.3. Exploitation Phase

When the $|E|$ is less than 1, the HHO algorithm enters the exploitation phase. There are four potential strategies for modeling the attacking stage according to the escaping behaviors of the prey and chasing strategies of Harris's hawks in real-world conditions. A random number that represents the chance of a prey successfully escaping is defined. When $r < 0.5$, it means that the prey escapes successfully. Conversely, when $r \geq 0.5$, it indicates that the prey does not escape successfully.

1. Soft besiege

When $0.5 \leq |E| < 1$ and $r \geq 0.5$, Harris's hawks will quietly encircle their prey softly to make the prey more exhausted before launching a sudden assault. This strategy is called soft besiege, which can be modeled as follows:

$$\vec{X}(t+1) = (\vec{X}_{prey}(t) - \vec{X}(t)) - E\left|J\vec{X}_{prey}(t) - \vec{X}(t)\right|, \tag{4}$$

where $J = 2(1 - r_5)$ represents the random jump strength of the prey over the procedure of escaping, and r_5 is a random number in the range of (0, 1). In each iteration, the value of J will change randomly to simulate the nature of prey motions.

2. Hard besiege

When $|E| < 0.5$ and $r \geq 0.5$, Harris's hawks hardly encircle their intended prey to finally launch a sudden assault. This strategy is called hard besiege, which can be modeled as follows:

$$\vec{X}(t+1) = \vec{X}_{prey}(t) - E\left|\vec{X}_{prey}(t) - \vec{X}(t)\right|, \tag{5}$$

3. Soft besiege with progressive rapid dives

When $0.5 \leq |E| < 1$ and $r < 0.5$, the prey possesses sufficient vitality to successfully escape and still a soft besiege is formed before launching a sudden assault. This procedure is more intelligent compared to the previous case. Here, the levy flight (LF) is introduced to mathematically model the current escape behaviors of the prey. The LF is used to emulate the real zigzag deceptive movements of the prey during the escape phase and the irregular, sudden, and rapid dives of the hawks around the prey. This strategy is called soft besiege with progressive rapid dives, which can be modeled as follows:

$$\vec{Y} = \vec{X}_{prey}(t) - E\left|J\vec{X}_{prey}(t) - \vec{X}(t)\right|, \tag{6}$$

Equation (6) bears resemblance to Equation (4), yet differences do exist. Subsequently, Harris's hawks will compare the result from Equation (6) with the previous dive to investigate its feasibility as a dive (a candidate solution). If it is unreasonable (indicating Harris's hawks consider the movements of the prey as deceptive), Harris's hawks will not use Equation (6) to finish diving. Instead, they will perform irregular, sudden, and rapid dives as they near the prey. According to the LF, the dive will be defined as follows:

$$\vec{Z} = \vec{Y} + \vec{S} \times LF(D), \tag{7}$$

where $\vec{Y}$ is calculated based on Equation (6), D is the dimension of problem space, $\vec{S}$ is a random vector with dimension, and LF represents the levy flight function, which can be calculated as follows:

$$LF(x) = 0.01 \times \frac{\mu \times \sigma}{|v|^{\frac{1}{\beta}}}, \tag{8}$$

$$\sigma = \left(\frac{\Gamma(1+\beta) \times \sin\left(\frac{\pi\beta}{2}\right)}{\Gamma\left(\frac{1+\beta}{2}\right) \times \beta \times 2^{\left(\frac{\beta-1}{2}\right)}}\right)^{\frac{1}{\beta}}, \tag{9}$$

where μ and v are random numbers between 0 and 1, and β is a constant value which is usually set to be 1.5.

In other words, when Harris's hawks discover that their prey is engaging in deceptive behavior, the action they take is adjusted according to their expected action, which is represented by Equation (6). The adjustment is achieved through the levy flight, as outlined in Equation (7). The selection of this particular diving mode is designed based on the fitness of the Harris Hawk's current position as follows:

$$\vec{X}(t+1) = \begin{cases} \vec{Y}, & if\ F(\vec{Y}) < F(\vec{X}(t)) \\ \vec{Z}, & if\ F(\vec{Z}) < F(\vec{X}(t)) \end{cases} \tag{10}$$

where $\vec{Y}$ and $\vec{Z}$ are derived from Equations (6) and (7), respectively. If we consider the optimization problem, then the $F(\cdot)$ is the objective function for that optimization problem.

4. Hard besiege with progressive rapid dives

When $|E| < 0.5$ and $r < 0.5$, the prey has not got enough energy to escape and Harris's hawks will build a hard besiege before launching a sudden assault to catch and kill the prey. Harris's hawks endeavor to shorten the distance between their average position and the escaping prey. Equation (10) is still used as the rule for the hard besiege condition. However, the $\vec{Y}$ is calculated using the new rules as follows:

$$\vec{Y} = \vec{X}_{prey}(t) - E\left|J\vec{X}_{prey}(t) - \vec{X}_m(t)\right|, \tag{11}$$

where $\vec{X}_m(t)$ is calculated by Equation (2).

2.2. HHO for GPR FWI

We develop a software package for GPR FWI to estimate SWC through HHO based on MATLAB 2020a. In this study, we only focus on inversion for SWC and layer thicknesses in the vadose zone by fixing attenuation factor. There are six steps for inverting SWC for GPR waveform data, which are as follows:

1. Define the parameters for the algorithm (population size, maximum number of iteration, upper and lower boundaries for the search space of model);
2. Initialize the parameters for the models (SWC and layer thicknesses);
3. Calculated the soil relative dielectric permittivity according to the Topp equation [58];
4. Conduct the forward modeling for all the initial models of the entire population;
5. Calculate the fitness for every model and update the initial models by HHO algorithm;
6. If reaching the maximum number of iterations or meeting the stopping criterion, the iterations will stop and output the best solution. If not, a new iteration will begin from stage 3. In this study, the stopping criterion can be expressed as follows:

$$\left\|\vec{d} - \vec{G}(\vec{m})\right\|_2 < tol, \tag{12}$$

where $\vec{d}$ represents the observed data, $\vec{m}$ indicates model parameters including the SWC (θ) and the layer thicknesses (z), $\vec{G}$ is a nonlinear forward operator, and tol stands for the predetermined threshold value. It means that the iteration will stop when the L_2 norm between the observed data and calculated data is lower than the predetermined value.

2.2.1. Objective Function

The purpose of inverse problem is to invert the model parameters based on observed data within a predetermined solution space. In general, the inverse problems are solved by an iteration algorithm. During the iteration, an objective function needs to be defined.

In this study, we first calculate the L_2 norm between the observed and calculated data. Then, we calculate the L_2 norm for the observed data. Finally, we build the objective function according to the equation as follows:

$$fitness = 100 \times \frac{\left\| \vec{E}_{obs} - \vec{E}_{cal} \right\|_2}{\left\| \vec{E}_{obs} \right\|_2}, \quad (13)$$

where $\vec{E}_{obs}$ and $\vec{E}_{cal}$ represent the observed data vector and the calculated data vector, respectively. Figure 2 shows the flow for inverting the SWC from GPR waveform inversion with HHO algorithm.

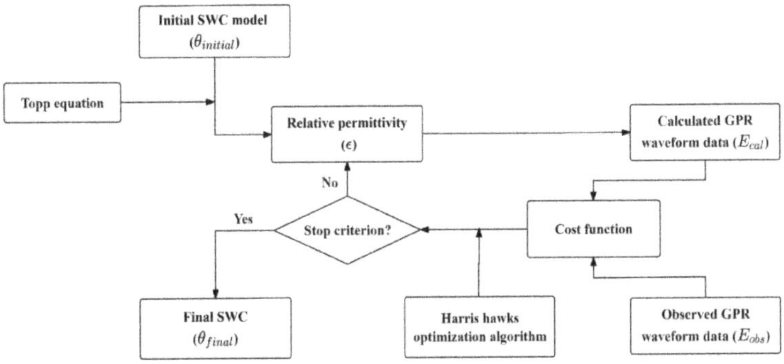

Figure 2. The flow for inverting the SWC from GPR waveform inversion with HHO algorithm.

2.2.2. Petrophysical Relationship

In order to obtain models of electrical properties for GPR forward modeling, we need a petrophysical relationship to convert the SWC to electrical properties. In this paper, we only focus on the relative dielectric permittivity by ignoring the conductivity. Therefore, we select the widely used Topp equation to calculate the relative dielectric permittivity from the volumetric SWC [58]. With the classical parameters, the Topp equation can be expressed as follows:

$$\varepsilon = 3.03 + 9.30 \times \theta + 146.00 \times \theta^2 - 76.70 \times \theta^3, \quad (14)$$

where ε stands for the relative dielectric permittivity of the soil, and θ represents the volumetric SWC. In this paper, we ignore the errors in the petrophysical relationship.

2.2.3. GPR Forward Modeling

In GPR inversion, the forward modeling is a great important link. At each iteration, the response of a subsurface model is required to be calculated. Subsequently, the parameters for the model are inverted by gradually narrowing the difference between the observed data and the calculated forward response. Maxwell's equations are the basic theory for the modeling of electromagnetic waves. Combining Maxwell's equations and constitutive relations, we can gain the propagation and diffusion equations of the electromagnetic waves, which are expressed as follows:

$$\Delta\vec{E} = \mu\sigma\frac{\partial\vec{E}}{\partial t} + \mu\varepsilon\frac{\partial^2\vec{E}}{\partial t}, \tag{15}$$

$$\Delta\vec{H} = \mu\sigma\frac{\partial\vec{H}}{\partial t} + \mu\varepsilon\frac{\partial^2\vec{H}}{\partial t}, \tag{16}$$

where $\vec{E}$ stands for the electrical field, $\vec{H}$ stands for the magnetic field, μ, σ, and ε represent the magnetic permeability, the electrical conductivity, and the dielectric permittivity of the medium, respectively, and t is the time. In this study, we only focus on the electric field. Therefore, we will not mention Equation (16) in the following section.

In this study, we assumed the electrical field to be a time-harmonic wave ($\vec{E} = \vec{E}_0 e^{-i\omega t}$), and so Equation (15) can be written as follows:

$$\Delta\vec{E} + k^2\vec{E} = 0, \tag{17}$$

where k is the complex wavenumber, ω is the angular frequency. Equation (17) is also called the Helmholtz equation. The k can be expressed as follows:

$$k^2 = \mu\omega(\varepsilon\omega + i\sigma), \tag{18}$$

When considering the attenuation (α), Equation (18) can be written as follows [59]:

$$k = \omega\sqrt{\mu(\varepsilon + i\frac{\sigma}{\omega})} = \beta + i\alpha, \tag{19}$$

where α and $\beta = \omega/V$ represent the absorption coefficient and phase velocity, respectively, and V stands for the velocity of electromagnetic wave in the medium.

In this study, we realize the forward modeling by using the solution of Equation (17) given by Banoas follows [59]:

$$E(z,\omega)\exp(ikz) = E(0,\omega)\exp(i\beta z)\exp(-\alpha z), \tag{20}$$

Let $E_0(\omega)$ represents the complex spectrum of the electrical source $e_0(t)$ at $z = 0$. Assuming a reflecting layer exists at depth of $z = z_1$, the complex spectrum resulting from a vertically incident wavelet, after it has traversed a homogeneous absorbing layer from $z = 0$, is given as follows:

$$E(\omega, z_1) = G(z_1)R(\omega)E_0(\omega)\exp\left(i\frac{\omega}{V}2z_1\right)\exp(-\alpha 2z_1)), \tag{21}$$

where $G(z_1)$ stands for the geometrical spreading and $R(\omega)$ is the reflection coefficient at $z = z_1$. α is equal to $\omega/2VQ$, where Q represents the quality factor. Finally, we can apply an inverse Fourier transform to Equation (21) to yield a GPR response [59].

3. Results and Discussions

3.1. Numerical Examples

In order to test the performance of the proposed inversion scheme with HHO algorithm, we designed three synthetic examples. In Model A, the soil was assumed to be homogeneous, there was only one kind of soil, and there was a water table at 0.85 m. For Model B and Model C, we assumed that the subsurface was divided into two layers, in other words, there were two types of soil. In Model B, the SWC increased from the surface to the boundary of the two types of soil, and then a sudden decrease occurred at about 0.5 m. Below the boundary, the SWC increased with the increasing of depth until it reached the water table at 0.85 m. In Model C, we assumed that the surface reached saturated first and the SWC decreased from the surface to the boundary of the two types of soil. Below the boundary, the trend was similar to the change in SWC in Model B. The SWC increased with the increasing of depth until the water table at 0.85 m. The parameters of the models are listed in Table 1. Figures 3–5 illustrate these three 1D surface models (blue solid lines in Figures 3, 4 and 5b) and their forward modeling responses (blue solid lines in Figures 3, 4 and 5a). For all the 1D forward modeling in this paper, the center frequency for the source was set to be 500 MHz.

Table 1. The soil property parameters and search space for Model A, Model B, and Model C.

Layer	Parameters of Model A		Search Space of Model A		Parameters of Model B		Search Space of Model B		Parameters of Model C		Search Space of Model C	
	SWC [1]	z [2]	SWC	z	SWC	z	SWC	z	SWC	z	SWC	z
1	0.05	0.15	0.03–0.08	0.08–0.23	0.37	0.15	0.19–0.56	0.08–0.23	0.07	0.15	0.04–0.11	0.08–0.23
2	0.05	0.20	0.03–0.08	0.10–0.30	0.21	0.20	0.10–0.31	0.10–0.30	0.18	0.20	0.09–0.27	0.10–0.30
3	0.07	0.10	0.03–0.10	0.05–0.15	0.25	0.10	0.13–0.38	0.05–0.15	0.31	0.10	0.15–0.46	0.05–0.15
4	0.10	0.20	0.05–0.16	0.10–0.30	0.07	0.20	0.03–0.10	0.10–0.30	0.06	0.20	0.03–0.08	0.10–0.30
5	0.20	0.10	0.10–0.30	0.05–0.15	0.20	0.10	0.10–0.30	0.05–0.15	0.16	0.10	0.08–0.23	0.05–0.15
6	0.43	0.10	0.22–0.65	0.05–0.15	0.43	0.10	0.22–0.65	0.05–0.15	0.43	0.10	0.22–0.65	0.05–0.15
7	0.43	0.10	0.22–0.65	0.05–0.15	0.43	0.10	0.22–0.65	0.05–0.15	0.43	0.10	0.22–0.65	0.05–0.15
8	0.43	∞ [3]	0.22–0.65	∞	0.43	∞	0.22–0.65	∞	0.43	∞	0.22–0.65	∞

[1] SWC is the abbreviation of soil water content. [2] z represents the layer thickness. [3] ∞ represents the infinite half space.

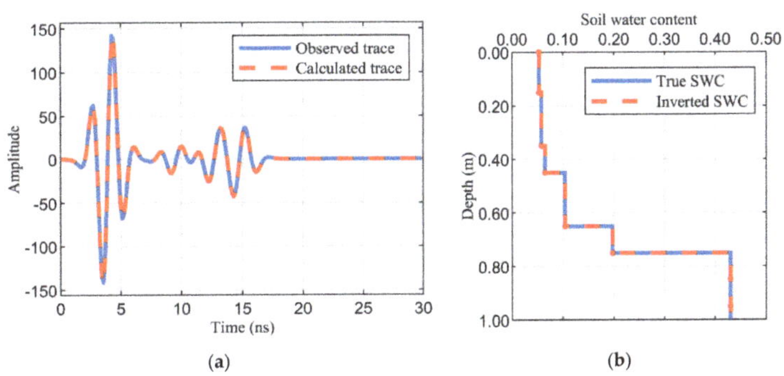

Figure 3. Inversion results of data without noise for Model A: (**a**) observed GPR trace (blue solid line) and calculated GPR response (red dashed line); (**b**) true SWC (blue solid line) and inverted SWC (red dashed line).

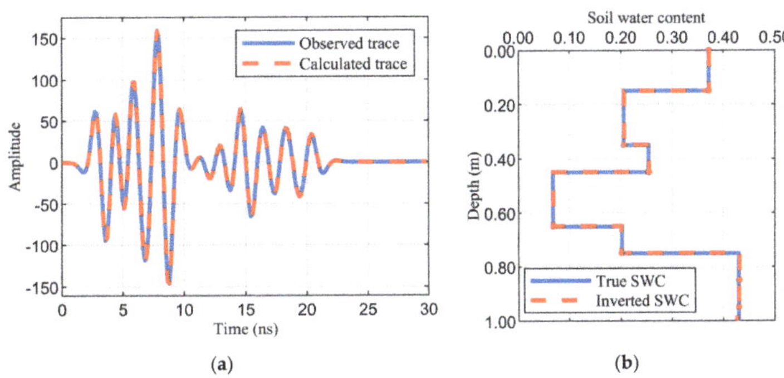

Figure 4. Inversion results of data without noise for Model B: (**a**) observed GPR trace (blue solid line) and calculated GPR response (red dashed line); (**b**) true SWC (blue solid line) and inverted SWC (red dashed line).

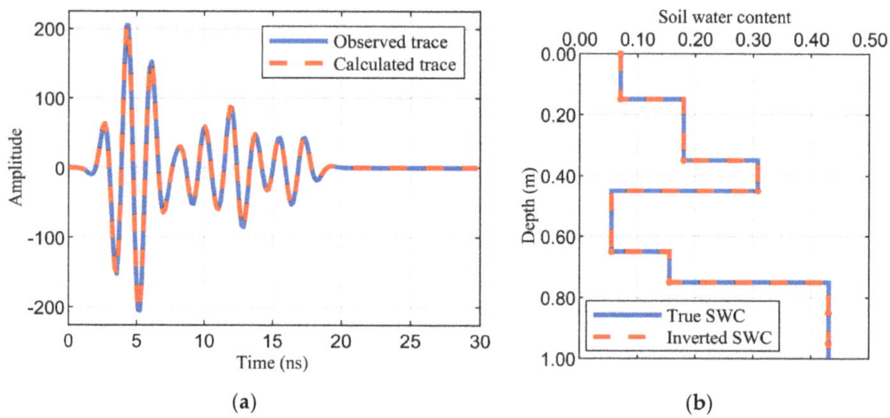

Figure 5. Inversion results of data without noise for Model C: (**a**) observed GPR trace (blue solid line) and calculated GPR response (red dashed line); (**b**) true SWC (blue solid line) and inverted SWC (red dashed line).

3.1.1. Synthetic Data Without Noise

The performance of the proposed inversion scheme is first investigated on synthetic data without noise. For all these three models, the population size is set to be $N = 10 \times D$, where D is the dimension of parameter space. Based on a few previous tests, the maximum number of iterations for data without noise is set to be 1000. The upper and lower boundaries for the search space depart 50% or more from the values of the true models for all the three models.

Figures 3–5 provide a comprehensive illustration of the detailed inversion results for Model A, B, and C, respectively, when there is no noise in the data. Figures 3, 4 and 5a show the fitness between the calculated traces (depicted by red dashed lines) that were derived from the final models inverted using the proposed scheme and the observed traces (represented by blue solid lines) originating from the synthetic models. It can be seen that the calculated traces fit with the observed traces very closely. The differences between the calculated traces and the observed traces for each model are so minimal that they are almost imperceptible, which indicates that the proposed inversion scheme is highly accurate and reliable for applying to data without noise. Figures 3, 4 and 5b present a detailed comparison of the inverted SWC (represented by red dashed lines) against the synthetic SWC model (depicted by blue solid lines). It is evident that all the three true models are almost accurately recovered by the proposed inversion scheme at each layer. These comprehensive results not only demonstrate the effectiveness of the method but also validate that the proposed inversion scheme possesses the ability to accurately invert the SWC and layer thicknesses directly from the GPR waveform data. From Tables 2–4, it can also be seen that the inverted SWC and layer thicknesses match with the synthetic SWC and layer thicknesses almost perfectly at every layer, which further demonstrate the accuracy and the capability of the proposed inversion scheme for estimating the SWC and layer thicknesses based on the GPR waveform data with HHO.

Table 2. The inversion results for Model A.

Layer	Model Parameters		Inversion Results for Data Without Noise		Inversion Results for Data with Noise	
	SWC [1]	z [2]	SWC	z	SWC	z
1	0.05	0.15	0.05	0.15	0.05	0.15
2	0.05	0.20	0.06	0.20	0.06	0.20
3	0.07	0.10	0.07	0.10	0.06	0.10
4	0.10	0.20	0.10	0.20	0.10	0.21
5	0.20	0.10	0.20	0.10	0.19	0.10
6	0.43	0.10	0.43	0.10	0.40	0.09
7	0.43	0.10	0.43	0.10	0.43	0.10
8	0.43	∞ [3]	0.43	∞	0.43	∞

[1] SWC is the abbreviation of soil water content. [2] z represents the layer thickness. [3] ∞ represents the infinite half space.

Table 3. The inversion results for Model B.

Layer	Model Parameters		Inversion Results for Data Without Noise		Inversion Results for Data with Noise	
	SWC [1]	z [2]	SWC	z	SWC	z
1	0.37	0.15	0.37	0.15	0.37	0.15
2	0.21	0.20	0.20	0.20	0.20	0.19
3	0.25	0.10	0.25	0.10	0.25	0.09
4	0.07	0.20	0.07	0.20	0.07	0.20
5	0.20	0.10	0.20	0.10	0.20	0.10
6	0.43	0.10	0.43	0.10	0.43	0.09
7	0.43	0.10	0.43	0.09	0.43	0.10
8	0.43	∞ [3]	0.43	∞	0.43	∞

[1] SWC is the abbreviation of soil water content. [2] z represents the layer thickness. [3] ∞ represents the infinite half space.

Table 4. The inversion results for Model C.

Layer	Model Parameters		Inversion Results for Data Without Noise		Inversion Results for Data with Noise	
	SWC [1]	z [2]	SWC	z	SWC	z
1	0.07	0.15	0.07	0.15	0.07	0.14
2	0.18	0.20	0.18	0.20	0.18	0.20
3	0.31	0.10	0.31	0.10	0.29	0.10
4	0.06	0.20	0.05	0.19	0.05	0.20
5	0.16	0.10	0.16	0.10	0.16	0.10
6	0.43	0.10	0.43	0.10	0.42	0.09
7	0.43	0.10	0.43	0.10	0.43	0.09
8	0.43	∞ [3]	0.43	∞	0.43	∞

[1] SWC is the abbreviation of soil water content. [2] z represents the layer thickness. [3] ∞ represents the infinite half space.

3.1.2. Synthetic Data with Noise

When dealing with real data that has been collected under real-world field conditions, it is an unavoidable fact that interference (including noise and clutter) is inevitable. In this study, we only consider the influence of the noise. Consequently, we deliberately introduced 10% white Gaussian noise to the GPR data from Models A, B, and C to further verify the performance of the proposed inversion scheme. After this addition of noise, we tested the proposed inversion scheme using the data with noise to ensure that our method remains reliable under a less-than-ideal condition.

Figures 6–8 illustrate the inversion results achieved through the proposed inversion scheme within a realistic setting. It becomes evident that the proposed inversion scheme still exhibits a robust performance. Although the data are contaminated with noise, the scheme is still capable of obtaining the acceptable solutions. In Figures 6, 7 and 8a, it can be clearly observed that the calculated traces (indicated by red dashed lines) derived from the inverted models still maintain a satisfactory alignment with the observed GPR traces (represented by blue solid lines). However, we can still see the impact of noise within the data. Specifically, when the amplitudes of the signal are diminished towards the end of the traces, it becomes challenging to identify the signal and the noise. Consequently, the fitness between the calculated traces and observed traces is observed to be less optimal towards the terminal parts of the traces compared to the earlier sections. This can also be demonstrated in Figures 6, 7 and 8b.

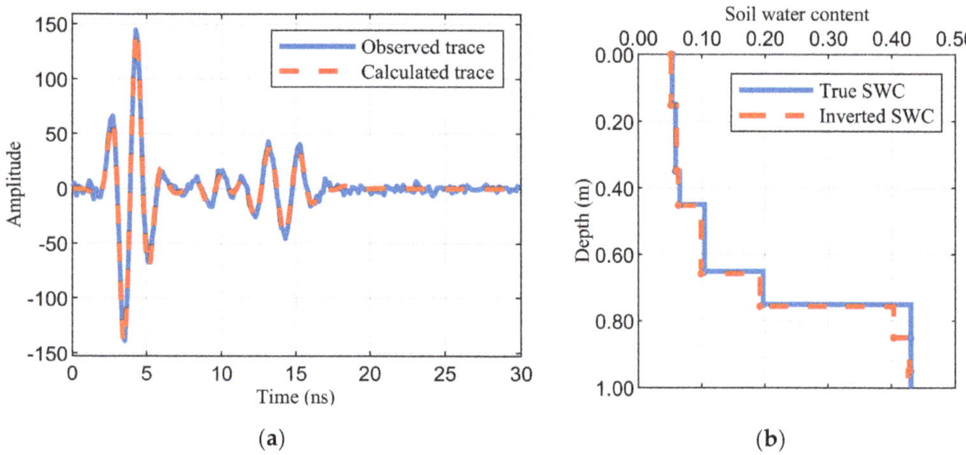

Figure 6. Inversion results of data with noise for Model A: (**a**) observed GPR trace (blue solid line) and calculated GPR response (red dashed line); (**b**) true SWC (blue solid line) and inverted SWC (red dashed line).

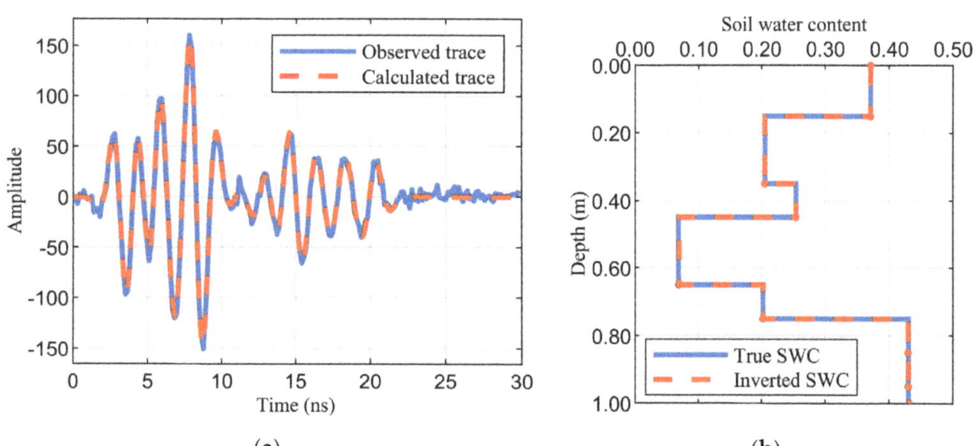

Figure 7. Inversion results of data with noise for Model B: (**a**) observed GPR trace (blue solid line) and calculated GPR response (red dashed line); (**b**) true SWC (blue solid line) and inverted SWC (red dashed line).

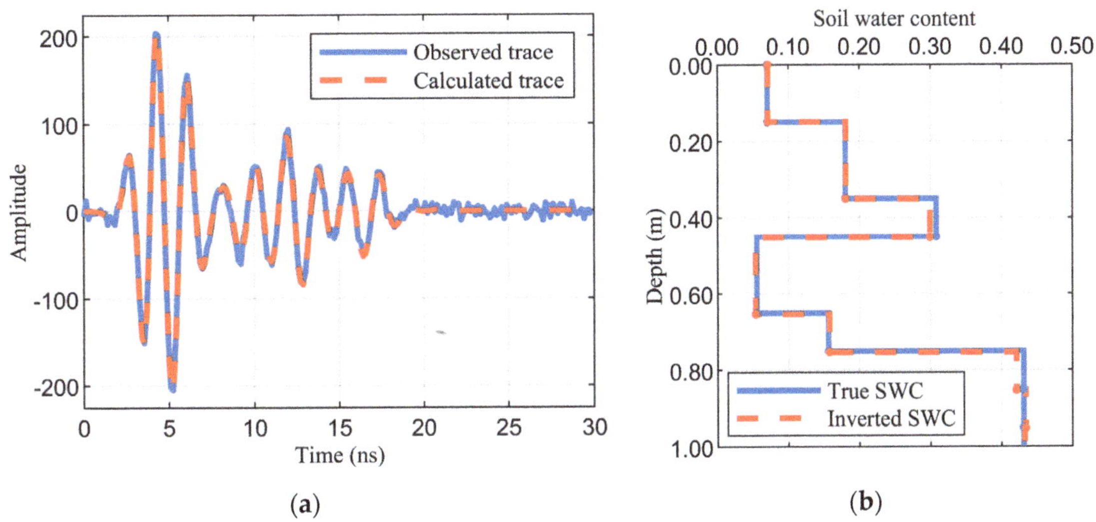

Figure 8. Inversion results of data with noise for Model C: (**a**) observed GPR trace (blue solid line) and calculated GPR response (red dashed line); (**b**) true SWC (blue solid line) and inverted SWC (red dashed line).

Figures 6, 7 and 8b present the results of SWC and layer thicknesses that have been inverted by the proposed inversion scheme based on the data with noise. It can be observed that the SWC and layer thicknesses still can be effectively and accurately recovered by the proposed inversion scheme even with the presence of noise. The inverted models (represented by red dashed lines) match well with the true models (indicated by blue solid lines) at each layer. Nevertheless, with the increases in the depth, the fitness becomes worse, particularly at the depths where the SWC almost reaches saturated. This trend corresponds to the end part of the GPR traces, which also suggests that the calculated traces do not align perfectly with the observed traces due to the disruptive effects of noise. Despite this, the overall results still serve to confirm the efficacy of the proposed inversion scheme in determining the SWC and the layer thicknesses by GPR waveform data, even if the data are impacted by noise. However, it is important to note that when compared to the inversion results derived from data without noise, the inversion results derived from noisy data show a discernible decrease in accuracy and reliability.

3.1.3. Analysis for Convergence Behavior

In order to examine the convergence behavior of HHO, Figure 9 displays the fitness evolution during HHO iterations using synthetic data both without and with noise from Model C. This offers significant understanding of the efficiency of the proposed inversion scheme. For data without noise, the fitness curves gradually converge to zero. The fitness curves gradually converge to a constant value for data with noise, demonstrating the resilience of the inversion scheme in the presence of disturbances. These findings indicate that the HHO algorithm simultaneously exhibits a high capacity for avoiding local optima and rapid convergence, which are essential characteristics for optimization problems. The ability to maintain a steady convergence rate even when faced with noisy data demonstrate the algorithm's versatility and potential for application.

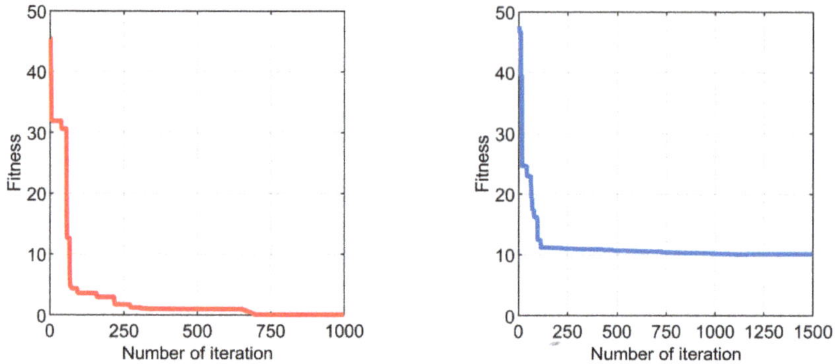

Figure 9. Fitness behavior of Model C using the HHO algorithm: (**a**) fitness behavior for data without noise; (**b**) fitness behavior for data with noise.

3.2. Comparisons with Other Algorithms

In order to verify the effectiveness evaluation of the proposed inversion scheme with HHO, we also performed a comparative analysis for the results of the HHO algorithm with other algorithms including the particle swarm optimization (PSO) and grey wolf optimizer (GWO) [53].

The synthetic data without and with noise which originated from Model C used in HHO inversion were, respectively, utilized to estimate the SWC and layer thicknesses from the GPR waveform data using PSO and GWO. For the sake of comparison and to ensure consistency, we applied the same search space and parameters, including the population size and the maximum number of iterations, as those that were used for the HHO inversion. Specifically, when conducting the inversion with the PSO algorithm, the inertia weighting factor was set to be $w = 0.8$, and the scaling factors were selected to be $c_1 = 1.8$ and $c_2 = 2.0$, respectively.

Figures 10 and 11 display the comparison of the inversion results obtained from three distinct optimization algorithms: HHO, PSO, and GWO. As depicted in Figure 10a, it can be observed that the calculated trace derived from the inversion results inverted by the HHO algorithm (represented by red solid line) exhibits a superior fit to the observed trace (indicated by green star line) in comparison to the calculated trace derived from the inversion results inverted by the PSO algorithm (represented by yellow solid line) and the GWO algorithm (represented by blue solid line). The red solid line almost perfectly aligns with the green star line at all significant peaks and valleys, demonstrating a high degree of precision in capturing the dynamic variations in the trace. In contrast, the blue solid line tracks the green star line with less accuracy than the red solid line, yet it still outperforms the yellow solid line in terms of following the observed trace. The phenomena suggests that the HHO algorithm is adept at accurately capturing the dynamic changes in the trace. When examining the entire trace, the red solid line consistently shows a higher degree of conformity with the shape of the trace compared to both the blue solid line and the yellow solid line, which indicates that the HHO algorithm possesses a higher approximation accuracy than both the GWO and PSO algorithm, highlighting its potential superiority in handling such optimization tasks. In Figure 11a, results similar to those in Figure 10a can be seen. For trace fitting, the HHO algorithm still exhibits a superior fit to the observed trace, even though the noise exists. Compared to both the blue and yellow solid lines, the red solid line matches better than the other two solid line and aligns with the green star line more accurately than the blue and yellow solid lines. This phenomena further demonstrates the potential superiority of the HHO algorithm although the existence of the noise.

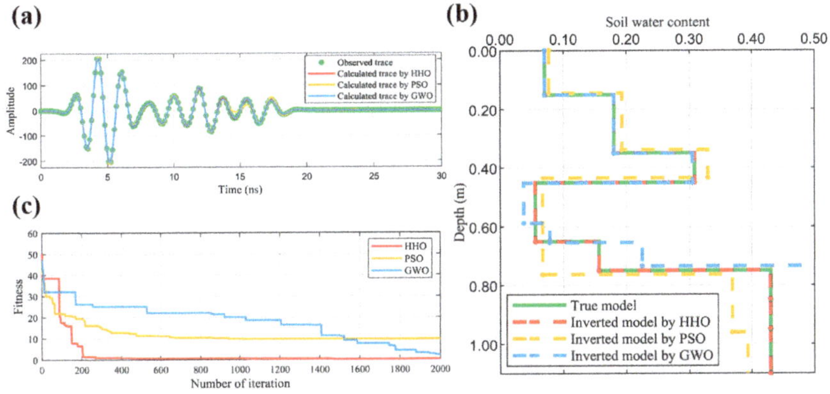

Figure 10. Comparison of inverted results of data without noise for Model C with HHO, PSO, and GWO: (**a**) comparison of calculated traces from inverted models; (**b**) comparison of fitness behavior; (**c**) comparison of inverted results.

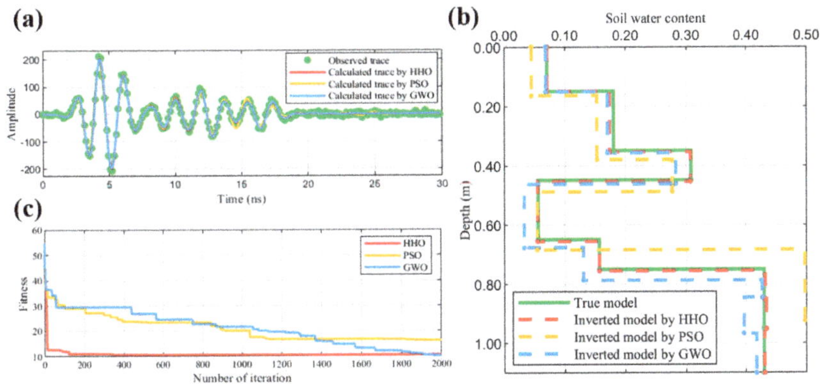

Figure 11. Comparison of inverted results of data with noise for Model C with HHO, PSO, and GWO: (**a**) comparison of calculated traces from inverted models; (**b**) comparison of fitness behavior; (**c**) comparison of inverted results.

Figure 10b shows a detailed comparison between the true SWC model and the SWC models inverted by using three different optimization algorithms: HHO, PSO, and GWO. Upon inspecting the degree of overlap among the four curves, it becomes evident that the red dashed line, representing the SWC inverted by the HHO, aligns more closely with the true SWC model. Conversely, the yellow dashed curve representing the SWC inverted by PSO and the blue dashed curve representing the SWC inverted by GWO exhibit more significant deviations from the true models. This finding further demonstrates the superior accuracy of the HHO algorithm in determining the SWC and layer thicknesses from GPR waveform data. Furthermore, this also implies that in terms of the correlation between SWC and depth, the SWC and layer thicknesses model inverted by the HHO algorithm is more capable of simulating real-world conditions with a higher level of precision. In Figure 11b, the inversion results are worse compared to the inversion results based on data without noise for the PSO and GWO algorithms. However, the inversion results of HHO still match closely with the true SWC model. This further demonstrates the superiority of the HHO algorithm for inverting SWC and layer thicknesses from data with noise.

At the same time, we also calculated the standard deviation (STD) and the overall average error (OAE) between the true model and the inverted model for these three optimization algorithms in Tables 5 and 6. After thorough analysis, it can be seen that the HHO algorithm exhibited the lowest values for both the STD and the OAE when compared to the other two algorithms. This result provides compelling evidence supporting the performance of the HHO algorithm in estimating the SWC and the layer thicknesses through the GPR waveform inversion. However, we can also see that the STD and OAE are all larger than that for data without noise.

Table 5. The comparison of STD and OAE of three algorithms for data without noise from Model C.

	HHO		PSO		GWO	
	STD [1]	**OAE** [2]	**STD**	**OAE**	**STD**	**OAE**
Model C	12.99%	0.03%	15.24%	6.86%	14.95%	4.29%

[1] STD is the standard deviation. [2] OAE is the overall average error.

Table 6. The comparison of STD and OAE of three algorithms for data with noise from Model C.

	HHO		PSO		GWO	
	STD [1]	**OAE** [2]	**STD**	**OAE**	**STD**	**OAE**
Model C	13.07%	0.17%	20.52%	12.97%	12.67%	1.56%

[1] STD is the standard deviation. [2] OAE is the overall average error.

Figures 10c and 11c display the comparison of the convergence behavior among these three optimization algorithms. It can be observed that the red solid line descends to a stable value more rapidly than the blue solid line and the yellow line, which means that the HHO algorithm can achieve convergence faster than the GWO and PSO algorithms. In the case of PSO in Figure 10c, the yellow solid line first reaches a stable value approximately after the 60th iteration. For GWO, the blue solid line nears stability around the 1900th iteration; however, the blue solid line can reach a lower value than the yellow solid line after approximately the 1500th iteration. In the context of optimization problems, a lower fitness value usually implies a solution that is closer to the optimum or a superior solution. The PSO algorithm may become stuck in a local minimum. This underscores that the GWO algorithm can gain more precise inversion results compared to the PSO algorithm, but the efficiency of the GWO is lower than that of the PSO. The red solid line demonstrates that the HHO not only reaches a stable value faster than the other two algorithms, after approximately the 400th iteration, but that it also attains a lower value compared to the GWO and the PSO. However, in Figure 11c, the yellow solid line first reaches a stable value after approximately the 1000th iteration. The existence of noise has a great influence on the PSO algorithm. The HHO and GWO algorithm both show good noise resistance. These results indicate that the HHO algorithm is capable of finding solutions of higher quality.

Overall, it is evident that the HHO algorithm performs better than both the GWO and the PSO algorithms regarding precision, model inversion, and convergence speed. We think that the mechanism of HHO is superior to that of GWO and PSO. First, the GWO is superior to PSO because the search of the GWO is based on a strict internal hierarchy, unlike the PSO which randomly initializes the particles of the population. And the grey wolf group is led by the three best leaders to approach the prey, which ensures that the GWO can better search for the true value [53]. For the HHO algorithm, it has a strong dynamic searching ability. By switching between the exploration phase and the development phase, the algorithm can effectively find the optimal solution in the solution space. In addition, the HHO introduced a variety of different hunting strategies, including multiple encirclement strategies, tracking, and raiding. This diversified search strategy enables HHO to better adapt to complex optimization problems and enhance the global search capability. Therefore, in the early stage, the HHO can achieve convergence faster and maintain a lower fitness value than the other two algorithms. This analysis offers significant insights for selecting the appropriate optimization algorithm.

3.3. Experimental Field Data Application

In order to assess the potential and practical use of the proposed inversion scheme, we implemented it on GPR data acquired at SCERES, an experimental site situated on the CNRS campus in Strasbourg-Cronenbourg, France. Figure 12 displays the underground structure of SCERES. It features a controlled artificial aquifer within a lined concrete basin. The edges of the basin are sealed to avoid any leakages to the outside. The inner dimensions of the basin are 24 m × 12 m × 3 m. The basin is equipped with the necessary tools for aquifer system and water table management. The basin is filled with three types of sand, and the specific parameters of the basin are detailed in Table 7. The water table lies roughly 0.85 m beneath the surface of the basin.

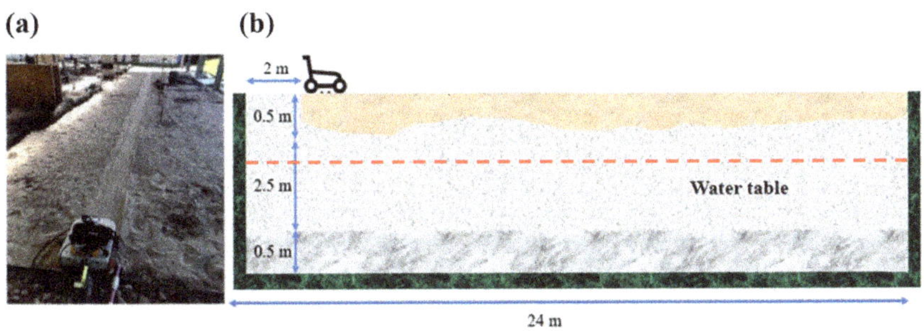

Figure 12. (**a**) Illustration for experimental site; (**b**) illustration for the subsurface structure of the experimental site (Modified by [60]).

Table 7. Parameters for the basin at the experimental site.

Layer	Thickness (m)	Porosity (%)
1	0.5	43
2	2.5	40
3	0.5	38

We used a MALA RAMAC system (which is produced by MALA company located at Stockholm, Sweden) equipped with 500 MHz antennas to collect GPR data across the basin, from one side to the other. Table 8 lists the technical index for the 500 MHz antennas. The entire length of the survey line is 24 m and we configured the system to collect one trace every 0.02 m. A DC filter and a shift of 20 samples are applied to the GPR profile. Figure 13 displays the processed GPR radargram [53]. We selected the 700th, 750th, 800th, and 850th traces as the observed traces for subsequent inversion based on experimental field data. To simulate synthetic GPR traces, it is essential to give a source wavelet. In this process, we first calculate the source wavelet by averaging the first arrivals from all traces, applying a Blackman window for weighting [61]. Subsequently, we employed a Butterworth band-pass filter to refine the source wavelet further. We took the 700th trace as an example. Figure 14 displays both the source wavelet and its corresponding spectrum.

Table 8. Technical indexes for the 500 MHz antennas.

Items	Technical Index
Frequency of antennas	500 MHz
Maximum detection depth	6 m
Radial resolution	0.05 m
Offset	0.18 m
Size	0.5 × 0.3 × 0.16 m
Weight	5.0 kg

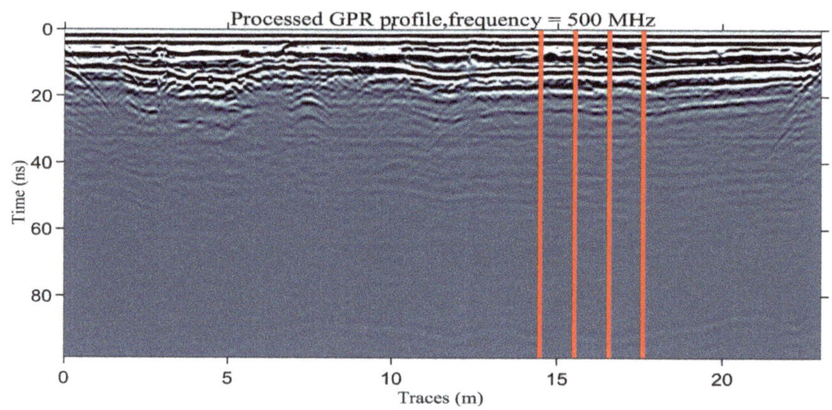

Figure 13. Processed observed GPR profile from experimental site. The red solid lines represent the position of the selected traces, which are regarded as the observed traces (the 700th, 750th, 800th, and 850th traces).

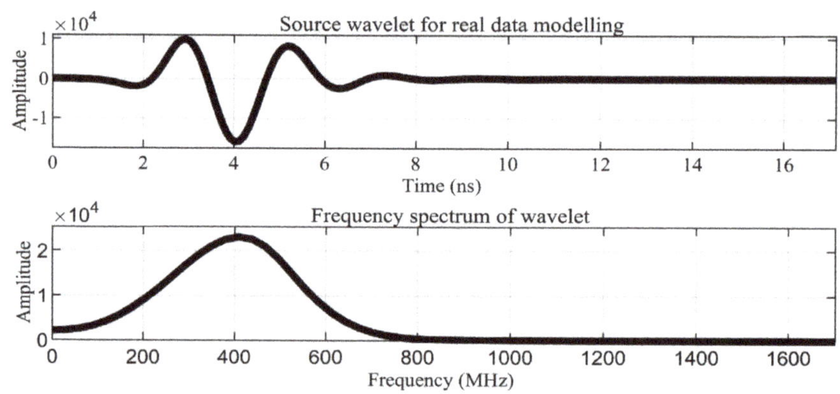

Figure 14. The source wavelet for real data modeling and the corresponding frequency spectrum for the 700th trace signal.

For the experimental field data, we utilized a comparable inversion scheme as that employed for the synthetic data. The parameters that required determination were the SWC and the layer thickness. The SWC measured by Sentek sensors and the depth of each Sentek sensor were used as prior information [62]. Based on this prior information, the upper and lower boundaries for the search areas were established. Table 9 outlines the search space and inversion results for the experimental field data. Throughout the inversion, we adopted an 11-layer subsurface configuration, based on the placement of the Sentek sensors, to execute the inversion using the proposed inversion scheme.

Table 9. The upper and lower limitations for search space and inversion results for field data.

Layer	Search Space Limitations		Inversion Results	
	SWC [1]	z [2]	SWC	z
1	0.04–0.11	0.08–0.23	0.07	0.18
2	0.07–0.21	0.05–0.15	0.08	0.15
3	0.10–0.30	0.05–0.15	0.22	0.05
4	0.13–0.38	0.05–0.15	0.32	0.07
5	0.04–0.05	0.05–0.15	0.04	0.08
6	0.04–0.11	0.05–0.15	0.10	0.14
7	0.11–0.32	0.05–0.15	0.14	0.08
8	0.20–0.43	0.05–0.15	0.37	0.08
9	0.20–0.43	0.05–0.15	0.21	0.12
10	0.20–0.43	0.05–0.15	0.36	0.05
11	0.20–0.43	∞ [3]	0.41	∞

[1] SWC is the abbreviation of soil water content. [2] z represents the layer thickness. [3] ∞ represents the infinite half space.

Figures 15–18 show the inversion results for the field data. From Figures 15, 16, 17 and 18a, it can be seen that all the calculated GPR traces (represented by red dashed line) derived from the results inverted by the proposed inversion scheme align quite well the observed GPR traces (represented by blue solid line). Figures 15, 16, 17 and 18b contrasts the inverted SWC and layer thicknesses obtained through the proposed inversion scheme, with the measured SWC by Sentek sensors. As illustrated in Figures 15, 16, 17 and 18b, the inverted models (indicated by red dashed line) based on the proposed inversion scheme correspond with the measured SWC captured by Sentek sensors (indicated by blue solid line). They exhibit a comparable trend. Until a depth of approximately 0.55 m (marking the boundary between the two sandy layers), the SWC increases with the increase in depth. Approximately at 0.55 m, the SWC diminishes abruptly, potentially due to the influence of the boundary and the alterations in the medium. In the subsequent layers, the SWC continues to ascend with the increase in depth until it reaches the water table. There is a slight variation in the inverted model around 0.85 m. This may stem from the noise being regarded as signals that were fitted during the inversion. Nevertheless, the overall trend remains similar. Additionally, Figures 15, 16, 17 and 18c demonstrate that all the fitness curves diminish at the beginning of the iterations, and then subsequently converge to a relatively constant value following the 1400th iteration. This convergence behavior is indicative of the robustness and stability of the inversion algorithm as it iteratively refines the calculated traces to better fit the observed traces. All inversion results further demonstrate the effectiveness of the proposed inversion scheme. However, we observed variations in the results obtained from different traces. We believe that the SWC exhibits not only vertical heterogeneity but also horizontal heterogeneity. Future studies should take this heterogeneity into account.

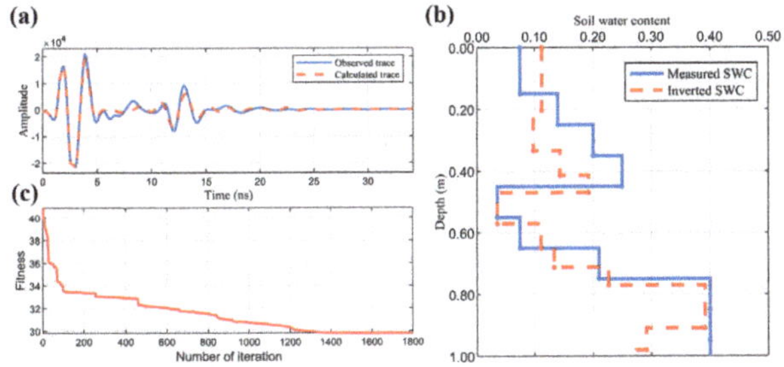

Figure 15. The inversion results of field data for 700th trace: (**a**) observed trace (blue solid line) and calculated trace (red dashed line); (**b**) measured SWC with Sentek (blue solid line) and inverted SWC (red dashed line); (**c**) fitness behavior of iteration.

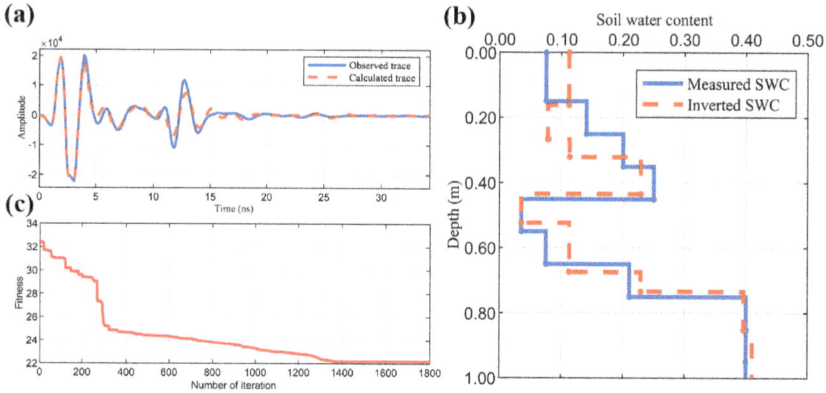

Figure 16. The inversion results of field data for 750th trace: (**a**) observed trace (blue solid line) and calculated trace (red dashed line); (**b**) measured SWC with Sentek (blue solid line) and inverted SWC (red dashed line); (**c**) fitness behavior of iteration.

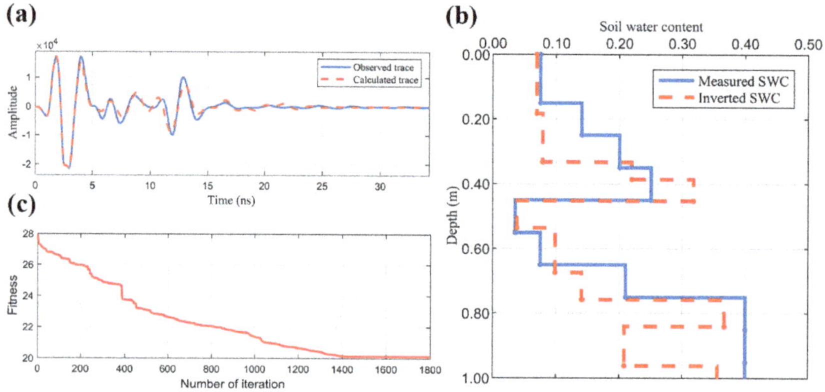

Figure 17. The inversion results of field data for 800th trace: (**a**) observed trace (blue solid line) and calculated trace (red dashed line); (**b**) measured SWC with Sentek (blue solid line) and inverted SWC (red dashed line); (**c**) fitness behavior of iteration.

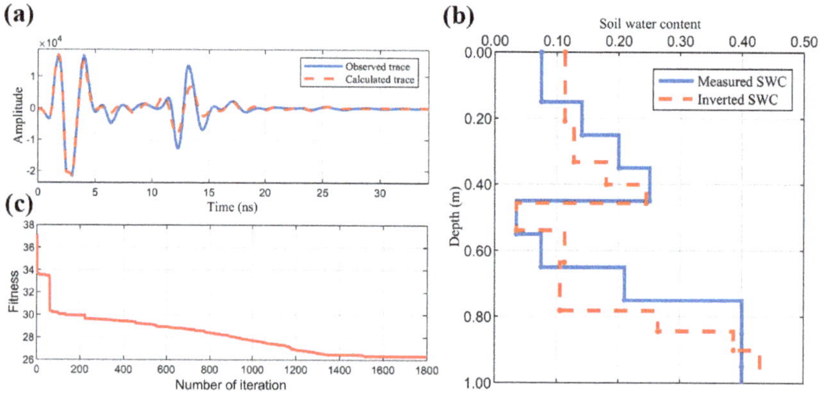

Figure 18. The inversion results of field data for 850th trace: (**a**) observed trace (blue solid line) and calculated trace (red dashed line); (**b**) measured SWC with Sentek (blue solid line) and inverted SWC (red dashed line); (**c**) fitness behavior of iteration.

4. Conclusions

This study proposed a novel and powerful GPR FWI scheme based on the HHO algorithm to estimate the SWC and the layer thicknesses with a fixing quality factor. The HHO algorithm is inspired by the social behavior of Harris's Hawks, mimicking their hunting behavior. Three main steps of exploration, the transition from exploration to exploitation, and exploitation, are used to realize the SWC and layer thicknesses estimation based on GPR waveform inversion. The proposed inversion scheme was first tested on three synthetic cases. Each case includes data without and with noise. In order to simulate more real-world scenarios where there is limited a priori information, we set wider boundaries for inversion search space. For further verification, we also compared the inversion results inverted from the HHO algorithm with the results from the PSO and GWO algorithms. The inversion results from the synthetic data demonstrate that the HHO algorithm is highly capable and converges rapidly to estimate the SWC and layer thicknesses from GPR waveform data. Subsequently, we further applied the proposed inversion scheme to field data collected from an experimental site at the CNRS campus in Strasbourg-Cronenbourg, France. For the experimental field data, we adopted the SWC measured by Sentek sensors and the depth of each Sentek sensor as a priori information. During the test with the experimental field data, the calculated trace closely matched the observed trace, and the inverted SWC showed a similar trend to the SWC measured by the Sentek sensors. The fitness curves from all tests on both synthetic and experimental field data demonstrates that the proposed scheme effectively avoids high local optima and achieves very fast convergence. Hence, we strongly suggest using the HHO algorithm for the SWC estimation based on GPR FWI.

This research is the initial endeavor to apply the HHO algorithm for GPR waveform inversion to invert the SWC and layer thicknesses. It also offers significant insights into the effectiveness of HHO in various geophysical inversions. In the future, we plan to improve our method to enhance the robustness and efficiency of HHO. At the same time, we also plan to consider the influence of more interference like clutter and further test the proposed inversion scheme on data with the interference of clutter. In addition, we will also continue to evaluate other cutting-edge algorithms to address the same issue. Furthermore, this technique can also be extended to challenges in other areas, including engineering, hydrology, and archeology, leveraging GPR technology.

Author Contributions: Conceptualization, H.Q. and M.Z.; methodology, H.Q., M.Z. and M.B.; software, H.Q., M.Z. and M.B.; validation, H.Q.; formal analysis, H.Q. and M.Z.; investigation, H.Q. and M.Z.; resources, H.Q.; data curation, H.Q. and M.Z.; writing—original draft preparation, H.Q.; writing—review and editing, M.Z. and M.B.; visualization, H.Q.; supervision, M.Z.; project administration, H.Q.; funding acquisition, H.Q. All authors have read and agreed to the published version of the manuscript.

Funding: This research was funded by the China Geological Survey Project, grant number DD20221638, the National Natural Science Foundation of China, grant number 42374125, the Chinese National Nonprofit Institute Research Grant, grant number AS2024J12, and the Chinese National Nonprofit Institute Research Grant, grant number AS2024J02.

Data Availability Statement: The raw data supporting the conclusions of this article will be made available by the authors on request.

Acknowledgments: The authors thank Nolwenn Lesparre, Benjamin Belfort, Pascal Friedmann, Jean-François Girard, François Lehmann, and Olivier Razakarisoa for helping us to acquire GPR data. This experiment was funded by the ITES laboratory via an internal call. The authors are very grateful to the anonymous reviewers for their valuable suggestions and comments that helped to improve the quality of the article.

Conflicts of Interest: The authors declare no conflicts of interest.

References

1. Vereecken, H.; Huisman, J.A.; Bogena, H.; Vanderborght, J.; Vrugt, J.A.; Hopmans, J.W. On the value of soil moisture measurements in vadose zone hydrology: A review. *Water Resour. Res.* **2008**, *44*, W00D06. [CrossRef]
2. Dobriyal, P.; Qureshi, A.; Badola, R.; Hussain, S.A. A review of the methods available for estimating soil moisture and its implications for water resource management. *J. Hydrol.* **2012**, *458–459*, 110–117. [CrossRef]
3. Blonquist, J.M.; Jones, S.B.; Robinson, D.A. Precise irrigation scheduling for turfgrass using a subsurface electromagnetic soil moisture sensor. *Agric. Water Manag.* **2006**, *84*, 153–165. [CrossRef]
4. Cai, G.; Vanderborght, J.; Langensiepen, M.; Schnepf, A.; Hüging, H.; Vereecken, H. Root growth, water uptake, and sap flow of winter wheat in response to different soil water conditions. *Hydrol. Earth Syst. Sci.* **2018**, *22*, 2449–2470. [CrossRef]
5. Robinson, D.A.; Campbell, C.S.; Hopmans, J.W.; Hornbuckle, B.K.; Jones, S.B.; Knight, R.; Ogden, F.; Selker, J.; Wendroth, O. Soil Moisture Measurement for Ecological and Hydrological Watershed-Scale Observatories: A Review. *Vadose Zone J.* **2008**, *7*, 358–389. [CrossRef]
6. Klotzsche, A.; Jonard, F.; Looms, M.C.; van der Kruk, J.; Huisman, J.A. Measuring Soil Water Content with Ground Penetrating Radar: A Decade of Progress. *Vadose Zone J.* **2018**, *17*, 180052. [CrossRef]
7. Huisman, J.A.; Hubbard, S.S.; Redman, J.D.; Annan, A.P. Measuring Soil Water Content with Ground Penetrating Radar: A Review. *Vadose Zone J.* **2003**, *2*, 476–491. [CrossRef]
8. Liu, X.; Chen, J.; Cui, X.; Liu, Q.; Cao, X.; Chen, X. Measurement of soil water content using ground-penetrating radar: A review of current methods. *Int. J. Digit. Earth* **2019**, *12*, 95–118. [CrossRef]
9. Zhang, M.; Feng, X.; Bano, M.; Xing, H.; Wang, T.; Liang, W.; Zhou, H.; Dong, Z.; An, Y.; Zhang, Y. Review of Ground Penetrating Radar Applications for Water Dynamics Studies in Unsaturated Zone. *Remote Sens.* **2022**, *14*, 5993. [CrossRef]
10. Ercoli, M.; Di Matteo, L.; Pauselli, C.; Mancinelli, P.; Frapiccini, S.; Talegalli, L.; Cannata, A. Integrated GPR and laboratory water content measures of sandy soils: From laboratory to field scale. *Constr. Build. Mater.* **2018**, *159*, 734–744. [CrossRef]
11. Grote, K.; Hubbard, S.; Rubin, Y. GPR monitoring of volumetric water content in soils applied to highway construction and maintenance. *Lead. Edge* **2002**, *21*, 482–504. [CrossRef]
12. Irving, J.; Knight, R.; Holliger, K. Estimation of the lateral correlation structure of subsurface water content from surface-based ground-penetrating radar reflection images. *Water Resour. Res.* **2009**, *45*, W12404. [CrossRef]
13. Klenk, P.; Jaumann, S.; Roth, K. Quantitative high-resolution observations of soil water dynamics in a complicated architecture using time-lapse ground-penetrating radar. *Hydrol. Earth Syst. Sci.* **2015**, *19*, 1125–1139. [CrossRef]
14. Lunt, I.A.; Hubbard, S.S.; Rubin, Y. Soil moisture content estimation using ground-penetrating radar reflection data. *J. Hydrol.* **2005**, *307*, 254–269. [CrossRef]
15. Moua, R.; Lesparre, N.; Girard, J.F.; Belfort, B.; Lehmann, F.; Younes, A. On the interest of ground penetrating radar data for the estimation of unsaturated soil parameters. *EGUsphere* **2022**, *2022*, 1–24. [CrossRef]
16. Pettinelli, E.; Di Matteo, A.; Beaubien, S.E.; Mattei, E.; Lauro, S.E.; Galli, A.; Vannaroni, G. A controlled experiment to investigate the correlation between early-time signal attributes of ground-coupled radar and soil dielectric properties. *J. Appl. Geophys.* **2014**, *101*, 68–76. [CrossRef]
17. Steelman, C.M.; Endres, A.L. An examination of direct ground wave soil moisture monitoring over an annual cycle of soil conditions. *Water Resour. Res.* **2010**, *46*, W11533. [CrossRef]
18. Yu, Y.; Klotzsche, A.; Weihermüller, L.; Huisman, J.A.; Vanderborght, J.; Vereecken, H.; van der Kruk, J. Measuring vertical soil water content profiles by combining horizontal borehole and dispersive surface ground penetrating radar data. *Near Surf. Geophys.* **2020**, *18*, 275–294. [CrossRef]
19. Huisman, J.A.; Snepvangers, J.J.J.C.; Bouten, W.; Heuvelink, G.B.M. Monitoring Temporal Development of Spatial Soil Water Content Variation: Comparison of Ground Penetrating Radar and Time Domain Reflectometry. *Vadose Zone J.* **2003**, *2*, 519–529. [CrossRef]
20. Allroggen, N.; van Schaik, N.L.M.B.; Tronicke, J. 4D ground-penetrating radar during a plot scale dye tracer experiment. *J. Appl. Geophys.* **2015**, *118*, 139–144. [CrossRef]
21. Bradford, J.H. Measuring Water Content Heterogeneity Using Multifold GPR with Reflection Tomography. *Vadose Zone J.* **2008**, *7*, 184–193. [CrossRef]
22. Strobbia, C.; Cassiani, G. Multilayer ground-penetrating radar guided waves in shallow soil layers for estimating soil water content. *Geophysics* **2007**, *72*, J17–J29. [CrossRef]
23. Hubbard, S.; Grote, K.; Rubin, Y. Mapping the volumetric soil water content of a California vineyard using high-frequency GPR ground wave data. *Lead. Edge* **2002**, *21*, 552–559. [CrossRef]
24. Huisman, J.A.; Sperl, C.; Bouten, W.; Verstraten, J.M. Soil water content measurements at different scales: Accuracy of time domain reflectometry and ground-penetrating radar. *J. Hydrol.* **2001**, *245*, 48–58. [CrossRef]
25. Iwasaki, T.; Kuroda, S.; Saito, H.; Tobe, Y.; Suzuki, K.; Fujimaki, H.; Inoue, M. Monitoring Infiltration Process Seamlessly Using Array Ground Penetrating Radar. *Agric. Environ. Lett.* **2016**, *1*, 160002. [CrossRef]

26. Saito, H.; Kuroda, S.; Iwasaki, T.; Sala, J.; Fujimaki, H. Estimating infiltration front depth using time-lapse multioffset gathers obtained from ground-penetrating-radar antenna array. *Geophysics* **2021**, *86*, WB51–WB59. [CrossRef]
27. Binley, A.; Cassiani, G.; Middleton, R.; Winship, P. Vadose zone flow model parameterisation using cross-borehole radar and resistivity imaging. *J. Hydrol.* **2002**, *267*, 147–159. [CrossRef]
28. Galagedara, L.W.; Parkin, G.W.; Redman, J.D.; Endres, A.L. Assessment of Soil Moisture Content Measured by Borehole GPR and TDR Under Transient Irrigation and Drainage. *J. Environ. Eng. Geophys.* **2003**, *8*, 77–86. [CrossRef]
29. Klotzsche, A.; Lärm, L.; Vanderborght, J.; Cai, G.; Morandage, S.; Zörner, M.; Vereecken, H.; van der Kruk, J. Monitoring Soil Water Content Using Time-Lapse Horizontal Borehole GPR Data at the Field-Plot Scale. *Vadose Zone J.* **2019**, *18*, 190044. [CrossRef]
30. Kowalsky, M.B.; Finsterle, S.; Rubin, Y. Estimating flow parameter distributions using ground-penetrating radar and hydrological measurements during transient flow in the vadose zone. *Adv. Water Resour.* **2004**, *27*, 583–599. [CrossRef]
31. Rucker, D.F.; Ferré, T.P.A. Near-Surface Water Content Estimation with Borehole Ground Penetrating Radar Using Critically Refracted Waves. *Vadose Zone J.* **2003**, *2*, 247–252. [CrossRef]
32. Looms, M.C.; Jensen, K.H.; Binley, A.; Nielsen, L. Monitoring Unsaturated Flow and Transport Using Cross-Borehole Geophysical Methods. *Vadose Zone J.* **2008**, *7*, 227–237. [CrossRef]
33. Wijewardana, Y.G.N.S.; Galagedara, L.W. Estimation of spatio-temporal variability of soil water content in agricultural fields with ground penetrating radar. *J. Hydrol.* **2010**, *391*, 24–33. [CrossRef]
34. Strobach, E.; Harris, B.D.; Dupuis, J.C.; Kepic, A.W. Time-lapse borehole radar for monitoring rainfall infiltration through podosol horizons in a sandy vadose zone. *Water Resour. Res.* **2014**, *50*, 2140–2163. [CrossRef]
35. Cheng, Q.; Su, Q.; Binley, A.; Liu, J.; Zhang, Z.; Chen, X. Estimation of Surface Soil Moisture by a Multi-Elevation UAV-Based Ground Penetrating Radar. *Water Resour. Res.* **2023**, *59*, e2022WR032621. [CrossRef]
36. Jadoon, K.Z.; Lambot, S.; Scharnagl, B.; van der Kruk, J.; Slob, E.; Vereecken, H. Quantifying field-scale surface soil water content from proximal GPR signal inversion in the time domain. *Near Surf. Geophys.* **2010**, *8*, 483–491. [CrossRef]
37. Jonard, F.; Weihermüller, L.; Vereecken, H.; Lambot, S. Accounting for soil surface roughness in the inversion of ultrawideband off-ground GPR signal for soil moisture retrieval. *Geophysics* **2012**, *77*, H1–H7. [CrossRef]
38. Lambot, S.; Rhebergen, J.; van den Bosch, I.; Slob, E.C.; Vanclooster, M. Measuring the Soil Water Content Profile of a Sandy Soil with an Off-Ground Monostatic Ground Penetrating Radar. *Vadose Zone J.* **2004**, *3*, 1063–1071. [CrossRef]
39. Lambot, S.; Weihermüller, L.; Huisman, J.A.; Vereecken, H.; Vanclooster, M.; Slob, E.C. Analysis of air-launched ground-penetrating radar techniques to measure the soil surface water content. *Water Resour. Res.* **2006**, *42*, W11403. [CrossRef]
40. Minet, J.; Bogaert, P.; Vanclooster, M.; Lambot, S. Validation of ground penetrating radar full-waveform inversion for field scale soil moisture mapping. *J. Hydrol.* **2012**, *424–425*, 112–123. [CrossRef]
41. Serbin, G.; Or, D. Ground-penetrating radar measurement of soil water content dynamics using a suspended horn antenna. *IEEE Trans. Geosci. Remote Sens.* **2004**, *42*, 1695–1705. [CrossRef]
42. Tran, A.P.; Ardekani, M.R.M.; Lambot, S. Coupling of dielectric mixing models with full-wave ground-penetrating radar signal inversion for sandy-soil-moisture estimation. *Geophysics* **2012**, *77*, H33–H44. [CrossRef]
43. Wu, K.; Rodriguez, G.A.; Zajc, M.; Jacquemin, E.; Clément, M.; De Coster, A.; Lambot, S. A new drone-borne GPR for soil moisture mapping. *Remote Sens. Environ.* **2019**, *235*, 111456. [CrossRef]
44. Tarantola, A. Inversion of seismic reflection data in the acoustic approximation. *Geophysics* **1984**, *49*, 1259–1266. [CrossRef]
45. Klotzsche, A.; Vereecken, H.; van der Kruk, J. Review of crosshole ground-penetrating radar full-waveform inversion of experimental data: Recent developments, challenges, and pitfalls. *Geophysics* **2019**, *84*, H13–H28. [CrossRef]
46. Kruk, J.v.d.; Liu, T.; Mozaffari, A.; Gueting, N.; Klotzsche, A.; Vereecken, H.; Warren, C.; Giannopoulos, A. GPR full-waveform inversion, recent developments, and future opportunities. In Proceedings of the 2018 17th International Conference on Ground Penetrating Radar (GPR), Rapperswil, Switzerland, 18–21 June 2018; pp. 1–6.
47. Slob, E.; Sato, M.; Olhoeft, G. Surface and borehole ground-penetrating-radar developments. *Geophysics* **2010**, *75*, 75A103–75A120. [CrossRef]
48. Lambot, S.; Slob, E.; Chavarro, D.; Lubczynski, M.; Vereecken, H. Measuring soil surface water content in irrigated areas of southern Tunisia using full-waveform inversion of proximal GPR data. *Near Surf. Geophys.* **2008**, *6*, 403–410. [CrossRef]
49. Lambot, S.; Slob, E.; Rhebergen, J.; Lopera, O.; Jadoon, K.Z.; Vereecken, H. Remote Estimation of the Hydraulic Properties of a Sand Using Full-Waveform Integrated Hydrogeophysical Inversion of Time-Lapse, Off-Ground GPR Data. *Vadose Zone J.* **2009**, *8*, 743–754. [CrossRef]
50. Busch, S.; van der Kruk, J.; Bikowski, J.; Vereecken, H. Quantitative conductivity and permittivity estimation using full-waveform inversion of on-ground GPR data. *Geophysics* **2012**, *77*, H79–H91. [CrossRef]
51. Lavoué, F.; Brossier, R.; Métivier, L.; Garambois, S.; Virieux, J. Two-dimensional permittivity and conductivity imaging by full waveform inversion of multioffset GPR data: A frequency-domain quasi-Newton approach. *Geophys. J. Int.* **2014**, *197*, 248–268. [CrossRef]

52. Wang, X.; Feng, D. Multiparameter Full-Waveform Inversion of 3-D On-Ground GPR With a Modified Total Variation Regularization Scheme. *IEEE Geosci. Remote Sens. Lett.* **2021**, *18*, 466–470. [CrossRef]
53. Zhang, M.H.; Feng, X.; Bano, M.; Liu, C.; Liu, Q.; Wang, X. Soil water content estimation by using ground penetrating radar data full waveform inversion with grey wolf optimizer algorithm. *Vadose Zone J.* **2025**, *24*, e20379. [CrossRef]
54. Heidari, A.A.; Mirjalili, S.; Faris, H.; Aljarah, I.; Mafarja, M.; Chen, H. Harris hawks optimization: Algorithm and applications. *Future Gener. Comput. Syst.* **2019**, *97*, 849–872. [CrossRef]
55. Ouyang, C.; Liao, C.; Zhu, D.; Zheng, Y.; Zhou, C.; Li, T. Integrated improved Harris hawks optimization for global and engineering optimization. *Sci. Rep.* **2024**, *14*, 7445. [CrossRef] [PubMed]
56. Tripathy, B.K.; Reddy Maddikunta, P.K.; Pham, Q.-V.; Gadekallu, T.R.; Dev, K.; Pandya, S.; ElHalawany, B.M.; Sarfraz, S. Harris Hawk Optimization: A Survey onVariants and Applications. *Comput. Intell. Neurosci.* **2022**, *2022*, 2218594. [CrossRef]
57. Hussien, A.G.; Abualigah, L.; Abu Zitar, R.; Hashim, F.A.; Amin, M.; Saber, A.; Almotairi, K.H.; Gandomi, A.H. Recent Advances in Harris Hawks Optimization: A Comparative Study and Applications. *Electronics* **2022**, *11*, 1919. [CrossRef]
58. Topp, G.C.; Davis, J.L.; Annan, A.P. Electromagnetic determination of soil water content: Measurements in coaxial transmission lines. *Water Resour. Res.* **1980**, *16*, 574–582. [CrossRef]
59. Bano, M. Modelling of GPR waves for lossy media obeying a complex power law of frequency for dielectric permittivity. *Geophys. Prospect.* **2004**, *52*, 11–26. [CrossRef]
60. Zhang, M.H.; Bano, M.; Feng, X.; Lesparre, N.; Girard, J.F.; Razakarisoa, O.; Belfort, B.; Lehmann, F.; Friedmann, P. Estimating water content of unsaturated sandy soils by GPR during a drainage experiment. *IOP Conf. Ser. Earth Environ. Sci.* **2021**, *660*, 012018. [CrossRef]
61. Guillemoteau, J.; Bano, M.; Dujardin, J.-R. Influence of grain size, shape and compaction on georadar waves: Examples of aeolian dunes. *Geophys. J. Int.* **2012**, *190*, 1455–1463. [CrossRef]
62. Jellali, S.; Muntzer, P.; Razakarisoa, O.; Schäfer, G. Large Scale Experiment on Transport of Trichloroethylene in a Controlled Aquifer. *Transp. Porous Media* **2001**, *44*, 145–163. [CrossRef]

remote sensing MDPI

Article

Adaptive Variational Mode Decomposition and Principal Component Analysis-Based Denoising Scheme for Borehole Radar Data

Ding Yang [1], Cheng Guo [1,*], Raffaele Persico [2], Yajie Liu [1], Handing Liu [1], Changjin Bai [1], Chao Lian [3] and Qing Zhao [1]

1 School of Resources and Environment, University of Electronic Science and Technology of China, Chengdu 611731, China; yuesibi@hotmail.com (D.Y.); 202222070302@std.uestc.edu.cn (Y.L.); 202122070203@std.uestc.edu.cn (H.L.); 202222070301@std.uestc.edu.cn (C.B.); zhaoq@uestc.edu.cn (Q.Z.)
2 Department of Engineering of the Environment and the Territory DIAM, University of Calabria, Via P. Bucci, Cubo 45 A, 87036 Rende, Italy; raffaele.persico@unical.it
3 School of Information Science and Engineering, Northeastern University, Shenyang 110819, China; 2210326@stu.neu.edu.cn
* Correspondence: guocheng@uestc.edu.cn

Abstract: To address the significant impact of noise on the target detection performance of borehole radar (BHR), a key type of ground-penetrating radar (GPR), a denoising scheme based on the whale optimization algorithm (WOA) for adaptive variational mode decomposition (VMD) and multiscale principal component analysis (MSPCA) is proposed. This study initially conducts the modal decomposition of BHR data using an improved adaptive VMD method based on the WOA; it then automatically selects modes meeting specific frequency band standards. The correlation coefficients between these modes and the original signal are computed, discarding weakly correlated modes before signal reconstruction. Finally, MSPCA further suppresses noise, yielding denoised BHR data. Simulations show that the proposed scheme increases the signal-to-noise ratio by 17.964 dB or higher, surpassing the more established denoising techniques of robust principal component analysis (RPCA), MSPCA, and empirical mode decomposition (EMD), and obtains the most favorable results in terms of the RMSE and MSE metrics. The experimental results demonstrate that the proposed scheme more effectively suppresses vertical and random noise signals in BHR data. Both the numerical simulations and experimental results confirm the effectiveness of this scheme in noise reduction for BHR data.

Keywords: borehole radar (BHR); denoising; adaptive variational mode decomposition (VMD); principal component analysis (PCA); whale optimization algorithm (WOA)

1. Introduction

GPR is widely used for geotechnical surveys, groundwater monitoring, polar glacier studies, lunar exploration, geological surveys, and buried object identification as a key tool in shallow surface geophysics [1–7]. BHR, identified as a key type of ground-penetrating radar (GPR), is employed as an electromagnetic detection tool for subsurface investigations. BHR is widely applied in mineral resource exploitation, civil engineering, geophysical exploration, and geological surveys [1,8]. Furthermore, BHR is utilized for the detection of underground structures [9,10], mapping of hydrological properties [11], monitoring of oil well perforations [12], and identification of tunnels [13]. Significant noise interference in BHR receiving signals originates from the complexity of the underground environment, thermal noise within electronic components, external electromagnetic interference, and the characteristics of wideband antennas [14–17]. This noise consists of coherent noise (clutter), which exhibits inter-trace correlation, and incoherent noise (random noise), which is randomly distributed across the data. Both types of noise significantly impact the data fidelity and the signal-to-noise ratio (SNR) [18,19]. Furthermore, the emission of high-frequency electromagnetic waves by the BHR system, in conjunction with the region's inherent filtering effects, exacerbates the nonlinearity and nonstationarity of the received signals [14,16,17,20,21]. Useful signals are masked by nonlinear and nonstationary noise, resulting in degraded target detection performance and impaired quality in data interpretation, migration imaging, and full-waveform inversion [22–24]. Consequently, the suppression of random noise continues to be a critical focus within BHR data processing research [25,26]. Regarding the suppression of random noise, the existing methods have been categorized into five main types, which are complementary when applied to BHR data. In subspace methods, clutter is suppressed through the decomposition of Bscan images into different subspaces. While theoretically straightforward and easy to implement, these methods encounter challenges in accurately estimating the noise subspace when the clutter energy is comparable to that of the target signal, resulting in unstable denoising performance. Low-rank and sparse matrix decomposition (LRSD) facilitates the separation of strong background energy from sparse targets by modeling the data as the combination of a low-rank background and sparse components. However, residual noise within the sparse matrix may be overlooked, and the performance remains highly sensitive to the hyperparameter selection, making the method prone to variations across different scenarios. Multiresolution methods process signals and noise across multiple scales and orientations, allowing for the preservation of local details while attenuating specific types of random noise. However, these methods exhibit limitations when dealing with highly nonstationary noise. Deep learning methods exhibit significant advantages in feature representation and adaptive modeling when sufficient labeled training data are available. However, their effectiveness is constrained by the requirement for large-scale labeled datasets, and their generalization performance remains unstable in new environments with distributional variations. Due to the nonstationary and nonlinear characteristics of BHR signals, modal decomposition enables adaptive signal decomposition into several intrinsic mode functions (IMFs). However, these methods are often constrained by mode mixing, endpoint effects, and the absence of rigorous mathematical frameworks for parameter selection. The necessity of the manual tuning of key parameters further restricts their effectiveness in real-world BHR data processing. Although these approaches complement each other to some extent, challenges remain, including noise-to-target energy similarity, the dependence on large training datasets, and mode mixing in decomposition. These limitations hinder the ability to achieve an optimal balance between denoising accuracy and signal fidelity, particularly in complex environments. To overcome these limitations, a BHR denoising strategy is proposed, integrating adaptive VMD with MSPCA. In particular, the whale optimization algorithm (WOA) is employed to optimize the VMD parameters, enabling the more precise decomposition of nonlinear and nonstationary signals. Residual noise is further suppressed using MSPCA in the multiscale domain. This scheme exploits the strengths of modal decomposition for nonstationary noise suppression while employing multiscale analysis for signal detail preservation. As a result, a superior balance is maintained between the BHR denoising effectiveness and weak target fidelity.

Denoising methods based on subspace decomposition, such as singular value decomposition (SVD), are applied to BHR data to enhance the correspondence with strata resistivity and improve the stratigraphic classification capabilities [10]. Other subspace-based techniques, including independent component analysis (ICA) [27] and principal component analysis (PCA) [28,29], are extensively utilized for radar data denoising. However, these methods rely on accurate noise subspace estimation, assuming significant differences between noise and target responses, with noise primarily captured by dominant components. When the noise intensity is similar to that of the target signals, separation becomes challenging, potentially compromising noise suppression. Morphological component analysis (MCA) is also employed for radar data denoising by decomposing signals into clutter and target components, although its performance heavily depends on the selected dictionary [30].

Low-rank and sparse matrix decomposition (LRSD) methods are widely employed for radar noise suppression. For instance, robust non-negative matrix factorization (RNMF) models radar data as low-rank clutter and sparse target signals to suppress clutter [31], while robust principal component analysis (RPCA) achieves the superior suppression of noise compared to PCA [32–34]. Furthermore, other LRSD approaches, such as GO decomposition (GoDec) [35] and robust orthogonal subspace learning (ROSL) [36], represent GPR Bscan data as combinations of low-rank and sparse matrices, corresponding to clutter and target signals. However, their performance is sensitive to the hyperparameter selection, which balances low-rank and sparse components, with the optimal values varying by scenario. Additionally, focusing on dominant components may result in neglecting residual noise in the sparse matrix, reducing the effectiveness when noise is inadequately removed [34].

Multiresolution techniques are extensively applied in radar noise suppression. The wavelet transform (WT), for instance, is frequently utilized for multiscale GPR signal denoising [37–39]. The curvelet transform (CT) and nonlocal mean (NLM) filtering have been applied to random noise attenuation in seismic and GPR data [40]. The structural features of radar signal reflection layers are captured by CT across multiple scales and orientations while preserving the primary energy components. Furthermore, CT filtering is robust in noise suppression and ensures signal integrity [40]. Bakshi's multiscale principal component analysis (MSPCA) combines the strengths of PCA and WT for multivariate process monitoring and noise elimination [41–43], and it has been successfully applied to denoise GPR and GNSS signals [14,44]. However, the WT is most effective for stationary signals, limiting its application to highly nonlinear and nonstationary radar data [14,45]. Other multiresolution methods, such as the curvelet and shearlet transforms [39,46], are also applied but remain primarily effective for uniformly distributed random noise [47], struggling with nonstationary noise in BHR data.

Deep learning has been applied in radar noise suppression, including clutter removal neural networks (CR-Nets) [48], convolutional autoencoder (CAE) networks [49], and autoencoder-based methods [50]. Despite their notable advantages, these methods often rely on large training datasets and specific conditions for optimal performance. Limited generalization capabilities, especially in diverse scenarios, remain a major research focus [25]. Furthermore, obtaining clutter-free Bscan images of underground targets for training presents a persistent challenge in practical BHR experiments.

Modal decomposition methods have also been utilized in BHR denoising. Empirical mode decomposition (EMD) adaptively decomposes nonlinear, nonstationary signals into finite subcomponents, thereby highlighting anomalous geological features in BHR data [51]. To address intrinsic mode mixing, variants like ensemble EMD (EEMD) [52], complete ensemble EMD (CEEMD) [53], and improved CEEMD (ICEEMD) [54] have been proposed. However, issues such as mode mixing, endpoint effects, and insufficient mathematical foundations persist [19,47]. Variational mode decomposition (VMD) [55] resolves the challenges associated with empirical mode decomposition (EMD), including mode mixing, endpoint effects, and the absence of a well-defined mathematical framework. This is achieved through a mathematical optimization process, ensuring more precise, stable, and controllable decomposition results. Furthermore, signals are adaptively decomposed into intrinsic mode functions (IMFs) with a finite bandwidth by VMD, thereby effectively enhancing the target signals while suppressing noise. However, its reliance on manual parameter tuning, such as mode number selection, limits its broader applicability [56].

A denoising scheme for BHR data is proposed in this study, integrating adaptive VMD and MSPCA. The core parameters (K, α) of VMD are adaptively optimized through the WOA, followed by signal decomposition using VMD. The frequencies of the decomposed modes are evaluated to ensure alignment with the predefined range of the BHR mean spectrum. Modes with weak correlations are removed based on the Pearson correlation coefficient (PCC). Finally, MSPCA is employed to further suppress residual noise. The performance of the proposed scheme is evaluated against RNMF, EMD, RPCA, MSPCA, and WOA-based Adaptive VMD, demonstrating superior noise suppression capabilities.

The main innovations of this study are as follows. (1) First Application of VMD for BHR Denoising and Reconstruction: Nonlinear and nonstationary BHR signals are decomposed into intrinsic mode functions (IMFs) with finite bandwidths, effectively enhancing target signals masked by noise. Additionally, this method resolves inherent issues in EMD, such as mode mixing and ambiguous physical interpretation. (2) Adaptive VMD: To improve the accuracy and robustness of VMD decomposition, the core parameters (K, α) are optimized using the WOA, which enables rapid convergence and global search capabilities. The maximum sample entropy $H(\cdot)$, recognized for its strong anti-interference capabilities, is employed as the optimization function. Specific frequency band modes are adaptively selected for reconstruction, facilitating the separation of major noise components from the target signals. The correlation between the selected modes and the original signal is evaluated, and highly correlated modes are reconstructed to enhance the denoising performance. These components collectively form the adaptive VMD BHR denoising algorithm. (3) Effective Handling of Nonstationary Signals with Adaptive VMD and MSPCA: Adaptive VMD processes nonstationary signals effectively, separating noise from target components. MSPCA combines the strengths of the WT and PCA, enabling the extraction of primary signal features across multiple scales. While the WT is more effective for stationary signals, it is limited in addressing nonstationary BHR data. Adaptive VMD addresses the nonlinear and nonstationary characteristics of BHR signals, and MSPCA subsequently suppresses residual noise and enhances weak signals. In summary, the proposed scheme achieves effective noise suppression, delivering optimal visualization and metric results.

2. Proposed Scheme

A denoising scheme for BHR data is proposed in this study, integrating adaptive VMD and MSPCA. The algorithmic foundations of this scheme are introduced in this section. Adaptive VMD incorporates WOA-based parameter optimization, frequency-domain filtering, and time-domain correlation-based mode selection. The optimization of the VMD parameters is initially conducted using the WOA, which improves the efficiency of VMD in extracting target signals. In the subsequent step, frequency bands and modes demonstrating a high degree of correlation with the original signal are chosen for reconstruction. MSPCA is then employed to suppress residual noise and to further diminish noise signals that resemble the target within the reconstructed Bscan. The detailed structure of the processing workflow is illustrated in Figure 1. Initially, BHR Bscan data, organized in M rows and N columns, are transformed into Ascan data of length N. This conversion simplifies the data processing required for the WOA optimization of the VMD parameters while preserving key data features. In this data compression method, $A_{i,j}$ denotes the value in the i-th row and j-th column of the BHR data, and $A_{\text{mean}}(j)$ is the mean value of the j-th column.

$$A_{\text{mean}}(j) = \frac{1}{M}\sum_{i=1}^{M} A_{i,j} \quad (j = 1, 2, \ldots, N) \tag{1}$$

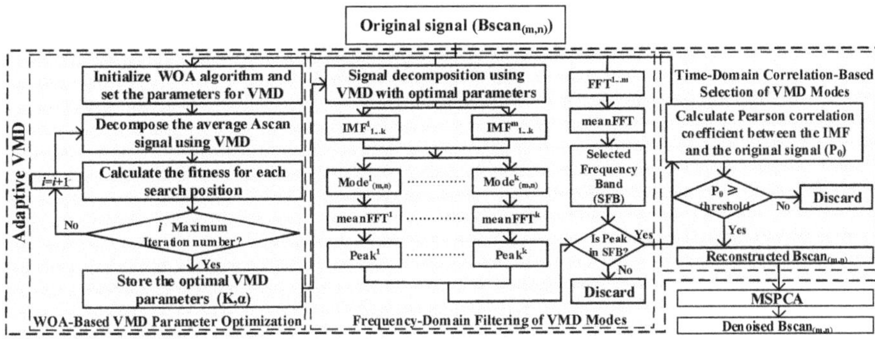

Figure 1. Processing workflow of the proposed scheme.

2.1. Adaptive VMD Algorithm

In this study, VMD is first applied for BHR data denoising, effectively highlighting target signals obscured by noise. In this application, the compressed Ascan data, A_{mean}, are decomposed by VMD into K modal functions, represented as u_k.

$$A_{\text{mean}} = \sum_{k=1}^{K} u_k \tag{2}$$

In VMD, the signal is decomposed by solving a variational problem that ensures minimal bandwidth for each modal function. This approach improves the accuracy and robustness of the decomposition. Through this decomposition process, the modal functions become more physically meaningful, allowing the more accurate representation of the original signal. In this study, ω_k denotes the center frequency of the K-th mode; Δ_n denotes the difference operator at discrete time n; $\delta[n]$ denotes the Kronecker Delta function at discrete time; and j denotes the imaginary unit. $\|\cdot\|_2$ denotes the L2 norm, also known as the Euclidean norm [55].

$$\min_{u_k,\omega_k}\left\{\sum_{k=1}^{K}\left\|\Delta_n\left[\left(\delta[n] + j\frac{1}{\pi n}\right) * u_k[n]\right]e^{-j\omega_k n}\right\|_2^2\right\} \text{ subject to } \sum_{k=1}^{K} u_k = A_{\text{mean}} \tag{3}$$

The optimization problem is formulated through the Lagrangian method by introducing a quadratic penalty term and Lagrange multipliers λ into the VMD formula. It is then solved using the Alternating Direction Method of Multipliers (ADMM). In this formulation, the Lagrange multiplier λ is used to adjust the constraints during the optimization process. Within the objective function, the penalty parameter α is used to determine the weight of the penalty term, balancing the reconstruction error with modal sparsity in the VMD decomposition. The formula is constructed with three key components. The modal bandwidth measure term minimizes the bandwidth of each decomposed mode, ensuring that the signal components are concentrated in the frequency domain. By reducing the bandwidth, the precision of the signal decomposition process is enhanced. The quadratic penalty term enhances the fidelity of signal reconstruction by penalizing deviations from the original signal. The Lagrange multiplier term ensures that the sum of all modal components strictly equals the original signal, thereby preserving signal completeness and integrity. It enforces strict reconstruction constraints throughout the optimization process. The combination of the quadratic penalty term and the Lagrange multiplier leverages the advantages of both. The quadratic penalty provides robust convergence properties, while the Lagrange multiplier enforces strict constraint satisfaction. Together, these components enable a balanced and stable optimization framework for the decomposition process [56].

$$\begin{aligned} L(u_k, \omega_k, \lambda) = \alpha \sum_{k=1}^{K} \left\| \Delta_n \left[\left(\delta[n] + j\frac{1}{\pi n} \right) * u_k[n] \right] e^{-j\omega_k n} \right\|_2^2 + \left\| A_{\text{mean}}[n] - \sum_{k=1}^{K} u_k[n] \right\|_2^2 \\ + \langle \lambda[n], A_{\text{mean}}[n] - \sum_{k=1}^{K} u_k[n] \rangle \end{aligned} \tag{4}$$

The update formula for the mode u_k is presented below:

$$\begin{aligned} u_k^{n+1}[n] = \arg\min_{u_k} \left\{ \alpha \left\| \Delta_n \left[\left(\delta[n] + \frac{j}{\pi n} \right) * u_k(n) \right] e^{-j\omega_k n} \right\|_2^2 \right. \\ \left. + \left\| A_{\text{mean}}[n] - \sum_{k=1}^{K} u_k[n] + \frac{\lambda(n)}{2} \right\|_2^2 \right\} \end{aligned} \tag{5}$$

The formula for the updating of ω_k is presented below:

$$\omega_k^{n+1} = \arg\min_{\omega_k} \left\{ \left\| \Delta_n \left[\left(\delta[n] + j\frac{1}{\pi n} \right) * u_k[n] \right] e^{-j\omega_k n} \right\|_2^2 \right\} \tag{6}$$

In VMD signal decomposition, the penalty parameter α and the number of mode decompositions K are predetermined and serve as crucial factors. A lower penalty parameter α reduces bandwidth constraints on the modal components, resulting in modal signals with a broader bandwidth. Conversely, an excessively high or low number of mode decompositions K can cause the over-decomposition or under-decomposition of the signal. Optimizing the VMD parameters through the whale optimization algorithm (WOA) enables the automatic selection of parameters suitable for the signal, enhancing the decomposition accuracy and robustness. The whale optimization algorithm (WOA) [57], inspired by the bubble-net hunting behavior of whales, is characterized by its rapid convergence and robust global search capabilities. The WOA dynamically adjusts the convergence factor to balance global exploration and local exploitation. This mechanism is employed to efficiently optimize complex, nonlinear problems and avoid local optima. In this study, the WOA is combined with variational mode decomposition (VMD) to adaptively optimize the penalty parameter α and the number of decompositions K. The global search and local exploitation properties of the WOA ensure the accurate optimization of both integer and continuous parameters, making it particularly suitable for tuning K and α. By leveraging the WOA's stochastic nature and linear convergence factor, the optimization process achieves improved stability and convergence speeds. Collectively, these attributes significantly enhance the decomposition performance of VMD.

The optimization of the variational mode decomposition (VMD) parameters (K, α) is performed using the whale optimization algorithm (WOA) [56,57]. The optimization process involves mapping the positions of the whales to the VMD parameter space, thereby enabling the adaptive adjustment of both the penalty parameter α and the number of modes K. Whale Position Representation: The position of each whale is denoted as $X[i] = [K_i, \alpha_i]$, where K_i refers to the number of decomposition modes and α_i corresponds to the penalty parameter. Population Initialization: A population of N whales is initialized randomly, with positions X_i constrained within predefined boundary limits. This ensures that all whale positions remain within the feasible optimization range. Parameter Updates During Iterations: In each iteration of the WOA, the positions $X[i]$ are updated according to the exploration and exploitation strategies embedded in the algorithm, allowing for the dynamic adjustment of both K and α. Following each update, boundary checks are conducted to ensure that the values of K and α remain within their valid ranges.

$$\{\vec{X}_i\}_{i=1}^{N} = \{(K_i, \alpha_i)\}_{i=1}^{N} \tag{7}$$

Prior to parameter optimization, the fitness function must be defined. The sample entropy, represented as $SampEn(m, r, u_k)$, is used to quantify the complexity and nonlinearity of a time series. The sample entropy is characterized by independence from the data length and strong resistance to interference, enabling stable calculations regardless of the data length. It is also insensitive to noise, allowing accurate complexity measurement despite disturbances. The sample entropy is defined using m as the embedding dimension and r as the tolerance to measure the similarity of subsequences within the time series. Here, $B^m(r)$ and $B^{m+1}(r)$ are the total counts of instances where the distances between subsequences are below r in dimensions m and $m + 1$, respectively. u_k is the input time series for which the sample entropy is being calculated [58].

$$SampEn(m, r, u_k) = -\ln\left(\frac{B^{m+1}(r)}{B^m(r)}\right) \tag{8}$$

The core parameters of VMD, specifically the number of decomposition modes K and the penalty factor α, are optimized through a fitness function derived from the sample entropy (SampEn). This fitness function quantifies the regularity of the intrinsic mode functions (IMFs) and is utilized to assess and enhance the quality of VMD signal decomposition. The primary objective is to minimize the local SampEn of the IMFs. Lower SampEn values signify greater regularity in the IMFs, which in turn improves the quality of signal decomposition. The fitness value calculation process is as follows. (1) Decomposition Step: VMD is applied with varying combinations of the number of decomposition modes K and penalty factor α to decompose the input signal into K IMFs. (2) Entropy Calculation: The SampEn value for each IMF is computed. (3) Fitness Selection: Since the appearance of targets in BHR echo signals is often accompanied by an increase in entropy, the maximum SampEn value among all IMFs is selected as the fitness value. The fitness value derived from the maximum SampEn represents the complexity and irregularity of the IMFs. Minimizing the fitness value ensures that the resulting IMFs are more regular, effectively suppressing noise while retaining the target signal components. This optimization strategy enhances the signal regularity and improves the interpretability of the decomposition results, underscoring its importance in applications such as target detection in BHR signals. The fitness value is calculated using the following formula:

$$\text{Fitness}_i = \max\left\{ SampEn(m, r, u_k)\right\}\Big|_{(K,\alpha)=(K_i,\alpha_i)} \tag{9}$$

The adjustment of the (K, α) parameters follows the update rules of the whale optimization algorithm (WOA). In this process, b represents a constant that determines the shape of the logarithmic spiral, while l is a randomly selected value within the range of $[-1, 1]$. In the WOA, the parameter p is a random probability value within $[0, 1]$. When $p < 0.5$, the encircling mechanism is applied; otherwise, if $p \geq 0.5$, spiral updating is employed. The current optimal solution is denoted by $\vec{X}^*(t)$, and the vectors $\vec{A}$ and $\vec{C}$ control the convergence factors [57].

$$\begin{aligned} \vec{X}_{i+1} &= \vec{X}_i^* - \vec{A} \cdot |\vec{C} \cdot \vec{X}_{\text{rand}} - \vec{X}|, & p < 0.5 \\ \vec{X}_{i+1} &= |\vec{X}_i^* - \vec{X}_i| \cdot e^{bl} \cdot \cos(2\pi l) + \vec{X}_i^*, & p \geq 0.5 \end{aligned} \tag{10}$$

The optimization process is repeated until the predefined maximum number of iterations, set to 16 by default, is reached. Upon the completion of the iterations, the optimal parameters are determined and retained for future application.

$$(K, \alpha) = \arg \max_{\{\vec{X}_i\}_{i=1}^{N}} \text{Fitness}_i \tag{11}$$

Once the optimal parameters (K, α) are obtained, the original BHR Bscan data are decomposed using VMD to generate the modal functions $\text{IMF}_k^{i,j}$. The Fourier transforms of the modes are then calculated to identify the peak frequency in the average Fourier spectrum. Subsequently, the average Fast Fourier Transform (FFT) of the original BHR Bscan data is computed. To determine the selected frequency band (SFB), the threshold is defined as the spectrum mean plus or minus one-fourth of the difference between the peak value and the spectrum mean. Modes within the selected frequency band are adaptively chosen for reconstruction, thereby separating the main noise components from the target signal and forming the reconstructed matrix $\hat{A}_{\text{SFB},i,k}$. The correlation coefficient $R_{i,k}$ between the selected mode $\hat{A}_{\text{SFB},i,k}$ and the i-th row of the corresponding original Bscan signal A_i is calculated to assess their relationship. In the formula for the correlation coefficient $R_{i,k}$, $\text{Cov}[A_i, \hat{A}_{\text{SFB},i,k}]$ is the covariance between them, and $\text{Var}[A_i]$ and $\text{Var}[\hat{A}_{\text{SFB},i,k}]$ are their respective variances [59].

$$R_{i,k} = \frac{\text{Cov}[A_i, \hat{A}_{\text{SFB},i,k}]}{\sqrt{\text{Var}[A_i] \times \text{Var}[\hat{A}_{\text{SFB},i,k}]}} \tag{12}$$

Modes with a correlation coefficient $R_{i,k}$ greater than the threshold $\text{Threshold}_{\text{COR}}$ are selected for reconstruction. The reconstructed signal is denoted by $\hat{A}_{\text{COR}}$, and $\hat{A}_{\text{SFB},i,k}$ represents the selected mode.

$$\hat{A}_{\text{COR}} = \sum_{k \in \{k_1, k_2, \ldots, k_m\}} \hat{A}_{\text{SFB},i,k} \quad \text{subject to } R_{i,k} > \text{Threshold}_{\text{COR}} \tag{13}$$

2.2. Multiscale Principal Component Analysis (MSPCA)

MSPCA leverages the complementary strengths of the WT and PCA to enable robust multiscale signal processing. The WT excels in decomposing signals across multiple scales, isolating high- and low-frequency components, while PCA identifies the principal features within these scales. Together, these methods facilitate the efficient extraction of meaningful signal components, even under challenging noise conditions. For BHR data, selecting appropriate MSPCA parameters is critical in maintaining signal fidelity and achieving effective noise suppression. The symmetric wavelet basis 'sym28' is employed due to its orthogonal and symmetric properties, which preserve the smoothness and essential features of BHR signals [38]. This choice is informed by the complex frequency components and significant noise inherent in BHR data. The high time–frequency resolution of 'sym28' enables the effective retention of the target signal characteristics while suppressing noise. To balance the computational complexity and feature extraction accuracy, the decomposition level is empirically set to five. This configuration prevents distortions from over-decomposition and ensures the clear separation of noise from the target signals. BHR signals are inherently nonstationary due to geological variations, subsurface dielectric constant differences, conductivity changes, and environmental interferences. Instrumental noise and subsurface interactions further contribute to their nonlinear and complex nature. Although the WT is highly effective in processing stationary signals, its performance diminishes when applied to nonlinear and nonstationary BHR data, reducing its capability to handle the complex geophysical noise patterns inherent in such signals. To address these limitations, the proposed methodology integrates adaptive VMD with MSPCA. Adaptive VMD is employed as the initial step to decompose the nonlinear and nonstationary BHR signals into intrinsic mode functions (IMFs) with a finite bandwidth. This decomposition mitigates the nonlinearity and nonstationarity of the signals. Subsequently, MSPCA is applied to refine the signal further, isolating critical features while suppressing residual noise. This sequential strategy combines the complementary capabilities of adaptive VMD and MSPCA to enhance the signal clarity and noise suppression. During the reconstruction phase, the WT decomposes the signal $\hat{A}_{\mathrm{COR}}$ into multiscale approximate coefficients $\mathbf{A}_j$, which represent overall trends, and detail coefficients $\mathbf{D}_j$, which capture finer details. Here, j denotes the decomposition level within the multiscale framework [41,43]. This integrated approach balances signal clarity and noise reduction, ensuring the reliable processing of BHR data.

$$\hat{A}_{\mathrm{COR}} = \sum_{j=1}^{J} (\mathbf{A}_j + \mathbf{D}_j) \tag{14}$$

The signal is decomposed using the discrete wavelet transform (DWT) through a high-pass filter $g[n]$ and a low-pass filter $h[n]$, which produces two types of coefficients: detail coefficients $D_j[n]$, capturing high-frequency information, and approximate coefficients $A_j[n]$, capturing the main trends in the signal at each decomposition level j.

$$A_j[n] = \sum_k \hat{A}_{\mathrm{COR}}[k] h[2n-k] \tag{15}$$

$$D_j[n] = \sum_k \hat{A}_{\mathrm{COR}}[k] g[2n-k] \tag{16}$$

After the approximate and detail coefficients are obtained across different scales, PCA is applied to the wavelet coefficients at each scale. This approach captures essential signal features across various frequency ranges, thereby enhancing noise suppression. At each wavelet decomposition level j, the covariance matrix $\mathbf{C}_j$ is calculated using the following formula. Here, $\mathbf{X}_j$ denotes the matrix of coefficients at that level, comprising both approximate and detail coefficients.

$$\mathbf{C}_j = \frac{1}{N} \mathbf{X}_j^T \mathbf{X}_j \tag{17}$$

The eigenvalue decomposition of the covariance matrix $\mathbf{C}_j$ is given by the following formula, where $\mathbf{P}_j$ is the eigenvector matrix and $\mathbf{\Lambda}_j$ is the diagonal matrix of eigenvalues:

$$\mathbf{C}_j = \mathbf{P}_j \mathbf{\Lambda}_j \mathbf{P}_j^T \tag{18}$$

In the PCA stage of MSPCA, high-frequency detail coefficient matrices, primarily containing noise, are discarded, while low-frequency approximate coefficient matrices, which capture primary trends and target features, are retained. The selection of principal components for the approximate coefficient matrices is guided by the cumulative variance criterion (CVC) [29], a method that sorts eigenvalues in descending order and accumulates their variance contributions until the cumulative contribution rate reaches 96%. This threshold ensures the retention of meaningful signal components while effectively reducing noise interference. By eliminating detail coefficient matrices and preserving key information within approximate coefficient matrices, the CVC enhances the fidelity of signal reconstruction, achieving a balance between signal integrity and noise suppression. The retained principal components, denoted as $k_{A,j}$ and $k_{D,j}$, correspond to approximate $\mathbf{A}_j^{\text{PCA}}$ and detail coefficient matrices $\mathbf{D}_j^{\text{PCA}}$, respectively. These components undergo an inverse PCA transformation, resulting in the reconstructed coefficient matrices $\mathbf{A}_j^{\text{recon}}$ and $\mathbf{D}_j^{\text{recon}}$. Finally, the denoised signal $\hat{A}_{\text{COR}}^{\text{recon}}$ is obtained through an inverse WT, completing the reconstruction process. This systematic approach demonstrates the effectiveness of MSPCA in processing BHR data, as it ensures robust noise suppression while preserving critical signal characteristics. The methodology leverages the complementary strengths of PCA and the WT to optimize signal reconstruction for nonstationary and noisy data environments.

$$\hat{A}_{\text{COR}}^{recon} = \sum_{j=1}^{J} (\mathbf{A}_j^{recon} + \mathbf{D}_j^{recon}) \tag{19}$$

3. Examples

3.1. Numerical Simulation Results

3.1.1. Case 1: Underwater Scenario with Multiple Rectangular Targets

The proposed method is applied to denoise BHR data, with the results compared to those of alternative methods. As shown in Figure 2, numerical simulations are conducted using gprMax [60], employing a homogeneous water model containing three iron plate targets, each measuring 20 cm in height and 2 cm in width. Targets 1 and 3 are positioned 60 cm from the BHR, while target 2 is located 50 cm away, with a distance of 60 cm between adjacent targets. The transmitting and receiving antennas, spaced 26 cm apart, are excited with a 230 MHz Ricker wavelet. A 100 ns time window with 3 cm step increments is used for BHR Bscan measurements.

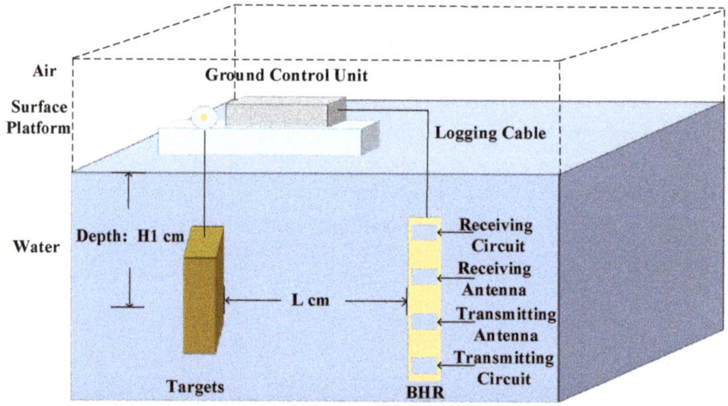

Figure 2. Experimental configuration for acquisition of target Bscan data through BHR.

As shown in Figure 3c, the RNMF method highlights the target signal to some extent, although its noise suppression remains limited. The clarity of the target signal is impacted by the significant noise remaining within the red box. RPCA is effective for sparse noise but exhibits limitations in suppressing highly random noise. These limitations result in incomplete noise suppression and hinder the accurate reconstruction of the signals, especially when the noise and target signals overlap. As shown in Figure 3d, significant residual noise is observed within the red box, demonstrating RPCA's reduced effectiveness in highly random noise environments. Although RPCA performs well in sparse noise environments, its efficiency diminishes significantly under conditions of prevalent random noise. RPCA struggles to fully isolate noise from the target signal, which compromises the clarity and fidelity of the reconstructed signal. As noted in Section 2.2, MSPCA integrates the complementary strengths of PCA and the WT. However, BHR signals are inherently nonstationary and nonlinear in nature. The WT is more effective for stationary signals, which constrains its applicability to BHR data. Additionally, MSPCA exhibits limited capabilities in suppressing random noise. As shown in Figure 3e, MSPCA effectively highlights weak echo signals, especially within the yellow box. However, residual noise remains significant in the red box, highlighting the limitations of its noise suppression capabilities.

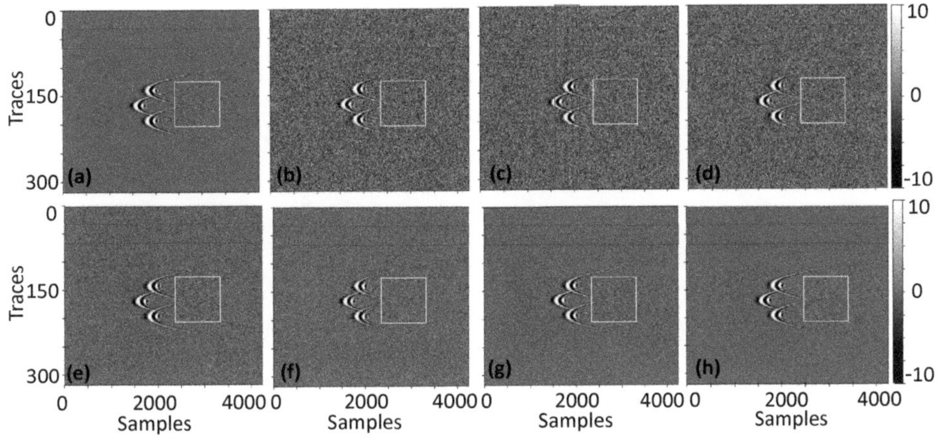

Figure 3. BHR-simulated Bscan data results. (**a**) Noise-free BHR-simulated Bscan data. (**b**) Noise-added Bscan data, −5.826 dB SNR. The denoised result for (**c**) RNMF-only, (**d**) RPCA-only, (**e**) MSPCA-only, (**f**) EMD-only, (**g**) the proposed adaptive VMD method, and (**h**) the proposed scheme (adaptive VMD + MSPCA).

EMD effectively decomposes nonlinear, nonstationary signals into intrinsic mode functions (IMFs), facilitating the enhancement of weak echo signals and achieving effective noise attenuation. However, the effectiveness of the method is constrained by mode mixing, which introduces high-frequency spike noise. This limitation diminishes its practical reliability and computational applicability, especially in scenarios involving complex signal interference. As depicted in Figure 3f, the residual noise patterns caused by mode mixing are evident, highlighting the challenges that EMD faces in maintaining signal clarity and fidelity. As demonstrated in Figure 3g, the adaptive VMD method excels in both noise suppression and highlighting weak echo signals, with the noise within the red box significantly reduced and the target signal within the yellow box appearing clearer. Noise and signals are effectively separated by adaptive VMD, making it suitable for the processing of nonstationary signals. As shown in Figure 3h, the proposed scheme (adaptive VMD + MSPCA) enhances both noise suppression and signal clarity. The noise within the red box is more effectively suppressed, and the target signal within the yellow box appears clearer and more distinct. The multiscale feature extraction advantages of adaptive VMD and MSPCA are combined in this scheme, resulting in more efficient noise suppression and signal enhancement. As described in Section 2.2, the sequential application of adaptive VMD and MSPCA is shown to achieve superior noise suppression and enhance weak signals. The effectiveness of the proposed scheme (adaptive VMD + MSPCA) in both noise suppression and target signal enhancement is validated through a comparative analysis of the denoising effects, demonstrating its suitability for the processing of noisy BHR signals.

A comprehensive comparison of the denoising performance, using metrics such as the signal-to-noise ratio (SNR), root mean square error (RMSE), and mean square error (MSE), is presented in Figure 4. These metrics are offered as critical insights into the effectiveness of the various denoising methods. The SNR quantifies the ratio of the signal amplitude to noise. Higher SNR values indicate improved signal clarity and reduced noise interference, which is essential for the accurate detection and interpretation of BHR data. For instance, an SNR increase of 19.088 dB, as achieved by the proposed method, results in significantly enhanced target signal interpretability in real-world applications, such as subsurface imaging and anomaly detection. The RMSE and MSE are quantified to indicate the fidelity of signal reconstruction. Lower RMSE and MSE values indicate effective noise suppression while preserving key signal features. For BHR applications, this ensures that the geological characteristics of the subsurface are accurately captured without distortion, enabling more reliable data interpretation.

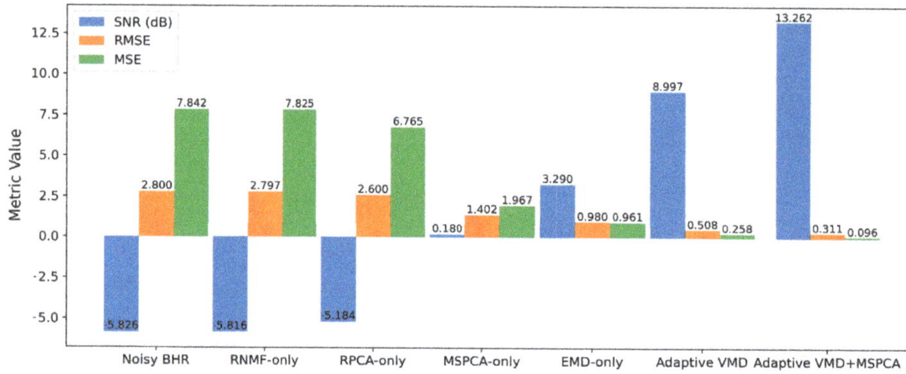

Figure 4. Comparisons of SNR, RMSE, and MSE are provided for various methods, including RNMF-only [31], RPCA-only [33], MSPCA-only [42], EMD-only [51], the proposed adaptive VMD method, and the proposed scheme (adaptive VMD + MSPCA) under numerical simulation data.

The initial SNR of the noisy Bscan data is −5.826 dB. After applying the various denoising methods, the following results are observed. RNMF: The SNR increases marginally to −5.816 dB, representing a slight improvement of 0.01 dB. This minimal improvement indicates RNMF's limited capabilities in distinguishing between noise and signals. RPCA: As described in Section 3.1.1, RPCA achieves an SNR improvement to −5.184 dB (approximately 0.642 dB increase). However, RPCA is less effective in suppressing high levels of random noise, leading to suboptimal results for BHR signal processing. MSPCA: Through the integration of PCA and the WT, MSPCA achieves a notable SNR improvement to 0.180 dB (approximately 6.006 dB increase). However, as highlighted in Section 2.2, MSPCA is more effective for stationary signals, restricting its applicability to the inherently nonstationary and nonlinear BHR signals. EMD: EMD achieves an SNR of 3.290 dB, an improvement of approximately 9.116 dB. While this indicates reasonable noise suppression, mode mixing issues, as described in Section 3.1.1, introduce high-frequency spike noise, which limits its effectiveness. Adaptive VMD: Adaptive VMD tunes the frequency bandwidth of modal components, resulting in a substantial improvement in the SNR to 8.997 dB (an increase of approximately 14.823 dB). This demonstrates its effectiveness in suppressing noise and enhancing the signal clarity. Proposed Method (Adaptive VMD + MSPCA): The proposed method combines the strengths of adaptive VMD and MSPCA, resulting in an SNR improvement to 13.262 dB, corresponding to a substantial increase of approximately 19.088 dB. This significant enhancement not only suppresses noise but also enhances the visibility of weak signals, making it the most effective method for BHR signal processing. Additionally, its superior performance in terms of the RMSE and MSE metrics further validates the method's robustness.

From the perspective of the simulation data, the increase in the SNR correlates with more distinct target features and reduced noise interference. For instance, as shown in Figures 3h and 5h, at an SNR improvement of 19.088 dB, previously obscured signals emerge more clearly, significantly enhancing the visibility and interpretability of geological features such as targets or stratigraphic boundaries. The reduction in the RMSE and MSE ensures higher accuracy in signal reconstruction, minimizing the likelihood of false positives or missed detections. In real-world scenarios, these improvements result in tangible benefits in BHR data processing. These include enhanced target detectability, as subsurface features become more distinguishable, aiding geophysical surveys and infrastructure inspections; increased data reliability, as lower noise levels and higher reconstruction fidelity reduce the risk of misinterpretation, supporting better decision-making in resource exploration and environmental monitoring; and greater applicability in complex environments, as the proposed method's ability to handle nonstationary signals ensures robust performance across diverse geological conditions.

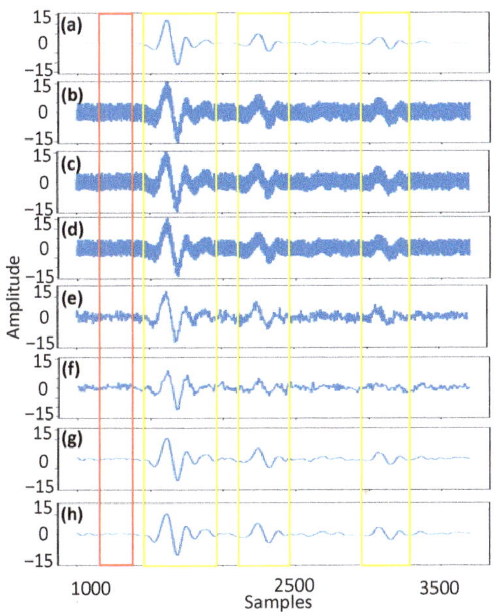

Figure 5. (**a**) Original noise-free BHR simulation data for the 160th Ascan. (**b**) Noise-added Ascan data. The denoised result for (**c**) RNMF-only, (**d**) RPCA-only, (**e**) MSPCA-only, (**f**) EMD-only, (**g**) the proposed adaptive VMD method, and (**h**) the proposed scheme (adaptive VMD + MSPCA).

Figure 5 presents the denoising results of the 160th simulated Ascan signal. The region subjected to significant random noise is delineated by the red box in Figure 5. This simulated noise is introduced to test the effectiveness of the proposed denoising method under high-interference conditions. In Figure 5, the yellow box encompasses the target signal and the tail edges of the target reflections. This area demonstrates how the proposed denoising method effectively enhances the target signal and its tail edges, even in the presence of substantial noise. The target signal is enhanced to some extent by RNMF and RPCA, but the noise is not effectively suppressed. As noted in Section 2.2, MSPCA integrates the complementary strengths of PCA and the WT and is suitable for multiscale signal processing. As seen from the target signal in the yellow box, weak echo signals are effectively highlighted by MSPCA. However, its noise suppression capabilities are weak in highly random noise environments .

As described in Section 3.1.1, while EMD highlights weak echo signals and achieves good noise suppression to some extent, its application is constrained by mode mixing, which introduces high-frequency spike noise. This limitation is evident as the noise within the red box remains inadequately suppressed. The high-frequency noise within the red box persists, and the signal quality is affected. The noise suppression and signal enhancement are significantly improved by adaptive VMD, resulting in a higher SNR. The noise within the red box is significantly reduced, and the target signal within the yellow box is clearer. Good separation of the noise and signal is demonstrated. The noise suppression and signal enhancement are further improved by our proposed method, which combines adaptive VMD and MSPCA. The noise within the red box is further suppressed, and the target signal within the yellow box becomes clearer and more distinct. The multiscale feature extraction advantages of adaptive VMD and MSPCA are combined in this scheme, demonstrating the best performance in terms of noise suppression and signal enhancement.

The effectiveness of the proposed method is assessed using synthetic BHR data at various noise levels. The SNR of the input data with noise varies from −25 dB to 0 dB. In Figure 6, the performance of various algorithms is shown in terms of the output SNR, MSE, and RMSE across the considered input SNR levels. The algorithms under evaluation include RNMF, RPCA, MSPCA, EMD, adaptive VMD, and the proposed adaptive VMD + MSPCA scheme. As depicted in Figure 6a, the output SNR improves for all algorithms as the input SNR increases. Specifically, the adaptive VMD method demonstrates a significant improvement in the SNR, with increases spanning from 14.4 dB to 14.8 dB. When integrated with the MSPCA method, the proposed adaptive VMD + MSPCA scheme attains even greater SNR improvements, ranging from 18.4 dB to 24.5 dB. This improvement is particularly pronounced under low input SNR conditions (e.g., −25 dB to −15 dB), where noise suppression poses significant challenges. The increase in the SNR results in more distinguishable target features in the BHR data. With a higher SNR, the signal becomes significantly clearer, reducing the interference from background noise. This enables the better detection and identification of subsurface features that may have been obscured in low-SNR conditions. Through the improvement in the SNR, the signal previously concealed by the noise is made significantly more visible, thereby greatly enhancing the target's detectability. For instance, in geological survey applications, an enhanced SNR helps in distinguishing between different geological layers or anomalies, improving the reliability of the results. The adaptive VMD method optimizes the number of modal components and their frequency bandwidths, thereby effectively separating noise from the signal. This optimization is critical in processing nonstationary signals, as frequently encountered with BHR data. Nonstationary signals, which vary over time, necessitate adaptive techniques to track changes in the signal's characteristics. The adaptive VMD method's ability to adaptively adjust the frequency bandwidths of modal components ensures that the signal is better separated from the noise, leading to an overall improvement in the signal quality. When combined with the multiscale feature extraction capabilities of MSPCA, the adaptive VMD + MSPCA scheme provides the best output SNR across all input SNR conditions. The multiscale nature of MSPCA allows for enhanced feature extraction at various resolutions, further improving the accuracy of signal detection and noise suppression. This combination proves highly effective in processing nonstationary BHR signals .

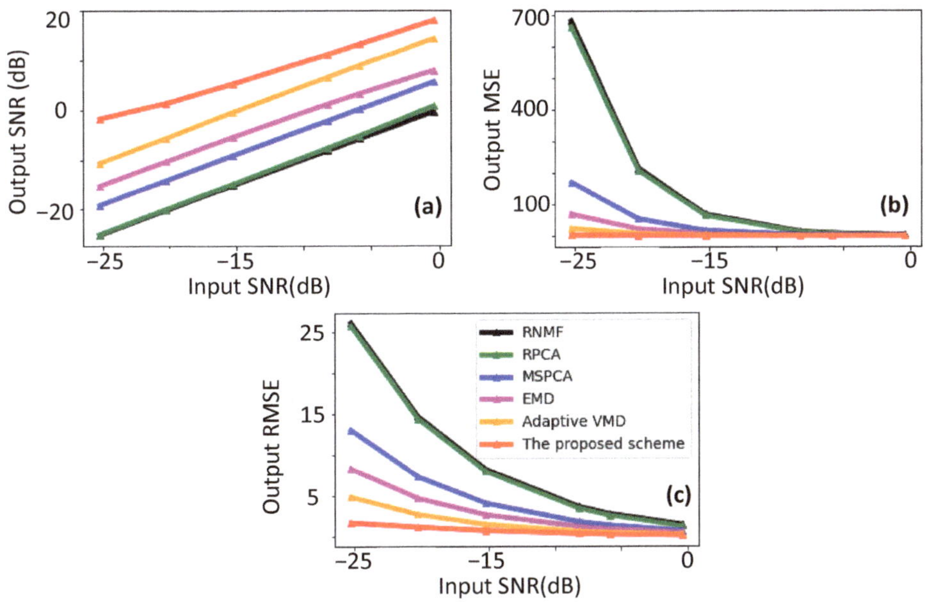

Figure 6. Output (**a**) SNR, (**b**) MSE, and (**c**) RMSE of the input noisy data with varied SNRs.

In Figure 6b,c, the adaptive VMD + MSPCA scheme is shown to exhibit the best performance in terms of both the output MSE and RMSE. The MSE and RMSE are essential metrics in assessing the accuracy of the reconstructed signal. A reduction in these values indicates the more accurate recovery of the signal, with minimal distortion or noise artifacts. The proposed method achieves the lowest MSE and RMSE values across all algorithms tested, indicating that it effectively removes noise while preserving the key features of the original BHR signal. The reduction in the MSE and RMSE, alongside the improvement in the SNR, ensures that the processed signal is both clearer and more accurate. Lower MSE and RMSE values mean that the denoised signal more closely matches the true subsurface conditions, reducing the risk of false positives or missed detections in subsequent analysis. A comparison of the denoising effects across different methods and input SNR levels reveals that the best performance in terms of noise suppression and signal enhancement is achieved by the adaptive VMD and the adaptive VMD + MSPCA schemes. These results validate the effectiveness of the proposed methods in denoising BHR signals. Specifically, the proposed adaptive VMD + MSPCA scheme not only maximizes the SNR but also minimizes the error between the original and denoised signals, making it the most effective method for practical BHR data processing.

3.1.2. Case 2: Underwater Scenario with Multiple Circular Targets

As shown in Figure 2, numerical simulations were conducted using gprMax [60], a specialized tool designed for the modeling of ground-penetrating radar scenarios. The simulation details align with those outlined for Case 1 in Section 3.1.1, with the primary difference being that the targets in Case 2 comprise five circular objects instead of three rectangular ones. The circular targets are positioned such that targets 1, 3, and 5 are 60 cm from the BHR, while targets 2 and 4 are 50 cm away. The center-to-center distance between adjacent targets is 80 cm. This modification creates a more intricate subsurface structure, establishing a robust benchmark for the assessment of the denoising performance of the proposed methods under diverse conditions. Figure 7 depicts the simulated Bscan data for BHR processed using various denoising techniques. The Bscan data for Case 2 were processed using several denoising techniques, as in Section 3.1.1 for Case 1, and the results are illustrated in Figure 7. Figure 7c shows that the RNMF method partially enhances the target signal (highlighted in the yellow box); however, significant residual noise remains, particularly in high-interference regions (highlighted in the red box), as observed in Section 3.1.1 for Case 1. Figure 7d shows that the RPCA method is effective in handling sparse noise but struggles with random noise, resulting in residual interference that compromises the clarity of the target signal. Figure 7e shows that, while the MSPCA method enhances weak echo signals, its performance in noise suppression is limited in regions with high levels of random noise. Figure 7f shows that the EMD method achieves noise suppression but is affected by mode mixing, which introduces high-frequency spike noise, impairing its performance with nonstationary signals. Figure 7g shows that the adaptive VMD method significantly mitigates noise, leading to a clear enhancement in the target signal. Figure 7h shows that the combination of adaptive VMD and MSPCA achieves the best denoising performance, effectively suppressing the noise and significantly improving the clarity of the target signal, consistent with the results observed in Section 3.1.1 for Case 1.

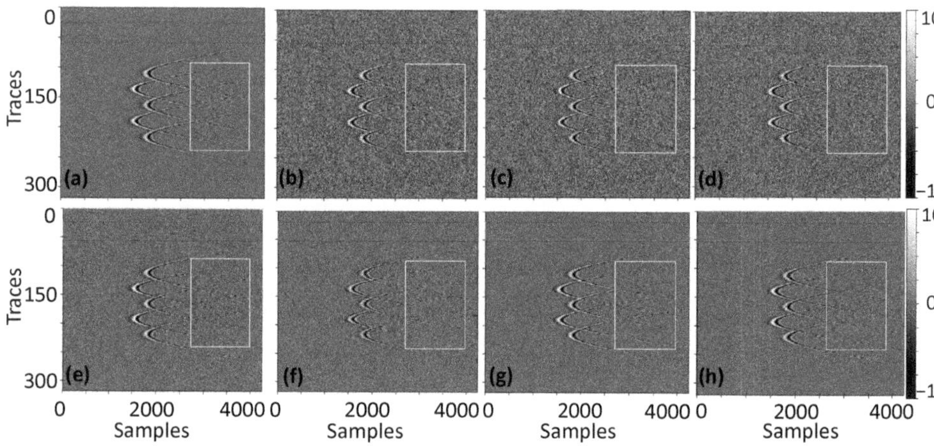

Figure 7. BHR-simulated Bscan data results. (**a**) Noise-free BHR-simulated Bscan data. (**b**) Noise-added Bscan data, −7.63 dB SNR. The denoised result for (**c**) RNMF-only, (**d**) RPCA-only, (**e**) MSPCA-only, (**f**) EMD-only, (**g**) the proposed adaptive VMD method, and (**h**) the proposed scheme (adaptive VMD + MSPCA).

Figure 8 summarizes the SNR, RMSE, and MSE results for the various methods applied to Case 2, similarly to those in Section 3.1.1 for Case 1. The initial SNR of the noisy Bscan data is −7.63 dB. After the application of the various denoising methods, the results were as follows. RNMF: A marginal improvement in the SNR, reaching −7.617 dB, with minimal reductions in the RMSE and MSE. RPCA: Improves the SNR to −7.084 dB but fails to effectively reduce the RMSE and MSE in the presence of high levels of random noise. MSPCA: Achieves a moderate SNR improvement to −1.611 dB, with corresponding reductions in the RMSE and MSE (1.216 and 1.478, respectively). EMD: Improves the SNR to 1.209 dB, with the RMSE and MSE further reduced to 0.878 and 0.772, respectively. Adaptive VMD: Outperforms all previous methods, with an SNR of 8.045 dB, RMSE of 0.400, and MSE of 0.160. Adaptive VMD + MSPCA: The proposed method achieves the highest SNR of 10.334 dB, the lowest RMSE (0.307), and the lowest MSE (0.094), thereby demonstrating its superior denoising capabilities. The adaptive VMD + MSPCA method provides superior denoising performance across multiple metrics, achieving the highest SNR and the lowest RMSE and MSE. These results reinforce the method's effectiveness, particularly in complex geophysical scenarios involving nonstationary and nonlinear signals, as demonstrated in Section 3.1.1 for Case 1. This method combines the adaptive frequency separation of VMD with the multiscale feature extraction of MSPCA, leading to improved noise suppression and clearer signal enhancement across various conditions.

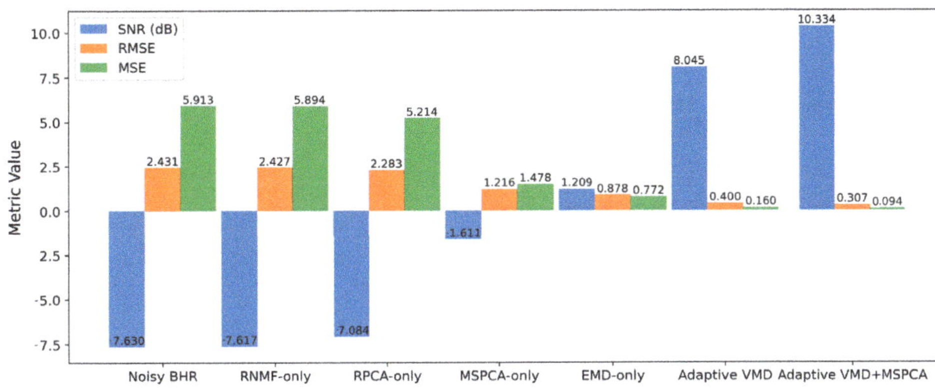

Figure 8. Comparisons of SNR, RMSE, and MSE are provided for various methods, including RNMF-only [31], RPCA-only [33], MSPCA-only [42], EMD-only [51], the proposed adaptive VMD method, and the proposed scheme (adaptive VMD + MSPCA), under numerical simulation data.

Figure 9 presents the denoising results for the 160th simulated Ascan signal, similarly to those in Section 3.1.1 for Case 1. The region affected by significant random noise is delineated by the red box, which was introduced to evaluate the effectiveness of the proposed denoising methods under high-interference conditions. The yellow box highlights the target signal and the trailing edges of its reflections, demonstrating how the proposed denoising method enhances both the target signal and its trailing edges, even in the presence of significant noise. The RNMF method (Figure 9c) and RPCA method (Figure 9d) marginally enhance the target signal but fail to suppress the noise effectively. The MSPCA method (Figure 9e) integrates the complementary strengths of PCA and the WT, making it suitable for multiscale signal processing. As seen in the yellow box, weak echo signals are effectively highlighted by MSPCA, but its noise suppression capabilities are limited, particularly in environments with high levels of random noise, as noted in Section 3.1.1 for Case 1. As discussed in Section 3.1.1 for Case 1, the EMD method (Figure 9f) provides good noise suppression and enhances weak echo signals but is constrained by mode mixing, which introduces high-frequency spike noise. As a result, the noise within the red box remains inadequately suppressed, and the signal quality is compromised. The noise suppression and signal enhancement are significantly improved by the adaptive VMD method (Figure 9g), resulting in a higher SNR. The noise within the red box is significantly reduced, and the target signal in the yellow box becomes clearer, demonstrating the effective separation of the noise and signal. The proposed method, adaptive VMD + MSPCA (Figure 9h), further improves the noise suppression and enhances the signal quality. The noise in the red box is more effectively suppressed, and the target signal within the yellow box becomes clearer and more distinct, further enhancing the signal clarity. By combining the multiscale feature extraction capabilities of adaptive VMD and MSPCA, this method demonstrates superior performance in terms of both noise suppression and signal enhancement, as confirmed in Section 3.1.1 for Case 1.

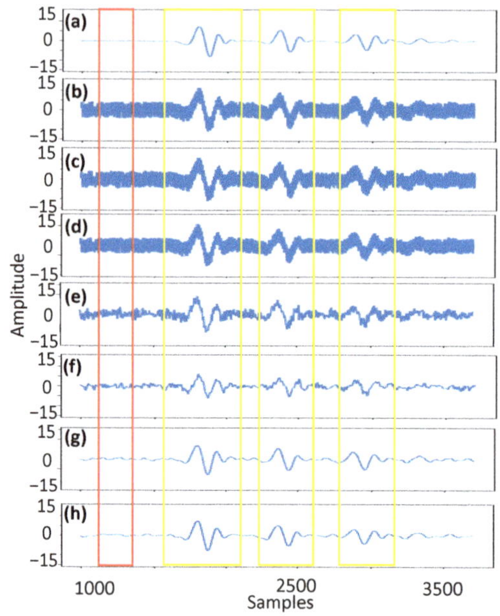

Figure 9. (**a**) Original noise-free BHR simulation data for the 160th Ascan. (**b**) Noise-added Ascan data. The denoised result for (**c**) RNMF-only, (**d**) RPCA-only, (**e**) MSPCA-only, (**f**) EMD-only, (**g**) the proposed adaptive VMD method, and (**h**) the proposed scheme (adaptive VMD + MSPCA).

Figure 10 illustrates the performance of various denoising algorithms under different input SNRs, using three metrics, the output SNR, MSE, and RMSE, as demonstrated in Section 3.1.1 for Case 1. The input SNR is varied from −25 dB to 0 dB to simulate varying levels of noise interference, providing a comprehensive evaluation of the algorithms' effectiveness. The methods evaluated include RNMF, RPCA, MSPCA, EMD, the proposed adaptive VMD method, and the combined adaptive VMD + MSPCA scheme. Figure 10a shows the output SNR as a function of the input SNR. As the input SNR increases, all algorithms exhibit improved performance. However, the adaptive VMD and adaptive VMD + MSPCA methods outperform the other methods across all SNR levels, especially under low-SNR conditions (−25 dB to −15 dB). The adaptive VMD method achieves significant SNR improvements by optimizing the modal components and suppressing the noise effectively. Furthermore, the adaptive VMD + MSPCA scheme enhances these gains, leveraging multiscale feature extraction and adaptive decomposition, with output SNR improvements exceeding 17.307 dB at 0 dB and up to 25.068 dB at lower SNR levels. Figure 10b presents the MSE results for the different input SNRs. Lower MSE values indicate better noise suppression and signal reconstruction. Among all methods, the adaptive VMD + MSPCA scheme consistently achieves the lowest MSE, significantly outperforming other methods, particularly at lower input SNRs. This superior performance is attributed to its integrated approach, combining adaptive frequency optimization and PCA to comprehensively reduce residual noise. Figure 10c shows the RMSE results, highlighting the robustness of the proposed methods in maintaining the signal's integrity while suppressing noise. Again, the adaptive VMD + MSPCA scheme achieves the lowest RMSE, demonstrating its ability to retain critical signal features while minimizing distortion. In contrast, RNMF and RPCA show limited noise suppression capabilities, particularly under high-noise conditions, resulting in higher RMSE values. Overall, Figure 10 confirms the effectiveness of the adaptive VMD and adaptive VMD + MSPCA schemes in denoising

BHR signals. As demonstrated in Section 3.1.1 for Case 1, adaptive VMD is particularly effective in handling nonstationary signals, while the combination of adaptive VMD and MSPCA provides the best overall performance. These results validate the superiority of the proposed methods in suppressing noise and enhancing the signal quality, confirming their suitability for geophysical applications involving noisy BHR data.

A comprehensive analysis of Cases 1 and 2 confirms the consistency and superiority of the adaptive VMD + MSPCA denoising scheme. The two cases are represented by different target layouts and noise environments; however, the adaptive VMD + MSPCA scheme consistently shows high denoising efficiency and effective signal enhancement across all conditions. Denoising Performance Consistency: In both the simple underwater scenario with targets (Case 1) and the more complex multitarget setup (Case 2), the adaptive VMD + MSPCA scheme significantly improves the SNR and reduces noise interference, thereby enhancing the detectability and interpretability of the target signal. Performance Consistency: In both cases, the scheme exhibits superior denoising performance and signal enhancement, particularly under low-SNR conditions. The adaptive VMD + MSPCA combination outperforms other traditional methods, such as RPCA, MSPCA, and EMD, in terms of its signal processing capabilities. Signal Processing Accuracy: The proposed adaptive VMD + MSPCA scheme significantly improves the SNR and optimizes the performance in terms of the RMSE and MSE metrics, indicating its ability to retain key signal features while reducing data distortion. Algorithm Adaptability: The introduction of adaptive VMD enhances the algorithm's adaptability in processing nonstationary signals. This ensures that the algorithm performs efficiently in diverse and complex environments. This comprehensive validation of both Cases 1 and 2 confirms the superior performance of the adaptive VMD + MSPCA denoising scheme. This method shows consistent advantages across various scenarios and noise environments, confirming its broad application potential in complex underground target detection and BHR data processing. This scheme improves the quality of BHR signals and enhances the clarity of target signals, providing reliable technical support for practical applications such as geological exploration and resource surveying.

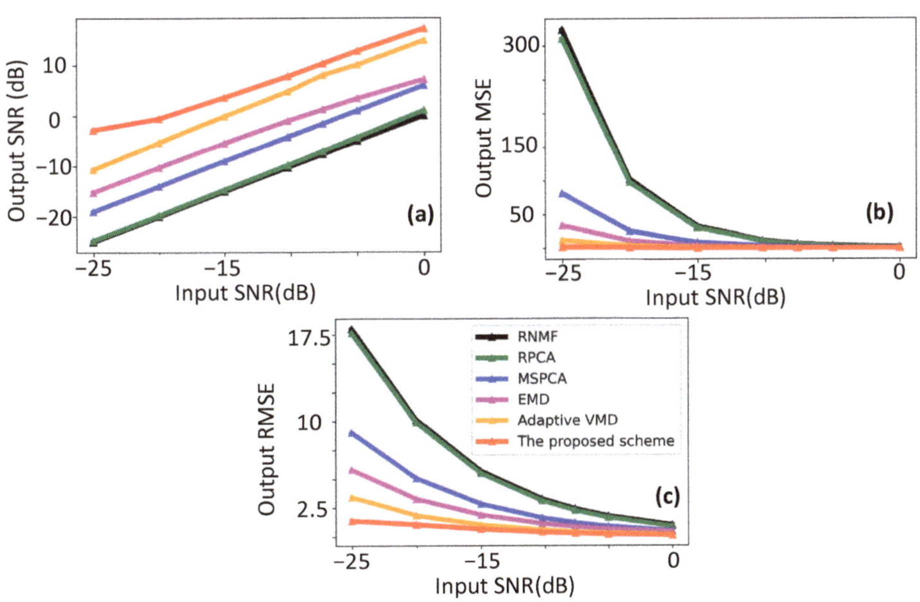

Figure 10. Output (**a**) SNR, (**b**) MSE, and (**c**) RMSE of the input noisy data with varied SNR.

3.2. Experimental Results

Figure 11 presents a complete diagram of the BHR equipment. The total length of the BHR equipment is approximately 1481.4 cm, with separation of 330 cm between the transmitting and receiving antennas. The equipment consists of the receiving circuit, receiving antenna, transmitting antenna, and transmitting circuit, with the transmitting circuit installed and connected sequentially at the bottom end. The electric cable connects the ground control unit to the underwater BHR. The ground control unit, wireless data transmission system, and mechanical structure, along with the electric cable, control the operating state of the underwater BHR and receive radar data. The BHR underwater experiment, conducted in Qiandao Lake, Hangzhou, China, is designed to validate the performance of the proposed scheme. A pulse BHR with a 230 MHz antenna is used to detect underwater iron plate targets. The target, a metal plate measuring 200 cm in length, 100 cm in width, and 5 cm in thickness, is positioned 100 cm from the instrument. Ascans are collected at 3 cm intervals, each consisting of 633 sampling points at a frequency of 2.5 GHz. The overall delay of the Ascan is set to 45 ns according to the ground control unit settings, and each Ascan is repeated and stacked 60 times at each position to form the final Ascan. This results in a Bscan composed of 371 Ascans. Basic processing steps, including background removal, are initially applied. The experimental setup is shown in Figure 2.

Figure 12 is presented, illustrating the denoising performance of various algorithms under different noise conditions, with the target signal marked in yellow and the noise highlighted within the red boxes in Figure 12a. In the two leftmost vertical red noise boxes, the RNMF (Figure 12b) and RPCA (Figure 12c) methods fail to effectively suppress the noise, leaving significant residual noise signals. Although MSPCA (Figure 12d) achieves some noise suppression, residual noise persists. The EMD (Figure 12e) method reduces the noise but introduces high-frequency spike noise due to mode mixing, as explained in Section 3.1.1. In contrast, the adaptive VMD method (Figure 12f) significantly reduces the noise, leading to a marked improvement in signal clarity. The adaptive VMD + MSPCA scheme (Figure 12g) nearly eliminates the residual noise, providing the most effective suppression. In the middle vertical red noise boxes, the RNMF and RPCA methods again leave substantial residuals. While MSPCA (Figure 12d) provides moderate noise suppression, some residual noise remains. The EMD method (Figure 12e) reduces the noise effectively, and adaptive VMD (Figure 12f) achieves substantial noise suppression. Finally, the adaptive VMD + MSPCA scheme (Figure 12g) outperforms the other methods, nearly eliminating the noise and significantly enhancing the clarity and resolution of the target signal.

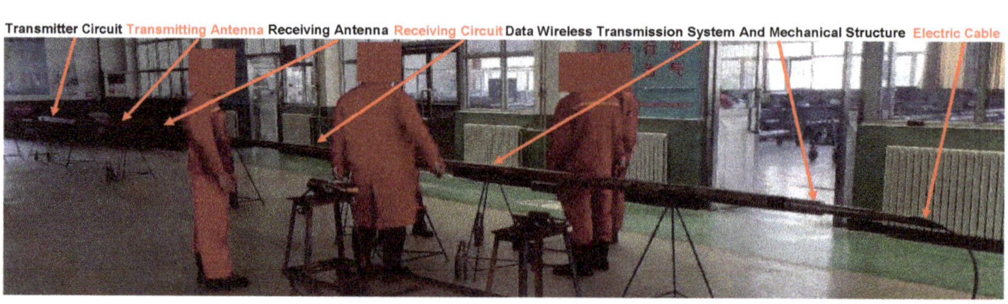

Figure 11. Complete structural diagram of the BHR system.

In Figure 12a, the bottom horizontal red noise box highlights the regions where poor noise suppression is exhibited by RNMF (Figure 12b), leaving significant residuals. Both RPCA (Figure 12c) and MSPCA (Figure 12d) show some noise reduction; however, their effectiveness is limited, particularly in this high-interference area. Effective noise reduction is achieved by EMD (Figure 12e), although some residual noise remains. In contrast, the adaptive VMD method (Figure 12f) shows significant noise reduction, leading to a marked improvement in signal clarity. Furthermore, the adaptive VMD + MSPCA scheme (Figure 12g) provides the most effective suppression, leaving almost no residual noise. When considering the upper and lower high-frequency red noise boxes on the right side of Figure 12a, RNMF (Figure 12b), RPCA (Figure 12c), and MSPCA (Figure 12d) exhibit limited suppression of the high-frequency noise. Noise reduction is achieved by EMD (Figure 12e), but this comes at the cost of introducing new high-frequency noise. The adaptive VMD method (Figure 12f) performs significantly better in reducing high-frequency noise, resulting in improved signal quality. The adaptive VMD + MSPCA scheme (Figure 12g) is the most effective in suppressing high-frequency noise, providing the best results in this context.

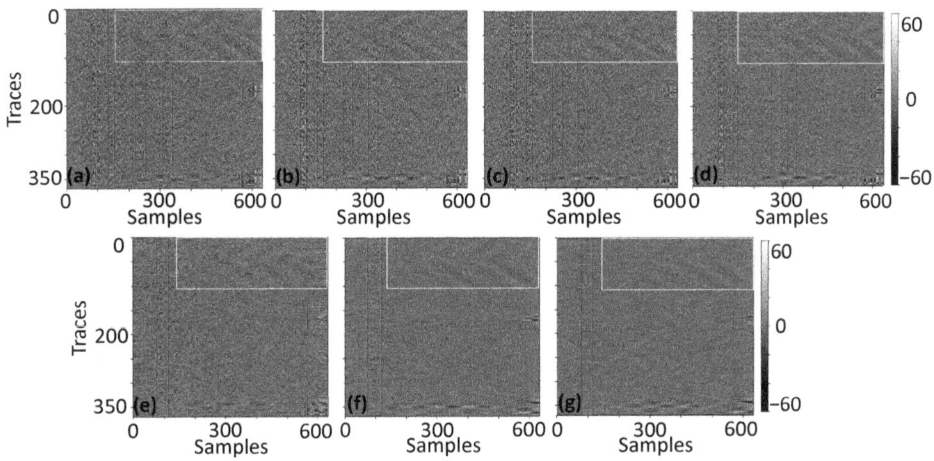

Figure 12. BHR experiment Bscan data results. (**a**) Bscan data processed through basic methods. The denoised result for (**b**) RNMF-only, (**c**) RPCA-only, (**d**) MSPCA-only, (**e**) EMD-only, (**f**) the proposed adaptive VMD method, and (**g**) the proposed scheme (adaptive VMD + MSPCA).

In the top yellow target signal area of Figure 12a, the RNMF (Figure 12b) and RPCA (Figure 12c) methods show limited effectiveness in enhancing the target signal, with substantial noise interference still present. MSPCA (Figure 12d) improves the target signal's clarity, but its noise suppression capabilities remain limited. The EMD method (Figure 12e) achieves effective noise reduction and some improvement in terms of target signal enhancement, but high-frequency noise remains a concern. In contrast, the adaptive VMD method (Figure 12f) significantly enhances the target signal's clarity while minimizing noise interference. The best results in terms of target signal enhancement with minimal noise interference are provided by the adaptive VMD + MSPCA scheme (Figure 12g), which is particularly well suited for the underwater uniform medium scenario.

The denoising effects of each algorithm indicate that the target signal's features are enhanced to some extent by RNMF (Figure 12b); however, its poor performance in terms of noise suppression complicates the distinction between the noise and signal. As discussed in Section 3.1.1, RPCA demonstrates effectiveness in managing sparse noise environments but is limited in suppressing high levels of random noise, which hinders effective noise suppression and signal separation (Figure 12c). MSPCA, as mentioned in Section 2.2, integrates the complementary strengths of PCA and the WT, enabling efficient multiscale signal processing. While weak echo signals are effectively highlighted by MSPCA, its noise suppression capabilities are limited in high-noise environments, making it more suitable for the processing of stationary signals (Figure 12d). EMD, as described in Section 3.1.1, successfully decomposes signals and suppresses noise (Figure 12e), but its practical application is limited by mode mixing, which can introduce high-frequency spike noise, thereby reducing its reliability in scenarios with complex signal interference. Adaptive VMD optimizes the number of mode components and their frequency bandwidths (Figure 12f), effectively separating noise from signals and providing excellent noise suppression and signal enhancement for nonstationary signals. The combination of adaptive VMD and MSPCA achieves the most effective noise suppression and signal enhancement (Figure 12g), with the denoised results closely resembling those of the underwater homogeneous medium scenario, thus validating the robustness of the proposed method. The sequential application of adaptive VMD and MSPCA, as detailed in Section 2.2, takes advantage of the frequency optimization provided by adaptive VMD and the multiscale feature extraction capabilities of MSPCA. Adaptive VMD is initially employed to process nonstationary signals, with MSPCA applied afterward for further optimization. This method effectively suppresses noise, enhances weak signals, and preserves critical features at various scales.

4. Conclusions

This study presents a denoising scheme for BHR signals, based on VMD and MSPCA. The core innovation of the scheme involves the application of the WOA to adaptively optimize the VMD parameters K and α, thereby achieving the automatic and efficient selection of decomposition modes and enhancing the adaptability of VMD in complex noise environments. Specifically, VMD is initially employed to decompose the nonlinear and nonstationary BHR signals into intrinsic mode functions (IMFs), effectively overcoming common issues such as mode mixing and the lack of clear physical interpretation associated with EMD. The WOA, in conjunction with the maximum sample entropy, is utilized to optimize the VMD parameters, facilitating precise mode selection and enhancing the separation between the noise and signal. By calculating the correlation between each mode and the original signal in both the time and frequency domains, modes exhibiting weak correlations are identified and discarded, effectively reducing the noise while preserving valuable signal information. Finally, MSPCA is applied to perform the multiscale processing of the signal, further extracting weak signals and suppressing residual noise to ensure the integrity of the original signal structure.

The proposed scheme is rigorously validated through numerical simulations and underwater BHR experiments. The simulations show that the proposed scheme increases the signal-to-noise ratio by 17.964 dB or higher, significantly outperforming traditional denoising methods. Additionally, the underwater BHR experiments confirm the scheme's practical applicability, particularly its effectiveness in suppressing vertical and random noise, commonly encountered in geophysical exploration, as well as its robustness in handling nonstationary noise environments. This method is especially suitable for geophysical exploration scenarios characterized by complex noise environments and nonstationary signals. By enhancing the SNR and reducing noise, this method facilitates the detection of geological features, subsurface structures, and resource distributions, thereby significantly improving the data quality and interpretability.

Despite its notable effectiveness, this scheme presents the following limitations. High computational intensity: The multistage processing involving VMD, MSPCA, and the WOA demands greater computational resources compared to simpler methods such as mean filtering or RNMF. Risk of over-denoising: The overlapping of noise characteristics with the signal may lead to the inadvertent removal of valuable signal components. To address these limitations and broaden its applicability, future research will concentrate on reducing the computational overhead by developing more efficient algorithms for parameter optimization and mode decomposition to decrease the processing time; enhancing noise–signal differentiation by improving the scheme's ability to distinguish between noise and signals, thereby minimizing the risk of over-denoising; and expanding the application range by adapting the scheme for use in heterogeneous geological environments and on various types of radar and geophysical data.

Author Contributions: Conceptualization, D.Y., C.G., and Q.Z.; methodology, D.Y., C.G., and Y.L.; supervision, R.P. and Q.Z.; validation, C.G., Y.L., H.L., and C.B.; formal analysis, D.Y. and C.L.; investigation, D.Y. and Y.L.; resources, D.Y., C.G., and Q.Z.; data curation, D.Y. and H.L.; writing—original draft preparation, D.Y.; writing—review and editing, C.G., R.P., Y.L., H.L., C.B., C.L., and Q.Z.; visualization, D.Y. and C.G.; project administration, C.G. and Q.Z.; funding acquisition, C.G. and Q.Z. All authors have read and agreed to the published version of the manuscript.

Funding: This work was supported by the National Key Research and Development Program (Grant No. 2023YFB3905501).

Data Availability Statement: Certain portions of the data are being used in ongoing related research and projects; premature disclosure may disrupt the progression of these studies or affect the outcomes of associated experiments. To ensure research integrity and orderly project advancement, the feasibility of data disclosure will be evaluated after all related research and projects are completed. For interested researchers, reasonable data requests will be assessed by the corresponding author, and the necessary data support will be provided within permissible boundaries.

Conflicts of Interest: The authors declare no conflicts of interest.

References

1. Tong, J.; Guo, J.; Huo, J.; Jiao, J.; Guo, C.; Zhang, H.; Liu, Y.; Liu, S.; Zhao, Q. A Compact Nonsymmetric Wideband Dipole Antenna for Borehole Radar Application. *IEEE Trans. Antennas Propag.* **2021**, *69*, 4323–4331. [CrossRef]
2. Persico, R.; Marasco, F.; Morelli, G.; Esposito, G.; Catapano, I. A posteriori insertion of information for focusing and time–depth conversion of ground-penetrating radar data. *Geophys. Prospect.* **2023**, *71*, 1756–1771. [CrossRef]
3. Persico, R.; Morelli, G. Combined migrations and time-depth conversions in GPR prospecting: Application to reinforced concrete. *Remote Sens.* **2020**, *12*, 2778. [CrossRef]
4. Persico, R.; Morelli, G.; Esposito, G.; Catapano, I.; Capozzoli, L.; De Martino, G.; Yang, D. An innovative time-depth conversion for the management of buried scenarios with strong discontinuities. *J. Appl. Geophys.* **2024**, *105435*. [CrossRef]
5. Yang, D.; Guo, C.; Wang, F.; Xie, L.; Zhang, K.; Jin, Y.; Zhang, S.; Zhao, Q. PICGAN: Conditional adversarial neural network-based permittivity inversions for ground penetrating radar data. *J. Appl. Geophys.* **2023**, *217*, 105164. [CrossRef]
6. Kong, Q.; Ye, S.; Liang, X.; Li, X.; Liu, X.; Fang, G.; Si, G. A clutter removal method based on the FK domain for ground-penetrating radar in complex scenarios. *Remote Sens.* **2024**, *16*, 935. [CrossRef]
7. Luo, W.; Lee, Y.H.; Jian, X.; Hao, T. A New Method for GPR Clutter Suppression Based on Stationary Graph Signals Processing. *IEEE Trans. Geosci. Remote Sens.* **2024**, *60*, 1–13. [CrossRef]
8. Liu, S.; Wu, J.; Zhou, J.; Zeng, Z. Numerical Simulations of Borehole Radar Detection for Metal Ore. *IEEE Geosci. Remote Sens. Lett.* **2010**, *8*, 308–312. [CrossRef]
9. Huo, J.; Zhou, B.; Mason, I.M.; Zhao, Q. An Application of Reflected Guided Borehole Radar Waves to Detection of Subhorizontal Structures. *IEEE Trans. Geosci. Remote Sens.* **2021**, *59*, 7201–7210. [CrossRef]
10. Zhang, H.; Wang, J.; Huo, J. SVD Description to Subsurface Strata Based on Borehole Radar. In Proceedings of the International Applied Computational Electromagnetics Society Symposium (ACES-China), Hangzhou, China, 15–18 August 2023; pp. 1–3.
11. Deiana, R.; Cassiani, G.; Kemna, A.; Villa, A.; Bruno, V.; Bagliani, A. An Experiment of Non-Invasive Characterization of the Vadose Zone via Water Injection and Cross-Hole Time-Lapse Geophysical Monitoring. *Near Surf. Geophys.* **2007**, *5*, 183–194. [CrossRef]
12. Oloumi, D.; Pettersson, M.I.; Mousavi, P.; Rambabu, K. Imaging of Oil-Well Perforations Using UWB Synthetic Aperture Radar. *IEEE Trans. Geosci. Remote Sens.* **2015**, *53*, 4510–4520. [CrossRef]
13. Wada, K.; Karasawa, S.; Kawata, K.; Ebihara, S. Fractures and Rock Properties Estimated by 3D Directional Borehole Radar. In Proceedings of the International Workshop on Advanced Ground Penetrating Radar (IWAGPR), Florence, Italy, 7–10 July 2015; pp. 1–4.

14. Hao, T.; Jing, L.; He, W. An Automated GPR Signal Denoising Scheme Based on Mode Decomposition and Principal Component Analysis. *IEEE Geosci. Remote Sens. Lett.* **2023**, *20*, 1–5. [CrossRef]
15. Li, R.; Zhang, H.; Chen, Z.; Yu, N.; Kong, W.; Li, T.; Wang, E.; Wu, X.; Liu, Y. Denoising Method of Ground-Penetrating Radar Signal Based on Independent Component Analysis with Multifractal Spectrum. *Measurement* **2022**, *192*, 110886. [CrossRef]
16. Li, Y.; Yang, X.; Lan, T.; Liu, R.; Qu, X. Parameter Inversion by a Modified Reflected Signal Reconstruction Method for Thin-Layered Media. *IEEE Antennas Wirel. Propag. Lett.* **2022**, *21*, 958–962. [CrossRef]
17. Xie, L.; Zhao, Q.; Ma, C.; Liao, B.; Huo, J. U-Net: Deep-Learning Schemes for Ground Penetrating Radar Data Inversion. *J. Environ. Eng. Geophys.* **2020**, *25*, 287–292. [CrossRef]
18. Zhang, G.; Cui, F.; Zhang, X.; Cheng, Q.; Wang, R.; Zhang, M. Research on Underground Pipeline Spatial Positioning Method Based on Multi-Offset Ground Penetrating Radar. *Meas. Sci. Technol.* **2024**, *36*, 015140. [CrossRef]
19. Cheng, Q.; Cui, F.; Chen, B.; Dong, G.; Wang, R.; Zhang, G.; Fu, J. Attenuation of Non-Stationary Random Noise in Ground Penetrating Radar Data Based on Time-Varying Filtering. *Measurement* **2024**, *236*, 115169. [CrossRef]
20. Zhang, X.; Qin, Y.; Hu, Z.; Hu, X.; Feng, X.; Chai, Y. GPR Energy Attribute Slices Based on Multivariate Variational Mode Decomposition and Teager–Kaiser Energy Operator. *Remote Sens.* **2022**, *14*, 4805. [CrossRef]
21. He, X.; Wang, C.; Zheng, R.; Li, X. GPR Image Noise Removal Using Grey Wolf Optimisation in the NSST Domain. *Remote Sens.* **2021**, *13*, 4416. [CrossRef]
22. Liu, H.; Long, Z.; Tian, B.; Han, F.; Fang, G.; Liu, Q. Two-Dimensional Reverse-Time Migration Applied to GPR with a 3-D-to-2-D Data Conversion. *IEEE J. Sel. Top. Appl. Earth Obs. Remote Sens.* **2017**, *10*, 4313–4320. [CrossRef]
23. Dai, Q.; Li, Y.; Jiang, T.; Sun, Q. A New B-Scan Interpretation Model in Complex Environments. *IEEE Trans. Instrum. Meas.* **2022**, *71*, 1–15. [CrossRef]
24. Feng, D.; Liu, Y.; Wang, X.; Zhang, B.; Ding, S.; Yu, T.; Li, B.; Feng, Z. Inspection and Imaging of Tree Trunk Defects Using GPR Multifrequency Full-Waveform Dual-Parameter Inversion. *IEEE Trans. Geosci. Remote Sens.* **2023**, *61*, 1–15. [CrossRef]
25. Zhao, Y.; Yang, X.; Qu, X.; Lan, T.; Gong, J. Clutter Removal Method for GPR Based on Low-Rank and Sparse Decomposition with Total Variation Regularization. *IEEE Geosci. Remote Sens. Lett.* **2023**, *20*, 1–5. [CrossRef]
26. Gutierrez, S.; Vega, F.; González, F.A.; Baer, C.; Sachs, J. Application of Polarimetric Features and Support Vector Machines for Classification of Improvised Explosive Devices. *IEEE Antennas Wireless Propag. Lett.* **2019**, *18*, 2282–2286. [CrossRef]
27. Abujarad, F.; Omar, A. Comparison of Independent Component Analysis (ICA) Algorithms for GPR Detection of Non-Metallic Land Mines. In Proceedings of the Image and Signal Processing for Remote Sensing XII, Stockholm, Sweden, 13–15 September 2006; Volume 6365, pp. 362–373.
28. Chen, G.; Fu, L.; Chen, K.; Boateng, C.D.; Ge, S. Adaptive Ground Clutter Reduction in Ground-Penetrating Radar Data Based on Principal Component Analysis. *IEEE Trans. Geosci. Remote Sens.* **2019**, *57*, 3271–3282. [CrossRef]
29. Brown, J. Choosing the Right Number of Components or Factors in PCA and EFA. *JALT Test. Eval. SIG Newsl.* **2009**, *13*, 2.
30. Temlioglu, E.; Erer, I. Clutter Removal in Ground-Penetrating Radar Images Using Morphological Component Analysis. *IEEE Geosci. Remote Sens. Lett.* **2016**, *13*, 1802–1806. [CrossRef]
31. Kumlu, D.; Erer, I. Improved Clutter Removal in GPR by Robust Nonnegative Matrix Factorization. *IEEE Geosci. Remote Sens. Lett.* **2020**, *17*, 958–962. [CrossRef]
32. Gao, Q.; Zhang, P.; Xia, W.; Xie, D.; Gao, X.; Tao, D. Enhanced Tensor RPCA and Its Application. *IEEE Trans. Pattern Anal. Mach. Intell.* **2020**, *43*, 2133–2140. [CrossRef]
33. Masarik, M.P.; Burns, J.; Thelen, B.T.; Kelly, J.; Havens, T.C. GPR Anomaly Detection with Robust Principal Component Analysis. In *Detection and Sensing of Mines, Explosive Objects, and Obscured Targets XX*; SPIE: Paris, France, 2015; Volume 9454, pp. 395–405.
34. Song, X.; Xiang, D.; Zhou, K.; Su, Y. Improving RPCA-Based Clutter Suppression in GPR Detection of Antipersonnel Mines. *IEEE Geosci. Remote Sens. Lett.* **2017**, *14*, 1338–1342. [CrossRef]
35. Xiang, D.; Zhou, K.; Su, Y. Fast Prescreening for GPR Antipersonnel Mine Detection via GO Decomposition. *IEEE Geosci. Remote Sens. Lett.* **2018**, *16*, 15–19. [CrossRef]
36. Shu, X.; Porikli, F.; Ahuja, N. Robust Orthonormal Subspace Learning: Efficient Recovery of Corrupted Low-Rank Matrices. In Proceedings of the IEEE Conference on Computer Vision and Pattern Recognition, Columbus, OH, USA, 23–28 June 2014; pp. 3874–3881.
37. Zhang, S.; Zhang, L.; Ling, T.; Fu, G.; Guo, Y. Experimental Research on Evaluation of Soil Water Content Using Ground Penetrating Radar and Wavelet Packet-Based Energy Analysis. *Remote Sens.* **2021**, *13*, 5047. [CrossRef]
38. Cui, H.; Song, G. Study of the Wavelet Basis Selections. In Proceedings of the International Conference on Computational and Information Science, Reading, UK, 28–31 May 2006; pp. 1009–1017.
39. Wang, X.; Liu, S. Noise Suppression and Direct Wave Arrivals Removal in GPR Data Based on Shearlet Transform. *Signal Process.* **2017**, *132*, 227–242. [CrossRef]
40. Oskooi, B.; Parnow, S.; Smirnov, M.; Varfinezhad, R.; Yari, M. Attenuation of random noise in GPR data by image processing. *Arab. J. Geosci.* **2018**, *11*, 1–10. [CrossRef]

41. Bakshi, B.R. Multiscale PCA with Application to Multivariate Statistical Process Monitoring. *AIChE J.* **1998**, *44*, 1596–1610. [CrossRef]
42. Sheriff, M.Z.; Mansouri, M.; Karim, M.N.; Nounou, H.; Nounou, M. Fault Detection Using Multiscale PCA-Based Moving Window GLRT. *J. Process Control* **2017**, *54*, 47–64. [CrossRef]
43. Sharma, L.N.; Dandapat, S.; Mahanta, A. Multichannel ECG Data Compression Based on Multiscale Principal Component Analysis. *IEEE Trans. Inf. Technol. Biomed.* **2012**, *16*, 730–736. [CrossRef] [PubMed]
44. Li, Y.; Chen, C.; Fang, R.; Yi, L. Accuracy Enhancement of High-Rate GNSS Positions Using a Complete Ensemble Empirical Mode Decomposition-Based Multiscale Multiway PCA. *J. Asian Earth Sci.* **2019**, *169*, 67–78. [CrossRef]
45. Jeng, Y.; Chen, C.-S. A Nonlinear Method of Removing Harmonic Noise in Geophysical Data. *Nonlinear Process. Geophys.* **2011**, *18*, 367–379. [CrossRef]
46. Terrasse, G.; Nicolas, J.M.; Trouvé, E.; Drouet, E. Application of the Curvelet Transform for Clutter and Noise Removal in GPR Data. *IEEE J. Sel. Top. Appl. Earth Obs. Remote Sens.* **2017**, *10*, 4280–4294. [CrossRef]
47. He, X.; Yan, H.; Wang, C.; Zheng, R.; Li, Y.; Li, X. Non-Stationary Random Noise Removal in Ground-Penetrating Radar Images by Using Self-Guided Filtering. *Digit. Signal Process.* **2022**, *129*, 103690. [CrossRef]
48. Sun, H.H.; Cheng, W.; Fan, Z. Learning to Remove Clutter in Real-World GPR Images Using Hybrid Data. *IEEE Trans. Geosci. Remote Sens.* **2022**, *60*, 1–14. [CrossRef]
49. Temlioglu, E.; Erer, I. A Novel Convolutional Autoencoder-Based Clutter Removal Method for Buried Threat Detection in Ground-Penetrating Radar. *IEEE Trans. Geosci. Remote Sens.* **2021**, *60*, 1–13. [CrossRef]
50. Ni, Z.; Ye, S.; Shi, C.; Li, C.; Fang, G. Clutter Suppression in GPR B-Scan Images Using Robust Autoencoder. *IEEE Geosci. Remote Sens. Lett.* **2020**, *19*, 1–5. [CrossRef]
51. Ma, C.; Zhao, Q.; Huo, J.; Xie, L. Borehole Radar Data Processing Based on Empirical Mode Decomposition. In Proceedings of the International Conference on Ground Penetrating Radar (GPR), Rapperswil, Switzerland, 18–21 June 2018; pp. 1–4.
52. Wu, Z.; Huang, N.E. Ensemble Empirical Mode Decomposition: A Noise-Assisted Data Analysis Method. *Adv. Adapt. Data Anal.* **2009**, *1*, 1–41. [CrossRef]
53. Li, J.; Liu, C.; Zeng, Z.; Chen, L. GPR Signal Denoising and Target Extraction with the CEEMD Method. *IEEE Geosci. Remote Sens. Lett.* **2015**, *12*, 1615–1619.
54. Colominas, M.A.; Schlotthauer, G.; Torres, M.E. Improved Complete Ensemble EMD: A Suitable Tool for Biomedical Signal Processing. *Biomed. Signal Process. Control* **2014**, *14*, 19–29. [CrossRef]
55. Liu, S.; Chen, Y.; Luo, C.; Jiang, H.; Li, H.; Li, H.; Lu, Q. Particle Swarm Optimization-Based Variational Mode Decomposition for Ground Penetrating Radar Data Denoising. *Remote Sens.* **2022**, *14*, 2973. [CrossRef]
56. Dragomiretskiy, K.; Zosso, D. Variational Mode Decomposition. *IEEE Trans. Signal Process.* **2013**, *62*, 531–544. [CrossRef]
57. Mirjalili, S.; Lewis, A. The Whale Optimization Algorithm. *Adv. Eng. Softw.* **2016**, *95*, 51–67. [CrossRef]
58. Richman, J.S.; Lake, D.E.; Moorman, J.R. Sample Entropy. In *Methods in Enzymology*; Elsevier: Amsterdam, The Netherlands, 2004; Volume 384, pp. 172–184.
59. Cohen, I.; Huang, Y.; Chen, J.; Benesty, J. Pearson Correlation Coefficient. In *Noise Reduct. Speech Process*; Springer: Berlin/Heidelberg, Germany, 2009; pp. 1–4.
60. Warren, C.; Giannopoulos, A.; Giannakis, I. gprMax: Open Source Software to Simulate Electromagnetic Wave Propagation for Ground Penetrating Radar. *Comput. Phys. Commun.* **2016**, *209*, 163–170. [CrossRef]

remote sensing

MDPI

Article

Two-Stage GPR Image Inversion Method Based on Multi-Scale Dilated Convolution and Hybrid Attention Gate

Mingze Wu, Qinghua Liu * and Shan Ouyang

School of Information and Communication, Guilin University of Electronic Technology, Guilin 541004, China; wumingze@mails.guet.edu.cn (M.W.); hmoysh@guet.edu.cn (S.O.)

* Correspondence: qhliu@guet.edu.cn

Abstract: Ground penetrating radar (GPR) image inversion is of great significance for interpreting GPR data. In practical applications, the complexity and nonuniformity of underground structures bring noise and clutter interference, making GPR inversion problems more challenging. To address these issues, this study proposes a two-stage GPR image inversion network called MHInvNet based on multi-scale dilated convolution (MSDC) and hybrid attention gate (HAG). This method first denoises the B-scan through the first network MHInvNet1, then combines the denoised B-scan from MHInvNet1 with the undenoised B-scan as input to the second network MHInvNet2 for inversion to reconstruct the distribution of the permittivity of underground targets. To further enhance network performance, the MSDC and HAG modules are simultaneously introduced to both networks. Experimental results from simulated and actual measurement data show that MHInvNet can accurately invert the position, shape, size, and permittivity of underground targets. A comparison with existing methods demonstrates the superior inversion performance of MHInvNet.

Keywords: ground penetrating radar; image inversion; dilated convolution; attention gate

1. Introduction

Ground-penetrating radar is a subsurface structure detection tool that offers portable and fast data acquisition technology, capable of providing dense and accurate high-resolution data [1]. It has been widely used in various fields such as rock strata detection [2], urban road disease detection [3], and building inspection [4]. GPR inversion establishes a mapping relationship between B-scans and electrical characteristics of underground media to obtain information on material structures and parameters in underground spaces.

Currently, GPR inversion methods can be divided into traditional and deep learning-based inversion methods. Traditional inversion methods include reverse time migration imaging, full waveform inversion (FWI), tomography, and others. Chen et al. [5] proposed a method that applies the normalized cross-correlation imaging condition to pre-stack migration imaging of ground-penetrating radar (GPR). Spatial high-pass filtering was used to suppress low-frequency noise generated during the cross-correlation process and reconstruct the spatial morphology and internal structural characteristics of the target. However, reverse time migration algorithms cannot provide electrical parameter information such as permittivity and conductivity of underground targets, which are crucial for identifying and detecting subsurface objects. Feng et al. [6] proposed a dual-parameter FWI method based on total variation regularization to achieve multi-scale inversion of GPR data. The dual-parameter inversion method provides reliable constraints, exhibits good noise adaptability, and can accurately invert the permittivity of underground targets. However, FWI requires multiple forward simulations to obtain the final inversion results, leading to high computational costs and long processing times. Irving et al. [7] proposed a causal linear model-based inverse Q-filtering algorithm derived from seismic tomography technology to eliminate wavelet dispersion and improve imaging resolution. However, this method relies on precise modeling and strict detection environments, limiting its practicality in real-world engineering applications. In summary, traditional methods suffer from slow computation speeds, low accuracy, and insufficient reliability.

In recent years, deep learning has gradually been applied to GPR inversion problems due to its rapid development and efficiency. Alvarez et al. [8] used three different deep neural networks (DNNs) to reconstruct underground permittivity distribution maps, demonstrating the feasibility of deep learning methods for GPR inversion tasks. Xie et al. [9] proposed a network called Ü-Net, which incorporates instance normalization layers into U-Net for inverting permittivity images. Liu et al. [10] proposed a network called GPRInvNet for tunnel lining defect inversion, which addresses the spatial alignment issue between B-scan images and permittivity images by designing a trace-to-trace encoder. Ji et al. [11] designed an inversion network named PINet, which further extracts global information from B-scan images using a global feature encoder comprising fully connected layers to improve inversion accuracy. Wang et al. [12] enhanced the reliability of reinforced concrete defect detection by introducing a multi-path encoder in the inversion network and using three types of GPR data simultaneously. Dai et al. [13] proposed a network called DMRF-UNet for permittivity inversion of underground targets under nonuniform soil conditions, cascading two U-Net-structured networks to eliminate clutter effects caused by soil heterogeneity during the inversion process.

Existing deep learning-based ground-penetrating radar (GPR) image inversion methods are mostly applied in ideal environments with homogeneous underground media. However, when inverting actually detected B-scan images, they are often influenced by noise and clutter interference in the actual environment, leading to a decrease in the accuracy and stability of the inversion results. To address the above issues, this study proposes a two-stage GPR image inversion network MHInvNet based on the U-Net structure [14]. The main contributions are as follows:

(1) A multi-scale dilated convolution (MSDC) module is developed. To better capture the correlation between adjacent A-scan signals in B-scan images, it is introduced into the network to extract and fuse features of different scales in B-scan images;

(2) A hybrid attention gate (HAG) module is proposed. Due to the skip connection operation in U-Net where encoder features are directly concatenated with decoder features, unnecessary interference information may be introduced to the decoder network. In order to enhance the network's denoising and inversion performance, it is introduced at the skip connection to highlight important features related to the target signal and suppress unnecessary noise features;

(3) Experiments with simulated data and real data obtained from actual environmental measurements demonstrate the effectiveness and superiority of the proposed network compared to existing methods.

2. Construction and Analysis of GPR Detection Model

Construct the GPR detection model as shown in Figure 1a, where the permittivities of the background medium and the target object in the model region are represented by ε_1 and ε_2, respectively. Move the GPR along the x-axis survey line direction, and the transmitting antenna Tx and receiving antenna Rx will measure multiple A-scan signals at a certain sampling interval. The A-scan reflects the temporal variation of the amplitude intensity of the reflected waves and indirectly reveals the reflection characteristics of electromagnetic waves propagating in the depth direction. By integrating and visualizing the obtained A-scan signals, B-scan image data can be produced. The B-scan illustrates the changes in the reflected waves in both the survey line direction and the depth direction, with the relationship between the A- and B-scans depicted in Figure 1b. The GPR model detection process described above can be simplified as a mathematical model $Y = H(X)$, where X represents the distribution of underground target media, H represents the physical propagation process of GPR electromagnetic signals, and Y represents the B-scans data obtained from the detection.

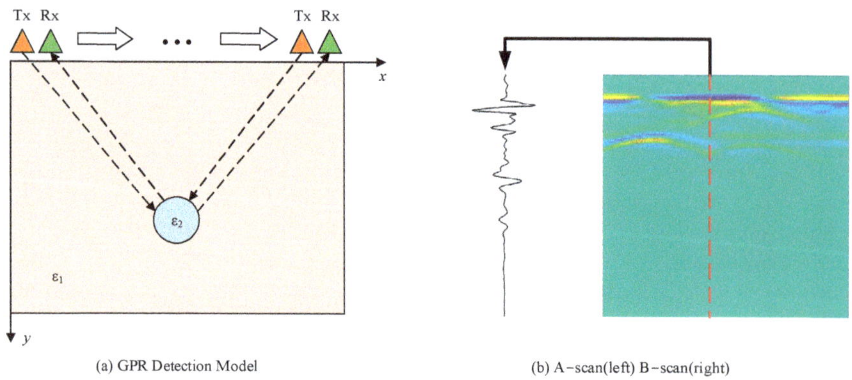

Figure 1. GPR detection model and data representation.

The GPR inversion problem involves the reverse reconstruction of physical information such as location, morphology, and permittivity of underground targets based on the acquired B-scan images. It can be simplified as the mathematical model $X = H^{-1}(Y)$, where H^{-1} is the inverse transformation of H, representing the inversion operator. The inversion aims to solve the inversion operator H^{-1}, which is a nonlinear problem. Deep learning provides a data-driven approach to learning the optimal H^{-1} by minimizing the differences between the predicted permittivity map and the actual permittivity map, thereby achieving a nonlinear mapping from B-scan images to permittivity maps, as shown in Figure 2.

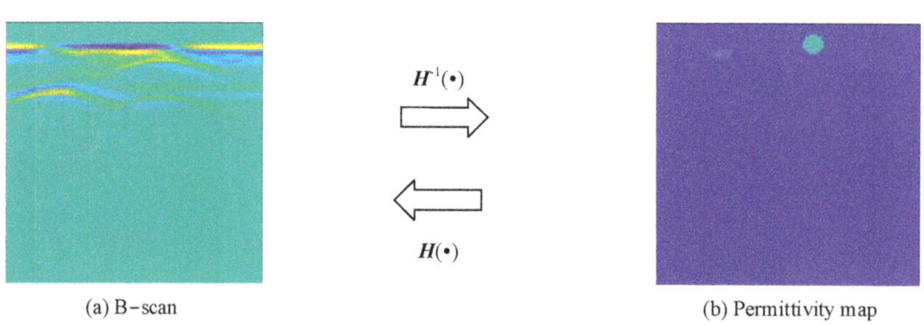

Figure 2. Inversion Process.

3. Network Structure

3.1. Overall Framework

The proposed network, MHInvNet, consists of two cascaded U-Nets, namely MHInvNet1 and MHInvNet2. The inversion process can be divided into two stages: first, the noisy B-scan images are denoised to obtain B-scan images that only contain the target signals. Subsequently, the inversion is performed on the dual-channel image obtained by concatenating the original noisy B-scan images with the denoised B-scan images to reconstruct the permittivity distribution of the underground targets. The overall framework of the network is shown in Figure 3.

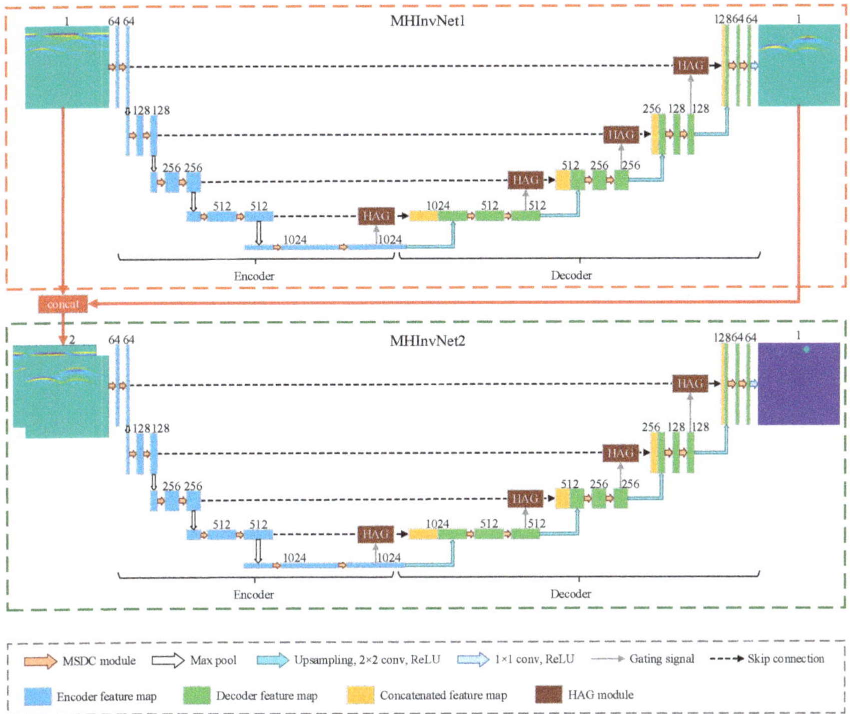

Figure 3. MHInvNet Structure.

MHInvNet1 is used for denoising the noisy input B-scans in the first stage to extract the target signal. MHInvNet1 consists of an encoder, a decoder, and skip connections with the introduction of the HAG module. The encoder consists of five feature extraction blocks, including four feature extraction blocks with downsampling operations comprising two MSDC modules and one max pooling layer, and one feature extraction block without downsampling operation consisting of two MSDC modules. The decoder consists of four upsampling blocks and a 1×1 convolutional layer with a rectified linear unit (ReLU) activation function, where each upsampling block consists of an upsampling layer and two MSDC modules. To compensate for information loss during encoding downsampling, skip connections are established between each layer of the encoder and decoder, with the introduction of the HAG module at the skip connections to help the network focus more on important feature information and suppress other interference.

MHInvNet2 is used for inverse processing of denoised B-scans in the second stage, converting B-scans into permittivity maps of underground targets. The network structure of MHInvNet2 is the same as MHInvNet1 overall, consisting of an encoder, decoder, and skip connections with the introduction of the HAG module. However, only the input part of the network is different. MHInvNet2 concatenates the denoised B-scans obtained from MHInvNet1 with the undenoised B-scans to obtain an image with a channel number of 2, which is used as the input image of the network. In turn, the network can focus more on features related to the target signal during the training process and avoid information loss in the B-scans caused by MHInvNet1's denoising process. This approach allows the network to learn and acquire more comprehensive feature information, enhancing the reconstruction capability of the permittivity for underground targets.

3.2. Multi-Scale Dilated Convolution Model

Scattering phenomena occur when electromagnetic waves propagate underground and encounter objects with different dielectric properties. The resulting scattered echoes reach the receiving antenna at different times. Therefore, in the B-scan data obtained from GPR detection, there is a certain correlation between adjacent A-scan signals. Additionally, due to the variety of underground targets, correlations between adjacent A-scans are not identical. To better capture the relationships between different A-scans, this paper proposes a Multi-Scale Dense Correlation (MSDC) module, whose structure is shown in Figure 4.

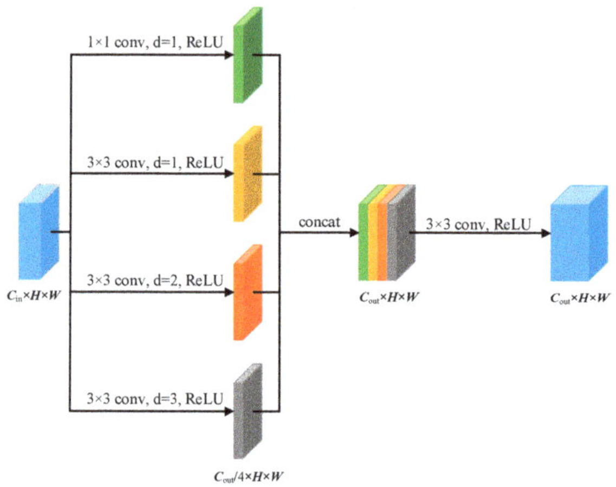

Figure 4. MSDC model.

Dilated convolution was originally applied to wavelet decomposition tasks [15], but now it is widely used in image processing fields such as semantic segmentation [16] and object detection [17]. Dilated convolution can change the size of the convolutional kernel receptive field by setting different dilation rates. This mechanism can expand the receptive field of the convolutional kernel without increasing additional parameters. The calculation formula for the equivalent standard convolutional kernel size is as follows:

$$K = (d - 1) \times (k - 1) + k \tag{1}$$

In the formula, K represents the size of the equivalent convolutional kernel, d is the dilation rate, and k is the size of the original convolutional kernel.

The MSDC module has four convolutional layer branches, including one branch with a convolution kernel size of 1×1 and three branches with convolution kernel sizes of 3×3. The dilation rate of the 1×1 convolution kernel is set to 1, and the dilation rates of the 3×3 convolution kernels are set to 1, 2, and 3, respectively. Using Formula (1), it can be seen that the equivalent convolution kernel sizes of the four convolution layer branches from top to bottom in Figure 4 are 1×1, 3×3, 5×5, and 7×7, respectively. The input feature map $\boldsymbol{F}_{\text{in}} \in \boldsymbol{R}^{C_{\text{in}} \times H \times W}$ first goes through the four convolution layer branches to capture feature information at different scales, while compressing the channel dimension to 1/4 of the output feature map size, obtaining the feature map $\boldsymbol{F}_{i \times i} \in \boldsymbol{R}^{C_{\text{out}}/4 \times H \times W}$, where i = 1, 3, 5, 7, represents the sizes of the equivalent convolution kernels corresponding to the four convolution layer branches, C_{in} represents the number of channels in the input feature map, C_{out} represents the number of channels in the output feature map, and H and W represent the height and width of the feature map. Then, concatenating $\boldsymbol{F}_{i\times i}$ results in the fused feature map $\boldsymbol{F}_c \in \boldsymbol{R}^{C_{out}\times H\times W}$. Finally, after passing through a 3×3 convolutional layer and ReLU activation function, the output feature map $\boldsymbol{F}_{out} \in \boldsymbol{R}^{C_{out}\times H\times W}$ is obtained. The specific implementation process can be described using mathematical formulas:

$$\boldsymbol{F}_{i\times i} = f_{\mathrm{D}}^{i\times i}(\boldsymbol{F}_{\mathrm{in}}),\ i = 1,3,5,7 \tag{2}$$

$$\boldsymbol{F}_{\mathrm{out}} = \mathrm{ReLU}(f^{3\times 3}(\mathrm{concat}(\boldsymbol{F}_{1\times 1}, \boldsymbol{F}_{3\times 3}, \boldsymbol{F}_{5\times 5}, \boldsymbol{F}_{7\times 7}))) \tag{3}$$

In the formula, $f_{\mathrm{D}}^{i\times i}$ represents the dilated convolution operation with an equivalent kernel size of $i\times i$, $f^{3\times 3}$ represents the convolution operation with a kernel size of 3×3, and concat represents the concatenation operation. In this study, both MHInvNet1 and MHInvNet2 utilize multi-scale dilated convolution modules for denoising and inversion, respectively.

3.3. Hybrid Attention Gate Model

When the encoder of the network downsamples to a deeper level, it extracts higher-level and more abstract feature information while losing some spatial information. By connecting the feature maps of the encoder and decoder through skip connections, it can compensate for lost spatial information. However, not all skip connections are useful. Since the lower-level features of the encoder may contain unwanted interference, this interference will negatively impact the effect of the skip connection fusion feature map. To highlight important features during the skip connection process and suppress unwanted interference, inspired by Attention U-Net [18] and ECANet [19], this study proposes a HAG module consisting of two parts: Channel attention gate (CGA) and Spatial attention gate (SGA), as shown in Figure 5.

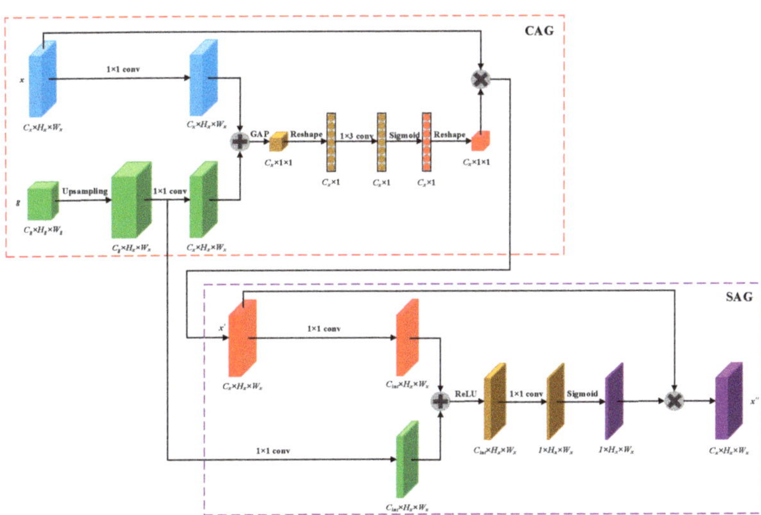

Figure 5. HAG model.

CAG is used to generate channel attention feature maps. Channel attention can capture the importance between different channels of feature maps to highlight those with important feature information. It has two inputs, the feature map $\boldsymbol{x} \in \boldsymbol{R}^{C_x\times H_x\times W_x}$ from the encoder skip connection and the gate signal $\boldsymbol{g} \in \boldsymbol{R}^{C_g\times H_g\times W_g}$ from the lower-level decoder, where C_x, H_x, W_x represent the number of channels, height, and width of the input feature map x, respectively, and C_g, H_g, W_g represent the number of channels, height, and width of the gate signal g. CAG first upsamples the gate signal g from the lower-level decoder to match the feature map sizes of x and g. Then, the upsampled x and g undergo convolution operations, and after being added, followed by global average pooling (GAP) to obtain the feature map $F_{\mathrm{GAP}} \in R^{C_x \times 1 \times 1}$. In this way, the context information in g adjusts x, enhancing the alignment weights of the two signals. Next, F_{GAP} is reshaped to obtain a one-dimensional feature $F_{\mathrm{1D}} \in R^{C_x \times 1}$, which then undergoes one-dimensional convolution, sigmoid, and reshape operations to produce the channel attention $F_{\mathrm{CA}} \in R^{C_x \times 1 \times 1}$. This local cross-channel interaction strategy avoids dimension reduction and can significantly reduce model complexity while maintaining performance. Finally, the output feature map $x' \in R^{C_x \times H_x \times W_x}$ weighted by the channel attention is obtained by multiplying F_{CA} by the input x. The specific implementation process can be described by the following mathematical formulae:

$$F_{\mathrm{GAP}} = f^{\mathrm{GAP}}(f^{1\times 1}(x + f^{\mathrm{UP}}(g))) \tag{4}$$

$$F_{1D} = \mathrm{Reshape}(F_{\mathrm{GAP}}) \tag{5}$$

$$F_{\mathrm{CA}} = \mathrm{Reshape}(\mathrm{Sigmoid}(f^{1\times 3}(F_{\mathrm{1D}}))) \tag{6}$$

$$\mathbf{x}' = \mathbf{x} \cdot F_{\mathrm{CA}} \tag{7}$$

In the equation, f^{GAP} represents global average pooling operation, f^{UP} represents upsampling operation, and $f^{1\times 1}$, $f^{1\times 3}$ represent convolution operations with kernel sizes of 1×1 and 3×3, respectively.

SAG is used to generate spatial attention feature maps. Spatial attention can capture the importance between different positions in the feature map to highlight positions with important feature information. It also has two inputs: the feature map $x' \in R^{C_x \times H_x \times W_x}$ weighted by channel attention from CAG and the gating signal $g \in R^{C_g \times H_g \times W_g}$ from the lower decoder. SAG first performs convolution on x' and upsampled g, adds them, and applies the ReLU activation function to obtain the intermediate feature map $F_{\mathrm{int}} \in R^{C_{\mathrm{int}} \times H_x \times W_x}$. This step also aims to adjust x' according to g, where C_{int} is the channel number of the input feature map F_{int}, usually set to half of C_x to reduce computation. F_{int} then undergoes another convolution and sigmoid activation function to generate the spatial attention $F_{\mathrm{SA}} \in R^{1 \times H_x \times W_x}$. Finally, F_{SA} is multiplied by the feature map x' to obtain the output feature map $x'' \in R^{C_x \times H_x \times W_x}$ after spatial attention weighting. The specific implementation process can be described by the following mathematical formulae:

$$F_{\mathrm{int}} = \mathrm{ReLU}(f^{1\times 1}(x' + f^{\mathrm{UP}}(g))) \tag{8}$$

$$F_{\mathrm{SA}} = \mathrm{Sigmoid}(f^{1\times 1}(F_{\mathrm{int}})) \tag{9}$$

$$\mathbf{x}'' = \mathbf{x}' \cdot F_{\mathrm{SA}} \tag{10}$$

HAG employs a combination of CAG and SAG in a series configuration. It first applies channel attention weighting to the input feature map, followed by spatial attention weighting to the channel-attended feature map. This approach highlights the important features in both the channel and spatial dimensions, thereby enhancing the overall feature extraction ability of the network. In this study, the HAG module is introduced into each layer's skip connection of both MHInvNet1 for denoising and MHInvNet2 for inversion.

3.4. Loss Function

The structural similarity index measure (SSIM) is a common image evaluation metric that comprehensively evaluates the similarity of two images in terms of brightness, contrast, and structural similarity. In deep learning, a variant of SSIM known as structural dissimilarity (DSSIM) is often used as a loss function, with the following expression:

$$\mathrm{SSIM}(x,y) = \frac{(2\mu_x\mu_y + c_1)(2\sigma_{x,y} + c_2)}{\left(\mu_x^2 + \mu_y^2 + c_1\right)\left(\sigma_x^2 + \sigma_y^2 + c_2\right)} \tag{11}$$

$$\mathrm{DSSIM}(x,y) = 1 - \mathrm{SSIM}(x,y) \tag{12}$$

In the equation, μ_x and μ_y represent the average values of images x and y, respectively; σ_x^2 and σ_y^2 represent the variances of images x and y, respectively; c_1 and c_2 are stability coefficients.

The total loss function of this study combines the losses of two networks, MHInvNet1 and MHInvNet2, to achieve simultaneous backpropagation of the two networks during the training process. Both MHInvNet1 and MHInvNet2 use DSSIM as the loss function. The definition of the total loss function is as follows:

$$l = l_1 + l_2 \tag{13}$$

In the equation, l represents the total loss function, and l_1 and l_2 represent the loss functions of MHInvNet1 and MHInvNet2, respectively.

4. Experimentation

4.1. Dataset and Evaluation Criteria

To validate the inversion performance and feasibility of the proposed network MHInvNet, this study used the open-source dataset provided by the authors of [13] to train and evaluate the network. The dataset used was simulated using the open-source software gprMax, with each dataset containing three images: the input noisy B-scan, the denoised B-scan, and the corresponding underground target permittivity map. A Peplinski mixed model was employed in the simulation process to simulate a real nonuniform soil environment for obtaining noisy B-scans with realistic dielectric and geometric properties. The specific simulation model size was 1.5 m wide and 0.5 m high. It had a background soil permittivity ranging from 3.82 to 9.99, and a target permittivity ranging from 2 to 32. The targets were randomly selected from circular, semi-circular, triangular, and rectangular shapes, with the radius of the circular and semi-circular shapes and distances from the three vertices of the triangles to the center ranging from 0.05 m to 0.08 m, and the widths and lengths of the rectangles ranging from 0.04 m to 0.06 m and 0.12 m to 0.16 m, respectively. The orientations of the semi-circular, triangular, and rectangular shapes were randomly chosen between 0° and 360°. The dataset consisted of 18,000 data sets, including 8000 single-target scene data and 10,000 double-target scene data, divided into training and testing sets in a ratio of 9:1. All images were normalized to a range of 0 to 1 and adjusted to a size of 128 × 128.

In order to evaluate the denoising performance and inversion performance of the network, three metrics including SSIM, mean absolute error (MAE), and mean square error (MSE) were adopted as evaluation criteria for denoising and inversion quality. SSIM can measure the structural similarity between two images, with a larger value indicating a higher similarity between the two images. MAE and MSE can measure the pixel-level errors between two images, with smaller values indicating minor errors between the two images.

4.2. Experimental Environment and Parameter Settings

The experimental environment was a Windows platform equipped with an NVIDIA TITAN RTX graphics card with 24 GB of graphics memory. The proposed MHInvNet network framework was implemented based on TensorFlow. An Adam optimizer was used for network training, with an initial learning rate set to 0.0001, and a batch_size set to 20. The network was trained for a total of 100 epochs, and the model with the lowest test loss was saved to obtain the best model.

4.3. Simulated Data Experiment

To verify the superior performance of the proposed network, it was compared with the network DMRF-UNet, which follows the same denoising and inversion process. All networks are configured with hyperparameters consistent with those described above. As shown in Table 1, #1 and #2 represent the denoising and inversion results of the networks. The denoising results of MHInvNet achieve SSIM, MAE, and MSE values of 0.99884, 0.16332, and 0.39140, respectively. The inversion results achieve SSIM, MAE, and MSE values of 0.99297, 0.21191, and 21.31944, respectively. Whether in denoising or inversion results, MHInvNet obtains the best scores in all three evaluation metrics. In terms of network model complexity, MHInvNet has 0.87 M fewer parameters than DMRF-UNet, which is attributed to the MSDC module employing dilated convolutions that reduce the computational burden associated with large-sized convolution kernels. Additionally, the HAG module utilizes strategies of local cross-channel interaction and compression of feature map channels to lower the module's complexity and parameter count. Regarding the time consumed to invert a B-scan image, MHInvNet has fewer parameters and requires slightly less time than DMRF-UNet. Overall, MHInvNet improves network performance while reducing model complexity.

Table 1. Comparison of denoising and inversion Performance of different networks.

Network		SSIM	MAE	MSE	Parameter/M	Time/ms
DMRF-UNet	#1	0.99773	0.79116	0.86150	12.41	23.9
	#2	0.99152	0.25348	23.26477		
MHInvNet	#1	0.99884	0.16332	0.39140	11.54	22.1
	#2	0.99297	0.21191	21.31944		

The denoised B-scan and permittivity map outputs are normalized to between −50 V/m and 75 V/m, and 0 to 32, respectively. Figure 6 shows the denoising and inversion results of different networks after normalization. Regarding denoising, as shown in Figure 6i–l, although DMRF-UNet can effectively remove the echoes of direct waves and clutter caused by soil inhomogeneity, there are still some pixel-level noise spots distributed laterally in the background of the B-scan after denoising. Figure 6q–t indicates that MHInvNet can effectively eliminate noise and unrelated target signals in the noisy B-scan, with almost no noise spots remaining in the background of the B-scan after denoising. In terms of inversion, as shown in Figure 6m–p, the target contours obtained by DMRF-UNet are blurry, with artifacts present, and the inversion effect for stacked targets is poor, sometimes leading to the inversion of false targets. In Figure 6u–x, the MHInvNet results are more accurate, showing clear boundaries between the inverted targets, closely resembling the shapes of the actual underground targets, and performing well when inverting stacked targets. Overall, MHInvNet demonstrates better performance in denoising and inversion. This improvement is attributed to the MSDC module, which extracts and integrates feature information at multiple scales, enabling the network to learn richer global features. Additionally, the HAG module provides attention constraints to the skip-connected feature maps, further enhancing important information in the feature maps and collectively improving the network's denoising and inversion performance.

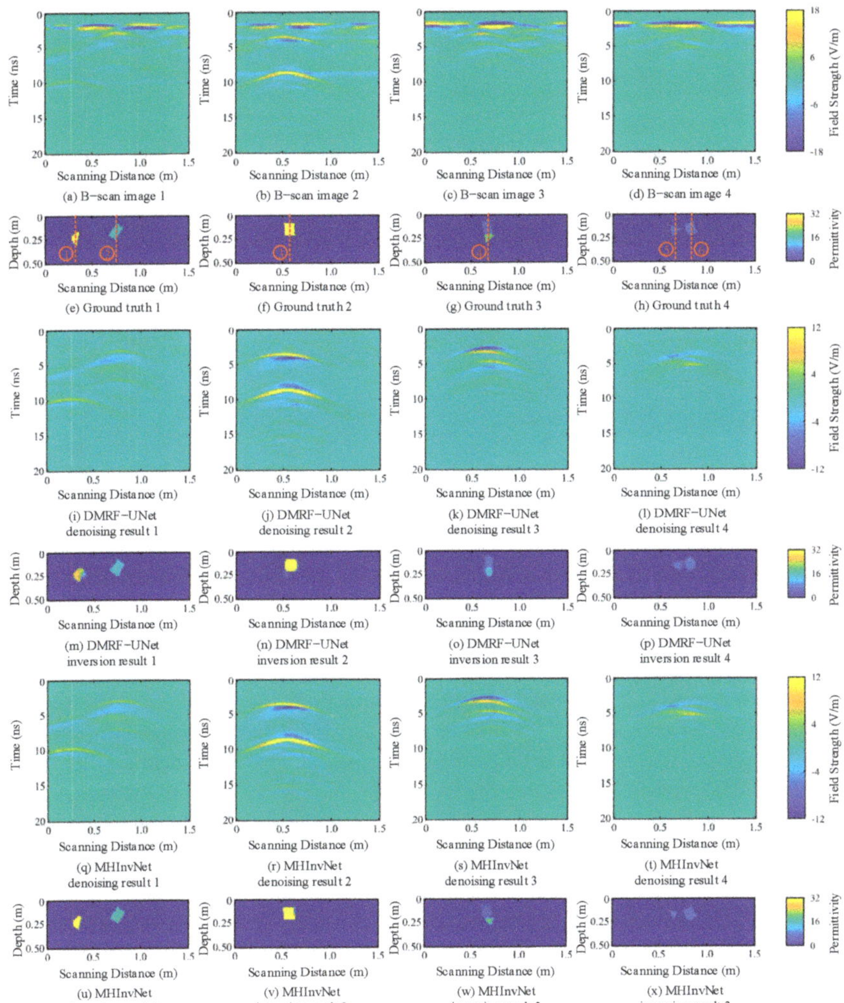

Figure 6. Denoising and inversion results of different networks.

To further validate the accuracy of the proposed network in the specific value inversion of permittivity, this study extracted the permittivity values along the six cutting lines marked from ① to ⑥ in Figure 6b for comparative analysis of the differences between the results inverted by DMRF-UNet and MHInvNet under the same cutting lines and the true permittivity model, as shown in Figure 7. It can be seen from Figure 7a–f that MHInvNet performs better in the specific value inversion of permittivity than DMRF-UNet. A further examination of Figure 7b,d,e,f reveals that the sizes and positions of the targets inverted by MHInvNet exhibit smaller deviations in the depth direction compared to the true model. Overall, the permittivity curve inverted by MHInvNet is closer to the true underground model curve, with only slight differences in the curve amplitudes.

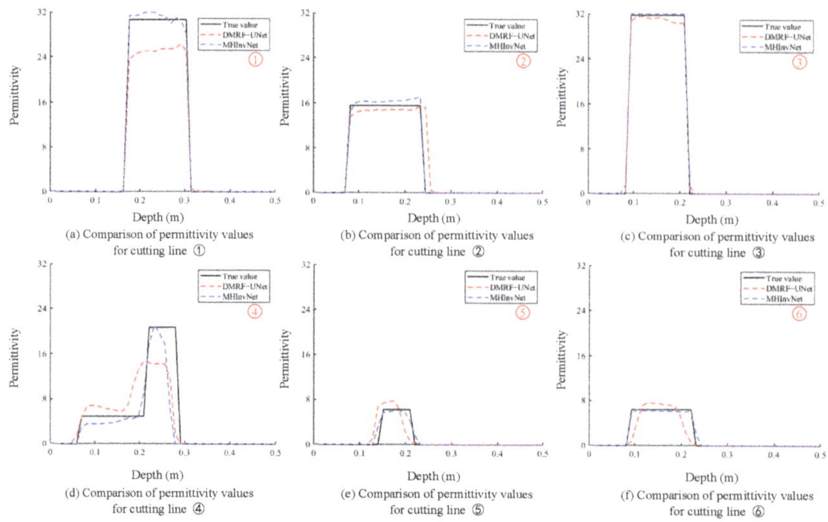

(a) Comparison of permittivity values for cutting line ①
(b) Comparison of permittivity values for cutting line ②
(c) Comparison of permittivity values for cutting line ③
(d) Comparison of permittivity values for cutting line ④
(e) Comparison of permittivity values for cutting line ⑤
(f) Comparison of permittivity values for cutting line ⑥

Figure 7. Inversion results of permittivity values at cutting lines ① to ⑥ using different networks.

4.4. Ablation Study

In order to demonstrate the effectiveness of the introduced MSDC module and HAG module, this paper compares and verifies four different network structures. They are MHInvNet_v1 without introducing any modules, MHInvNet_v2 with only the MSDC module, MHInvNet_v3 with only the HAG module, and the proposed network MHInvNet. The performance comparison of the four networks is shown in Table 2. MHInvNet_v1, without introducing any modules, obtained the worst performance indicators. The introduction of the MSDC module in MHInvNet_v2 enhances the network's feature extraction capability, resulting in improved performance indicators. The introduction of the HAG module in MHInvNet_v3 highlights the more important information in the feature map while improving the network's performance indicators. The simultaneous introduction of the MSDC and HAG modules in MHInvNet significantly enhances network performance, achieving the best performance indicators. It also demonstrates the effectiveness of the two modules in the denoising and inversion tasks of B-scan images.

Table 2. Performance comparison of ablation research.

Network		SSIM	MAE	MSE
MHInvNet_v1	#1	0.99838	0.79703	0.95108
	#2	0.99146	0.26318	27.65724
MHInvNet_v2	#1	0.99837	0.33972	0.39919
	#2	0.99282	0.22192	21.43576
MHInvNet_v3	#1	0.99871	0.27688	0.59640
	#2	0.99145	0.25849	27.74265
MHInvNet	#1	0.99884	0.16332	0.39140
	#2	0.99297	0.21191	21.31944

4.5. Visualization Study

In order to further understand the underlying working principles of the MSDC module and the HAG module, this study extracts the feature maps obtained from the convolutions of different downsampling layers in the inversion network for visual analysis, enabling a more intuitive observation of the effects of these two modules on enhancing network performance. For the MSDC module, feature maps are extracted from the convolution of the shallowest feature extraction layer of the inversion network. These shallow-layer feature maps are relatively close to the input images and contain information such as textures and edges, demonstrating the network's ability to capture image details. As shown in Figure 8, when compared to a network that does not incorporate the MSDC module, the extraction of feature maps from the same position and channel reveals that the MSDC module provides better performance in the feature extraction of target signals. It can clearly delineate the edges and contours of the target signals, preserve details related to them, and capture a greater amount of feature information.

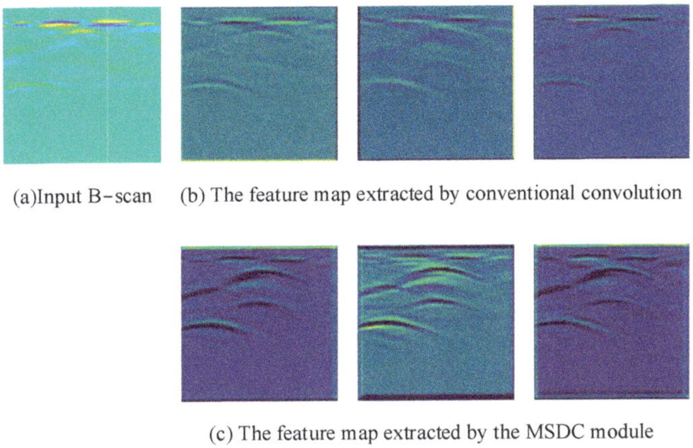

Figure 8. MSDC module visualization.

For the HAG module, since the feature maps obtained from the deepest convolution contain higher-level, more abstract semantic information, the attention matrix extracted by HAG can more distinctly highlight important semantic features. Therefore, the attention matrix output at the deepest skip connection of the inversion network is visualized. Figure 9b shows that the attention matrix effectively represents information about the important parts in the B-scan image. The brighter the color in the matrix, the greater the attention weight at that position, indicating that the corresponding features are more important. By adjusting the extracted attention matrix to the size of the input image and superimposing the two, an attention heatmap is generated, as shown in Figure 9c. The HAG module better highlights the features of the target signal in the B-scan, allowing the network to focus on learning important features.

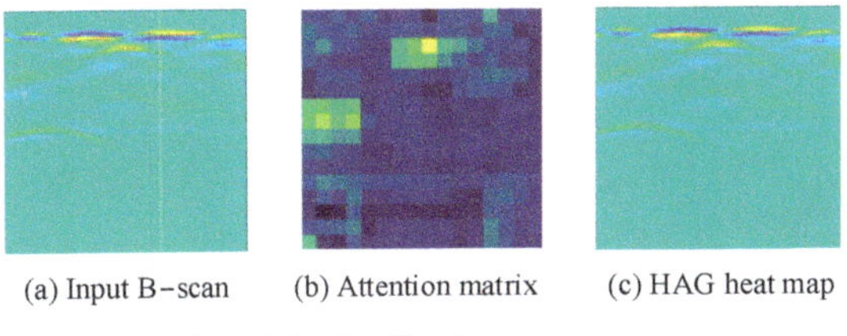

Figure 9. HAG module visualization.

4.6. Actual Measurement Data Experiment

In order to validate the effectiveness of the proposed network in denoising and inversion on actual measurement data, this study also used the actual open-source dataset provided by the authors of [13]. The data set was collected using commercial GSSI's Utility

Scan Pro GPR system on outdoor non-flat, nonuniform sand, and the center frequency of the GPR antenna was 400 MHz. Five different shapes, sizes, and relative permittivities of wooden objects were selected as buried objects in the sandy soil. The positions, depths, horizontal angles, and vertical angles of the objects buried underground were selected from ranges of 20 cm to 80 cm, 9 cm to 25 cm, 0° to 60°, and 0° to 60°, respectively. The collected data was normalized to between 0 and 1 and adjusted to a size of 128 × 128, with a total of 196 groups of data collected. Among them, 180 groups of data were selected as the training set for further training the network to fine-tune the weights obtained from training on simulated data. The remaining 16 groups of data were used as a test set to test the network. The network was trained for an additional 150 epochs on top of the previous training, with the remaining training parameters consistent with those described in Section 3.2.

Figure 10 shows the denoising and inversion results of different networks on actual measurement data. From Figures 10i–l and 10q–t, it can be seen that MHInvNet and DMRF-UNet perform similarly in denoising tasks. Although the clutter signals in the background are not completely removed, both can still effectively extract the target signal to reduce the adverse impact of clutter on the accuracy of target inversion. Figure 10m–p,u–x indicate that MHInvNet performs better in terms of the shape of the inverted target contour than DMRF-UNet, as the latter may produce irregular target shapes and artifacts in some cases. Overall, MHInvNet demonstrates better overall performance on actual measurement data.

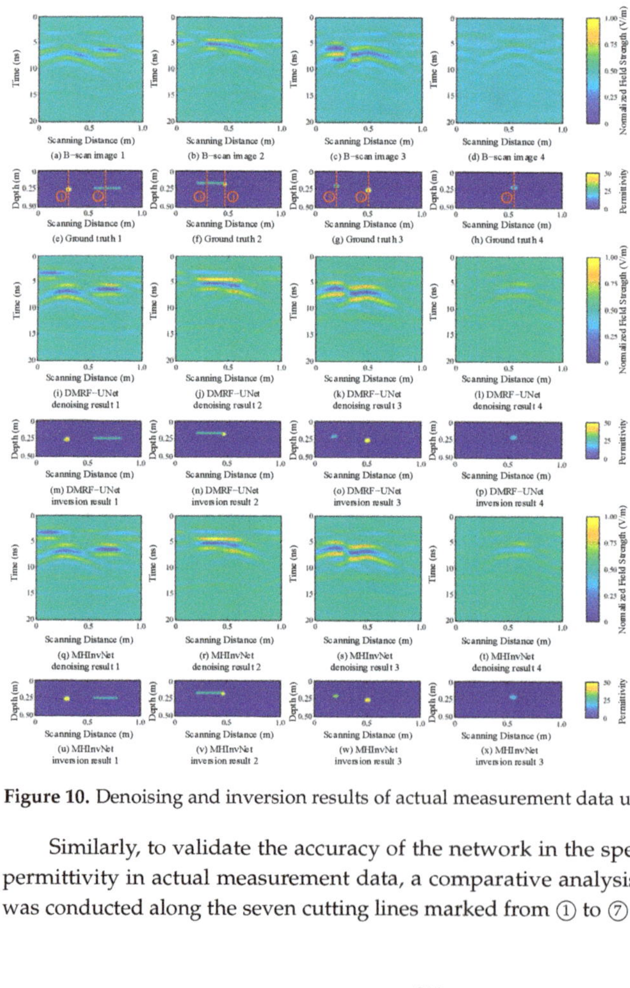

Figure 10. Denoising and inversion results of actual measurement data using different networks.

Similarly, to validate the accuracy of the network in the specific value inversion of permittivity in actual measurement data, a comparative analysis of permittivity values was conducted along the seven cutting lines marked from ① to ⑦ in Figure 10b, as shown in Figure 11. It can be observed that the permittivity values inverted by MHInvNet are more accurate than DMRF-UNet, with only a relatively large error in the inverted target permittivity along the cutting line ⑤. Further examination of Figure 11a,f,g reveals that the distribution of target permittivity inverted by DMRF-UNet is not uniform, exhibiting slight oscillations in the permittivity curve. By contrast, the permittivity curve of MHInvNet is relatively smoother and closer to the true underground model curve. Overall, MHInvNet achieves better permittivity inversion results along the seven cutting lines.

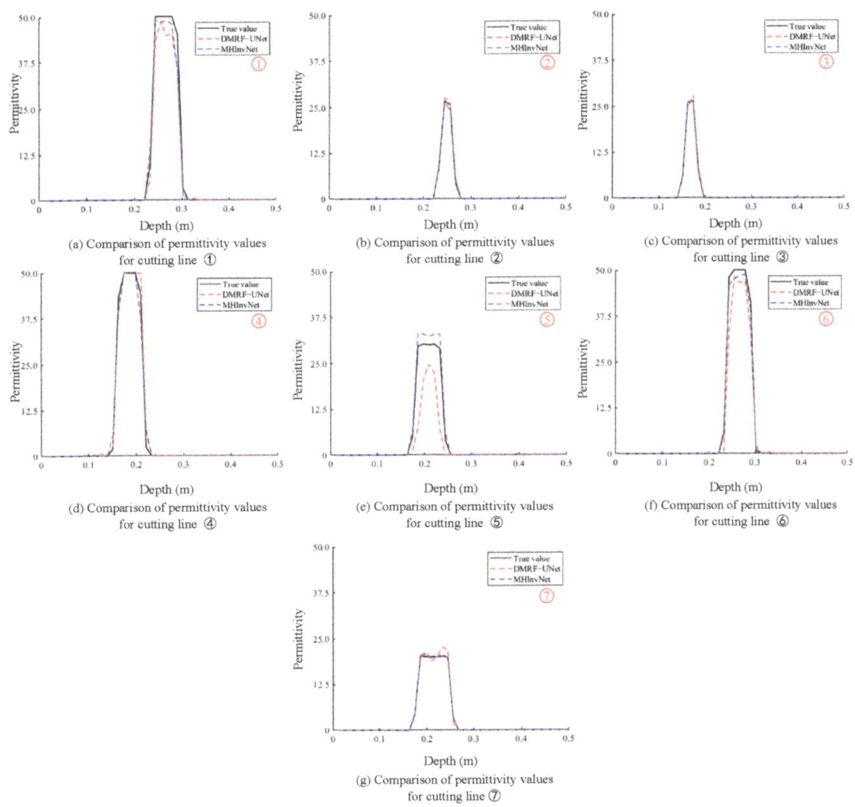

Figure 11. Inversion results of permittivity values at cutting lines ① to ⑦ of actual measurement data using different networks.

4.7. Measured Data Experiments in Different Geographical Environments

To further validate the network's generalization ability in inverting different measured data, we collected measured data from real scenarios that differ from those in Section 4.6 for experimentation. The experiment utilized a GPR from Y-Line, model YL-GPR400M, for data collection, with the antenna center frequency set at 400 MHz. The experimental scenario is depicted in Figure 12a, which shows a concrete wall with a thickness of 0.5 m, within which a solid steel bar with a diameter of 5 cm is embedded, located 0.2 m from the external surface of the wall. Behind the inner side of the wall is dry sand. Data acquisition was conducted along the external surface of the wall using radar, as shown in the experimental schematic in Figure 12b.

The collected data were preprocessed using the same method as the open-source measured datasets (i.e., normalization, resizing, and mean subtraction). Figure 13 presents the denoising and inversion results for the measured steel bar data. In Figure 13b,c, there are no significant differences in the denoised images obtained by MHInvNet and DMRF-UNet, as both methods effectively suppress the background clutter in the input B-scan. Figure 13e,f indicate that both MHInvNet and DMRF-UNet exhibit some deviation in the inversion at the target location, demonstrating horizontal and depth shifts compared to the actual target. Specifically, the target outline obtained by DMRF-UNet is unclear and irregularly shaped, with an uneven texture. It also produces nonexistent false targets. By contrast, MHInvNet performs target inversion relatively well. Although the dimensions of the inverted target are slightly smaller than those of the actual target, its overall performance still surpasses that of DMRF-UNet, demonstrating better generalization capability.

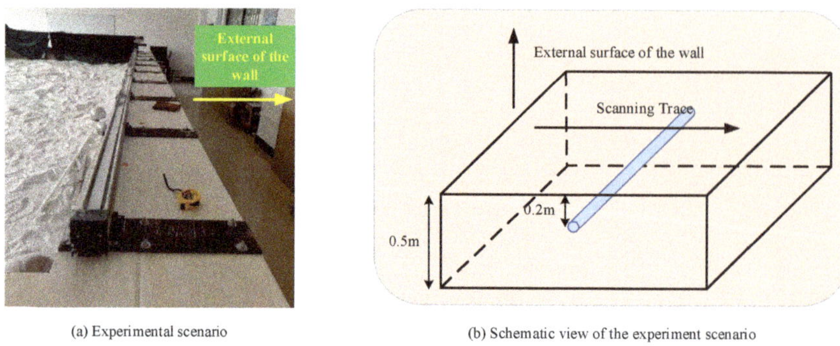

Figure 12. Measurement experiments of measured data in different geographical environments.

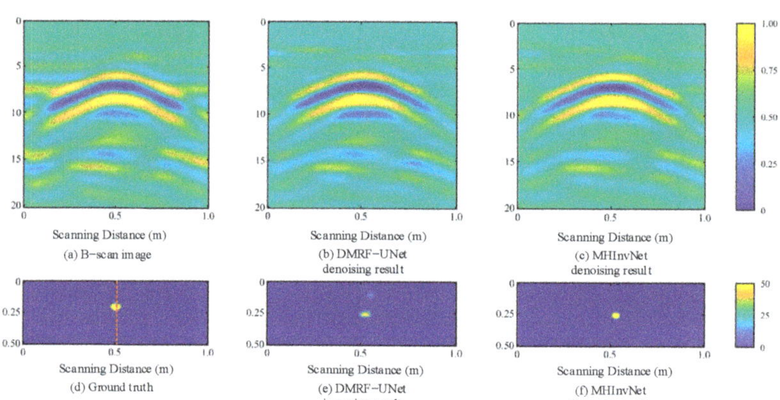

Figure 13. The denoising and inversion results for the measured data of steel bars by different networks.

The accuracy analysis of the permittivity inversion along the cutting lines shown in Figure 13d is presented in Figure 14. Due to the positional inversion deviations of MHInvNet and DMRF-UNet, the permittivity curves of both models do not align well with the true model curve in the horizontal coordinate direction. Regarding the values of permittivity, the maximum permittivity of the targets included in the measured dataset used for weight fine-tuning by both networks is 50. By contrast, the permittivity of the steel bar theoretically approaches infinity. Therefore, the correct value obtained for the permittivity of the steel bar during inversion is theoretically the maximum value that the network can invert, which is 50. It can be concluded that since the peak value of the permittivity curve obtained by MHInvNet reaches the theoretically correct value of 50, while that of DMRF-UNet is less than 50, the accuracy of MHInvNet's permittivity inversion is higher, showing good generalization capability in the specific inversion of permittivity values.

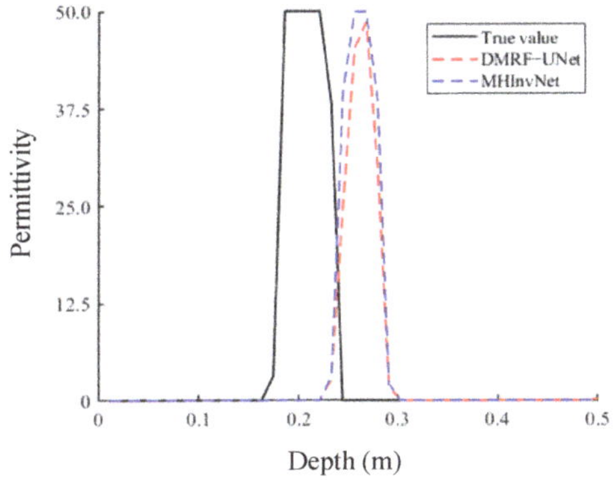

Figure 14. Inversion results of permittivity values for the measured data of steel bars by different networks.

5. Discussion

This study proposes an underground target permittivity inversion network MHInvNet for nonuniform soil environments to reduce the negative impact of clutter interference on inversion accuracy through a strategy of denoising followed by inversion. Introducing the MSDC and HAG modules enhanced network performance. Experimental validations were also conducted on both simulated and measured data. The results of the study are discussed as follows:

(1) The MSDC module consists of dilated convolutions with different receptive field sizes. Dilated convolutions allow for an increase in the receptive field of the convolution kernel without introducing extra parameters, thus reducing network complexity to some extent. Convolution kernels with larger receptive fields can capture global features with high-level semantic information in the image. By contrast, smaller receptive fields focus more on local detail features. MSDC learns the correlations of adjacent A-scans more comprehensively by extracting and fusing feature maps of different scales;

(2) The HAG module comprises CAG and SAG, which combine channel attention and spatial attention to weight the feature maps from skip connections. This process highlights important features related to the target signal while suppressing irrelevant features such as clutter and noise. By adjusting the input feature maps using gated signals from lower-level decoders, CAG and SAG further enhance the alignment of weights in feature maps of two different sizes, which guides the feature maps from the skip connection to some extent;

(3) Experimental verification using simulated and measured data shows that the proposed network achieves performance improvements in both denoising and inversion, outperforming existing inversion networks. To validate the generalization ability of the network, measured data from different geographical environments were collected for inversion. The results indicate that the proposed network can achieve high-quality inversions for measured data in various geographical contexts, demonstrating excellent generalization ability. Furthermore, visualization studies provide an in-depth analysis of the impact of the MSDC and HAG modules on network feature extraction, offering a clear understanding of their working principles;

(4) The proposed network can effectively invert the permittivity of targets in underground nonuniform background media, reducing noise and clutter interference to some extent. However, when faced with harsh and complex underground environments,

the collected B-scan images may contain strong interference signals. The noise and clutter intensity exceeds the capacity of the denoising network, rendering it unable to effectively denoise the B-scan and subsequently leading to failure in the inversion network. To overcome this limitation, it is necessary to study how to achieve high-quality inversion of B-scan images under low signal-to-noise ratio and low signal-to-clutter ratio conditions. Additionally, the types of underground targets that the proposed network can invert depend on the sample size of the training dataset; therefore, to achieve accurate inversions for more types and forms of underground targets, it is essential to collect higher-quality and more diverse data to build the dataset and enhance network learning.

6. Conclusions

This study proposes a two-stage GPR image inversion network MHInvNet based on the U-Net structure. Firstly, the first-stage network denoises the input B-scan images and then concatenates the denoised B-scan images with the undenoised B-scan images as inputs for the second-stage network for inversion, obtaining permittivity maps of underground targets. The MSDC module is introduced at both stages of the network to extract features from multiple scales. It merges them to capture the correlation between adjacent A-scans in the B-scan. The HAG module was also introduced to sequentially perform channel-wise attention weighting and spatial attention weighting on the feature maps of skip connections to reduce potential interference and enhance the network's ability to learn important feature information. Comparative experiments with existing methods show that the proposed network achieves good denoising and inversion effects on both simulated and actual measurement data. This study aims to invert the permittivity of targets buried in nonuniform soil environments. In future work, we will further research the inversion tasks of background media permittivity and more complex targets such as underground voids, cracks, and cavities.

Author Contributions: Conceptualization, M.W. and Q.L.; methodology, M.W. and Q.L.; validation, M.W. and Q.L.; writing—original draft, M.W.; writing—review and editing, Q.L. and S.O. All authors have read and agreed to the published version of the manuscript.

Funding: This work was supported by the National Natural Science Foundation of China under Grant 62361015 and the Guangxi Special Fund Project for Innovation-Driven Development under Grant AA21077008.

Data Availability Statement: The original data presented in the study is open available in Google Drive at https://drive.google.com/drive/folders/1s_C7Cfp5XlbWF-MjW1z0XIaiq-zpNltN?usp=sharing, access date 9 August 2024. The original author of the data production is Qiqi Dai.

Conflicts of Interest: The authors declare no conflicts of interest.

References

1. Solla, M.; Pérez-Gracia, V.; Fontul, S. A Review of GPR Application on Transport Infrastructures: Troubleshooting and Best Practices. *Remote Sens.* **2021**, *13*, 672. [CrossRef]
2. Tolooiyan, A.; Dyson, A.P.; Karami, M.; Shaghaghi, T.; Ghadrdan, M. Application of Ground Penetrating Radar (GPR) to Detect Joints in Organic Soft Rock. *Geotech. Test. J.* **2019**, *42*, 257–274. [CrossRef]
3. Wang, Z.; Wan, B.; Han, M. A Three-Dimensional Visualization Framework for Underground Geohazard Recognition on Urban Road-Facing GPRData. *ISPRS Int. J. Geo-Inf.* **2020**, *9*, 668. [CrossRef]
4. Urban, T.M.; Leon, J.F.; Manning, S.W.; Fisher, K.D. High resolution GPR mapping of Late Bronze Age architecture at Kalavasos-Ayios Dhimitrios, Cyprus. *J. Appl. Geophys.* **2014**, *107*, 129–136. [CrossRef]
5. Chen, D.P.; Dai, Q.W.; Feng, D.S.; Wang, H.H.; Zhang, B. Reverse time migration of ground penetrating radar based on normalized cross correlation imaging condition. *J. Cent. South Univ. (Sci. Technol.)* **2018**, *49*, 7. (In Chinese)

6. Feng, D.; Cao, C.; Wang, X. Multiscale Full-Waveform Dual-Parameter Inversion Based on Total Variation Regularization to On-Ground GPR Data. *IEEE Trans. Geosci. Remote Sens.* **2019**, *57*, 9450–9465. [CrossRef]
7. Irving, J.D.; Knight, R.J. Removal of wavelet dispersion from ground-penetrating radar data. *Geophysics* **2003**, *68*, 960–970. [CrossRef]
8. Alvarez, J.K.; Kodagoda, S. Application of deep learning image-to-image transformation networks to GPR radargrams for sub-surface imaging in infrastructure monitoring. In Proceedings of the 13th IEEE Conference on Industrial Electronics and Applications, Wuhan, China, 31 May–2 June 2018; pp. 611–616.
9. Xie, L.; Zhao, Q.; Ma, C.; Liao, B.; Huo, J. Ü-net: Deep-learning schemes for ground penetrating radar data inversion. *J. Environ. Eng. Geophys.* **2020**, *25*, 287–292. [CrossRef]
10. Liu, B.; Ren, Y.; Liu, H.; Xu, H.; Wang, Z.; Cohn, A.G.; Jiang, P. GPRInvNet: Deep learning-based ground-penetrating radar data inversion for tunnel linings. *IEEE Trans. Geosci. Remote Sens.* **2021**, *59*, 8305–8325. [CrossRef]
11. Ji, Y.; Zhang, F.; Wang, J.; Wang, Z.; Jiang, P.; Liu, H.; Sui, Q. Deep neural network-based permittivity inversions for ground penetrating radar data. *IEEE Sens. J.* **2021**, *21*, 8172–8183. [CrossRef]
12. Wang, Y.; Qin, H.; Miao, F. A Multi-Path Encoder Network for GPR Data Inversion to Improve Defect Detection in Reinforced Concrete. *Remote Sens.* **2022**, *14*, 5871. [CrossRef]
13. Dai, Q.; Lee, Y.H.; Sun, H.-H.; Ow, G.; Yusof, M.L.M.; Yucel, A.C. Dmrf-unet: A two-stage deep learning scheme for gpr data inversion under heterogeneous soil conditions. *IEEE Trans. Antennas Propag.* **2022**, *70*, 6313–6328. [CrossRef]
14. Ronneberger, O.; Fischer, P.; Brox, T. U-Net: Convolutional net-works for biomedical image segmentation. In Proceedings of the International Conference on Medical Image Computing and Computer-Assisted Intervention, Munich, Germany, 5–9 October 2015; pp. 234–241.
15. Holschneider, M.; Kronland-Martinet, R.; Morlet, J.; Tchamitchian, P. *A Real-Time Algorithm for Signal Analysis with the Help of the Wavelet Transform*; Springer: Berlin/Heidelberg, Germany, 1989; pp. 286–297.
16. Yu, F.; Koltun, V. Multi-Scale Context Aggregation by Dilated Convolutions. In Proceedings of the ICLR, San Juan, Puerto Rico, 2–4 May 2016.
17. Xu, Q.; Zhu, H.; Fan, H.; Zhou, H.; Yu, G. Study on detection of steel plate surface defects by improved YOLOv3 network. *Comput. Eng. Appl.* **2020**, *56*, 265–272.
18. Oktay, O.; Schlemper, J.; Folgoc, L.L.; Lee, M.; Heinrich, M.; Misawa, K.; Mori, K.; McDonagh, S.; Hammerla, N.Y.; Kainz, B.; et al. Attention U-Net: Learning where to look for the pancreas. *arXiv* **2018**, arXiv:1804.03999.
19. Wang, Q.; Wu, B.; Zhu, P.; Li, P.; Zuo, W.; Hu, Q. ECA-Net: Efficient channel attention for deep convolutional neural networks. In Proceedings of the IEEE/CVF Conference on Computer Vision and Pattern Recognition, Seattle, WA, USA, 14–19 June 2020; pp. 11534–11542.

remote sensing

MDPI

Article

Cross-Line Fusion of Ground Penetrating Radar for Full-Space Localization of External Defects in Drainage Pipelines

Yuanjin Fang *, Feng Yang, Xu Qiao, Maoxuan Xu, Liang Fang, Jialin Liu and Fanruo Li

School of Artificial Intelligence, China University of Mining and Technology (Beijing), Beijing 100083, China; yangf@cumtb.edu.cn (F.Y.); qiaoxu@cumtb.edu.cn (X.Q.); xmx@cumtb.edu.cn (M.X.); bqt1800405021@student.cumtb.edu.cn (L.F.); bqt2000405023@student.cumtb.edu.cn (J.L.); bqt2200405019@student.cumtb.edu.cn (F.L.)

* Correspondence: bqt1900405017@student.cumtb.edu.cn

Abstract: Drainage pipelines face significant threats to underground safety due to external defects. Ground Penetrating Radar (GPR) is a primary tool for detecting such defects from within the pipeline. However, existing methods are limited to single or multiple axial scan lines, which cannot provide the precise spatial coordinates of the defects. To address this limitation, this study introduces a novel GPR-based drainage pipeline inspection robot system integrated with multiple sensors. The system incorporates MEMS-IMU, encoder modules, and ultrasonic ranging modules to control the GPR antenna for axial and circumferential scanning. A novel Cross-Line Fusion of GPR (CLF-GPR) method is introduced to integrate axial and circumferential scan data for the precise localization of external pipeline defects. Laboratory simulations were performed to assess the effectiveness of the proposed technology and method, while its practical applicability was further validated through real-world drainage pipeline inspections. The results demonstrate that the proposed approach achieves axial positioning errors of less than 2.0 cm, spatial angular positioning errors below 2°, and depth coordinate errors within 2.3 cm. These findings indicate that the proposed approach is reliable and has the potential to support the transparency and digitalization of urban underground drainage networks.

Keywords: ground penetrating radar (GPR); pipeline robot; drainage pipeline; underground space safety

1. Introduction

Drainage pipelines are a critical component of urban infrastructure [1]. However, they are highly susceptible to aging, corrosion, and cracking due to water flow erosion and gas intrusion, which often lead to structural damage [2]. This damage allows external soil to infiltrate the pipeline, resulting in soil loss near the outer wall of the pipe [3]. The loss of external soil can cause the formation of cavities, loose soil, and water-saturated zones outside the pipeline. These external defects are a major contributing factor to urban road collapses [4,5]. The accurate localization of external cavities provides essential guidance for pipeline maintenance and repair, ensuring the safety of urban underground spaces.

Traditional methods for inspecting drainage pipelines primarily involve using devices equipped with closed-circuit television (CCTV) [6–9], sonar sensors [10], or multi-sensor fusion [11] technologies. However, these methods are limited to detecting internal pipeline defects and cannot identify or locate external defects. Moreover, in practical applications, cameras are often rendered ineffective due to unpredictable water flow within the pipeline, which can submerge or erode the camera, resulting in blurred images. Consequently, these methods fail to adequately inspect drainage pipelines.

Ground Penetrating Radar (GPR) is an effective, non-invasive technology widely used to detect underground cavities [12–14], especially beneath roadways. Typically, GPR systems are dragged across the surface to identify subsurface defects [15–17]. However, this method is most effective for detecting shallow anomalies. The technology faces limitations due to the trade-off between resolution and depth. High-frequency GPR offers high resolution but limited depth penetration, whereas low-frequency GPR penetrates deeper but lacks the resolution to detect small defects. To overcome these challenges, Liu et al. [18] proposed a collaborative GPR detection method. This approach combines multiple survey lines on the ground with manually adjusted radar angles within the pipeline to gather comprehensive data. By integrating GPR data from both the ground and the pipeline, this method improves the accuracy of shallow pipeline detection. However, it remains unsuitable for drainage pipelines buried at depths of 3–10 m [3]. These pipelines are positioned beneath other underground utilities at significant depths, resulting in severe signal attenuation. Interference from intersecting pipelines further hinders detection accuracy.

To effectively identify defects outside drainage pipelines, it is necessary to deploy GPR inside the pipelines. Due to safety concerns, this process must be conducted using robotic systems. The earliest research on robots equipped with GPR dates back to 2006, when Koo et al. [19] developed a wheeled pipeline robot designed to carry GPR for defect detection. Building on this, Csaba Ekes et al. [20,21] introduced the concept of Pipe Penetrating Radar (PPR), a GPR-based approach designed specifically for pipelines. They proposed mounting multiple GPR sensors on a robot to conduct multi-line surveys, enabling the detection of defects on the top and sides of the pipeline and expanding the inspection range. In more recent studies, Xu et al. [22] enhanced this concept by designing a three-channel adaptive GPR pipeline robot in 2023. The robot is equipped with three GPR sensors positioned at 120-degree intervals around the pipeline circumference. By collecting GPR data along three axial lines, the system enables defect detection in drainage pipelines and has been successfully applied to urban drainage inspections. Further improvements were made in 2024, when Xu et al. [23] developed a correction algorithm to enhance the robot's stability during inspections. Despite these advancements, significant limitations remain. Current technology does not provide full coverage of the space outside the pipeline, and its precision is insufficient for determining the exact spatial coordinates of external defects.

In conclusion, existing methods for detecting external defects in drainage pipelines primarily rely on axial surveys along single or multiple lines. These approaches have limited accuracy and cannot provide precise spatial coordinates of the defects. To address these challenges, this study introduces an advanced pipeline inspection robot system that integrates GPR with multiple sensors. The robot is designed to adapt to pipelines of varying diameters through an adjustable mechanism. Ultrasonic sensors ensure that the antenna maintains close contact with the inner wall, while MEMS-IMU and encoder modules control the robot's posture and antenna angles. This enables the robot to perform multi-line scans both along the pipeline axis and around its circumference. The cross-sectional GPR data collected are then processed using a novel CLF-GPR method, allowing for the accurate localization of external defects in the surrounding soil. The system underwent rigorous testing through physical simulations and was validated using a full-scale, 1:1 drainage pipeline model that mimics real-world conditions. The results confirm the system's feasibility and effectiveness in addressing the limitations of current detection methods.

2. Methods

2.1. Principle of GPR Detection

Figure 1 illustrates the application of GPR for detecting external defects around drainage pipes. As shown in Figure 1a, such defects can result in road subsidence. A pipeline robot equipped with GPR is deployed to inspect the interior of the drainage pipe. The GPR transmitter emits high-frequency electromagnetic pulses, which are reflected by underground anomalies, such as defects, as depicted in Figure 1b. The receiver captures these reflected signals, allowing for the accurate detection and localization of external defects surrounding the drainage pipe.

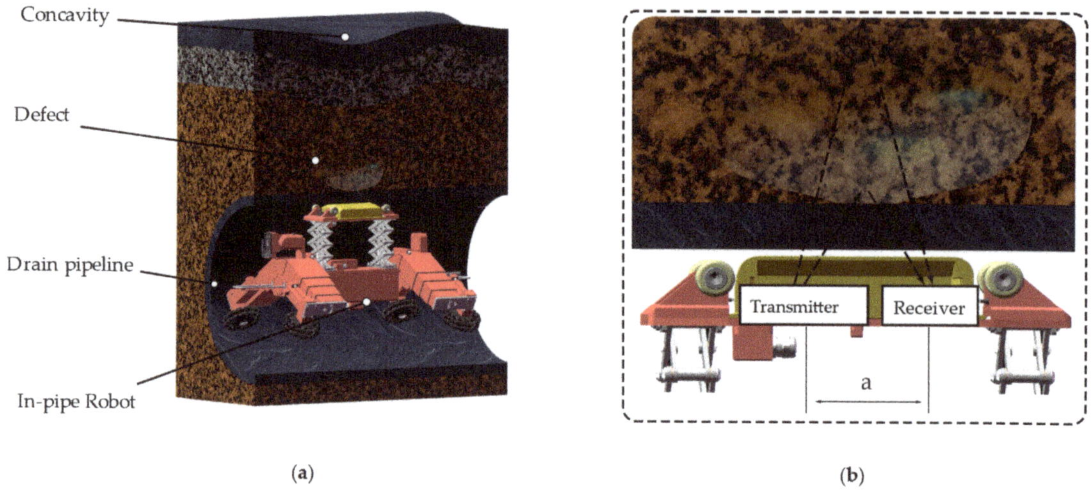

Figure 1. Schematic diagram of GPR detection principle for drainage pipelines. (**a**) Actual working conditions. (**b**) Detection principle.

The geological structure and the distance d from the observation point can be calculated using Equation (1) [24]:

$$d = \frac{\sqrt{(vt)^2 - a^2}}{2}, \tag{1}$$

where v is the propagation speed of the electromagnetic waves in a medium with dielectric constant ε_r, and a is the distance between the transmitter and the receiver, as labeled in Figure 1b. If c represents the speed of light, then

$$v = \frac{c}{\sqrt{\varepsilon_r}}. \tag{2}$$

2.2. GPR Pipeline Robot System

To address the limitations of existing GPR methods for detecting external defects around drainage pipes, this study proposes a GPR-based inspection robot system specifically designed to adapt to the operational conditions of drainage pipelines, as illustrated in Figure 2.

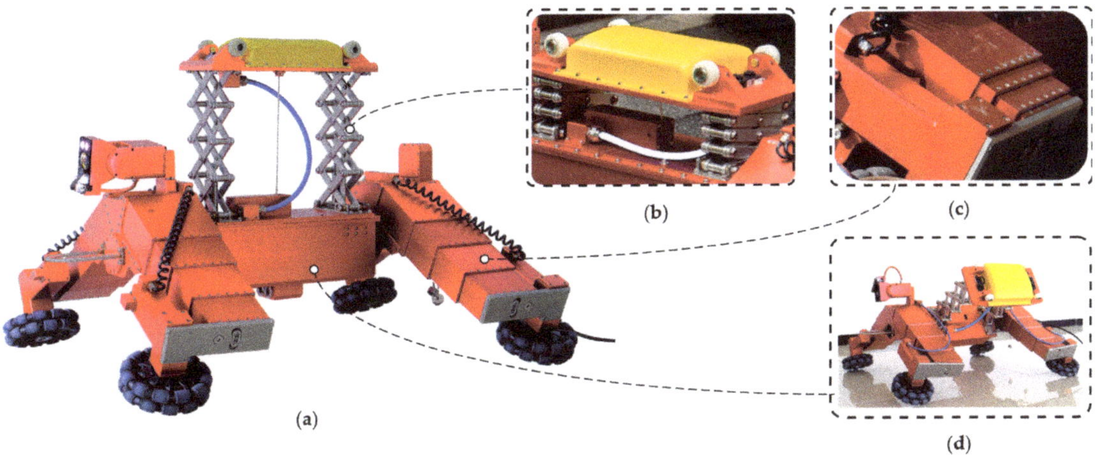

Figure 2. Diagram of GPR pipeline robot. (**a**) Extended state of prototype robot. (**b**) Collapsed state of telescopic antenna module. (**c**) Collapsed state of telescopic driving module. (**d**) Circular scanning state of GPR pipeline robot.

2.2.1. Sensing and Control

The robot system features four adjustable drive wheels and a telescoping antenna, enabling it to adapt to drainage pipelines of varying diameters. The antenna can rotate to perform a full 360-degree scan of the space around the pipeline. Three MEMS IMU modules are integrated into the front and rear telescopic driving modules and the telescopic antenna module, along with encoder and odometer modules, to calculate the robot's motion parameters in real time, control the tilt angle of the GPR antenna, and adjust the robot's posture. Additionally, two ultrasonic distance sensors are mounted on the front and rear of the telescopic antenna module to measure and maintain the GPR antenna's proximity to the pipe wall. The key specifications of each sensor are summarized in Table 1.

Table 1. Main parameters of each sensor.

Sensor	Sensor Model (Manufacturer, Country)	Data Type	Precision
MEMS-IMU	JY931 (Wit Motion, China)	Roll, pitch, yaw	Roll and pitch: 0.2°; yaw: 1°
Ultrasonic distance sensor	A02 (Best Sensor, China)	Distance	±(1 + s × 0.3%)
Odometer	E6A2-CW3C (Omron, Japan)	Distance	0.01 m

Additionally, the telescopic antenna module is equipped with 750 MHz Ground Penetrating Radar developed by the China University of Mining and Technology (Beijing, China).

2.2.2. Working Process

As shown in Figure 3, the operational setup includes a robotic support module positioned on the ground, which is connected to the robot via a cable. This module provides power and facilitates data and command transmission. An integrated odometer calculates the robot's travel distance by monitoring the cable's length. Inside the pipeline, the robot autonomously navigates while its GPR antenna adjusts to various angles to perform axial scans along multiple measurement lines. This approach ensures nearly complete coverage of the external space surrounding the drainage pipe. Upon detecting abnormal signals, the robot records the corresponding odometer coordinates and performs a circular scan of the pipeline's cross-section at the specified location. By integrating axial and circular data, the system achieves the precise detection and localization of external defects.

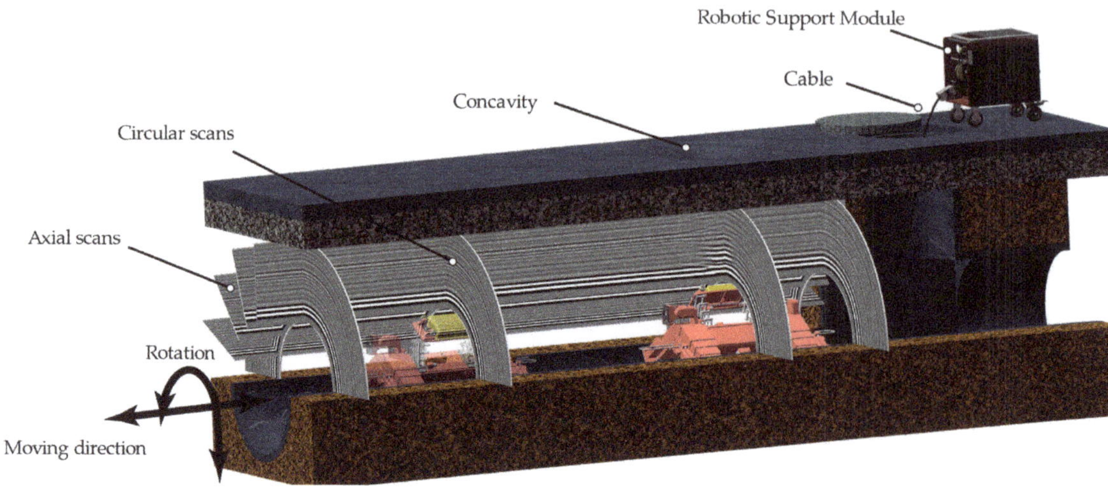

Figure 3. Operational principle of GPR pipeline robot system.

In this study, for each GPR trace recorded by the GPR mounted on the pipeline robot, the corresponding mileage information and timestamp and all motion parameters are logged. These motion parameters include the three-axis acceleration of both the robot and the antenna, attitude angles, mileage, the distance between the antenna and the pipe wall, water flow velocity, and the extension lengths of the robot's antenna and telescopic arm.

The GPR signals are triggered in two ways:

- Distance-triggered:

As the robot moves along the pipeline axis, the cable car on the ground remains stationary. The transmission cable extends as the robot moves forward and retracts as it moves backward. GPR signals are emitted and received based on the distance traveled. The robot triggers a GPR signal after covering a predefined distance, as measured by the odometer within the robotic support module. All robot parameters are recorded simultaneously with each GPR trace.

- Time-triggered:

In this mode, the robot remains stationary while the GPR antenna performs a constant-speed circular scan of the pipeline. GPR signals are triggered at fixed time intervals, and all robot parameters are recorded together with each GPR trace.

2.2.3. Definition of Coordinate Systems

As shown in Figure 4a, the robot coordinate system $R({}^Rx, {}^R\theta, {}^R\rho)$ is defined with the rotation center of the GPR antenna as the origin. The robot coordinate system moves along with the robot as it progresses through the pipeline and the coordinate system's axes are as follows:

- x-Axis: It is aligned with the axial direction of the drainage pipeline, with the positive direction corresponding to the robot's forward movement.
- ρ-Axis: It is perpendicular to the GPR antenna, extending outward from the pipeline (positive direction).

- θ: It is the angle between the ρ-axis and the vertical direction. From a rear perspective, a clockwise rotation is considered positive. In the figure, the θ angle is shown as negative.

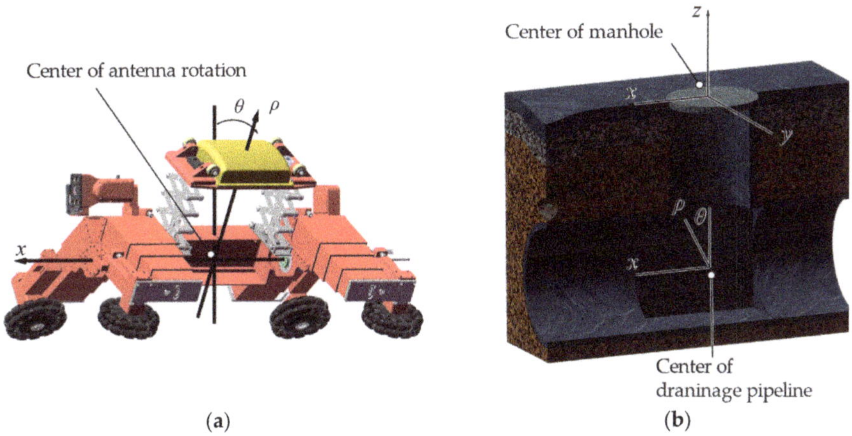

Figure 4. Definition of coordinate system. (**a**) Robot coordinate system. (**b**) Pipeline coordinate system and ground coordinate system.

The pipeline coordinate system, as shown in Figure 4b, differs from the robot coordinate system in that its origin is fixed at the starting center of the drainage pipeline. Since the pipeline robot can adjust its size to fit the pipeline, its rotation center may not always align with the pipeline's geometric center when viewed in a cross-section.

Lastly, the ground coordinate system, as shown in Figure 4b, is defined with its origin at the center of the manhole located at the starting point of the detected pipeline segment. In this system,

x-Axis: It points in the forward direction of the robot's movement (positive direction).

z-Axis: It points vertically upward (positive direction).

The y-axis is determined using the right-hand rule, ensuring a consistent, right-handed coordinate system.

These coordinate systems collectively provide a comprehensive framework for describing the robot's movements and the pipeline geometry during inspection operations.

For a point p in space, its position can be represented as ${}^{R}p$ in the robot coordinate system R, ${}^{S}p$ in the pipeline coordinate system S, and ${}^{G}p$ in the ground coordinate system. The transformation from the robot coordinate system to the pipeline coordinate system can be expressed as Equation (3):

$$\begin{cases} {}^{S}\rho = \sqrt{{}^{R}\rho_2 + \Delta h_1^2 - 2 \cdot {}^{R}\rho \cdot \Delta h_1 \cdot \cos {}^{R}\theta} \\ {}^{S}\theta = {}^{R}\theta + \arcsin\left(\frac{\Delta h_1}{{}^{R}\rho} \sin {}^{R}\theta\right) \\ {}^{S}x = \chi + w \end{cases}, \quad (3)$$

where (${}^{S}x$,${}^{S}\theta$,${}^{S}\rho$) represent the respective coordinate values of point p in the pipeline coordinate system; (${}^{R}x$,${}^{R}\theta$,${}^{R}\rho$) represent the respective coordinate values of point p in the robot coordinate system. Δh_1 represents the offset between the robot coordinate system and the pipeline coordinate system; χ represents the distance traveled by the robot; w represents the initial distance along the pipeline axis between the origin of the robot coordinate system and the pipeline coordinate system.

The transformation between the pipeline coordinate system and the ground coordinate system can be expressed as

$$^{G}p = C_S^G \cdot {}^{S}p - D_S^G \tag{4}$$

where

$$C_S^G = \begin{bmatrix} 1 & 0 & 0 \\ 0 & -\sin{}^{S}\theta & 0 \\ 0 & 0 & \cos{}^{S}\theta \end{bmatrix} \text{ and } D_S^G = \begin{bmatrix} 0 \\ 0 \\ -\Delta h_2 \end{bmatrix}.$$

Δh_2 is the vertical distance between the origin of the robot coordinate system and the origin of the ground coordinate system, and this distance can be measured before deployment into the pipeline. During the movement and scanning of the pipeline robot within the drainage pipeline, each recorded GPR data trace is associated with the corresponding antenna rotation angle $^{R}\theta$. The robot's axial displacement χ along the pipeline axis can be obtained using the odometer on the robotic support module. The offset between the robot coordinate system and the pipeline coordinate system can be calculated using the robot's posture data provided by the three onboard MEMS IMU sensors. Since all the parameters required for the coordinate transformation are known, determining the coordinates of a defect in the robot coordinate system allows its position relative to the ground to be calculated with precision.

2.3. *Cross-Line Fusion of GPR (CLF-GPR) Method*

The CLF-GPR method consists of two main processes:

- Cross-Signal Fusion

This process involves integrating GPR signals collected from multiple scanning lines (both axial and circumferential) to enhance the consistency and accuracy of the defect detection data. By combining these signals, the method reduces noise and highlights key features, enabling the better identification of defect boundaries.

- CLF-GPR Resolution

In this step, the fused signals are processed and analyzed to determine the spatial coordinates of defects. This involves advanced algorithms that utilize the fused data to calculate the precise location and dimensions of anomalies, ensuring accurate mapping of defects in the pipeline environment.

These two processes work together to enhance the accuracy and reliability of defect detection using the CLF-GPR approach.

2.3.1. Cross-Signal Fusion

GPR pulses exhibit non-stationary and non-uniform propagation behavior [25], resulting in inherently weak signals that require amplification. However, GPR signals are highly susceptible to interference. For example, standing wave interference, often distributed horizontally along the axial direction, can be modeled as constant-value noise. Mean filtering techniques are applied to mitigate this type of interference.

While removing constant-value noise, random noise components become more prominent and are further amplified during the signal gain process. These random noise signals are isolated and exhibit broadband, high-frequency characteristics. To suppress such noise, bandpass filtering is applied to each GPR signal trace collected by the pipeline robot.

Mechanical vibrations during axial and circumferential scanning introduce additional noise, which appears as inter-trace variations [26,27]. To address this, inter-trace filtering is applied to maintain data consistency. Wavelet transform, a widely used method for analyzing non-stationary time series, is employed for noise suppression. By utilizing translation and scaling, wavelet transform enables a multi-scale analysis of frequency characteristics at different spatial positions. In this study, wavelet transform is applied to two-dimensional inter-trace GPR signal sequences rather than individual traces, enabling more robust noise suppression during both axial and circumferential scans.

A key challenge arises from the variable number of GPR traces, which may not conform to a power-of-two structure. To overcome this, the Maximum Overlap Discrete Wavelet Transform (MODWT) [28,29] is utilized. MODWT offers flexibility in handling non-standard trace counts while preserving signal integrity during decomposition and reconstruction.

After filtering, GPR images of defects show strong reflections along the defect boundaries [30], creating high-energy contrast with surrounding areas. This enhanced contrast facilitates boundary extraction using advanced techniques such as edge detection [31] and Hough transform [32]. Following axial scans, circumferential scans are performed at identified defect locations. By integrating data from axial and circumferential scans with gyroscope measurements, fusion is achieved through correlation coefficient optimization. This comprehensive fusion process generates detailed three-dimensional cross-sectional views of defects, enabling precise localization and characterization.

This methodology provides a robust framework for defect detection and spatial analyses in drainage pipelines, effectively addressing challenges related to signal noise and spatial data fusion.

2.3.2. CLF-GPR Resolution

During the inspection process, the corresponding mileage information, timestamps, and motion parameters are recorded. By identifying defects in the GPR signal map and associating them with specific GPR traces, the defect's coordinates in the robot coordinate system can be calculated using the parameters linked to those traces. These coordinates are subsequently transformed into the pipeline coordinate system and, finally, into the ground coordinate system.

During axial scanning, the GPR antenna detects defect signals, and the Cross-Signal Fusion method is employed to determine the cross-section where the defect is located. The axial coordinate ${}^{S}x$ of the defect within the pipeline is calculated as the product of the number of GPR traces and the distance between consecutive traces.

After the cross-sectional position of the external defect is identified, the robot system moves to the corresponding location for circumferential scanning. The angular position ${}^{R}\theta$ of the defect relative to the robot coordinate system is then determined. However, since the rotation center of the GPR antenna may not align with the pipeline's geometric center, ${}^{S}\theta$, the defect's angular position in the pipeline coordinate system, cannot be directly equated to ${}^{R}\theta$.

During the circumferential scanning process of the GPR, energy variations are observed when detecting defects, allowing them to be distinguished from reflections originating from the pipe wall. The difference in sampling points between the defect signal and the pipe wall reflection, denoted as Δn, is illustrated in Figure 5a; the GPR signal corresponding to the defect forms a hyperbolic pattern, with Δn_{center} representing the sampling point difference at the defect center. The travel time Δt corresponding to a specific point on the hyperbola can be expressed as

$$\Delta t = \frac{T_{\text{wnd}}}{N_{\text{sample}}}\Delta n, \tag{5}$$

where T_{wnd} is the acquisition time window, and N_{sample} is the total number of sampling points. When $\Delta n = \Delta n_{\text{center}}$, the distance d_{center} between the defect and the antenna can be calculated using Equation (1). The conventional approach directly uses d_{center} as an estimate for the defect location D. However, in drainage pipeline inspections, where defects are often close to the pipe wall, the value of Δn_{center} is typically small. As a result, even minor counting errors can significantly affect the outcome. Assume at this point that the distance from the antenna to the robot's rotation center is l_{center}, and the tilt angle is ${}^{R}\theta_{\text{center}}$. Figure 5b illustrates the geometric relationship during circumferential scanning: P represents the defect location outside the pipe; ${}^{S}O$ and ${}^{R}O$ are the pipe center and robot's rotation center, respectively; when the antenna is aligned with the defect, D represents the actual distance from the GPR antenna to the defect; ${}^{R}\theta$ denotes the tilt angle of the antenna; L represents the actual distance from the antenna to the robot's coordinate origin. When the GPR antenna detects the defect but is not directly below it at a given moment i, the distance from the antenna to the defect d_i can be calculated using Equation (1). The distance from the antenna to the rotation center l_i is obtained from the robot's encoder. The tilt angle at that moment, ${}^{R}\theta_i$, is measured by the JY931 sensor, manufactured by Wit Motion, China.

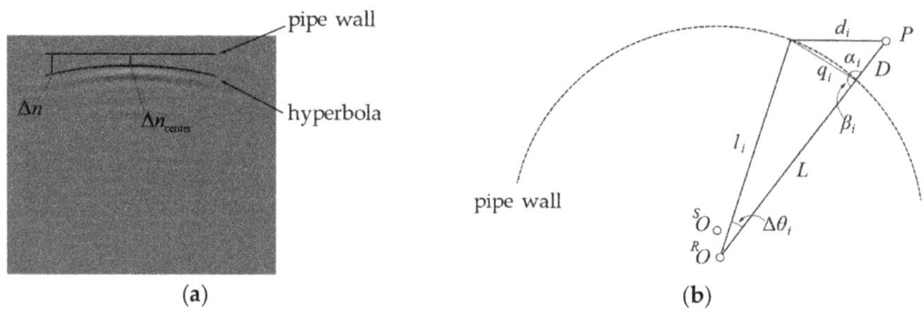

Figure 5. Geometric diagram of (**a**) Δn and (**b**) circumferential scan.

We can then use d_{center}, ${}^{R}\theta_{\text{center}}$, and l_{center} as estimates for D, ${}^{R}\theta$, and L, respectively, and i.e., $D = d_{\text{center}}$, ${}^{R}\widetilde{\theta} = {}^{R}\theta_{\text{center}}$, and $\widetilde{L} = l_{\text{center}}$. Therefore, $\Delta{}^{R}\widetilde{\theta}_i = {}^{R}\theta_i - {}^{R}\theta_{\text{center}}$, and based on the geometric relationships, the equation system can be written as

$$\begin{cases} q_i^2 = l_i^2 + L^2 - 2l_i L\cos\Delta^{R}\theta_i \\ \cos\beta_i = \frac{q_i^2+L^2-l_i^2}{2q_i L} \\ \cos\alpha_i = -\cos\beta_i \\ d_i^2 = q_i^2 + D^2 - 2q_i D\cos\alpha_i \end{cases} \tag{6}$$

where the q_i, α_i, and β_i are intermediate variables. By solving the above system of Equation (6), the estimated distance between the defect and the pipe wall $\widetilde{D}$ can be calculated as

$$\widetilde{D}_i = -\widetilde{L} + l_i\cos\Delta\widetilde{\theta}_i + \sqrt{d_i^2 - l_i^2\sin^2\Delta\widetilde{\theta}_i} \tag{7}$$

In the CLF-GPR method, N GPR signal traces detecting the defect are selected and processed individually to calculate their respective $\widetilde{D}_i$ values. These calculated values are then averaged to provide a more robust and reliable estimate. This averaging process reduces the errors associated with single-point measurements, thereby enhancing the overall accuracy and reliability. By incorporating the antenna length, a more accurate distance estimate is obtained as

$${}^{R}\rho = \widetilde{D} + l_0 = \frac{1}{N}\sum_{i}^{N}\widetilde{D}_i + l_0 \tag{8}$$

The final coordinates of the defect in both the pipeline coordinate system and the ground coordinate system can be obtained from Equations (3) and (4).

3. Laboratory and Field Testing

To assess the effectiveness of this technology, experiments were conducted in both laboratory and field settings.

3.1. Laboratory Testing

The laboratory experiments were designed to simulate drainage pipeline conditions and evaluate the system's performance under controlled settings.

3.1.1. Setup and Initial Configuration for Laboratory Testing

As shown in Figure 6, a glass bottle filled with water was used as a simulated defect outside an 800 mm diameter PVC pipe selected for the experiment. The GPR system was configured with a 25-nanosecond time window, with a spacing of 0.000624 m between consecutive GPR traces. The GPR antenna was horizontally positioned 32 cm from the rear wheel axle of the robot. At the experiment's outset, the rear wheel was positioned 6 cm from the pipe opening, and the GPR's rotation center was vertically offset by 16 cm from the pipe center. The simulated defect was placed 55 cm from the pipe opening, outside the PVC pipe, at a 10-degree angle. Due to the symmetry of the drainage pipeline, control groups with different defect angles were deemed unnecessary. Four experimental sets were conducted, with their specific configurations detailed in Table 2.

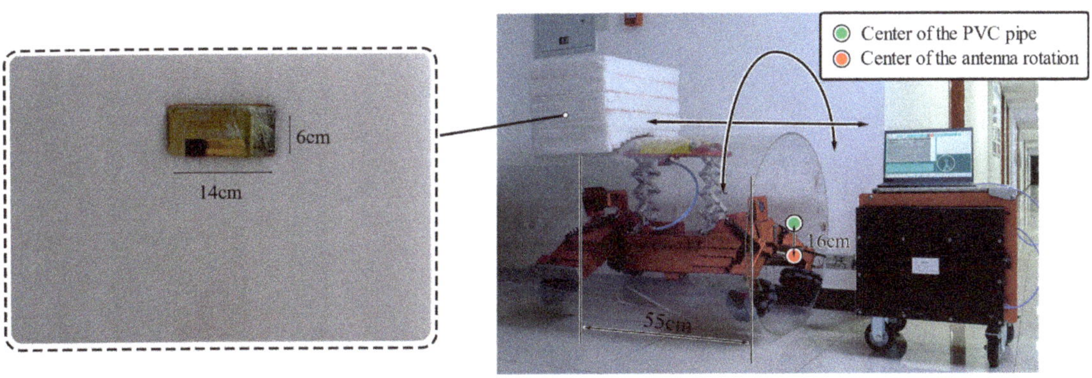

Figure 6. Diagram of simulated defect and experimental setup.

Table 2. The specific setup of laboratory testing.

Number	s_x	s_θ	Distance to Pipe Wall	Size
1#	55 cm	10°	3 cm	14 cm × 6 cm × 6 cm
2#	55 cm	10°	10 cm	14 cm × 6 cm × 6 cm
3#	55 cm	10°	17 cm	14 cm × 6 cm × 6 cm
4#	55 cm	10°	24 cm	14 cm × 6 cm × 6 cm

3.1.2. Experiment Execution and Implementation of CLF-GPR Method

We elaborate on the details of experiment 1. The antenna was positioned at 0 degrees, indicating that it was oriented directly upward. The pipeline robot was controlled to move along the axial direction of the pipe. The raw GPR signal captured during this movement is presented in Figure 7a.

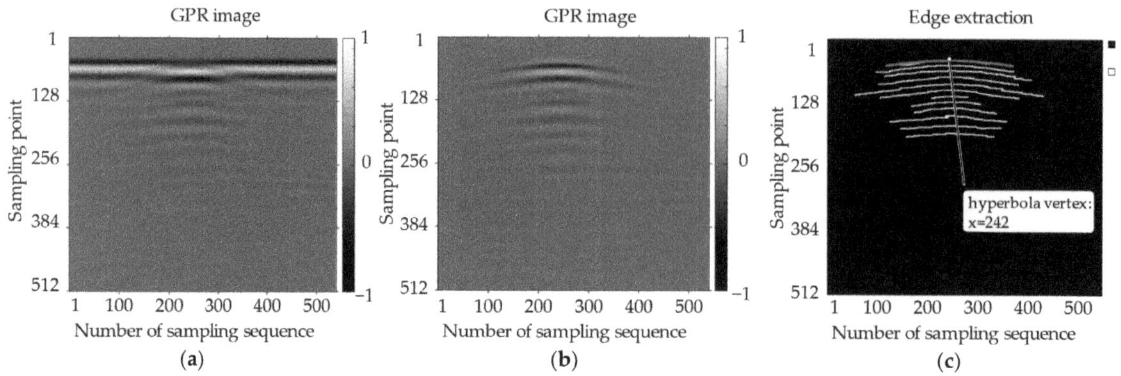

Figure 7. Processing of axial scan in experiment 1#: (**a**) Raw image of circumferential GPR signal; (**b**) GPR image after denoising and reconstructed; (**c**) edge detection and center point determination.

The GPR image underwent background noise reduction and wavelet reconstruction, allowing for the removal of interference signals, including standing waves, noise, and reflections from the antenna casing and pipe wall. Based on the proposed CLF-GPR method, during the experiment, traces 1–20 were selected as background traces to perform background noise removal on the raw GPR signals, effectively eliminating standing wave interference. Subsequently, a bandpass filter was applied to focus on the frequency band around the central frequency of the antenna, effectively suppressing interference and clutter from frequency ranges outside the GPR's operational bandwidth. The resulting lateral signals were further processed using MODWT with the db2 wavelet basis, effectively reducing inter-trace noise caused by vibrations and other disturbances. The processed GPR image after filtering and reconstruction is presented in Figure 7b. This process effectively eliminated standing waves and reduced noise and other interference signals.

To determine the axial position of the defect, the vertex of the hyperbola needed to be identified. The edge of the filtered and reconstructed GPR signal was extracted, and a Hough transform was applied, resulting in the hyperbola shown in Figure 7c. The vertices corresponding to this hyperbola were identified at GPR trace 242. The GPR signal at this trace corresponds to the cross-section of the pipe where the simulated defect was placed. By combining the trace spacing (0.000624 m) with the trace number, the forward movement of the robot was calculated to be 15.1 cm.

Additionally, the horizontal distance from the robot's GPR antenna center to the rear wheel axle is 32 cm, and the distance from the rear wheel to the pipe opening is 6 cm. Using these measurements and applying Equation (3), the position of the defect can be accurately calculated as

$$^{S}x = \chi + w = 53.1 \text{ cm}.$$

After determining the cross-sectional location of the simulated defect, the pipeline robot is controlled to return to this position and perform a circumferential scan with the antenna. The GPR signals collected at this section are shown in Figure 8a, which represents the raw GPR data.

The raw signal undergoes filtering to eliminate noise, as well as the influence of antenna casing and pipe wall reflections. After processing and reconstruction, the GPR signal is displayed in Figure 8b. Next, the edges of the processed GPR image are extracted and a Hough transform is applied to identify the hyperbolic shape, as shown in Figure 8c. The vertices corresponding to this hyperbola are located at GPR trace 256. At this point, the robot's GPR antenna tilt angle $^{R}\theta$ is measured to be 9.18 degrees, as shown in Figure 8c.

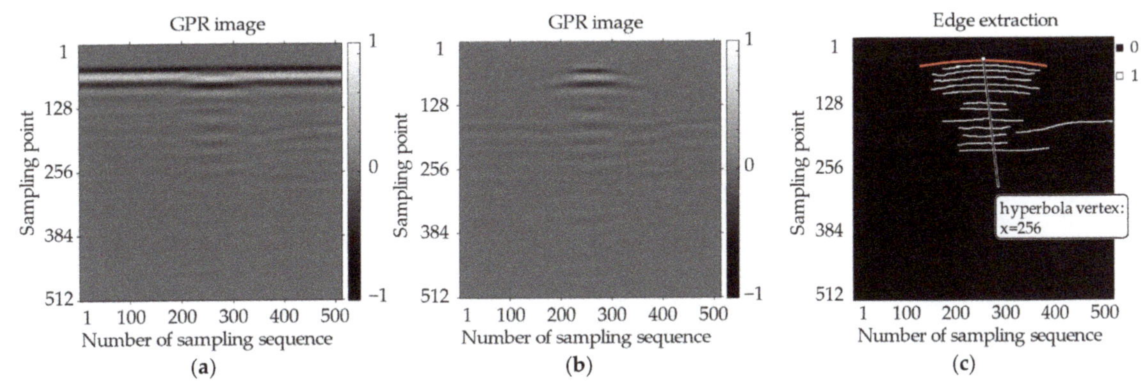

Figure 8. Processing of circular scan in experiment 1#: (**a**) Raw image of circumferential GPR signal; (**b**) GPR image after denoising and reconstructed; (**c**) edge detection and center point determination.

Using Equation (3), the tilt angle in the pipeline coordinate system ${}^{S}\theta$ can be calculated as

$$^{S}\theta = {}^{R}\theta + \arcsin\left(\frac{\Delta h_1}{^{R}\rho}\sin{}^{R}\theta\right) = 11.65°.$$

To calculate ${}^{S}\rho$ in the pipeline coordinate system, GPR signals from the initial few sampling traces are selected (highlighted by black box in Figure 9a), and their detailed waveforms are plotted, as shown in Figure 9b. Since the GPR antenna is always in direct contact with the pipe wall, the reflected signals from both the antenna casing and the pipe wall are merged. As a result, the first prominent peak in the waveform corresponds to the location of the pipe wall. These peaks are highlighted with a red line in Figure 9. Notably, these peaks align with the 50th sampling point, allowing for an accurate determination of the pipe wall's position at the 50th sampling point.

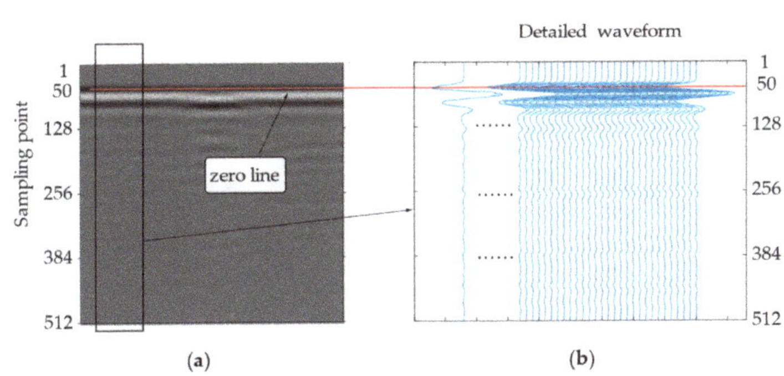

Figure 9. Zero-point calibration for GPR image; (**a**) raw image of GPR signal; (**b**) detailed waveforms of initial GPR signal traces.

As shown in Figure 10a, a CLF-GPR image for experiment 1 is generated based on the results above. The red dots indicate the locations of the water-filled bottle, identified through the Hough transform. To minimize errors, a series of GPR signals is selected from regions where the GPR signal shows significant energy changes (i.e., regions where the simulated defect can be detected). The corresponding antenna tilt angles for these signals are marked as points in Figure 10b. The filtered and reconstructed signals for each GPR trace are displayed in Figure 10c. The first prominent peak in each GPR trace is used to calculate the defect location, as indicated by the red lines in Figure 10c. These peaks correspond to sampling points between the 60th and 62nd, with five such peaks marked in Figure 10c.

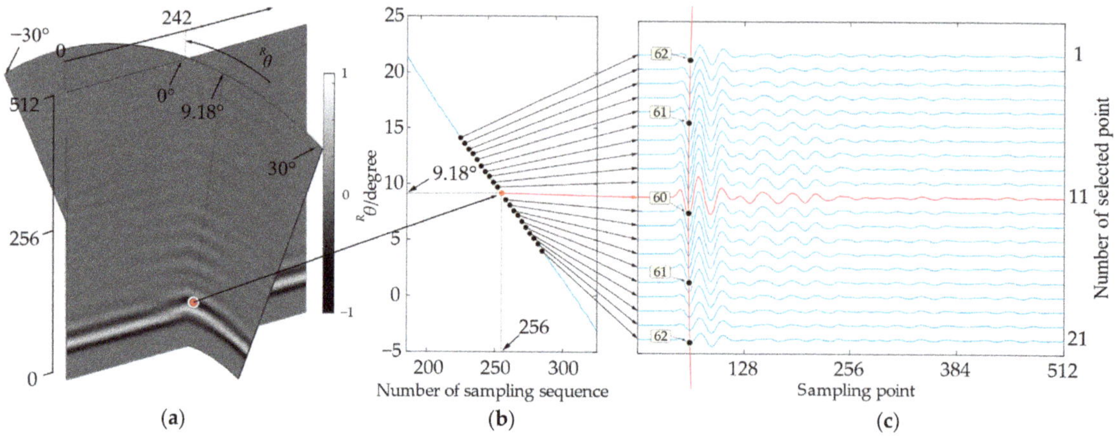

Figure 10. CLF-GPR resolution; (**a**) the CLF-GPR image of experiment 1#; (**b**) antenna angles corresponding to different GPR traces selected; (**c**) filtered and reconstructed GPR signals selected.

According to Equations (5) and (1), the travel time and the distance from the defect can be calculated for each selected GPR signal trace. Furthermore, using Equation (7), the estimated distance between the defect and the pipe wall can be determined, as shown in Figure 11.

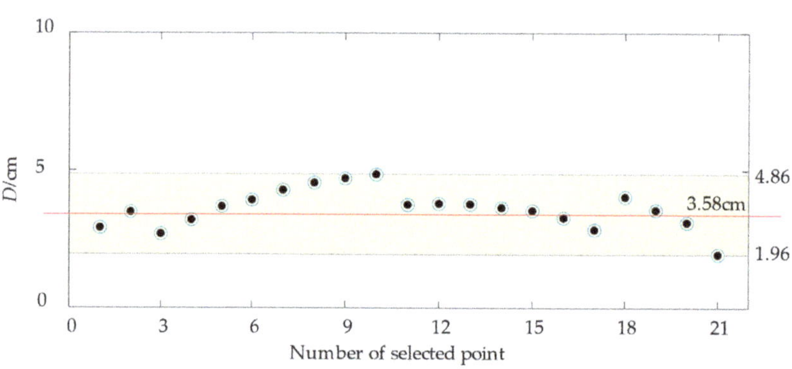

Figure 11. The results of the estimated distance between the simulated defect and inner pipe wall.

Using the results in Figure 11, the value of ${}^{S}\widetilde{\rho}$ can be calculated by Equation (8), and, i.e.,

$$^{S}\widetilde{\rho} = \frac{1}{N}\sum_{i}^{N} \widetilde{D}_i + l_0 = 43.58 \text{ cm}.$$

By incorporating the pipeline radius of 40 cm, the distance from the simulated defect to the inner pipe wall can be calculated as 3.58 cm.

For experiments 2 through 4, the calculations were conducted, and the corresponding processes and results, similar to those of experiment 1, are illustrated in Figure 12a–l.

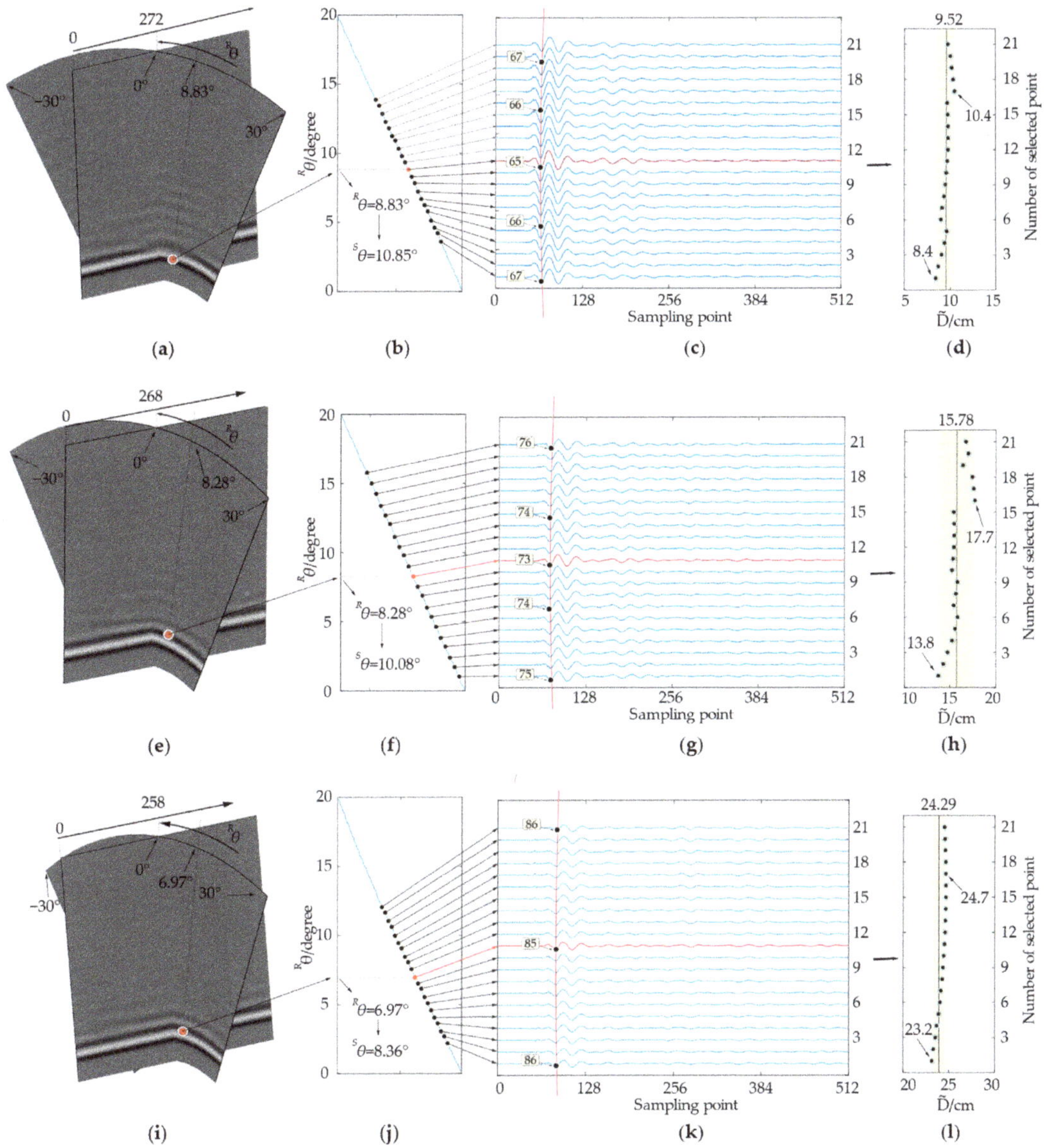

Figure 12. The CLF-GPR image and the results. (**a**), (**e**), and (**i**) are CLF-GPR images of 2#, 3#, and 4#, respectively; (**b**), (**f**), and (**j**) show the antenna angles corresponding to different GPR traces selected in 2#, 3#, and 4#, respectively; (**c**), (**g**), and (**k**) illustrate the GPR signals selected after filtering and reconstructing in 2#, 3#, and 4#, respectively; (**d**), (**h**), and (**l**) are the results of the estimated distance between the simulated defect and inner pipe wall in 2#, 3#, and 4#, respectively.

3.1.3. The Results

The coordinates obtained from the four experiments using the CLF-GPR method are summarized in Table 3.

Table 3. The results of experiment 1#~4#.

Experiment Number	Calculated Coordinates	Actual Coordinates
1#	(53.1 cm, 11.65°, 43.58 cm)	(55 cm, 10°, 44 cm)
2#	(55.0 cm, 10.85°, 49.52 cm)	(55 cm, 10°, 51 cm)
3#	(54.7 cm, 10.08°, 55.78 cm)	(55 cm, 10°, 58 cm)
4#	(54.1 cm, 8.39°, 64.29 cm)	(55 cm, 10°, 64 cm)

3.2. Field Testing

3.2.1. Setup and Initial Configuration for Field Testing

The experimental setup and procedure are illustrated in Figure 13. The drainage pipeline used in the experiment was designed and constructed as a 1:1 scale model, fully adhering to the standards for urban underground drainage systems.

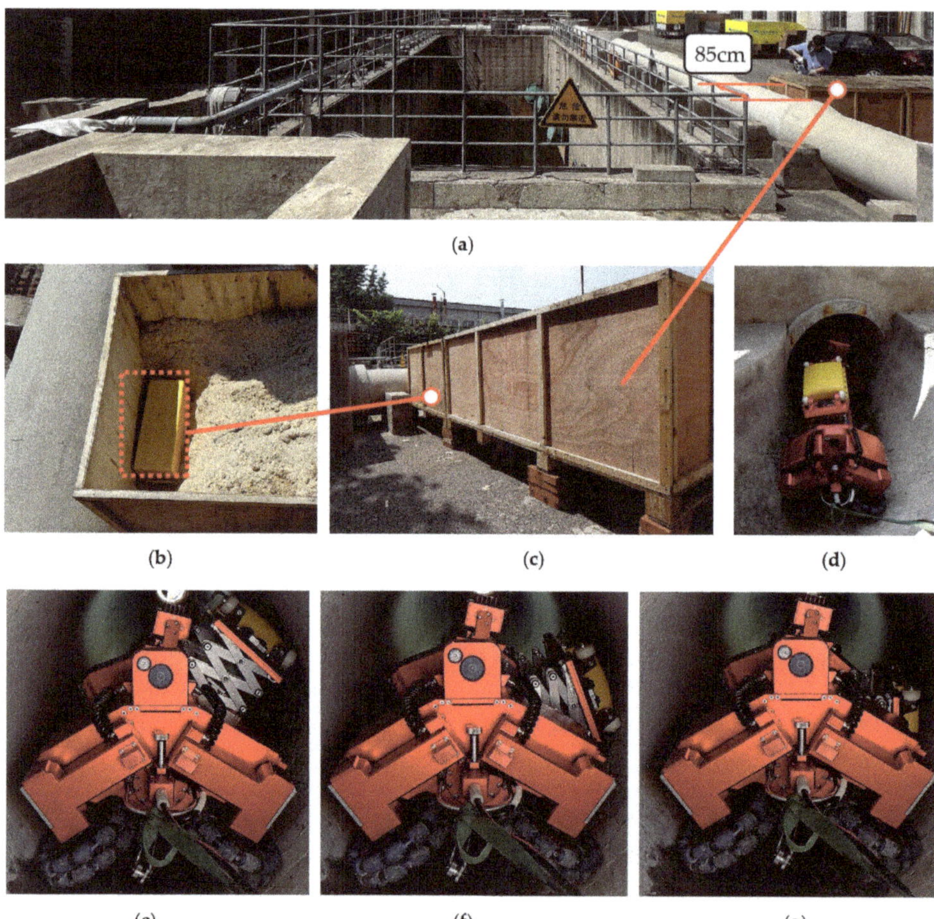

Figure 13. Filed testing. (**a**) Field overview. (**b**) Simulated defect location; (**c**) wooden box placement; (**d**) Robot Advancement; Antenna Swing to (**e**) angle 1, (**f**) angle 2, and (**g**) angle 3.

The pipeline is made of concrete, with each section measuring 2.3 m in length. The experimental segment has a diameter of 600 mm, as shown in Figure 13a.

A wooden box filled with sand was used to simulate the underground environment, with a water-filled cavity embedded in the sand to simulate an external defect, as illustrated in Figure 13b. The wooden box was positioned on one side of the concrete pipe, as shown in Figure 13c. Due to the symmetrical design of drainage pipelines, placing the wooden box on the top or the side of the pipe yields equivalent results. For safety reasons, the box was placed on the side of the pipe. The defect was positioned 85 cm from the start of the first pipe section. With a pipe wall thickness of 10 cm and a wooden box thickness of 4 cm, the total distance from the defect to the inner pipe wall was 14 cm. The defect was oriented at a 90-degree angle.

The pipeline robot was first controlled to move axially along the pipeline, as shown in Figure 13d. The GPR antenna was adjusted to measure at various angles. After detecting the external defect, the antenna assembly was rotated to further refine the defect's position, as illustrated in Figure 13e–g. The time window was set to 25 nanoseconds, and the GPR trace spacing was configured to 0.005 m.

3.2.2. CLF-GPR Resolution and Results

The results are shown in Figure 14. Figure 14a displays the raw GPR signal slices from both axial and circumferential scans. Using the proposed CLF-GPR method to process the axial scan GPR signals, the defect was located at GPR trace 170. Given a trace spacing of 0.005 m, the calculated distance from the defect to the pipe section is 85.0 cm.

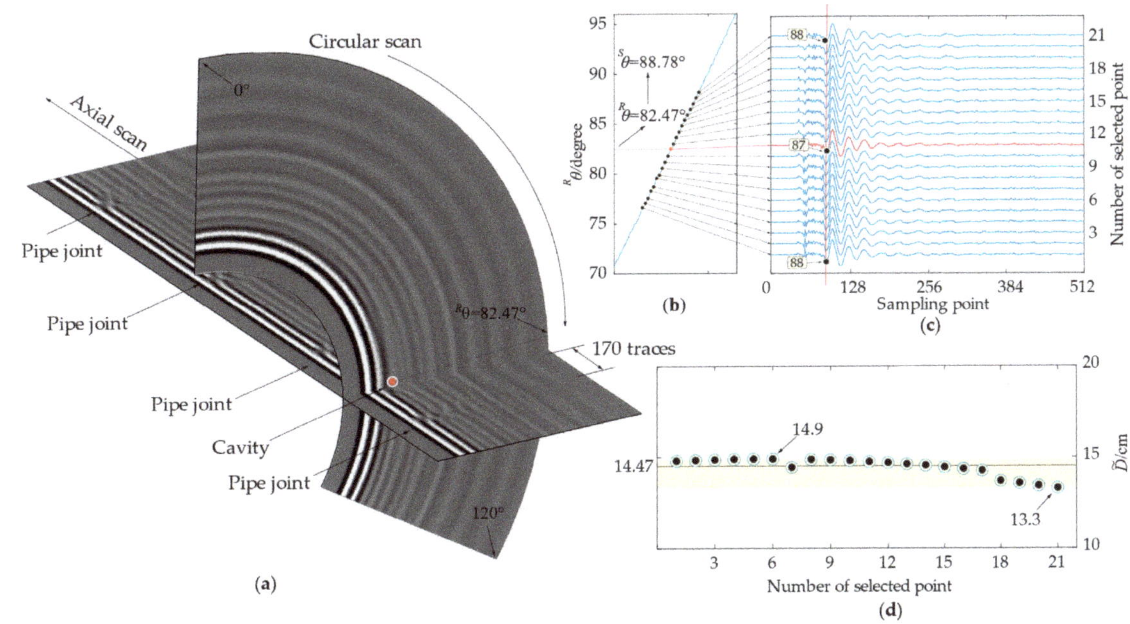

Figure 14. The CLF-GPR resolution of the experiment; (**a**) the CLF-GPR image; (**b**) antenna angles corresponding to different GPR traces selected; (**c**) filtered and reconstructed GPR signals selected; (**d**) the estimated distance between the simulated defect and inner pipe wall.

For the circumferential scan GPR signal, the CLF-GPR method identified the peak of the hyperbolic signal, corresponding to a GPR antenna tilt angle of 82.47 degrees. Using Equation (3), the tilt angle in the pipeline coordinate system was calculated as 88.78 degrees, showing an error of 1.22 degrees from the actual value of 90 degrees.

To refine the analysis, 21 points near the hyperbolic peak in the axial scan were selected, as shown in Figure 14b. The defect reflection peaks for these selected points were all located between the 87th and 88th sampling points, as shown in Figure 14c. The pipe wall reflection peak was at the 50th sampling point (consistent with Figure 8). Using the CLF-GPR method with Equations (1), (5), and (7), the distances between the simulated defect and the inner pipe wall were calculated for all 21 points, and their scatterplot is shown in Figure 14d. The average distance between the simulated defect and inner pipe-wall was calculated as 14.47 cm, with an error of 0.47 cm from the actual value of 14 cm.

In summary, the coordinates of the simulated defect in the pipe coordinate system can be expressed as $(85.0\ \text{cm},\ 88.78^{\circ},\ 14.47\ \text{cm})$, and the actual coordinates are $(85\ \text{cm},\ 90^{\circ},\ 14\ \text{cm})$.

4. Discussion

4.1. Analysis of Results

The performance of the proposed method is evaluated by analyzing the testing results from three key perspectives: the axial displacement, angle error, and depth error. The analysis includes both laboratory and field tests, with a focus on identifying errors and investigating their underlying causes to ensure accuracy and reliability in practical applications.

4.1.1. Axial Displacement Error

The axial displacement results of each group of experiments are shown in Table 4.

Table 4. The axial displacement results.

Experiment	Calculated Displacement	Actual Position	Error	Relative Error
1#	53.1 cm	55 cm	−1.9 cm	3.5%
2#	55.0 cm	55 cm	0.0 cm	0.0%
3#	54.7 cm	55 cm	−0.3 cm	0.5%
4#	54.1 cm	55 cm	−0.9 cm	1.6%
field testing	85.0 cm	85 cm	0.0 cm	0.0%

From the testing results, the absolute error of axial displacement ranged between 0 and 2 cm. These errors can be attributed to the fact that the robot's initial position was not exactly 6 cm from the pipe opening at the start of each trial.

In actual inspection processes, GPR signals near pipe joints are highly distinct in GPR images. Since each section of a drainage pipeline has a standardized length, these pipe joint signals can be utilized to compensate for and correct mileage information. In real-world inspections, the axial component ${}^{S}x$ of the pipeline coordinate system in Equation (3) can be redefined using two parameters:

$$ {}^{S}x = (M, x'), \tag{9} $$

where M represents the number of pipe sections, and x' denotes the distance traveled within the current pipe section. This approach effectively reduces cumulative errors in actual inspection processes. This approach was adopted in the field testing, which effectively reduced the cumulative error.

4.1.2. Angle Error

The angle results of each group of experiments are shown in Table 5.

Table 5. The angle results.

Experiment	Calculated Angle	Actual Angle	Error
1#	11.65°	10°	1.65°
2#	10.85°	10°	0.85°
3#	10.08°	10°	0.08°
4#	8.39°	10°	−1.61°
field testing	88.78°	90°	−1.22°

The absolute angle errors for all experiments were within 2 degrees. These errors may arise due to several factors: the angle is calculated using the robot's tilt angle in its own coordinate system, and the CLF-GPR method is used to compute the pipeline coordinate system. Deformations in the pipeline or misalignments in the robot's posture can affect the accuracy of these parameters. During laboratory experiments, it was challenging to ensure that the simulated defect was placed at precisely the intended angle. However, in real drainage pipeline inspections, an error within 2 degrees is acceptable.

Additionally, during circumferential scanning, gyroscope data revealed that the defect's detectable range in the GPR image was within 20 degrees. Therefore, during repeated axial scans, the angular interval between scans should be less than 20 degrees to ensure comprehensive coverage.

4.1.3. Depth Error

The depth results of each group of experiments are shown in Table 6.

Table 6. The depth results.

Experiment	Calculated Depth	Actual Depth	Error	Relative Error
1#	43.58 cm	44 cm	−0.42 cm	1.0%
2#	49.52 cm	51 cm	−1.48 cm	2.9%
3#	55.78 cm	58 cm	−2.22 cm	3.8%
4#	64.29 cm	64 cm	0.29 cm	0.5%
field testing	44.47 cm	44 cm	0.47 cm	1.1%

The relative errors of the depth coordinate were all controlled within 4%. Calculating the average of multiple points improves the accuracy of results by significantly reducing the impact of random errors caused by point selection. However, the number of selected points should not be excessive. For instance, in experiment 1, we initially selected 21 points for the calculation of $\widetilde{D}$. In the following analysis, we will vary the number of selected points, N, and evaluate the impact of this variation on the mean square error (MSE) of the calculated $\widetilde{D}$. The MSE can be calculated as

$$\mathrm{MSE} = \frac{1}{N}\sum_{i=1}^{N}\left(\widetilde{D} - D_i\right)^2, \tag{10}$$

and the MSE versus the number of selected points N, using data from experiment 1, is shown in Figure 15.

As illustrated in Figure 15, the MSE remains relatively stable as the number of points increases. However, when the number of points exceeds 68, the MSE rises sharply. Selecting too many points for the calculation significantly increases the MSE. This is because an excessive number of points may extend beyond the actual defect area, thereby introducing inaccuracies into the analysis.

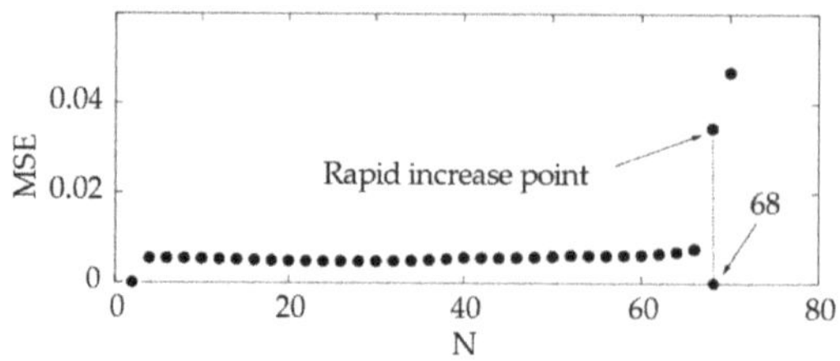

Figure 15. Variation in MSE with the number of selected points.

To mitigate this issue, incorporating a standard error test after defect localization is both meaningful and essential. This approach ensures the accuracy of results by identifying and addressing errors introduced by overextending the selection of points.

4.2. Future Research

The proposed GPR pipeline robot system, combined with the CLF-GPR method, has strong scalability for real-world urban drainage systems. Its modular design enables the replacement of GPR antenna modules with different central frequencies, allowing it to adapt to various real-world scenarios. Additionally, its telescopic structure ensures compatibility with urban drainage systems of varying pipe diameters. However, the system still faces challenges, including variability in environmental conditions, instability under extreme pipeline environments, and the complexities associated with non-standard pipelines. To address these challenges in practical detection scenarios, future research should focus on the following key areas.

4.2.1. Improving Performance in Varying Environmental Conditions

The accuracy and reliability of GPR signals are significantly influenced by environmental factors such as soil type and moisture content. For example, high soil conductivity can result in substantial signal attenuation, thereby reducing the effective range of defect detection. While the current technology demonstrates the ability to localize defects in external pipelines, its performance may deteriorate under extreme environmental conditions, highlighting the need for further optimization or the incorporation of complementary techniques.

Building on existing findings, future studies should further investigate the response characteristics of GPR in drainage pipeline inspections. This includes examining how GPR interacts with defects of varying types, sizes, and underground water content.

4.2.2. Enhancing Mechanical and Operational Stability

The system relies on the precise movement of the GPR antenna for both axial and circumferential scanning. However, in real-world pipeline environments, external factors such as water flow impacts and sediment accumulation on pipe walls often introduce significant vibrations and slippage. These disturbances can compromise the consistency and accuracy of GPR signals, leading to challenges in defect localization that are not entirely mitigated by the CLF-GPR method.

Future research should focus on analyzing the effects of these disturbances on GPR signal characteristics. Additionally, robust reconstruction methods should be developed to mitigate their impact, thereby enhancing the reliability and accuracy of defect detection.

4.2.3. Adapting to Non-Standard Pipeline Configurations

The robot system and CLF-GPR method are primarily designed for standard drainage pipelines. However, pipelines with non-standard dimensions or irregular geometries pose significant challenges for accurate defect localization. For example, drainage pipes may deform into elliptical shapes due to external compression, while factors such as corrosion, silt accumulation, and irregular inner surfaces further complicate boundary conditions. These variations in boundary conditions can adversely affect both angular and depth measurements, ultimately reducing the accuracy of defect detection.

Addressing these challenges will require the integration of additional sensors and advancements in algorithms. Future research should focus on improving the system and method through mechanical enhancements, advanced sensing technologies, and refined algorithms to enable adaptation to non-standard pipeline geometries and sizes.

5. Conclusions

This study proposed a novel drainage pipeline inspection robot system based on GPR, integrated with multiple sensors. Through multi-sensor fusion, the system enables the robot to navigate complex drainage pipeline environments and perform comprehensive spatial scanning using GPR. Building on this foundation, a CLF-GPR method was developed to achieve the accurate localization of external defects in drainage pipelines, including azimuthal angles that were previously undetectable.

The experimental results demonstrated high precision, with axial displacement errors within 2 cm, azimuthal angle errors within 2 degrees, and depth coordinate errors controlled within 2.3 cm. Comprehensive physical simulations, along with field validations conducted in a 1:1-scale drainage pipeline model, confirmed the feasibility and effectiveness of this technology.

These findings address critical challenges in detecting and localizing external defects in drainage pipelines, offering a reliable solution for enhancing the safety of urban underground spaces. This research represents a significant contribution to advancing the transparency, digitalization, and sustainable management of urban drainage infrastructure.

Author Contributions: Y.F. contributed to the research idea, methodology, and model design. He also wrote the main portion of the manuscript. F.Y. validated the whole integration scheme, and was responsible for the thesis revision proofreading. X.Q., L.F. and J.L. assisted with the GPR inspection device data acquisition. M.X. and F.L. assisted with the signal analysis. All authors have read and agreed to the published version of the manuscript.

Funding: This research was funded by the National Key Research and Development Program of China, grant numbers 2021YFC3090303 and 2021YFC3090304.

Data Availability Statement: Restrictions apply to the availability of these data. Data were obtained from the China University of Mining and Technology (Beijing) and are available from the authors with the permission of the China University of Mining and Technology (Beijing).

Conflicts of Interest: The authors declare no conflict of interest.

References

1. Zeng, X.; Wang, Z.; Wang, H.; Zhu, S.; Chen, S. Progress in Drainage Pipeline Condition Assessment and Deterioration Prediction Models. *Sustainability* **2023**, *15*, 3849. [CrossRef]
2. Guo, S.; Shao, Y.; Zhang, T.; Zhu, D.Z.; Zhang, Y. Physical Modeling on Sand Erosion around Defective Sewer Pipes under the Influence of Groundwater. *J. Hydraul. Eng.* **2013**, *139*, 1247–1257. [CrossRef]
3. Noshahri, H.; Van Der Meijde, M.; Olde Scholtenhuis, L. GPR Surveys in Enclosed Underground Sewer Pipe Space. *Tunn. Undergr. Space Technol.* **2022**, *129*, 104689. [CrossRef]
4. Wang, K.; Zhang, J.; Gao, G.; Qiu, J.; Zhong, Y.; Guo, C.; Zhao, W.; Tang, K.; Su, X. Causes, Risk Analysis, and Countermeasures of Urban Road Collapse in China from 2019 to 2020. *J. Perform. Constr. Facil.* **2022**, *36*, 04022054. [CrossRef]
5. Wang, Y.; Shi, G.; Tian, X.; Li, C.; Cheng, H. Experimental Study on City Road Collapse under Vibrating Load. *Shock Vib.* **2020**, *2020*, 6074658. [CrossRef]

6. Chuang, T.-Y.; Sung, C.-C. Learning and SLAM Based Decision Support Platform for Sewer Inspection. *Remote Sens.* **2020**, *12*, 968. [CrossRef]
7. Wang, Y.; Li, P.; Li, J. The Monitoring Approaches and Non-Destructive Testing Technologies for Sewer Pipelines. *Water Sci. Technol.* **2022**, *85*, 3107–3121. [CrossRef]
8. Fang, X.; Li, Q.; Zhu, J.; Chen, Z.; Zhang, D.; Wu, K.; Ding, K.; Li, Q. Sewer Defect Instance Segmentation, Localization, and 3D Reconstruction for Sewer Floating Capsule Robots. *Autom. Constr.* **2022**, *142*, 104494. [CrossRef]
9. Betgeri, S.N.; Matthews, J.C.; Vladeanu, G. Development of Comprehensive Rating for the Evaluation of Sewer Pipelines. *J. Pipeline Syst. Eng. Pract.* **2023**, *14*, 04023001. [CrossRef]
10. Li, C.; Li, H.; Chen, K. Convolutional Point Transformer for Semantic Segmentation of Sewer Sonar Point Clouds. *Eng. Appl. Artif. Intell.* **2024**, *138*, 109456. [CrossRef]
11. Li, C.; Chen, K.; Li, H.; Luo, H. Multisensor Data Fusion Approach for Sediment Assessment of Sewers in Operation. *Eng. Appl. Artif. Intell.* **2024**, *132*, 107965. [CrossRef]
12. Lee, J.S.; Nguyen, C.; Scullion, T. A Novel, Compact, Low-Cost, Impulse Ground-Penetrating Radar for Nondestructive Evaluation of Pavements. *IEEE Trans. Instrum. Meas.* **2004**, *53*, 1502–1509. [CrossRef]
13. Li, H.; Chou, C.; Fan, L.; Li, B.; Wang, D.; Song, D. Robotic Subsurface Pipeline Mapping with a Ground-Penetrating Radar and a Camera. In Proceedings of the 2018 IEEE/RSJ International Conference on Intelligent Robots and Systems (IROS), Madrid, Spain, 1–5 October 2018; pp. 3145–3150.
14. Li, F.; Yang, F.; Xie, Y.; Qiao, X.; Du, C.; Li, C.; Ru, Q.; Zhang, F.; Gu, X.; Yong, Z. Research on 3D Ground Penetrating Radar Deep Underground Cavity Identification Algorithm in Urban Roads Using Multi-Dimensional Time-Frequency Features. *NDT E Int.* **2024**, *143*, 103060. [CrossRef]
15. Allred, B.J. A GPR Agricultural Drainage Pipe Detection Case Study: Effects of Antenna Orientation Relative to Drainage Pipe Directional Trend. *J. Environ. Eng. Geophys.* **2013**, *18*, 55–69. [CrossRef]
16. Liang, X.; Hu, D.; Li, Y.; Zhang, Y.; Yang, X. Application of GPR Underground Pipeline Detection Technology in Urban Complex Geological Environments. *Geofluids* **2022**, *2022*, 7465919. [CrossRef]
17. Wang, M.; Hu, D.; Chen, J.; Li, S. Underground Infrastructure Detection and Localization Using Deep Learning Enabled Radargram Inversion and Vision Based Mapping. *Autom. Constr.* **2023**, *154*, 105004. [CrossRef]
18. Liu, H.; Chen, J.; Zhang, X.; Dai, D.; Cui, J.; Spencer, B.F. Collaborative Imaging of Subsurface Cavities Using Ground-Pipeline Penetrating Radar. *IEEE Geosci. Remote Sens. Lett.* **2024**, *21*, 1–5. [CrossRef]
19. Koo, D.-H.; Ariaratnam, S.T. Innovative Method for Assessment of Underground Sewer Pipe Condition. *Autom. Constr.* **2006**, *15*, 479–488. [CrossRef]
20. Ékes, C.; Neducza, B.; Takacs, P. Pipe Penetrating Radar Inspection of Large Diameter Underground Pipes. In Proceedings of the 15th International Conference on Ground Penetrating Radar, IEEE, Brussels, Belgium, 30 June–4 July 2014; pp. 368–371.
21. Ékes, C. Sinkhole Locating and Corrosion Quantification with Pipe Penetrating Radar. In Proceedings of the Pipelines 2017, Phoenix, AZ, USA, 6–9 August 2017; American Society of Civil Engineers: Reston, VA, USA, 2017; pp. 1–11.
22. Xu, M.; Yang, F.; Yan, R. Design and Implementation of Three-Channel Drainage Pipeline Ground Penetrating Radar Device. *Sensors* **2023**, *23*, 4525. [CrossRef] [PubMed]
23. Xu, M.; Yang, F.; Fang, Y.; Li, F.; Yan, R. Research on Time Series-Based Pipeline Ground Penetrating Radar Calibration Angle Prediction Algorithm. *Sensors* **2024**, *24*, 379. [CrossRef]
24. Liu, J.; Tang, X.; Yang, F.; Qiao, X.; Li, F.; Peng, S.; Huang, X.; Fang, Y.; Xu, M. Study on the Identification Method of Planar Geological Structures in Coal Mines Using Ground-Penetrating Radar. *Remote Sens.* **2024**, *16*, 3990. [CrossRef]
25. Feng, D.; Liu, S.; Yang, J.; Wang, X.; Wang, X. The Noise Attenuation and Stochastic Clutter Removal of Ground Penetrating Radar Based on the K-SVD Dictionary Learning. *IEEE Access* **2021**, *9*, 74879–74890. [CrossRef]
26. Wang, S.; Al-Qadi, I.L. Impact and Removal of Ground-Penetrating Radar Vibration on Continuous Asphalt Concrete Pavement Density Prediction. *IEEE Trans. Geosci. Remote Sens.* **2022**, *60*, 5106414. [CrossRef]
27. Sui, X.; Leng, Z.; Wang, S.; Zhang, S. Removal of the GPR Antenna Vibration Effect through Height Shifting Correction for Enhanced Multi-Layer Asphalt Pavement Thickness Determination. *Autom. Constr.* **2024**, *161*, 105335. [CrossRef]
28. Seo, Y.; Choi, Y.; Choi, J. River Stage Modeling by Combining Maximal Overlap Discrete Wavelet Transform, Support Vector Machines and Genetic Algorithm. *Water* **2017**, *9*, 525. [CrossRef]
29. Nigam, R.; Kumar, R.; Singh, S.K. Numerical and Experimental Investigation on Crack Detection in a Beam Using Maximal Overlapping Discrete Wavelet Transform. *Struct. Health Monit.* **2024**, *23*, 2942–2962. [CrossRef]
30. Dai, Y.; Xie, F. Detection and Automatic Identification of Voids by Ground Penetrating Radar in Pipes Based on Machine Learning. *Urban Geotech. Investig. Surv.* **2021**, *z1*, 21–33. [CrossRef]

31. Sugashini, T.; Balakrishnan, G. Visually Impaired Object Segmentation and Detection Using Hybrid Canny Edge Detector, Hough Transform, and Improved Momentum Search in YOLOv7. *Signal Image Video Process.* **2024**, *18*, 251–265. [CrossRef]
32. Chen, H.; Cohn, A.G. Probabilistic Robust Hyperbola Mixture Model for Interpreting Ground Penetrating Radar Data. In Proceedings of the 2010 International Joint Conference on Neural Networks (IJCNN), Barcelona, Spain, 18–23 July 2010; pp. 1–8.

remote sensing

MDPI

Article

Effect of the Outer Pipe on Reducing Direct Coupling of the Thin Borehole Radar Probe in Thick Water-Filled Borehole

Satoshi Ebihara *, Raiki Masui, Koki Koyama, Yuki Tsujikawa and Yuto Nishida

Department of Electrical and Electronic Engineering, Osaka Electro-Communication University, Neyagawa 572-8530, Japan; ee19a079@oecu.jp (R.M.); mf21a004@oecu.jp (K.K.); mf21a010@oecu.jp (Y.T.); ee19a066@oecu.jp (Y.N.)

* Correspondence: ebihara@osakac.ac.jp

Abstract: We propose an outer pipe to reduce a direct wave in a thin single-hole borehole radar probe in a thick water-filled borehole. The outer pipe replaces the medium, such as water inside the borehole, with low-permittivity materials, such as air and plastics. According to numerical calculations, the cylindrical water layer makes the direct wave from the transmitting loop antenna to the receiving one have significant power and narrow frequency bandwidth. This is caused by the low attenuation of the TE_{01} surface wave when there is a cylindrical water layer. The MoM analysis showed that wearing the outer pipe on the radar probe decreased the direct wave's power more than the reflected wave from the subsurface objects, improving the detection of that reflected wave. We realized the radar system with the outer pipe by attaching the two acrylic pipes with different diameters. With this outer pie, we conducted field experiments to estimate the position of metal ore near the borehole in skarn with the loop antenna array type borehole radar. The direct wave having oscillation prevented the detection of the reflected wave from the sphalerite vein in the time domain without the outer pipe. However, attaching the outer pipe highlighted that reflected wave.

Keywords: directional borehole radar; loop array antenna; direct wave; Sommerfeld integral

1. Introduction

Japan must select a final site for a deep geological repository of high-level radioactive waste. It is necessary to have a detailed understanding of the three-dimensional groundwater flow pathways in the rocks within a one hundred-meter radius of the disposal site, i.e., a near-field scale solute transport model [1]. For this purpose, physical exploration would be required to understand the three-dimensional distribution of permeable fractures to make the discrete fracture network model [2]. This information would also be required to preserve historical and cultural heritage [3]. In addition to this field, for example, in mineral resource exploration, it is necessary to delineate the three-dimensional contours of mineral deposits [4]. In the early stage of these explorations, we observe the drilling core and conduct various logging. Then, we need remote sensing to measure the objects away from the borehole. Borehole radar is one type of remote sensing for this measurement.

The borehole radar is one form of ground-penetrating radar. An antenna in a borehole radiates electromagnetic waves whose frequencies are between about 10 MHz and 500 MHz [5]. In a cross-hole borehole radar, a transmitter and receiver are set in different boreholes. A direct wave's arrival time and amplitude in the cross-hole borehole radar supply a topographical image between the two boreholes. The full wave inversion gives more precise information in the subsurface. In single-hole borehole radar, the transmitter

https://doi.org/10.3390/rs17010100

and receiver are in the same borehole. The most significant advantage is that this measurement is possible with just one borehole. In this radar measurement, we acquire a direct wave arriving at the receiver first in the time domain. The arrival time and amplitude of the direct wave are related to the electromagnetic property around the medium around the antenna [6]. Reflected waves from subsurface objects reach the receiver after the direct wave arrives in the time domain. These objects include fracture zones of faults [7] and lithological boundaries. The borehole radar delineated the McConnell nickel sulfide deposit [8] and the gold reef called the Ventersdorp Contact Reef in [5] (p. 478). The computer simulation confirmed that metal ore should be the target of the borehole radar [9].

The single-hole borehole radar typically uses one dipole antenna for the transmitter and another for the receiver in the same borehole. In this case, the radar is omnidirectional, and the subsurface object's position estimation is limited to the two-dimensional space of depth and distance. Extending it to the three-dimensional space requires replacing the dipole antenna with a directive antenna in a borehole. There are several ideas for this directive antenna, including an array antenna. Array-type directional borehole antennas date back 30 years [10]. The principle of directivity with the array measures the arrival time difference among the array elements in a borehole. This arrival difference is usually less than 1 ns, and we often introduce a vector network analyzer for the precise arrival time measurement. After the proposal to introduce the array antenna, many authors presented essential knowledge, including electromagnetic phenomena around the antenna [11], radar system, and signal processing [12] for the array-type directional borehole radar. The array-type directional borehole typically uses a vertical dipole antenna in a vertical borehole as an array antenna element [13]. This means that the radiated and observed wave polarization is vertically polarized. Adding the other wave polarization, i.e., a horizontally polarized wave, would introduce radar polarimetry. For this purpose, several horizontal loop antennas arranged vertically were recently proposed [14]. These loop array antennas with a horizontal loop plane in a vertical borehole radiate and are sensitive to horizontally polarized waves.

The diameters of the boreholes for natural resource exploration would be about 10 cm or more. In this exploration, the drilling depth is sometimes considerable, more than 100 m, and the boreholes may be used for extracting natural resources. These applications require large borehole diameters. On the other hand, the diameters of the boreholes for civil engineering exploration, such as the detection of foundation piles, might be less than 10 cm. The existence of boreholes with various diameters means that the smaller the diameter of the radar probe is, the larger the number of boreholes that are possible for measurement. This situation encourages us to develop the borehole radar probe as thin as possible [15]. However, a significant difference between the diameter of the thin borehole radar probe and the large borehole diameters makes a cylindrical insulation layer around the borehole radar probe. The deep borehole is usually filled with fresh water, which is the medium of that cylindrical layer. The permittivity of the water is much larger than air and rock, and this layer may influence the borehole radar signals. Many authors sometimes call this phenomenon the borehole effect. The larger thickness of the insulation layer would make the borehole effect larger. For example, the slight eccentricity of the borehole radar probe in a borehole excites the propagating mode, such as HE_{11} mode. It sometimes influences the direct wave in the single-hole borehole radar measurements [16]. The wave in single-hole borehole radar can propagate firmly along a metallic logging cable, and the ferrite loading may reduce it [17]. In the array-type directional borehole, the borehole effects often change the resonances among the dipole array antenna [11,18] and distort the arrival times of the incoming wave from outside the borehole [13,19,20]. Some researchers positively made use of the mode propagating along the borehole. The numerical simulation was conducted for dielectric logging exploiting the propagation mode [21]. The propagation mode can be used to estimate subhorizontal structures [22]. The mode propagating along a conductor in a borehole was investigated for forward-looking events [23].

Subsurface objects, i.e., radar targets, are sometimes near the borehole that includes the radar probe. In this case, a direct wave propagating along the borehole arrives at the receiver first, after which the reflected wave approaches the receiver in the time domain in the single-hole borehole radar. These two waveforms are sometimes superimposed in the time domain, which prevents the detection of the reflected waves [16]. In this case, the signal processing method, such as the moving average subtraction in Equation (7.9) in [5], is helpful to remove the direct wave numerically. This signal processing assumes that the electromagnetic properties of the medium do not vary in the depth direction, and the arrival time of the reflected wave well changes with the antenna's depth. However, there is sometimes disagreement between this assumption and the actual situation. which worsens the array-type directional borehole radar's directivity capability. This is because this radar requires precise arrival time measurements. In this situation, it would be preferable to suppress the direct wave by changing hardware, such as the antenna configuration, rather than the signal processing.

Controlling the borehole effect with a change of the radar probe design may improve the borehole radar's performance. Here are a few critical experimental reports. Tronicke and Holliger developed a unique packer system, which replaces the medium in a borehole with nitrogen in the limited depth range of seven meters [24]. The experimental results of the difference between the water-filled and gas-filled boreholes agreed with the numerical simulation. A similar experiment was conducted in [25], in which they pumped water from the borehole and compared the cross-hole borehole radar signals between the water-filled and air-filled boreholes. These two pieces of research aimed to confirm that the theoretical prediction on the borehole effects was correct. However, we should notice that these experimental results imply that we can modify the borehole effect on the radar signals by transforming the medium between the radar probe and the borehole's wall without changing the antennas and medium inside the radar probe.

When the difference between the radar probe's and the borehole's diameters is prominent in the single-hole borehole radar, this might influence the radar signals. This influence would be particularly significant if the medium inside the borehole is water. Remembering the unique packer system's success, inserting a pipe covering the radar probe can remove the medium inside the borehole and change the radar signal. It is crucial to be mindful of whether this pipe's insertion improves or worsens the detection of the reflection from radar targets such as the metal ore and fractures in the rock. Under these conditions, a meticulous investigation of the effect of removing the medium inside the borehole is essential.

In this paper, we propose an outer pipe to improve the detection of a reflected wave from a subsurface object. Section 2 introduces the directional borehole radar with a directive loop array antenna. In Section 3, we claim that replacing a water layer with an air and plastic layer reduces the direct wave with a point source and observation model. In Section 4, we investigate the influence of using antennas instead of the point source and observation on the effect of the outer pipe. Section 5 shows the impact of the outer pipe with the numerical simulation in the borehole radar measurements. Section 6 provides the field experiment results of single-hole measurements with a developed outer pipe.

This article is a revised and expanded version of the conference paper [26].

2. The Loop-Array Type Directional Borehole Radar Surrounded by the Outer Pipe

Figure 1a introduces the single-hole borehole radar probe. The transmitter and receiver are loop antennas, unlike dipole antennas. Each loop antenna is on the plane perpendicular to the borehole axis. The loop antenna radiates and senses the horizontally polarized wave in a vertical borehole, like the slotted antenna on a conducting cylinder [27]. The receiver is a directive antenna with four loop antennas [14]. The arrival times of an incoming wave to the array estimate the direction of arrival (DOA), and the DOA estimation algorithm was presented. The angle of the DOA includes the elevation and azimuth angles. The feeding lines feed the loop antenna elements, extending from the centered conducting cylinder. This centered conducting cylinder consists of a thin part near the antenna elements and a thick part including electronic circuits. If the distance between the thick part and the antenna elements is significant, we may ignore the wave scattering from this thick part [14]. In this paper, we ignore the existence of the centered conducting cylinder in the numerical model in Sections 3–5 and return to this issue in the field experiment in Section 6.2.

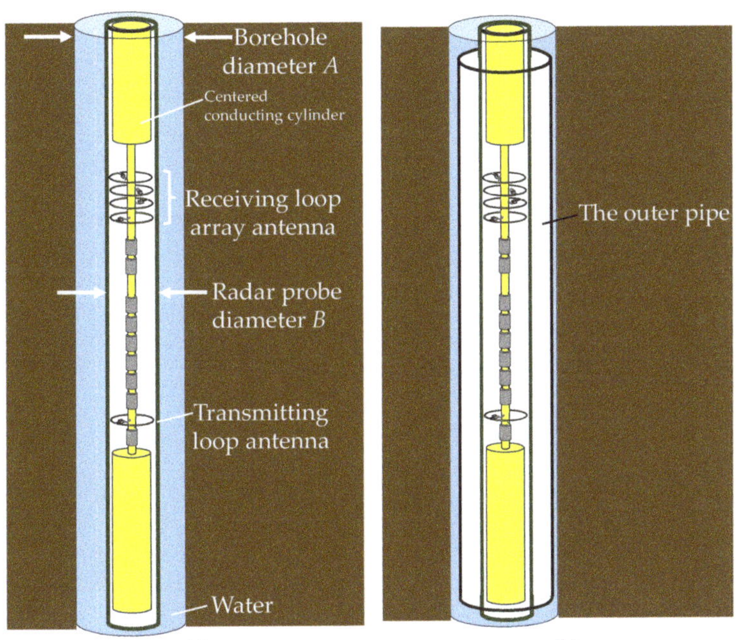

(a) (b)

Figure 1. The directional borehole radar with the loop array antenna and the outer pipe. (**a**) The thin radar probe in the thick borehole has the condition that the borehole diameter A is much larger than the radar probe diameter B. (**b**) Proposing the outer pipe replacing the medium, such as water, to be attached to the outside of the radar probe.

The diameter of the radar probe used in [14] was 5.7 cm. Suppose the borehole diameter is about 6.7 cm, which corresponds to the inner diameter of polyvinyl chloride (PVC) pipe, called VP65 in Japan. In that case, the thickness of the medium between the radar probe and the borehole diameter is 5 mm. Civil engineering used the borehole for this PVC pipe case. On the other hand, resources engineering needs thick boreholes, such as with a 10 cm inner diameter for the PVC case, called VP100. In this case, the thickness of the medium is 2.15 cm. If this medium is fresh water, this water's high permittivity might influence the direct wave. In this paper, we propose attaching an outer pipe made of material with low permittivity, such as acrylic and air. This pipe will replace the original medium inside the borehole with the outer pipe, as shown in Figure 1b.

3. Analysis of a Direct Wave with the Point Source

As stated in Section 2, the antennas have cylindrical layers around them, similar to a cylindrical dielectric waveguide. The optical and microwave theory says a mode may propagate in the cylinder axis direction, and the TM_{0n}, TE_{0n}, HE_{mn}, and EH_{mn}, where m and n are integers, modes may propagate along the waveguide [28,29]. This excitation of the mode significantly influences the direct wave from the transmitter to the receiver in the borehole radar measurement. The TM_{01} and HE_{11} modes will be excited dominantly if the wave source is a vertical dipole in a vertical borehole [16]. In the case of the loop antenna, as in this paper, the dominant mode would be the TE_{0n} mode since the transmitting loop antenna would excite the axially symmetric and concentric electric field concerning the z-axis. The most fundamental mode of the TE mode is TE_{01} mode. This section simplifies the transmitting and receiving antennas to the point ones and numerically analyzes the effect of replacing the medium with the outer pipe on the direct wave.

3.1. Formulation

Figure 2 shows the model used to calculate the direct wave's field. We consider rectangular (x, y, z) and cylindrical (ρ, ϕ, z) coordinate systems, which share the same origin. N cylindrical layers have cylindrical boundaries at $\rho = a_i$ $(i = 1, 2, \ldots, N-1)$. These layers simulate borehole boundaries in the borehole radar. The values ε_i and μ_i are the permittivity and permeability in the ith region. All permeability is assumed to be μ_0, the one in a vacuum in this paper. We consider the Hertzian dipole $\mathbf{J} = Il\boldsymbol{\phi}$ to be at (r', ϕ', z'), where the unit vector $\boldsymbol{\phi}$ is directed in the ϕ direction, I is current, and l is the length of the Hertzian dipole. This horizontally directed Hertzian dipole will be part of a transmitting loop in Section 4. The ϕ-component of the total electric field at (ρ, ϕ, z) radiated by the Hertzian dipole is defined as $E_{1\phi}^{total}$, which is decomposed by

$$E_{1\phi}^{total} = E_{1\phi}^{direct} + E_{1\phi}^{ref}, \tag{1}$$

where the direct field $E_{1\phi}^{direct}$ corresponds to the Hertzian dipole's field in a homogeneous media by having the same electromagnetic properties as the ones in the first layer with $\rho < a_1$; see (2.3.1) in [30] to obtain this equation, for example. The reflected field $E_{1\phi}^{ref}$ corresponds to the reflections from the cylindrical boundaries. Note that the total field $E_{1\phi}^{total}$ corresponds to the direct wave from the Hertzian dipole, and this direct wave consists of the direct field $E_{1\phi}^{direct}$ and reflected field $E_{1\phi}^{ref}$. We assume the Hertzian dipole radiates a sine wave with the angle frequency ω. We interpret the total electric field's ϕ-component as the electric field on the receiving loop antenna in Section 4. Referring to (3.3.6) in [30], the reflected field can be broken down into cylindrical and wavenumber harmonics $E_{1\phi}^{(n,k_z)}$ as

$$\begin{aligned} E_{1\phi}^{ref} &= \sum_{n=-\infty}^{\infty} e^{in(\phi-\phi')} \int_{-\infty}^{\infty} E_{1\phi}^{(n,k_z)} dk_z \\ &= \sum_{n=-\infty}^{\infty} e^{in(\phi-\phi')} \int_{-\infty}^{\infty} \frac{1}{k_{1\rho}^2} \left\{ ik_z \frac{in}{\rho} E_{1z}^{(n,k_z)} - i\omega\mu_1 \frac{\partial}{\partial\rho} H_{1Z}^{(n,k_z)} \right\} dk_z \end{aligned} \tag{2}$$

Table 1. The materials in the numerical model in this paper.

Layer Number i	Radius a_i	With the Outer Tube	Without the Outer Tube
1	2.35 cm	Air: $\varepsilon_1 = \varepsilon_0$	
2	2.85 cm	FRP: $\varepsilon_2 = 5\varepsilon_0$	
3	4.5 cm	Plastic: $\varepsilon_3 = 3\varepsilon_0$	Water: $\varepsilon_3 = \varepsilon_4$
4	5 cm	Water: ε_4 is given by (5) in [16] with $\alpha = 0.09$, $\tau_0 = 9.4 \times 10^{-12}$ s, $\varepsilon_\infty = 1.8\varepsilon_0$, $\varepsilon_{\text{low}} = 81\varepsilon_0$, $\sigma_{\text{dc}} = 0$. For example, $\varepsilon_4 = 80.9\varepsilon_0 + i0.0041/\omega$ at 100 MHz.	
5		Rock: ε_5 is given by (6) in [16], with $\sigma_{\text{dc}} = 5 \times 10^{-3}$ S/m, $\chi_r = 1.7$, $n = 0.61$, $\varepsilon_\infty = 8.7\varepsilon_0$ and $\omega_r = 2\pi \times 100$ MHz. Also see [31,32]. For example, $\varepsilon_5 = 10.4\varepsilon_0 + i0.012/\omega$ at 100 MHz.	

The constant ε_0 is the permittivity in a vacuum.

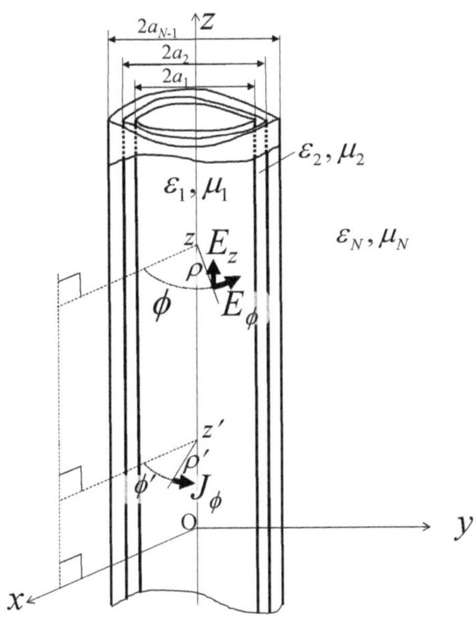

Figure 2. The model of the point source. The materials and radii of the layers are in Table 1.

This paper assumes and suppresses the time factor $\exp(-i\omega t)$. The value $k_{i\rho}$ is $\sqrt{k_i^2 - k_z^2}$, where k_i is the wave number of the plane wave in the ith layer, and k_z is the z-component of the wave number in all regions. The values $(E_{1z}^{(n,k_z)}, H_{1z}^{(n,k_z)})$ are given in Appendix A. We used Equation (3.1.4a) in [30] by transforming the above equations from the first to the second line.

If all the radii a_i ($i = 1, 2, \ldots, N-1$) are much smaller than the plane wave's wavelength in the outermost layer, the value at the right hand in (2) when the order n is zero is dominant. This condition is common in borehole radar measurements. For example, the borehole diameter a_{N-1} is 10 cm, much smaller than the wavelength 63 cm with Re $\varepsilon_N = 10\varepsilon_0$ for hard rock and 40 cm with Re $\varepsilon_N = 25\varepsilon_0$ for wet soil. From these insights, we define the value $E_{1\phi}^{ref,(n=0)}$ as the representative one $E_{1\phi}^{ref}$. Extracting the value at the right hand with $n = 0$ in (2), we have

$$\begin{aligned} E_{1\phi}^{ref,(n=0)} &= \int_{-\infty}^{\infty} E_{1\phi}^{(0,k_z)} dk_z \\ &= \int_{-\infty}^{\infty} -\frac{Il\omega\mu_1}{8\pi} e^{ik_z(z-z')} \frac{1}{k_{1\rho}^2} \left\{ \frac{\partial}{\partial\rho} J_n(k_{1\rho}\rho) \right\} \left\{ \frac{\partial}{\partial\rho'} J_n(k_{1\rho}\rho') \right\} R_{1,2}^{(2,2)} dk_z \end{aligned} \quad (3)$$

where the reflection coefficient $R_{1,2}^{(2,2)}$ is the value in the second row and second column cell of $\tilde{\overline{\mathbf{R}}}_{1,2}$, which is the 2×2 generalized reflection matrix and given by (3.2.8) in [30]. The function $J_n(\cdot)$ is the Bessel function. The right hand on the first and second rows in (3) includes the Sommerfeld integral. The singular points of the integrand $E_{1\phi}^{(0,k_z)}$ on the complex k_z plane contribute to the value of $E_{1\phi}^{ref,(n=0)}$ [16,30].

3.2. Observing the Impact of the Outer Tube on the Direct Wave

We observe the numerical values of the integrand on the wave number plane to see the physical phenomena when the outer tube covers the radar probe and not. A similar analysis was conducted in [16]. The main difference between that prior work and the one in this paper is the source and the observed electric field component. Table 1 shows the numerical values for each layer of the model. We assume the borehole is drilled in hard, dry rock and filled with water. The first layer is air, housed by the second layer of fiber-reinforced plastic vessel (FRP). The difference between the case with the outer tube and without is an electric property in the third layer. The outer tube removes the water and forms the third layer, which is the case with the outer tube. The actual material of the outer tube would be air and acrylic, as shown in Section 5. That is why the relative permittivity of the third layer with the outer tube is slightly smaller than that of the second layer and more significant than the first. The outermost layer is assumed to be hard, dry rock.

Figure 3 shows the absolute value of the integrand in the righthand side in (3). We calculated the two cases. Figure 3a shows one case when with the outer pipe. In this wave number plane, two branch points are associated with the wave numbers k_1 and k_4 in the innermost and outermost layers, respectively. In this case, the singular points, including the wavenumbers k_2, k_3, and k_4 in the middle layers, are removed [33,34]. The most critical matter is that no surface wave pole is near the real axis. The maximization of the integrand value corresponds to the surface wave pole, and there is no maximization. We interpret this fact as no mode that propagates well with the outer tube. Figure 3b shows the case without the outer tube. There is a surface wave pole at $k_z = 17.7 + 2.8i$, which might correspond to the TE_{01} mode. This implies that this mode propagates with little attenuation in the direction of the z-axis.

Figure 4a shows the total field $E_{1\phi}^{total}$, direct field $E_{1\phi}^{direct}$, and reflected field $E_{1\phi}^{ref}$ of (1) when the outer tube is in the third layer. At $z = 12$ and 52 cm, the amplitude of the reflected field is low, and the amplitudes of the total and direct fields are close to each other. Due to this fact, at these positions, the phase of the total field is closer to that of the direct field rather than the reflected wave. This means the reflected field does not affect the total field component well at these positions. At $z = 30$ and 75 cm, the amplitude of the reflected field is the largest, where the phase of the total field is closer to that of the reflected field than the direct field. As seen in all the positions, the larger amplitude of the direct and reflected fields depends on position z, which means the two components' amplitudes are at the same level. The amplitude of the total field is about 68 dB at $z = 10$ cm and 45 dB at $z = 78$ cm. This implies the direct wave's amplitude decreases by 23 dB when the direct wave propagates at a distance of 63 cm. Figure 4b shows the case when the outer tube is absent, i.e., the third layer is water. The amplitude of the reflected wave is always more significant than that of the direct field by about 15 dB at the position where 15 cm < z, which causes the total and reflected fields to be almost the same in amplitude and phase. This is because the TE_{01} surface wave's pole is near the real axis in the wavenumber plane in Figure 3b, and this mode included in the reflected field propagates with little attenuation. The amplitude of the direct wave decreases by 8 dB between the positions of z = 10 and 78 cm, and this attenuation is much smaller by 15 dB than the case with the outer tube.

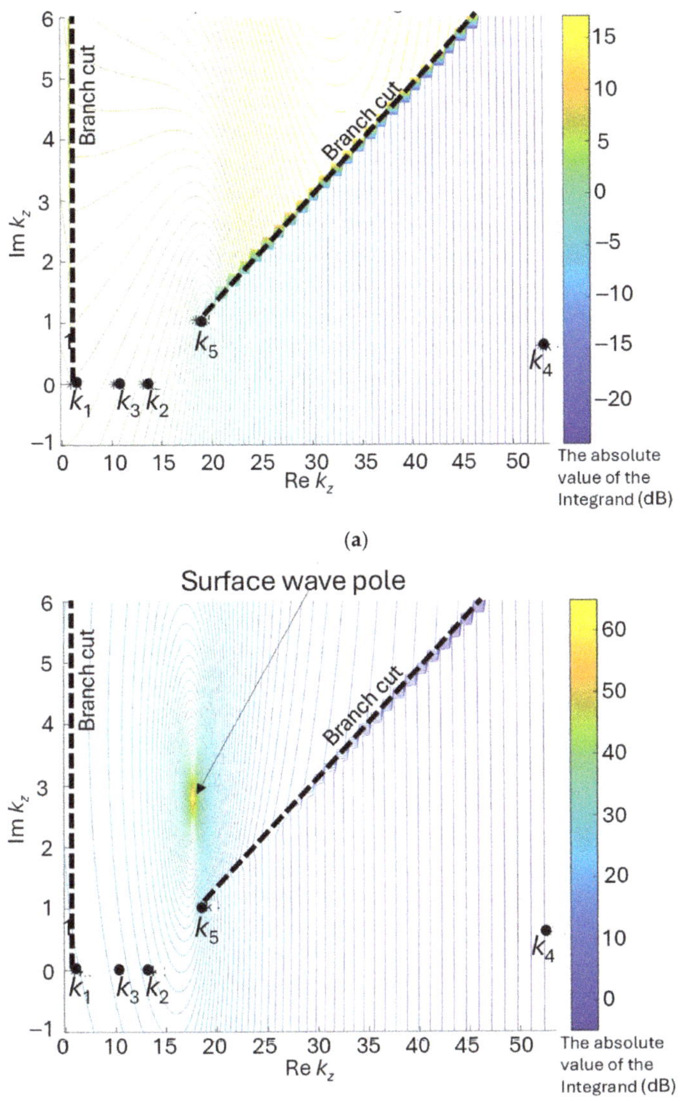

Figure 3. The amplitude $20\log_{10}\left|E_{1\phi}^{(0,k_z)}\right|$ of the integrand of the Sommerfeld integral in $E_{1\phi}^{ref,(n=0)}$ of (3). The frequency is 280 MHz. The current source parameters I = 1 A and l = 1 m are used. The source and observation points are located at (ρ' = 1.8 cm, ϕ' = 0, z' = 0) and (ρ = 1.8 cm, ϕ = 0, z = 0), respectively. (**a**) With the cover. (**b**) Without the cover.

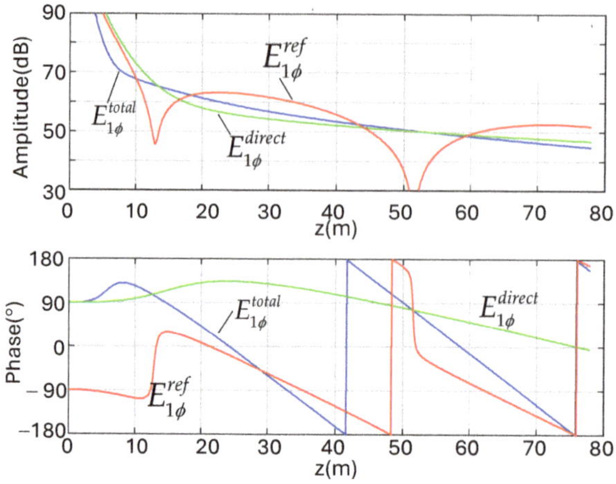

(**a**)

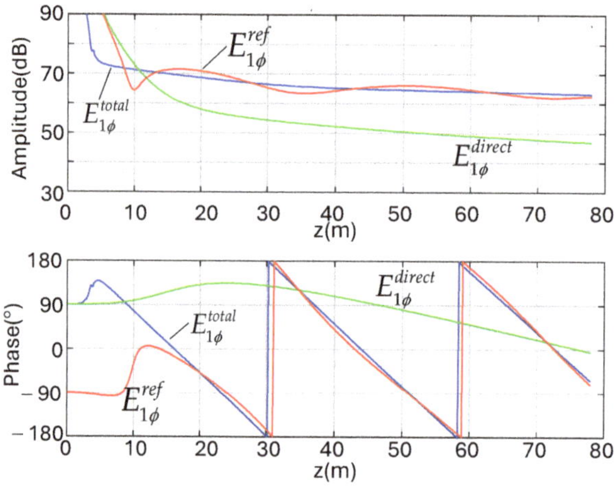

(**b**)

Figure 4. The amplitude and phase of $E_{1\phi}^{total}$, $E_{1\phi}^{direct}$, and $E_{1\phi}^{ref}$ of (1) at (ρ = 1.8 cm, $\boldsymbol{\phi}$ = 0, z). The amplitude of electric field E is calculated by $20\log_{10}|E|$ [dB]. The frequency, current source parameters, and source position are identical to those in Figure 3. (**a**) With the outer tube. (**b**) Without the outer tube.

4. Numerical Investigation with the Antennas

In this section, we model and incorporate antennas into the electromagnetic field calculations. We replace the Hertzian dipole in Section 3 with a transmitting loop antenna, which will be used in a borehole. We also put another loop antenna for the receiver. Figure 5 shows the antenna model. The transmitter is a loop antenna, and the receiver is an array of loop antennas. The method of moments (MoM) [35] outputs the voltage of the feeding point for a direct wave from one transmitter to the receiver. We coded the MoM for the borehole radar by modifying Green's functions, including the reflected fields of (2). This inclusion in the MoM considers the borehole effect and a similar attempt was made for dipole antennas in [16,36]. The frequency domain data output from the MoM is windowed with a band-pass windowing function, whose center frequency is f_{center}. Then, we apply an inverse Fourier transform to the windowed data and obtain the time domain signals.

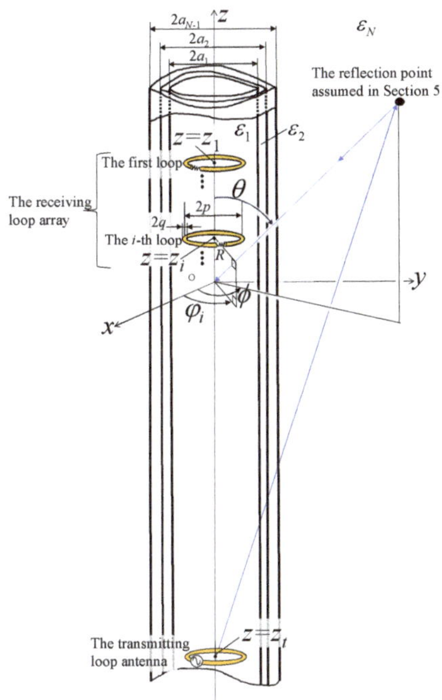

Figure 5. The antenna model. The materials and radii of the layers are in Table 1. The transmitting loop antenna's feeding point is directed to the x-axis. The ith receiving loop antenna feeding point at $z = z_i$ is directed to the azimuth angle φ_i. All the loop antenna's feeding points are terminated by the resistance R. We assume that a radar target and its reflection point exist only in Section 5 rather than Section 4.

4.1. Direct Wave from the Transmitting to the Receiving Antennas

Figure 6a shows the amplitude received at the feeding point of the receiving loop antenna in the frequency domain. This corresponds to the direct wave signal's amplitude in borehole radar. The amplitudes with and without the outer tube are close to each other at frequencies below 100 MHz, while that without the outer tube is larger by about 45 dB than that with the outer tube at 300 MHz. We find that the absence of the outer tube increases the direct wave's power in higher frequencies. Figure 6b shows the time domain signals at the feeding point of the receiver when a pulse in Figure 6c is excited at the transmitting antenna's feeding point. The amplitude of the direct wave without the outer tube is more than five times larger than that with the outer tube. Furthermore, the oscillated wave continues until 50 ns. If the outer tube is not attached to the radar probe, a direct wave with a large amplitude and oscillation will influence the detection of a reflected wave from a radar target, such as a fracture or a vein. There are the following reasons why the outer pipe significantly reduces direct wave power. One is the mode propagating along the cylindrical layers attenuated when the outer pipe is added, as seen in Section 3. Another possible reason is that the outer pipe might change the loop antenna's antenna efficiency. In Section 4.2, we will investigate the effect of the outer pipe on the antenna efficiency.

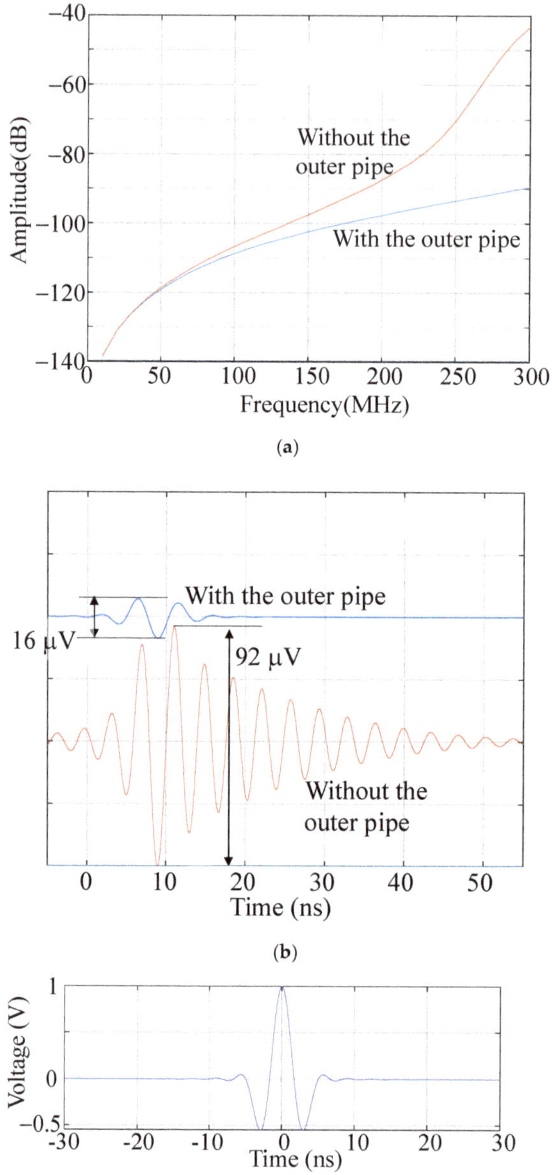

Figure 6. The direct wave is received at the receiver's feeding point. One receiving loop antenna. The loop antenna parameters are $z_t = 0$, $z_1 = 76$ cm, $2p = 3.6$ cm and $2q = 2$ mm, $R = 250\ \Omega$. (**a**) The received amplitude in the frequency domain. The value $20\log_{10}|V|$ [dB] is shown when the voltage is V. (**b**) The voltages at the receiver's feeding point in the time domain. (**c**) The voltage excited at the transmitter's feeding point in the time domain. The center frequency f_{center} is 150 MHz.

4.2. Consideration of Antenna Efficiency

The existence of the outer pipe might affect the antenna efficiency. Referring to p. 65 [37], the total antenna efficiency e_0 is a product of the reflection efficiency e_r, conduction efficiency e_c, and dielectric efficiency e_d, i.e.,

$$e_0 = e_r\, e_c\, e_d. \tag{4}$$

The antenna radiation efficiency e_{cd}, which equals $e_c\, e_d$, is caused by the conductive and dielectric losses in the medium, such as the water in the borehole and the material of the outer pipe. If the conductivity of the medium inside the borehole is large, we cannot ignore its influence on the borehole radar signals [22,38]. However, in this paper, we assume the medium inside the borehole is fresh water, of which conductivity is small, as seen in the field test site in Section 6.2. In this case, the existence of the outer pipe would not significantly influence the antenna radiation efficiency, considering the difference in the loss of water and the outer pipe's material around the antenna is small. This paper concentrates on reflection efficiency. This corresponds to the mismatch between the feeding line's impedance and the antenna input impedance.

Figure 7 shows the calculated input impedance of the loop antenna. At frequencies over 200 MHz, the difference in the resistance and reactance between the outer pipe's existence and its absence is significant by about 10 Ω. The difference in the inhomogeneity between the cases with the outer tube and without causes this difference in the antenna input impedance at frequencies over 200 MHz. Figure 8 shows the antenna's return loss RL, defined as $-20 \log |\Gamma|$ with the reflection coefficient Γ. The reflection coefficient Γ corresponds to the variable ρ defined in (10) of [39] or [40]. The difference between the cases with the outer pipe and without is about 0.6 dB at 300 MHz. This means the antenna efficiency with the outer pipe is slightly better than without the outer pipe. This amount would affect the power of receiving both a direct wave signal and a reflected wave from a subsurface object, a radar target. However, this antenna efficiency variation will not be essential in detecting the reflected wave since the mode in Section 4.1 is more important than the antenna efficiency variation, as mentioned in Section 5.

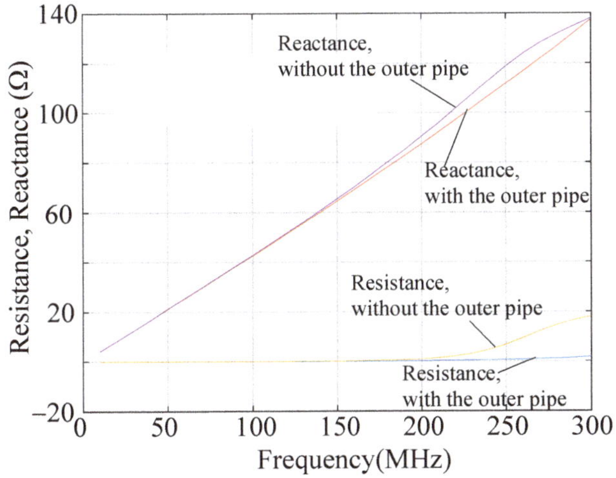

Figure 7. Input impedance of the antenna. The antenna configuration, such as q and p, is the same as in Figure 6.

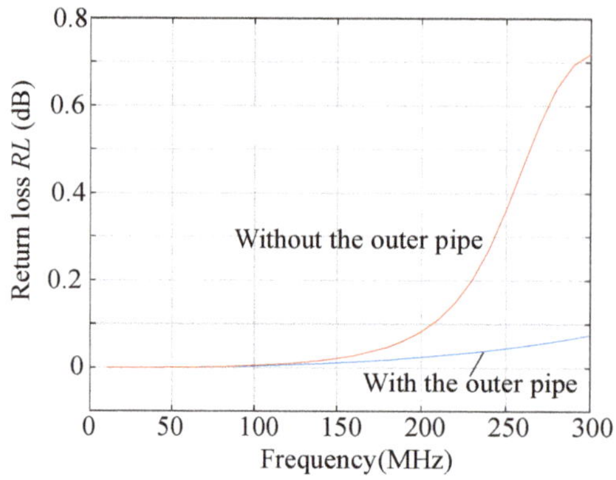

Figure 8. Return loss RL. The antenna configuration, such as q and p, is the same as in Figure 6. We assume that the feeding line's characteristic impedance Z_0 is 50 Ω.

5. Numerical Simulation of the Borehole Radar, Demonstrating the Effect of the Outer Pipe

In borehole radar measurement, the reflected wave from a radar target, like a subsurface object, reaches the receiver after the direct wave arrives in the time domain. If this radar target position is close to the borehole, the arrival times of the direct and reflected waves are almost the same. Furthermore, the power of the reflected wave is usually smaller than that of the direct wave. These facts decrease the detectability of the reflected wave from the radar target close to the borehole. This section affirms that the cylindrical layer simulating the outer tube highlights only the reflected wave from a radar target, reducing the power of the propagation mode of the direct wave.

Figure 9 shows the model for the numerical simulation. One set of cylindrical layers exists around the D-axis, simulating the borehole's center axis. The r-axis is perpendicular to the D-axis and extends from $D = D_0$ on the D-axis. The innermost layer includes one transmitting and four receiving loop antennas, as in Figure 5. The top and first loop antenna is 21 cm above the position at $D = D_r$. The directions of the ρ-axis in Figure 9 and the x-axis in Figure 5 are the same, and the D-axis in Figure 9 and the z-axis in Figure 5 are overlapped. We assume an infinitely large planar conducting interface as a radar target. This interface intersects the D-axis on the D-axis at $D = D_0$. This model is similar to Figure 5 in [20], except for the antenna configurations and the medium's electromagnetic parameters. The MoM with the modified Green's functions, including the borehole effects, outputs the voltages of the receiving loop antennas' feeding points. The inverse Fourier transform after windowing outputs the time domain signals in the same manner as in Figure 6b.

Figure 10a,b shows each antenna position's calculated time domain signals at $D = D_r$. These figures include the direct and reflected waves from the planer interface. In Figure 10a,b, the direct wave arrives at about 9 ns. The direct and reflected waves are superimposed in both figures at depths between $D_r = -806$ cm and -842 cm. For reference, Figure 10c,d shows the reflected waves only, in which the direct wave is removed from the waveforms in Figure 10a,b, and amplitudes are normalized, respectively. However, when the depth D_r is below -848 cm in Figure 10a, the two waves are separated, and the reflected wave arrives after the direct wave. The arrival times of these reflected waves get later when the depth D_r gets deeper since the antennas move away from the intersection at $D_0 = -700$ cm in Figure 9. The array of the receiving loop antenna measures this direct wave with a slight arrival time difference, while it measures the reflected waves with little arrival time difference. This is because the difference in the wave propagation directions affects the arrival times among the array antennas.

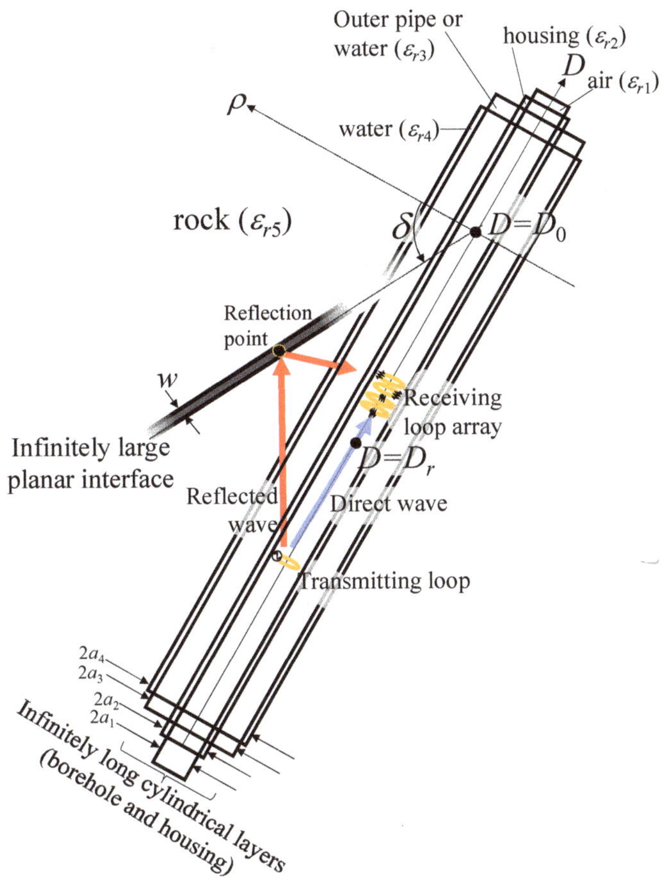

Figure 9. The model for the MoM. D_0= −700 cm. δ = 70°. The planar interface is a perfect conductor with w = 1 mm. The antenna parameters in Figure 5 are z_t = −57 cm, z_i = 21 − (I − 1) × 3 cm (i = 1, 2, 3, 4) cm, φ_1 = 0°, φ_2 = 180°, φ_3 = 270°, and φ_4 = 90°. $2p$ = 3.6 cm and $2q$ = 2 mm, and R = 250 Ω. The origin's position in Figure 5 is located at (D = D_r, r = 0) in this figure. The x-axis and z-axis in Figure 5 are directed in the same direction as the D-axis and r-axis, respectively, in this figure.

In the case without the outer tube in Figure 10b, we can see only the direct wave since the power of the direct wave is much larger than that of the reflected wave. The direct wave's arrival time does not change much with the depth. The wave oscillates since the mode's power is enormous at high frequencies and the frequency bandwidth is narrow. According to the discussion in Figures 7 and 8, the outer pipe affected the antenna efficiency. This would increase the amplitudes of both the direct and reflected waves without the outer tube compared with the outer tube. However, attaching the outer tube made the reflected waves detectable, as shown in Figure 10. This implies the outer tube suppressed the direct wave's propagation mode rather than affecting the antenna efficiency. The mode propagating along the cylindrical layers, such as TE_{01}, significantly increases the direct wave's power in the case without the outer tube.

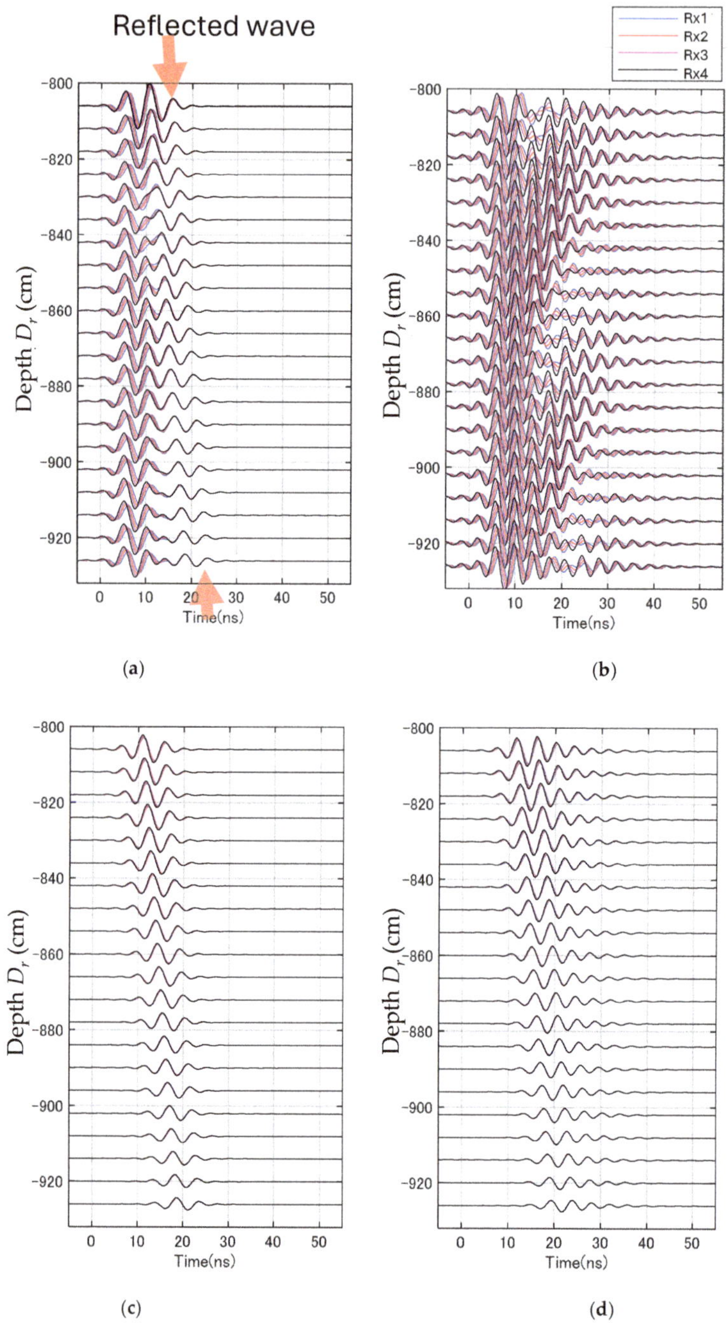

Figure 10. The simulated time domain signals. The signal color means the array antenna number used; see the legend in the upper figures in Figure 11. (**a**) With the outer tube. The direct wave and reflected wave. (**b**) Without the outer tube. The direct wave and reflected wave. (**c**) With the outer tube. The reflected wave only. (**d**) Without the outer tube. The reflected wave only.

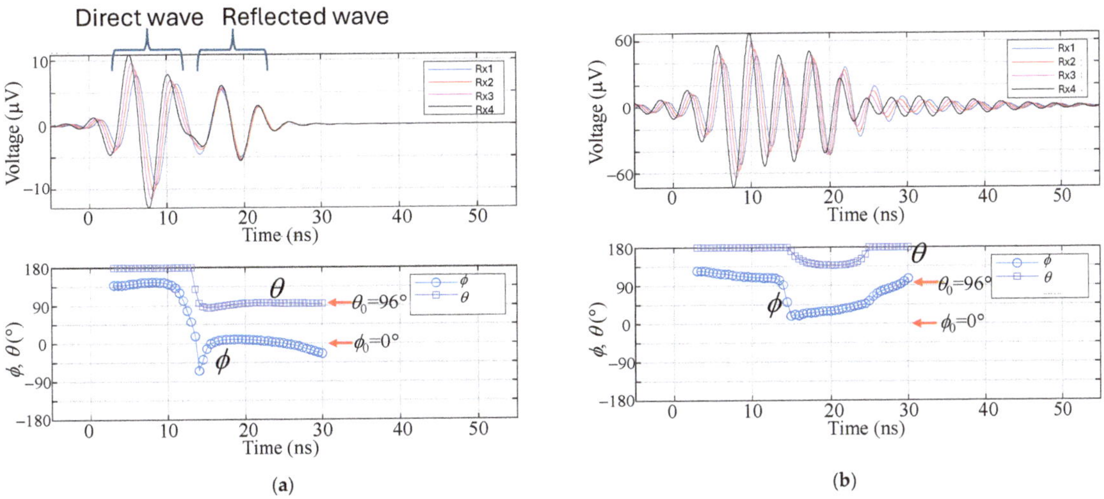

Figure 11. The estimated directions of the azimuth angle ϕ and elevation angle θ at the depth of $D_r = -902$ cm. Analyzing the time domain loop array signals in the computer simulation. The array signals are in the upper figures, and the estimated DOA is in the lower figures. The azimuth angle ϕ_0 and elevation angle θ_0 calculated by the GOs are 0° and 96°, respectively, and are also shown. In the DOA estimation algorithm, the velocity $v = v_0/\sqrt{8}$ was used as v in (3) of [14], where v_0 is the velocity in the vacuum. The legend in the upper figures shows the relationship between the signal color and the antenna number. For example, "Rx1" means the first receiving loop antenna. (**a**) With the outer tube. (**b**) Without the outer tube.

When a plane wave is incident on the receiving loop array antenna, it can estimate the wave's DOA, as we stated in Section 2. This means the receiving loop array can calculate the azimuth angle ϕ and elevation angle θ of the directions of the reflection point in Figure 5. This is based on the assumption that the distance between the antenna positions and the reflection point is much larger than the operating wavelength, i.e., holding a far-field assumption. Figure 11 shows an example of the reflected wave's estimated DOA. This way of showing is similar to Figure 11 in [14]. The direct wave arrives at about 8 ns in Figure 11a. At these times, the estimated elevation angles are 180°, which agrees that the direct wave propagates in the positive direction of the D-axis. The elevation and azimuth angles of the reflected wave at about 20 ns are about 90° and 0°, respectively. Geometrical optics (GOs) [13,41], a high-frequency method, can also estimate the directions according to the gematrical arrangement of the antennas and planar interface. The red arrows with θ_0 and ϕ_0 indicate these GOs' estimations in the lower two figures in Figure 11. The radar estimations and GOs are almost agreed upon in the lower figure (a). This means that the radar can detect the reflected waves from the radar target when with the outer pipe. However, Figure 11b, when without the outer pipe, shows that the elevation angles estimated by the radar are almost 180° at all times, corresponding to the direction of not the reflected wave but the direct wave. The azimuth angles of the radar's estimation range between 18° and 125°. These estimations differ from the GOs' estimation, indicated by the red arrows with θ_0 and ϕ_0. This implies that the power of the direct wave is much larger than that of the reflected wave and spreads in the time domain, which causes the overlap of the two waves.

6. Field Experiments with the Borehole Radar with the Outer Pipe

6.1. Making the Outer Pipe

We designed the outer pipe with the insight of the computer simulation in Section 5. The diameter of the loop-array type directional borehole radar probe is 57 mm, while the inner diameter of the borehole at the field experiment site is 10 cm. These values are similar to those in computer simulation. We set the outer and inner diameters of the outer pipe to be 90 and 57 mm, as shown in Figure 12. The length of the outer pipe is about 2 m, which covers all antennas. The inner material of the designed outer pipe is air, which is trapped between the two pipes made of acrylic. We expect this designed outer pipe to work as the infinitely long dielectric cylindrical layer outside the radar probe. It is easy to put on the outer pipe on the radar probe's vessel and take off it.

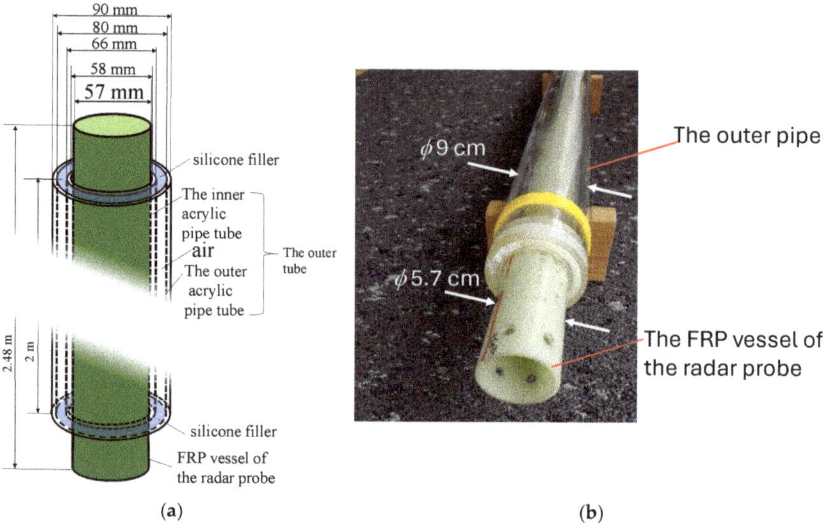

Figure 12. The outer pipe that was made in this paper. (**a**) The size of the outer pipe. (**b**) The picture of the outer pipe.

6.2. Field Experiments

We conducted field experiments at the Nakatatsu mine in Fukui Prefecture, Japan. The host rock is composed of skarns. A fracture zone is observed in the gallery of this site, and we drill a borehole that penetrates this fracture zone. With this borehole, we made the three-dimensional image using the dipole array directional borehole radar in past research [42]. We observed that the boring cores obtained at shallower depths than 7 m included much sphalerite mineral. This means that sphalerite veins were distributed in the skarns at these depths. One possible cross-section of the test site is the one in Figure 13. There is a sphalerite vein and the fracture zone near the borehole. This paper sets the interface between the sphalerite vein and the host rock as our radar target. This would be the beginning to delineate the three-dimensional contours of the mineral deposits. Figure 14 shows the picture when we insert the radar probe into the borehole.

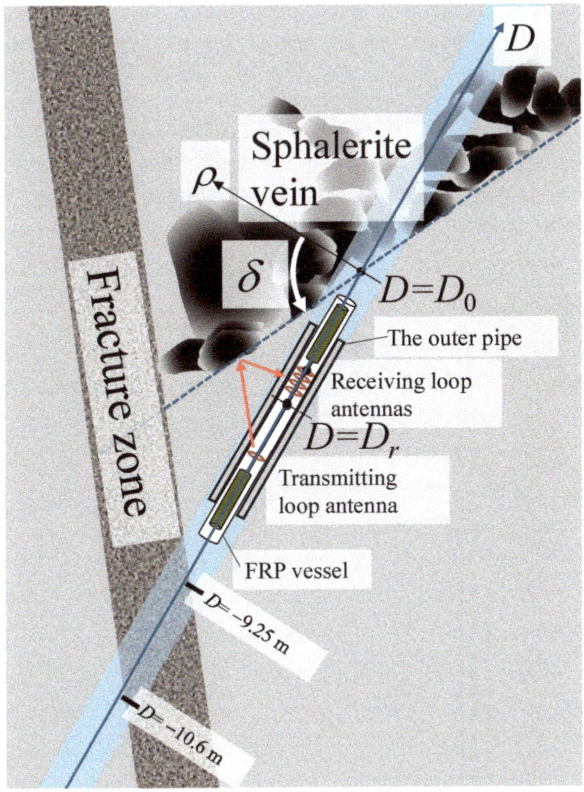

Figure 13. The borehole radar probe with the outer tube in Nakatatsu mine, Japan. The borehole, sphalerite vein, and fracture zone.

Figure 14. Insertion of the radar probe into the borehole, August 2022. Without the outer tube.

Figure 15 shows the used antennas. This antenna's configuration is similar to the model in the computer simulation. The main difference between the actual antenna and the model is the existence of the feeding lines to the antennas in the experiment, as we stated in Section 2. A thin-centered conducing cylinder includes the feeding lines and exists at the center axes of the radar probe. The distance between the thick-centered conducting cylinder and the antenna elements is over 15 cm, enough to ignore the scattering from the thick-centered conducting cylinder to the antenna elements. The receiver consists of the four loop array elements, as shown in Figure 14 in [14]. A vector network analyzer measures the transmission between the feeding points of the transmitting loop antenna and one of the receiving loops in the frequency domain between 0.3 and 300 MHz. In a similar manner to the process of the frequency domain data in the computer simulation, we obtain time domain signals.

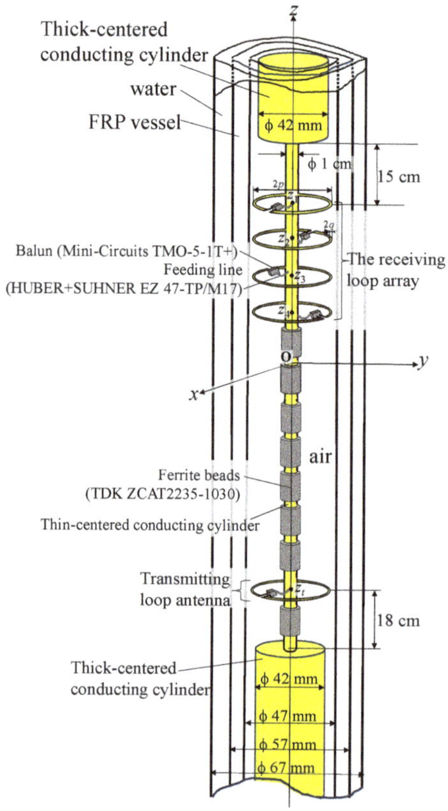

Figure 15. The antennas used in the field experiment. The antenna size parameters are similar to those used in Figure 9.

Figure 16a shows the power spectrum of the transmission function between the transmitter and receiver. Although it includes both the components of the direct waves and the reflected waves from the subsurface objects, the direct wave's component would be larger than the reflected wave's. This enables us to compare this figure and Figure 6a. The difference in the power between the case with the outer tube and without is 36 dB at 300 MHz in Figure 16a, while the difference in the power in Figure 6 was 40 dB. The outer pipe reduced power in the experiment and computer simulation by almost the same values. This implies the effect of the outer pipe on the reduction of the direct wave's power also in the experiment, as we expected in the computer simulation. This situation is also evident in the time domain signals in Figure 16b. The direct wave arrives at about 10 ns in both cases with and without the outer tube in this figure. We find that the amplitude of the direct wave without the outer tube is much larger than that with the outer tube. There is an oscillated wave just after the direct wave arrives in the case without the outer tube, which was observed also in the case without the outer tube in Figure 6b.

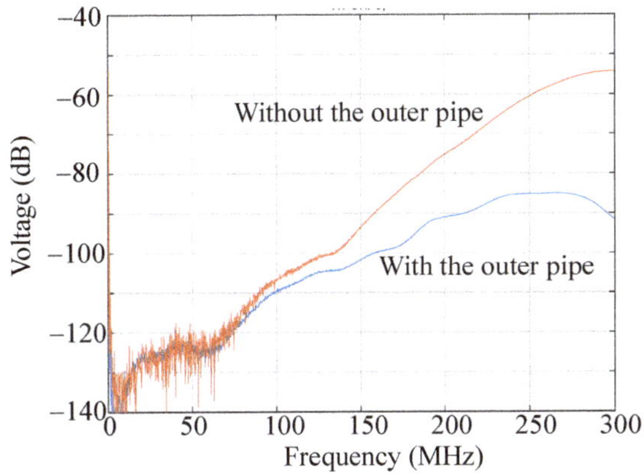

(a)

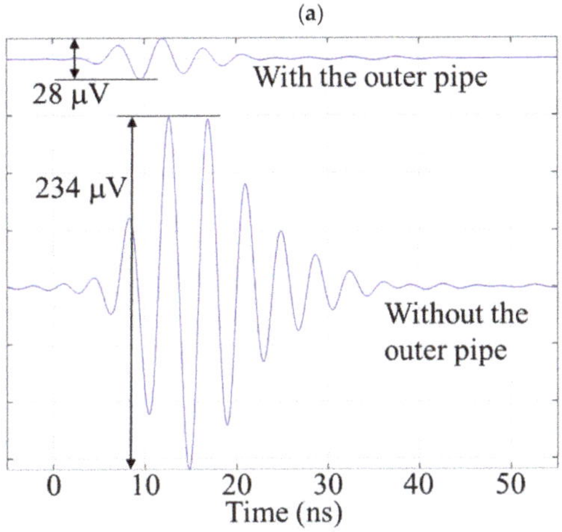

(b)

Figure 16. The example of the measured data at the depth of $D = -800$ cm for the case with the outer tube and $D = -804$ cm for the one without. The receiver is the first receiving array element. (**a**) The power spectrum of the transmission function between the transmitter and receiver. (**b**) The received time domain signals when the excited voltage wave at the feeding point of the transmitter was the one in Figure 6c.

Figure 17a shows the received time domain signals in the case with the outer tube. The amplitudes of the signals are normalized at each depth. One pulse at about 10 ns is separated into two when the depth gets deeper. One is the direct wave arriving at about 10 ns at all depths. The other is the reflected wave from the interface between the sphalerite vein and the host rock, and its arrival times get late when the depths from 800 cm to 920 cm get deeper. The characteristics of the two waves in Figure 17a are similar to those of the direct and reflected waves in Figure 10a. Figure 17b shows the received time domain signals in the case without the outer tube. We find one wave at depths from −796 cm to −980 cm, whose waveforms do not change despite the depth change. This wave corresponds to the direct wave. It is difficult to find the reflected waves from the sphalerite vein since the amplitude of the direct wave is large and its energy spread in the time domain. This situation is similar to that in Figure 10b.

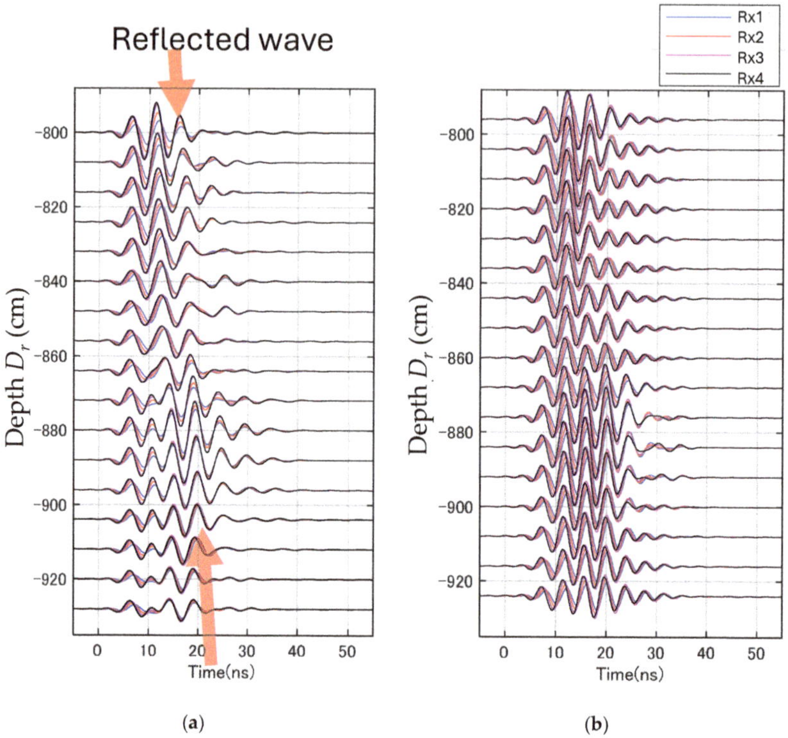

Figure 17. The measured time domain signals. The signal color means the array antenna number used; see the legend in the upper figures in Figure 18. (**a**) With the outer tube. (**b**) Without the outer tube.

Figure 18a shows the DOA estimation results at a depth of about 9 m. We expect the direct and reflected waves to be about 10 ns and 20 ns, respectively, according to Figures 11a and 17a. The estimated elevation angles vary slightly from 166° at 3 ns to 129° at 10.5 ns. This is when the direct wave arrives since the estimated direction is directed almost to the transmitter, and these times are close to the arrival time of the direct wave in Figure 11. From 12 ns to 13 ns, the estimated elevation and azimuth angle rapidly change, which implies another wave's arrival. After 14 ns, the estimated elevation angles range between 108° and 51° and the estimated azimuth angle is between −28° and 36°. If we assume that the sphalerite vein exists, as in Figure 13, the GOs estimate the DOA with the red arrows with θ_0 and ϕ_0. These angles are inside the limited angles by the colored broken lines. This implies that the direct and reflected waves are separated clearly in the case of the radar probe with the outer tube.

In the case without the outer tube in Figure 18b, the estimated elevation angle is 180° until about 9 ns, and the smallest elevation angle is 118° at 25 ns, even after 10 ns in Figure 18b. These elevation angles are much above the GOs' estimated elevation angle θ_0 of 96°. This is because much of the wave energy is from a direct wave. This makes it difficult to detect the reflected waves from the sphalerite vein in the case without the outer tube.

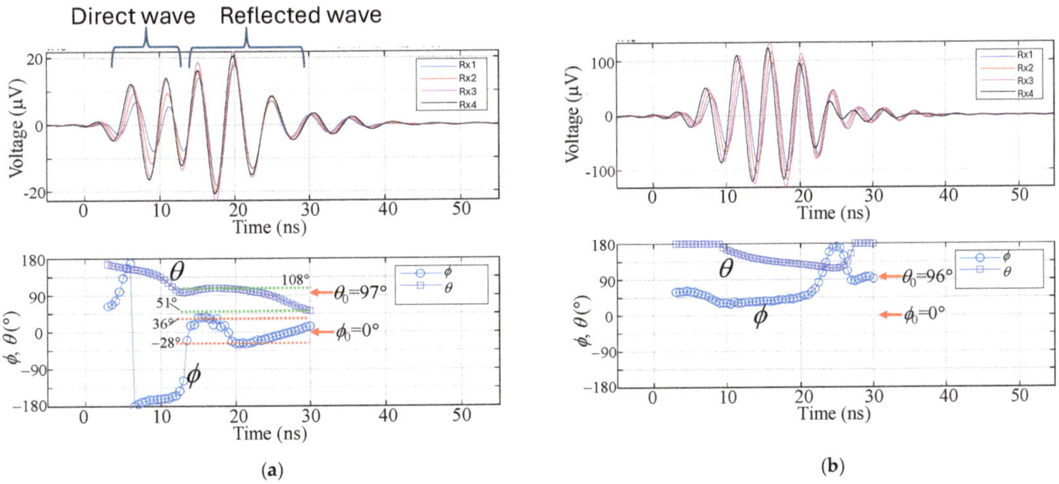

Figure 18. The estimated directions of the azimuth angle ϕ and elevation angle θ. The measured results. The azimuth angle ϕ_0 calculated by the GOs is 0, which is also shown. In the DOA estimation algorithm, $v = v_0/\sqrt{8}$. The legend in the upper figures shows the relationship between the signal color and the antenna number, as in Figure 11. (**a**) With the outer tube at a depth of D_r = −896 cm. The elevation angle θ_0 estimated by the GOs is 97°. The colored broken lines show the estimated angle range at the time expected for the reflected wave. The green broken lines show the maximum and minimum elevation angles at 51° and 108°, respectively. The red broken lines correspond to the maximum and minimum azimuth angles at −28° and 36°, respectively. (**b**) Without the outer tube at a depth of D_r = −900 cm. The GOs estimated θ_0 = 96°.

7. Conclusions

We investigated the effect of an outer tube on the reduction of a direct wave in single-hole directional borehole radar with loop antennas. Our numerical electromagnetic analysis confirmed the water cylindrical layer, whose thickness is about 2 cm, the mode, such as TE_{01} mode, propagating along the borehole occurs at frequencies above 200 MHz. The MoM analysis concluded that the mode makes the direct wave's power significant and its frequency bandwidth narrow. Although this direct wave prevented the borehole radar from detecting a reflected wave from a subsurface object near the borehole, attaching the outer tube resolved this problem. This outer tube consists of a hollow cylinder on the radar probe's vessel, pushing the water layer away. The numerical simulation showed the outer tube reduced the direct wave's power by more than 40 dB. According to the MoM analysis, the outer tube reduces the direct wave's power more than the reflected wave from a conducting sheet that simulated a radar target. According to the numerical simulation's insight, we made the outer tube of two plastic cylinders. We carried out the loop-array directional borehole radar with the outer tube in a water-filled borehole in Skarns. The radar target was an interface between the sphalerite vein and the host rock. The diameter of the borehole was 10 cm, while that of the radar probe vessel was 5.7 cm. Although the direct wave's power in the case without the outer tube is considerable, and it was hard to detect the reflected wave from the radar target, applying the outer tube enabled us to reduce the direct wave's power and highlight the reflected wave. The estimation of the DOAs of the waves agreed that the outer tube separates the direct and reflected waves well.

This paper examined the outer tube's effect on the loop antennas only, while most borehole radars use dipole antennas. It is interesting to investigate the outer tube's impact on them. We expect a similar manner to the one shown in this paper for the dipole antenna case. The essential surface wave in the dipole antenna case would be TM_{01} mode, referring to [16]. Future work should focus on whether the outer tube works to reduce this surface wave.

The materials for the outer tube should be optimized. This paper's numerical simulation showed that the outer tube may decrease antenna efficiency. By observing these factors, such as the reduction of the direct wave and the antenna efficiency, we may tune the electromagnetic properties and sizes of the outer tube.

An existing signal processing method, such as moving average subtraction, can numerically reduce direct wave power. Comparing this signal processing to its usage of the devices presented in this paper would be an interesting topic.

Author Contributions: Conceptualization, formal analysis, supervision, writing—review and editing, project administration and funding acquisition, S.E.; methodology, visualization, software, investigation, resources, and data curation, writing—original draft preparation, R.M., K.K., Y.T. and Y.N. All authors have read and agreed to the published version of the manuscript.

Funding: This work was financially supported by the Japan Society for the Promotion of Science (JSPS) KAKENHI Grant Number 24K08328.

Data Availability Statement: Due to the nature of this research, the participants did not agree to share their data publicly, so supporting data are unavailable.

Acknowledgments: The authors thank Kazuhiko Yamanaka of Mitsui Mining & Smelting Co., Ltd. for his permission to experiment with the Nakatatsu mine. We appreciate Seiichi Ikeda at Nakatatsu Mine for his assistance with the experiments in September 2022.

Conflicts of Interest: The authors declare no conflicts of interest.

Appendix A. The Point Source Fields in a Cylindrically Layered Medium

We observe total electric and magnetic fields' z-components $(E_{1z}^{total}, H_{1z}^{total})$ at (r, ϕ, z) in the innermost layer, which forms the direct wave. Each of these fields is the summation of a direct field $(E_{1z}^{direct}, H_{1z}^{direct})$ and reflected field $(E_{1z}^{ref}, H_{1z}^{ref})$. Referring to (3.3.20) in [30], we write down the reflected field by the Hertzian dipole directed in the ϕ direction like

$$\begin{bmatrix} E_{1z}^{ref} \\ H_{1z}^{ref} \end{bmatrix} = \sum_{n=-\infty}^{\infty} e^{in(\phi-\phi')} \int_{-\infty}^{\infty} \begin{bmatrix} E_{1z}^{(n,k_z)} \\ H_{1z}^{(n,k_z)} \end{bmatrix} dk_z, \quad \text{(A1)}$$

where

$$\begin{bmatrix} E_{1z}^{(n,k_z)} \\ H_{1z}^{(n,k_z)} \end{bmatrix} = \frac{iIl}{4\pi\omega\varepsilon_1} e^{ik_z(z-z')} J_n(k_{1\rho}\rho) J_n(k_{1\rho}\rho') \tilde{\tilde{\mathbf{R}}}_{1,2} \frac{i}{2} \begin{bmatrix} -\frac{nk_z}{\rho'} \\ i\omega\varepsilon_1 \frac{\partial}{\partial\rho'} \end{bmatrix}. \quad \text{(A2)}$$

Note that the differential operator $\frac{\partial}{\partial\rho'}$ acts to its left.

References

1. Fujiyama, T.; Kaku, K. Current status of the geological disposal programme and an overview of the safety case at the pre-siting stage in Japan. *Rock Mech. Bull.* **2023**, *2*, 100062. [CrossRef]
2. Tóth, E.; Hrabovszki, E.; Schubert, F.; Tóth, T.M. Discrete fracture network (DFN) modelling of a high-level radioactive waste repository host rock and the effects on its hydrogeological behaviour. *J. Struct. Geol.* **2022**, *156*, 104556. [CrossRef]
3. Zhao, Q.; Liu, S.; Yuan, W.; Lu, Q.; Tian, S.; Ni, J. Fracture System Assess in Beishan Grottos, Chongqing, China Based on Multiple Geophysical Data. *IEEE Access* **2024**, *12*, 117665–117677. [CrossRef]
4. Wang, Y.; Chen, J.; Jia, D. Three-Dimensional Mineral Potential Mapping for Reducing Multiplicity and Uncertainty: Kaerqueka Polymetallic Deposit, QingHai Province, China. *Nat. Resour. Res.* **2020**, *29*, 365–393. [CrossRef]
5. Daniels, D.J. *Ground Penetrating Radar*, 2nd ed.; IET: London, UK, 2004.

6. Liu, S.; Sato, M. Electromagnetic logging technique based on borehole radar. *IEEE Trans. Geosci. Remote Sens.* **2002**, *40*, 2083–2092. [CrossRef]
7. Kulich, J.; Bleibinhaus, F. Fault Detection with Crosshole and Reflection Geo-Radar for Underground Mine Safety. *Geosciences* **2020**, *10*, 456. [CrossRef]
8. Fullagar, P.K.; Livelybrooks, D.W.; Zhang, P.; Calvert, A.J.; Wu, Y. Radio tomography and borehole radar delineation of the McConnell nickel sulfide deposit, Sudbury, Ontario, Canada. *Geophysics* **2000**, *65*, 1920–1930. [CrossRef]
9. Liu, S.; Wu, J.; Zhou, J.; Zeng, Z. Numerical Simulations of Borehole Radar Detection for Metal Ore. *IEEE Geosci. Remote Sens. Lett.* **2011**, *8*, 308–312. [CrossRef]
10. Sato, M.; Tanimoto, T. A shielded loop array antenna for a directional borehole radar. In Proceedings of the Fourth International Conference GPR, Geological Survey of Finland, Rovaniemi, Finland, 8–13 June 1992; Special Paper 16. pp. 323–327.
11. Wang, W.; Liu, S.; Shen, X.; Zheng, W. Analysis of the Borehole Effect in Borehole Radar Detection. *Sensors* **2020**, *20*, 5812. [CrossRef]
12. Ni, J.; Liu, S.; Han, X.; Lu, Q.; Tian, S.; Zhao, Q. 3-D Directional Borehole Radar Imaging Based on Echo Separation. *IEEE Trans. Geosci. Remote Sens.* **2024**, *62*, 5912612. [CrossRef]
13. Ebihara, S.; Toyozawa, K.; Hasegawa, T.; Tsujikawa, Y.; Koyama, K. Influence of a Horizontally Polarized Wave Incidence on DOA Estimation by a Dipole Array Antenna in a Borehole. *IEEE Trans. Geosci. Remote Sens.* **2024**, *62*, 1000116. [CrossRef]
14. Ebihara, S.; Matsumoto, M. Loop Array Antenna in a Borehole for Directivity to a Horizontally Polarized Wave. *IEEE Trans. Geosci. Remote Sens.* **2022**, *60*, 2005310. [CrossRef]
15. Wang, L.; Bao, X.; Liu, C.; Yan, W.; Jiang, B.; Wang, Y.; Si, L.; Schreurs, D.; Shi, Z.; Zhang, A. Highly Compact Time-Domain Borehole Radar with Time-Varying Gain Equivalent Sampling. *IEEE Trans. Microw. Theory Technol.* **2024**, *72*, 6107–6119. [CrossRef]
16. Ebihara, S.; Sasakura, A.; Takemoto, T. HE_{11} mode effect on direct wave in single-hole borehole radar. *IEEE Trans. Geosci. Remote Sens.* **2011**, *49*, 854–867. [CrossRef]
17. Cho, J.; Jung, J.; Kim, S.; Kim, S. Suppression of Borehole-Guided Waves Supported by the Connection Cable of a Single-Borehole Monostatic Pulse Radar. *IEEE Trans. Geosci. Remote Sens.* **2014**, *51*, 3431–3438. [CrossRef]
18. Ebihara, S.; Yamamoto, T. Resonance analysis of a circular dipole array antenna in cylindrically layered media for directional borehole radar. *IEEE Trans. Geosci. Remote Sens.* **2006**, *44*, 22–31. [CrossRef]
19. Irving, J.D.; Knoll, M.D.; Knight, R.J. Improving crosshole radar velocity tomograms: A new approach to incorporating high-angle travel time data. *Geophysics* **2007**, *72*, J31–J41. [CrossRef]
20. Ebihara, S.; Kotani, S.; Fujiwara, K.; Kimura, Y.; Shimomura, T.; Uchimura, R. Consideration of Oblique Incidence in 3-D Imaging of a Planar Interface with a Circular Dipole Array in an Air-Filled Borehole. *IEEE J. Sel. Top. Appl. Earth Obs. Remote Sens.* **2020**, *13*, 3711–3722. [CrossRef]
21. Wang, B.; Huang, Z.; Xing, L.; Lao, L.; Wei, Z.; Ge, X. Quantitative Analysis of Guided Wave in Dielectric Logging Through Numerical Simulation. *IEEE Geosci. Remote Sens. Lett.* **2022**, *19*, 8001905. [CrossRef]
22. Huo, J.; Zhou, B.; Mason, I.M.; Zhao, Q. An Application of Reflected Guided Borehole Radar Waves to Detection of Subhorizontal Structures. *IEEE Trans. Geosci. Remote Sens.* **2021**, *59*, 7201–7210. [CrossRef]
23. Mao, L.; Zhou, B. Simulation and Analysis of Conductively Guided Borehole Radar Wave. *IEEE Trans. Geosci. Remote Sens.* **2017**, *55*, 2646–2657. [CrossRef]
24. Tronicke, J.; Holliger, K. Effects of gas- and water-filled boreholes on the amplitudes of crosshole georadar data as inferred from experimental evidence. *Geophysics* **2004**, *69*, 1255–1260. [CrossRef]
25. Ebihara, S.; Hashimoto, Y. MoM Analysis of Dipole Antennas in Crosshole Borehole Radar and Field Experiments. *IEEE Trans. Geosci. Remote Sens.* **2007**, *45*, 2435–2450. [CrossRef]
26. Ebihara, S.; Masui, R.; Koyama, K.; Tsujikawa, Y.; Nishida, Y. Adding the outer pipe to a thin borehole radar probe in a thick water-filled borehole in rock to reduce direct coupling. *J. Phys. Conf. Ser.* **2024**, *2887*, 012049. [CrossRef]
27. Miwa, T.; Sato, M.; Niitsuma, H. Enhancement of reflected waves in single-hole polarimetric borehole radar measurement. *IEEE Trans. Antennas Propag.* **2000**, *48*, 1430–1437. [CrossRef]
28. Collin, R.E.; Chang, H. Field Theory of Guided Waves. *Phys. Today* **1961**, *14*, 50–51. [CrossRef]
29. Snyder, A.W.; Love, J.D. *Optical Waveguide Theory*; Springer: Boston, MA, USA, 1983.
30. Chew, W.C. *Waves and Fields in Inhomogeneous Media*; IEEE Press: New York, NY, USA, 1995.
31. Hollender, F.; Tillard, S. Modeling ground-penetrating radar wave propagation and reflection with the Jonscher parameterization. *Geophysics* **1998**, *63*, 1933–1942. [CrossRef]
32. Jonscher, A. The 'universal' dielectric response. *Nature* **1977**, *267*, 673–679. [CrossRef]
33. Chew, W. The singularities of a Fourier-type integral in a multicylindrical layer problem. *IEEE Trans. Antennas Propag.* **1983**, *31*, 653–655. [CrossRef]
34. Ebihara, S.; Inoue, Y. Analysis of Eccentered Dipole Antenna for Borehole Radar. *IEEE Trans. Geosci. Remote Sens.* **2009**, *47*, 1073–1088. [CrossRef]

35. Harrington, R.F. *Field Computation by Moment Methods*; Wiley: New York, NY, USA, 1993.
36. Ebihara, S.; Kimura, Y.; Shimomura, T.; Uchimura, R.; Choshi, H. Coaxial-Fed Circular Dipole Array Antenna with Ferrite Loading for Thin Directional Borehole Radar Sonde. *IEEE Trans. Geosci. Remote Sens.* **2015**, *53*, 1842–1854. [CrossRef]
37. Balanis, C.A. *Antenna Theory*, 3rd ed.; John Wiley & Sons, Inc.: Hoboken, NJ, USA, 2005.
38. Mason, I.M.; Bray, A.J.; Sindle, T.G.; Simmat, C.M.; Cloete, J.H. The Effect of Conduction on VHF Radar Images Shot in Water-Filled Boreholes. *IEEE Geosci. Remote Sens. Lett.* **2008**, *5*, 304–307. [CrossRef]
39. Fischer, B.E.; LaHaie, I.J. On the Definition of Return Loss. *IEEE Antennas Propag. Mag.* **2013**, *55*, 172–174. [CrossRef]
40. Bird, T.S. Definition and Misuse of Return Loss. *IEEE Antennas Propag. Mag.* **2009**, *51*, 166–1670. [CrossRef]
41. Balanis, C.A. *Balanis' Advanced Engineering Electromagnetics*; Wiley: New York, NY, USA, 1989.
42. Ebihara, S.; Kawai, H.; Wada, K. Estimating the 3D position and inclination of a planar interface with a directional borehole radar. *Near Surf. Geophys.* **2013**, *11*, 185–196. [CrossRef]

remote sensing

MDPI

Article

The Ground-Penetrating Radar Image Matching Method Based on Central Dense Structure Context Features

Jie Xu [1,2], Qifeng Lai [1,*], Dongyan Wei [1], Xinchun Ji [1,3], Ge Shen [1] and Hong Yuan [1]

[1] Aerospace Information Research Institute, Chinese Academy of Sciences, Beijing 100094, China; xujie22@mails.ucas.ac.cn (J.X.); weidy@aircas.ac.cn (D.W.); jixc@aircas.ac.cn (X.J.); shenge@aircas.ac.cn (G.S.); yuanhong@aircas.ac.cn (H.Y.)

[2] School of Electronic, Electrical and Communication Engineering, University of Chinese Academy of Sciences, Beijing 100049, China

[3] School of Electronics and Information, Northwestern Polytechnical University, Xi'an 710072, China

* Correspondence: laiqf@aircas.ac.cn

Abstract: Subsurface structural distribution can be detected using Ground-Penetrating Radar (GPR). The distribution can be considered as road fingerprints for vehicle positioning. Similar to the principle of visual image matching for localization, the position coordinates of the vehicle can be calculated by matching real-time GPR images with pre-constructed reference GPR images. However, GPR images, due to their low resolution, cannot extract well-defined geometric features such as corners and lines. Thus, traditional visual image processing algorithms perform inadequately when applied to GPR image matching. To address this issue, this paper innovatively proposes a GPR image matching and localization method based on a novel feature descriptor, termed as central dense structure context (CDSC) features. The algorithm utilizes the strip-like elements in GPR images to improve the accuracy of GPR image matching. First, a CDSC feature descriptor is designed. By applying threshold segmentation and extremum point extraction to the GPR image, stratified strip-like elements and pseudo-corner points are obtained. The pseudo-corner points are treated as the centers, and the surrounding strip-like elements are described in context to form the GPR feature descriptors. Then, based on the feature description method, feature descriptors for both the real-time image and the reference image are calculated separately. By searching for the nearest matching point pairs and removing erroneous pairs, GPR image matching and localization are achieved. The proposed algorithm was evaluated on datasets collected from urban roads and railway tracks, achieving localization errors of 0.06 m (RMSE) and 1.22 m (RMSE), respectively. Compared to the traditional Speeded Up Robust Features (SURF) visual image matching algorithm, localization errors were reduced by 86.6% and 95.7% in urban road and railway track scenarios, respectively.

Keywords: ground-penetrating radar; shape context; image matching; localization

1. Introduction

In the fields of intelligent transportation and automotive driving, particularly in complex urban environments, there is a pressing demand for high-precision and highly reliable navigation and localization systems. Currently, various vehicle localization methods are available, such as the Global Navigation Satellite System (GNSS), Inertial Navigation System (INS), Light Detection and Ranging (LiDAR)-based algorithms, and vision-based algorithms. However, relying solely on single localization algorithms cannot ensure continuous and reliable positioning.

For instance, GNSS fails to provide accurate and reliable location information in GNSS-denied environments like urban canyons and tunnels [1,2]. INS can achieve autonomous positioning through gyroscopes and accelerometers over short periods but suffers from position drift over extended durations [3]. Vision-based localization methods are unable to achieve positioning under challenging lighting conditions. LiDAR-based algorithms work by emitting multiple laser beams to measure distances and create environmental point cloud maps for positioning [4,5]. While LiDAR is unaffected by lighting conditions, it encounters difficulties in degraded environments, such as tunnels, where it also fails to achieve reliable localization [6].

Through the above analysis, the following conclusion can be drawn: single localization algorithms cannot provide continuous, reliable, and accurate positioning. Ground-Penetrating Radar (GPR), which senses subsurface environmental information, offers a natural advantage over traditional algorithms in adverse weather and low-visibility scenarios. When applied to localization, GPR serves as a valuable complement to existing localization algorithms.

GPR is a non-invasive technology that utilizes electromagnetic waves to detect underground targets. By analyzing variations in reflection time and intensity, the structure of underground targets and the properties of the medium can be determined [7]. The detection process is illustrated in Figure 1. As an efficient subsurface exploration technique, GPR can rapidly and accurately detect underground targets without the need for surface contact. This technology has found widespread application in areas such as road maintenance, geological surveys, and battlefield demining [8–10]. Recently, researchers have initiated studies on using GPR to assist with vehicle localization [11].

Figure 1. Vehicle-borne GPR detects underground information.

Research on GPR-based localization algorithms predominantly utilizes a matching mechanism, requiring the construction of a reference database in which GPR images are compared to determine vehicle location. Selecting stable and unique features is essential for the performance of such systems. Cornick et al. from MIT pioneered the study of vehicle localization using GPR image matching [11]. They used the normalized cross-correlation (NCC) as the similarity evaluation strategy for GPR images. By fusing a Localizing Ground-Penetrating Radar (LGPR) system with GPS/INS systems, they achieved decimeter-level or even centimeter-level localization accuracy. Subsequently, MIT's Ort et al. expanded upon the LGPR system to achieve the localization of autonomous vehicles under various conditions, such as clear weather, rain, snow, and nighttime, demonstrating the GPR system's strong environmental adaptability [12]. However, full image matching methods require substantial storage and computational resources, limiting their practical applicability. Consequently, researchers have shifted focus to extracting features from GPR images and using these features to perform similarity matching instead of utilizing the entire image. Baikovitz et al. employed a ResNet-18 autoencoder to capture high-dimensional deep learning features from GPR images [13]. Zhang et al. from the National University of Defense Technology (NUDT) used the real and imaginary parts of the Log-Gabor filter to convolve with GPR images and extract phase symmetry features, which demonstrated localization accuracy within 5 pixels across asphalt, cement, and brick roads. However, the operation speed is slow [14]. Ni et al. from the Aerospace Information Research Institute, Chinese Academy of Sciences, applied Faster-RCNN to extract hyperbolic features from GPR images, creating a feature map for matching-based localization. However, due to the sparse distribution and indistinct nature of underground hyperbolic features, there was insufficient localization continuity [15]. Zhang et al. from NUDT further developed TSVR-Net based on SVR-Net, using both B-Scan and D-Scan as network inputs to enhance localization stability. The results showed localization accuracy within 0.1 m. However, compared to the original data, the map storage and computational requirements for network training increased [16].

In summary, most existing GPR image matching algorithms are direct adaptations of visual image matching algorithms. Due to the low resolution of GPR images and the absence of well-defined geometric features, conventional image feature extraction methods are inadequate for effectively describing GPR image characteristics, leading to unsatisfactory matching results. GPR images commonly exhibit strip-like structures, which represent abstract representations of actual underground formations. These features are not only largely invariant over time but also distinct across different locations, making them ideal for matching. Shape context, a descriptor used to capture the outlines of shapes, is not constrained by the specific form of the target and can accurately reflect the distribution of sampled points along a contour. This method has been widely applied in object recognition and image matching [17–19].

This paper draws on the concepts of shape context and graphical statistics to propose a novel method for extracting Central Dense Structure Context (CDSC) features. By applying threshold segmentation and extremum point extraction to GPR images, stripe structures and pseudo-corner points are obtained. These pseudo-corner points serve as centers for contextual descriptions of the surrounding stripe structures, forming GPR feature descriptors. Based on these descriptors, the feature sets of the target image and the reference image are calculated. Furthermore, a GPR image matching method based on these features is designed for vehicle positioning. Experimental results demonstrate that the proposed algorithm performs well in both positioning accuracy and computational efficiency.

The remainder of this paper is organized as follows: Section 2 analyzes the characteristics of underground structural features. Section 3 presents the algorithm for CDSC feature extraction and the feature matching positioning method. Section 4 provides experimental analysis and accuracy assessment. Section 5 offers conclusions and a discussion.

2. Analysis of the Characteristics of Underground Structures

The construction materials of roads primarily consist of concrete, asphalt, gravel, and steel reinforcement, providing a rich variety of information about the subsurface. Additionally, cables and pipelines are often buried beneath roads, which can serve as distinctive underground features. Moreover, road defects such as voids and insufficient compaction are pervasive and can also be regarded as unique characteristics of the roadway. A single radar pulse from ground-penetrating radar (GPR) generates an A-scan, as shown in Figure 2a. A continuous sequence of A-scans along the driving direction forms a two-dimensional underground profile, referred to as a B-scan, as shown in Figure 2b. Combining multiple B-scans perpendicular to the driving direction results in a three-dimensional C-scan, as shown in Figure 2c. Compared to B-scans, C-scans not only provide underground information along the driving direction but also capture underground information perpendicular to the driving direction. Therefore, C-scans have an advantage when dealing with shifts in the vehicle's trajectory. The richness of information increases with the scanning dimensions, but this also introduces significant challenges for data processing. To balance the trade-off between the amount of information and processing time, when vehicles travel within the same lane, B-scans are sufficient to meet the localization requirements. B-scan data were selected as the source for matching and positioning in this paper.

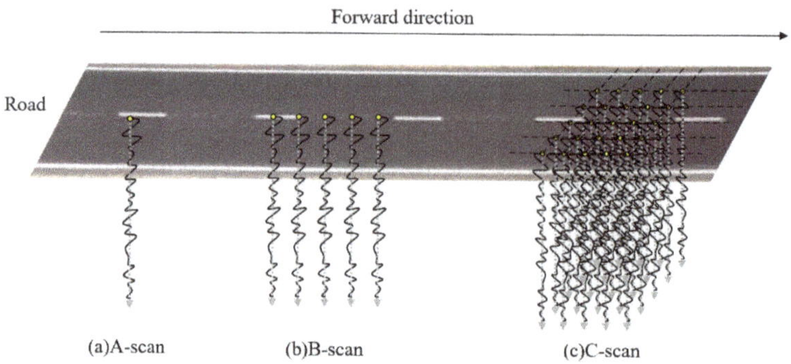

Figure 2. Three data types of ground-penetrating radar. (**a**) A-scan: the single-channel data signal acquired by GPR. (**b**) B-scan: data acquired by the GPR antenna through continuous scanning in the direction of movement. (**c**) C-scan: data composed of multiple B-scans.

The underground structural features of the road are represented in GPR B-scan images as alternating bright and dark stripe textures, as shown in the red box in Figure 3, These stripes are horizontally distributed across the image. Additionally, higher-intensity spots appear on each stripe, as indicated by the yellow box in Figure 3. The effective extraction of these features for location is the primary focus of this paper.

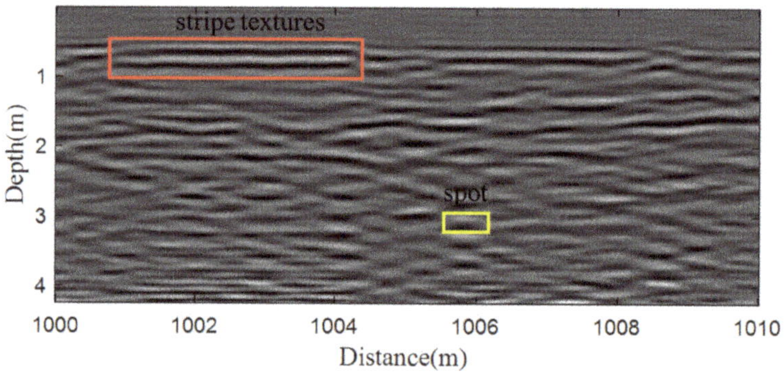

Figure 3. Stripe and spot features in GPR images.

These spots are typically circular or elliptical in shape, with pixel intensities that are similar within the spot but distinct from the surrounding neighborhood. The spots generally correspond one-to-one with underground structures and exhibit a degree of stability over time and space. To verify the stability of GPR image features under varying environmental conditions, six datasets were collected along the same road on three different days: 25 June 2023, 14 July 2023, and 17 July 2023. The dates 25 June 2023 and 17 July 2023 were sunny days, while 14 July 2023 was rainy. As shown in Figure 4, two datasets were collected each time along the same trajectory, labeled Track 1 and Track 2. This is evident in the yellow boxes in Figure 4, where similar underground features are observed across all datasets. A comparison of datasets a, b, e, and f reveals that even after a 20-day interval, the underground structural images remain largely unchanged and stable, thus meeting the requirements for positioning. Similar underground features can be observed in the two GPR images from datasets c and d, although the similarity has decreased. This is because rainfall altered the dielectric properties of the soil, which weakened the signal's penetration ability. However, the underground features are still measurable.

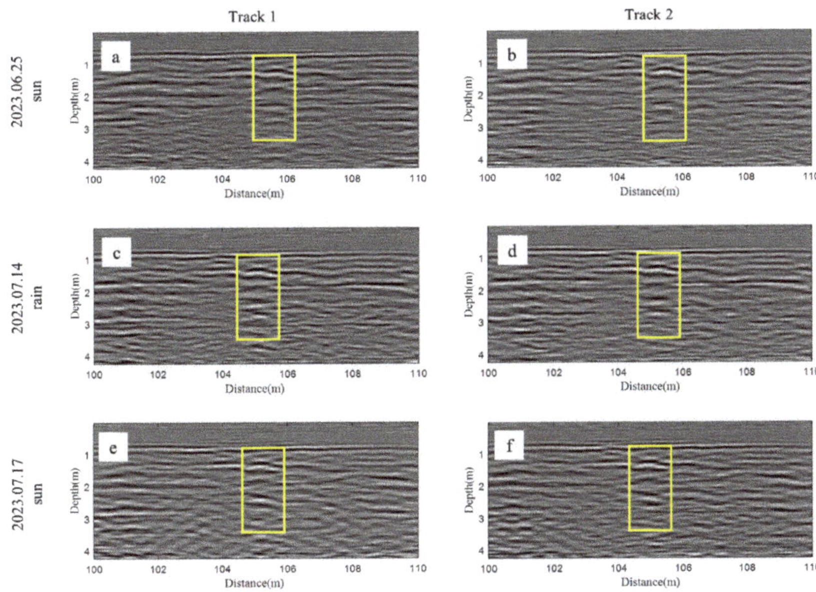

Figure 4. Data were collected from the same road segment at different times and under different weather conditions using GPR. Subfigure (**a**,**b**,**e**,**f**) are collected in sunny days, and subfigure (**c**,**d**) are collected in rainy days.

In summary, using GPR images for positioning offers the following advantages:

(1) The underground features are rich and highly correlated with spatial locations, providing coverage across most conventional roads, which can serve as natural geographic information for vehicle positioning.

(2) These underground features are long lasting and stable, with minimal influence from external factors such as weather, resulting in strong adaptability to various positioning environments.

3. Matching Method Based on Center Dense Structure Context Features

This section first introduces the basic process of the method proposed in this paper, which first needs to build a feature reference database in advance, and then match the collected GPR images with the pre-built feature reference database and output the current geographical location, as shown in Figure 5.

GPR detects underground structures by transmitting and receiving electromagnetic waves. However, during vehicle movement, maintaining a constant speed is challenging, which inevitably leads to distance scaling issues in GPR images. To address this, spatial resampling and noise reduction filtering must be applied as part of the preprocessing operations. In the proposed algorithm, when constructing a reference database, the preprocessed images are first transformed from image coordinates to spatial position coordinates before extracting CDSC features. The extracted image features are then stored in a specific format to form the feature reference database.

The feature reference database needs to contain location information *pos*, direction *dir*, and descriptor *dsc*:

$$Data_{base} = \{pos, dir, dsc\} \tag{1}$$

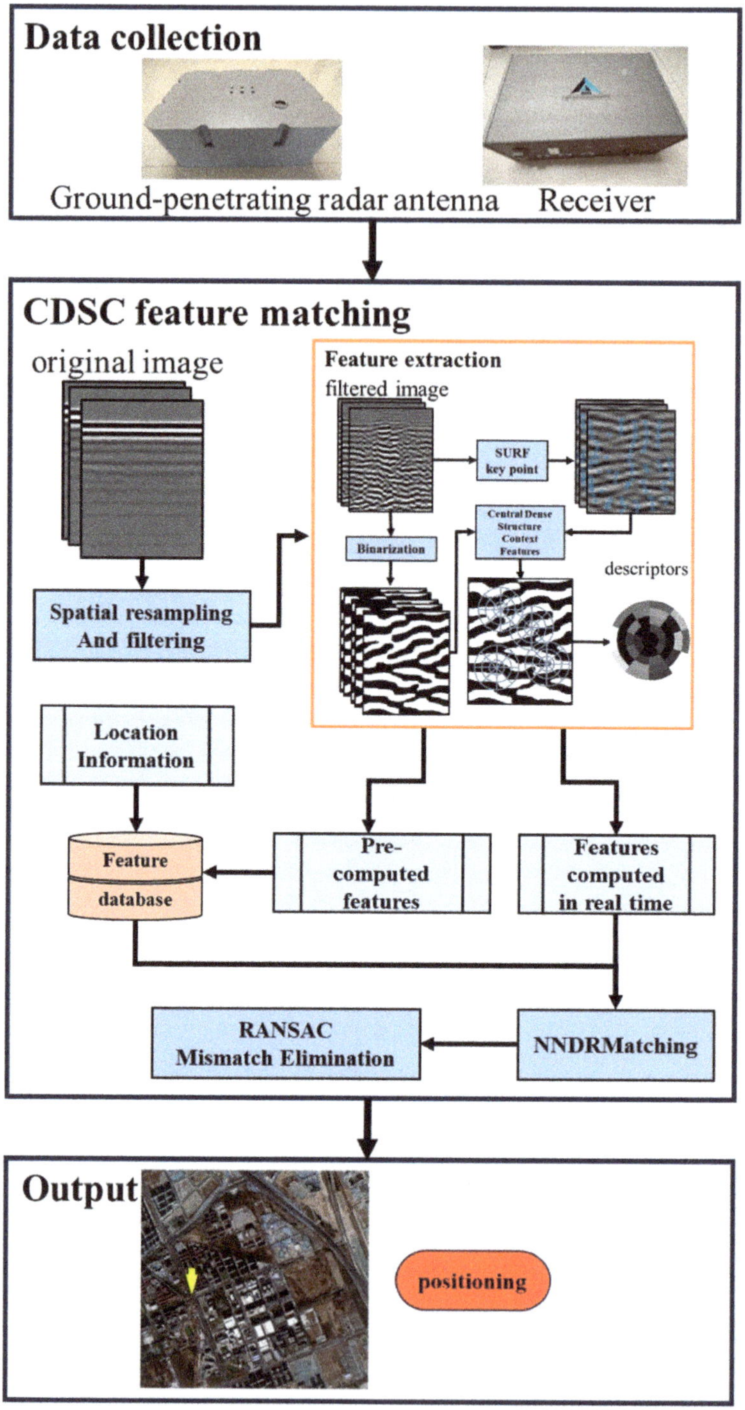

Figure 5. Algorithm flow.

During the positioning stage, the vehicle-mounted GPR captures the GPR images to be matched. These images undergo the same processing steps, including spatial resampling, noise reduction filtering, and CDSC feature extraction, to obtain the feature sequence for matching.

$$Data_{current} = \{dir, dsc\} \tag{2}$$

The similarity measurement method and mismatch elimination technique are used to determine the optimal matching position of the feature sequence within the feature reference database, ultimately yielding the positioning results.

3.1. SURF-Based GPR Image Corner Detection

GPR images are affected by factors such as jitter and noise. However, since spots represent strong underground reflective targets, they can be reliably detected by GPR, making them ideal image features. Feature point extraction has been extensively studied in visual navigation, with classic methods including ORB [20], SIFT [21], and SURF [22]. ORB combines the efficiency of FAST and the reliability of the Harris algorithm to detect feature points utilizing the intensity differences between the center and surrounding pixels. However, its performance in detecting spot features in GPR images is suboptimal. SIFT detects local extrema in a DoG pyramid and uses the Hessian matrix of local intensity values as a filter condition for extracting feature points. While SIFT performs well in detecting spots, the construction of the DoG pyramid and the calculation of the Hessian matrix are computationally intensive, resulting in longer detection times. SURF, an improvement on SIFT, is a fast and robust feature point matching algorithm that excels in detecting GPR spot features compared to the other two methods. The algorithm leverages 2D Haar wavelet responses, integral images, and scale-space techniques, providing strong robustness to image rotation, translation, scaling, and noise, while also offering improved speed over SIFT.

The performance of the aforementioned feature point detection algorithms on GPR images is illustrated in Figure 6, with results from SIFT, ORB, and SURF shown from left to right. Among the three algorithms, ORB performs the worst, detecting only a few spots and generating a significant number of false feature points. SIFT and SURF detect a larger number of spot features. However, SURF identifies more feature points with a more uniform distribution, resulting in a more complete description of the image. Therefore, SURF was selected as the feature point detection algorithm in this study.

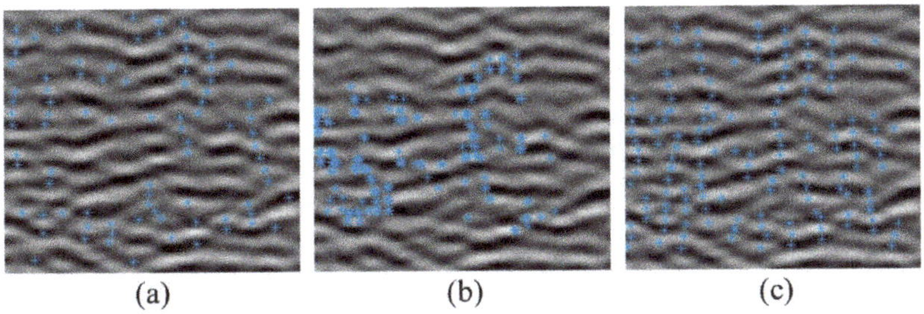

Figure 6. Comparison of feature point extraction results. (**a**) The SITF algorithm was used to extract feature points in GPR images. (**b**) The ORB algorithm was used to extract feature points in GPR images. (**c**) The SURF algorithm was used to extract feature points in GPR images.

3.2. Central Dense Structure Context Feature Extraction

The inconsistencies between roadbed layers and material differences result in alternating bright and dark stripes in GPR images. These structural variations within the road naturally provide descriptive information for pseudo-corner points. Therefore, this paper proposes the CDSC algorithm, based on the shape context algorithm, specifically targeting the stripe structures in GPR images.

3.2.1. Shape Context Algorithm

Shape context can describe arbitrary shapes and is widely used in contour matching, as shown in Figure 7. The contour line is uniformly sampled to obtain a set of contour points. For any given point in the contour point set, a polar coordinate system is established centered on that point. The number of contour points falling into small regions around that point is then counted. This process is repeated until each point in the contour point set has been accounted for, resulting in the SC feature.

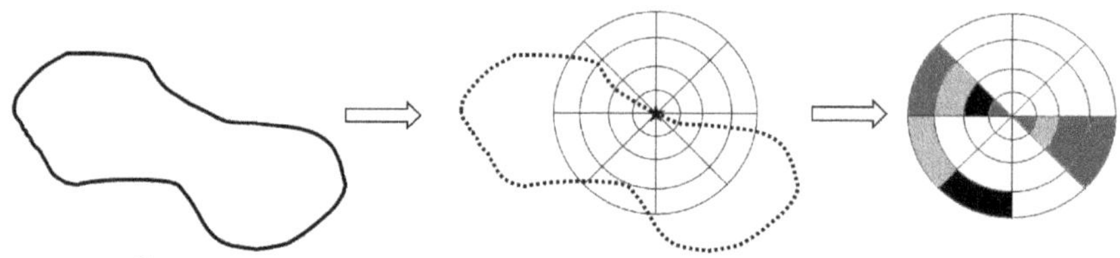

Figure 7. Shape context algorithm feature extraction.

The main steps are as follows: first, the image contour is uniformly sampled. Suppose there are N sample points. A point, $p_i(x_i, y_i)$, is selected, and the relative spatial position, including direction θ and distance ρ, between the remaining $N-1$ sample points $p_j(x_j, y_j)$ and $p_i(x_i, y_i)$ is calculated.

$$\begin{aligned} \rho &= \sqrt{(x_i - x_j)^2 + (y_i - y_j)^2} \\ \theta &= \arctan(y_i - y_j / x_i - x_j) \end{aligned} \quad (3)$$

A polar coordinate system is constructed around each sampling point, and the system is divided into $M \times N$ grids based on distance $\Delta\rho$ and directional $\Delta\theta$ intervals.

$$\begin{aligned} M &= R / \Delta d \\ N &= 360^\circ / \Delta\theta \end{aligned} \quad (4)$$

In this formula, R represents the local neighborhood range of the sampling point. The number of sampling points falling into each grid is counted to obtain the shape context feature histogram. The statistical rules for constructing the histogram are as follows:

$$h_i(l) = \#\{p_j \neq p_i : (p_j - p_i) \in bin(l)\} \quad (5)$$

In the equation, $h_i(l)$ represents the l-th component of the histogram for the i-th contour point.

The shape context algorithm requires uniform sampling of the contours in the original image. However, the stripe structures in GPR images are neither independent nor closed shapes. Therefore, building on the concept of shape context, this paper proposes the CDSC feature extraction algorithm to extract features from these stripe structures.

3.2.2. Center Dense Structure Context Feature Extraction

The CDSC is an improved version of the shape context algorithm, specifically designed for feature extraction from threshold-segmented GPR images centered around pseudo-corner points. The CDSC extraction process is illustrated in Figure 8. Compared to traditional shape context, this approach addresses the challenge of describing non-independent and non-closed images. Additionally, the dense stripe structures provide more neighborhood information for the descriptor.

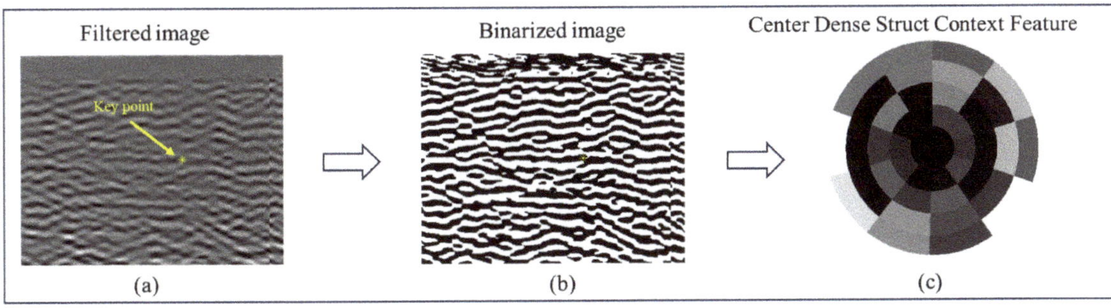

Figure 8. The steps of CDSC feature extraction. Descriptors are obtained by leveraging the stripes in the binarized image surrounding the pseudo-corner points. Subgraph (**a**) is the filtered image, subgraph (**b**) is the Binarized image, and subgraph (**c**) is the Center Dense Struct Context Feature.

The GPR image is segmented using the OTSU thresholding algorithm [23], which separates the stripe structures from the background, emphasizing large-scale structural features while ignoring local gradient details. The specific steps are as follows:

Assuming the GPR image $img(x,y)$ has L grayscale levels, with n pixels at the i-th grayscale level, the total number of pixels N in the image $img(x,y)$ is given by the following:

$$N = \sum_{i=1}^{L} n_i \tag{6}$$

The probability P_i of a pixel having an i-th grayscale level and the mean grayscale level μ of the entire image are given by the following:

$$P_i = \frac{n_i}{N}, \mu = \sum_{i=1}^{L} iP_i \tag{7}$$

By using a grayscale value k as the threshold to segment the image into A_0 and A_1, where A_0 consists of pixels with grayscale levels in the range $[1,k]$ and A_1 consists of pixels with grayscale levels in the range $[k+1, L-1)$, the probabilities P_{A0} and P_{A1} of A_0 and A_1 occurring, respectively, are given by the following:

$$P_{A0} = \sum_{i=0}^{k} P_i, P_{A1} = \sum_{i=k+1}^{L-1} P_i \tag{8}$$

The mean grayscale values E_{A0} and E_{A1} for A_0 and A_1, respectively, are given by the following:

$$E_{A0} = \sum_{i=0}^{k} iP_i|P_{A0}, E_{A1} = \sum_{i=k+1}^{L-1} iP_i|P_{A1} \tag{9}$$

The mean grayscale value E of the image $img(x,y)$ is given by the following:

$$E = \sum_{i=0}^{L-1} iP_i \tag{10}$$

The between-class variance for A_0 and A_1 is given by the following:

$$\theta^2(k) = P_{A0}(E - E_{A0})^2 + P_{A1}(E - E_{A1})^2 \tag{11}$$

The optimal threshold is the value of k that maximizes $\theta^2(k)$.

When the grayscale value of the original image $img(x, y)$ exceeds a predetermined threshold k, the pixel value is set to 1; otherwise, it is set to 0. This binarization process effectively reduces the complexity of the image and enhances the efficiency of subsequent operations.

$$img_{bw}(x,y) = \begin{cases} 1, img(x,y) \geq k \\ 0, img(x,y) < k \end{cases} \tag{12}$$

The GPR image is divided into two classes: the portion greater than the threshold, representing the stripe structures (non-zero points), is denoted as P_{STRUCT}, while the portion less than the threshold is considered the image background. The binarized GPR image emphasizes the existing stripe structures while ignoring minor local variations. This is because large-scale structural information is more stable compared to local gradient information.

Using spots as the center, the stripe structures are employed to describe the spots. Traditional shape context uses an "$N - 1$" feature construction method, where N contour points are used to describe a single point with $N - 1$ of these points. In contrast, this paper adopts a "1 + N" feature construction method, where a single center point is described by N stripe structure points. The specific steps are as follows:

Let $p_{s,i}$ be the i-th non-zero point in the image img_{bw}, $p_{s,i}(x_{s,i}, y_{s,i}) \in P_{STRUCT}$, with coordinates $(x_{s,j}, y_{s,j})$. In Section 3.1, the set of pseudo-corner points extracted from the GPR image is denoted as P_{SURF},$p_{c,j}(x_{c,j}, y_{c,j}) \in P_{SURF}$, with coordinates $(x_{c,j}, y_{c,j})$, serving as the center points for the feature descriptors.

Replace the contour points in Equation (5) with the SURF pseudo-corner points P_{SURF} and the stripe structure points P_{STRUCT}, respectively.

$$\begin{aligned} \rho &= \sqrt{(x_{c,i} - x_{s,j})^2 + (y_{c,i} - y_{s,j})^2} \\ \theta &= \arctan(y_{c,i} - y_{s,j} / x_{c,i} - x_{s,j}) \end{aligned} \tag{13}$$

Next, count the number of P_{STRUCT} points that fall into each grid within the neighborhood range of P_{SURF}, and obtain the feature vector $D(l)$.

$$D(l) = \#\{p_{s,j} \in bin(l)\} \tag{14}$$

3.3. Localization Algorithm Based on CDSC Features

As shown in Figure 9, nine pseudo-corner points are selected from each image to extract CDSC features. The extracted features from both images are observed to be very similar.

Therefore, CDSC descriptors are used as the feature representation for the images, and the similarity of CDSC features is used to describe the similarity between the two images. During the matching and positioning process, the features are extracted from the image to be matched. This region's features are then compared with a larger range of image features in the feature reference database. This approach helps minimize both false matches and missed matches. The specific process is illustrated in Figure 10.

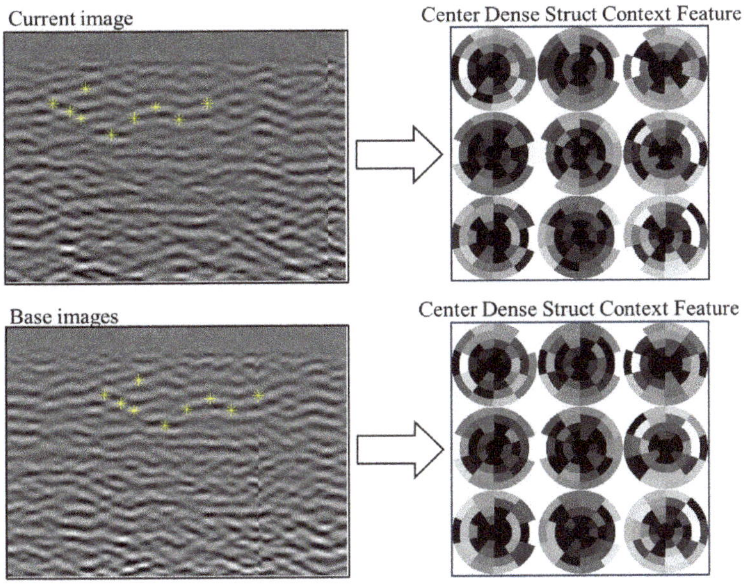

Figure 9. CDSC features were extracted from GPR images collected twice at the same location and compared. The nine feature points on each image on the left correspond to the nine features on the right.

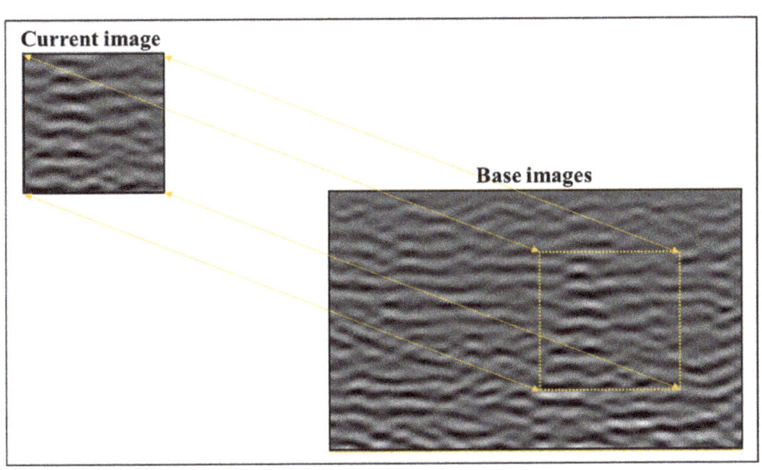

Figure 10. Image matching schematic diagram.

(1) CDSC Feature Matching

Nearest Neighbor Distance Ratio (NNDR) is a metric used to evaluate whether two feature descriptors are similar. When a match is correct, the distance between the two feature descriptors will be closer compared to the distance between mismatched descriptors. Therefore, the ratio of the distance to the nearest neighbor to the distance to the second nearest neighbor is calculated. If this ratio is below a certain threshold, the match is considered correct. The specific implementation of this method is as follows:

Let $D_{base} = [d_{b,1}, d_{b,2}, \ldots, d_{b,q}]$ be the feature vector from the reference database and $D_{current} = [d_{c,1}, d_{c,2}, \ldots, d_{c,q}]$ be the feature vector extracted from the image to be matched. The distance between the two descriptors is measured using the Euclidean distance:

$$dist(D_{base}, D_{current}) = \sqrt{\sum_{j=1}^{q} (d_{b,j}, d_{c,j})^2} \tag{15}$$

The ratio is the following:

$$d_{NNDR} = \frac{dist_1}{dist_2} \tag{16}$$

If d_{NNDR} is less than the specified threshold, it indicates a successful match between the two feature vectors.

(2) Elimination of False Matches

Each matching process yields a set of data points, but not all points in this set are correctly matched. Therefore, it is necessary to remove incorrect matches from the data set to improve matching accuracy. The Random Sample Consensus (RANSAC) algorithm can identify correct data from a set containing outliers by iteratively refining the dataset to exclude anomalous data [24]. Let the point set obtained from a single matching be denoted as follows:

$$\begin{aligned} P_{match} = \{ & [p_{match,b1}, p_{match,c1}], [p_{match,b2}, p_{match,c2}], \\ & \ldots, [p_{match,bn}, p_{match,cn}] \} \end{aligned} \tag{17}$$

In this context, $p_{match,bi}(px_{match,bi}, py_{match,bi})$ represents the points in the feature database with coordinates $px_{match,bi}$ and $py_{match,bi}$, while $p_{match,ci}$ denotes the points in the image to be matched. In the RANSAC algorithm, correct data points are classified as inliers (P_{in}), and mismatched points are classified as outliers (P_{out}). The first step is to find the optimal homography matrix H:

$$\begin{gathered} s\begin{bmatrix} px_{match,bi} \\ py_{match,bi} \\ 1 \end{bmatrix} = H\begin{bmatrix} px_{match,ci} \\ py_{match,ci} \\ 1 \end{bmatrix} \\ H = \begin{bmatrix} h_{11} & h_{12} & h_{13} \\ h_{21} & h_{22} & h_{23} \\ h_{31} & h_{32} & h_{33} \end{bmatrix} \end{gathered} \tag{18}$$

Here, s is the scale parameter. Typically, $h_{33} = 1$ is set to 1 to normalize the matrix H. Thus, the matrix has eight unknowns, and solving it requires four pairs of matching points. The algorithm selects four pairs of sample points from the data set, ensuring that the points are not collinear, computes the homography matrix, and evaluates the cost function J:

$$\begin{aligned} J = \sum_{i=1}^{n} & \left(px_{match,bi} - \frac{h_{11}px_{match,ci} + h_{12}py_{match,ci} + h_{13}}{h_{31}px_{match,ci} + h_{32}py_{match,ci} + h_{33}} \right)^2 \\ & + \left(py_{match,bi} - \frac{h_{21}px_{match,ci} + h_{22}py_{match,ci} + h_{23}}{h_{31}px_{match,ci} + h_{31}py_{match,ci} + h_{33}} \right)^2 \end{aligned} \tag{19}$$

When the cost function is minimized, the final output is the set of matched points:

$$\begin{aligned} P_{RANSAC} = \{ & [p_{r,b1}, p_{r,c1}], [p_{r,b2}, p_{r,c2}], \\ & \ldots, [p_{r,bn}, p_{r,cn}] \} \end{aligned} \tag{20}$$

4. Experiments and Results

4.1. Data and Experimental Scenarios

To evaluate the performance of the proposed algorithm in different scenarios, localization verification experiments were designed for two typical vehicle positioning environments: urban roads and railway tracks. During the experiments, the vehicle repeatedly collected two sets of data along a predetermined trajectory, which were used as the image set to be matched and the reference image set. An odometer provided mileage marker information for the images. After manually calibrating the mileage markers corresponding to the GPR data, the mileage marker error was controlled to be within 0.1 m. The key parameters of the ground-penetrating radar (GPR) used in this study are shown in Table 1. The detection depth of the GPR is closely related to the antenna's center frequency. According to highway and railway subgrade design specifications [25,26], the depth of artificial structures beneath the pavement typically does not exceed 2 m, providing ample feature information for matching and localization. According to the working principle of GPR, 300 Hz GPR achieves a detection depth of 3~4 m [27]. The chosen GPR has a scanning rate of 500 Hz, and with the vehicle traveling at 120 km/h, the scanning interval is better than 0.08 m. This resolution meets the requirements for sub-meter level localization, ensuring effective data collection and maintaining continuity and resolution even at high speeds.

Table 1. GPR parameters.

Parameters	Parameter Values
Frequency	300 MHz
Number of sample points	512
Scanning rate	500 Hz
Detection depth	3~4 m

The original output images from the GPR record both effective echoes and various interference signals, making them unsuitable for direct image matching and localization. Therefore, preprocessing techniques such as Singular Value Decomposition (SVD) and filtering are applied to the GPR images [7] to highlight the structural features within. A comparison of the GPR images before and after preprocessing is shown in Figure 11. Panel (a) displays the original output image, while panel (b) shows the image after filtering. The results indicate that the subsurface structural distribution in the filtered image is more pronounced, which aids in subsequent feature extraction. Using interpolation functions to discretely sample the continuously distributed original GPR data, images with uniform sampling intervals were obtained.

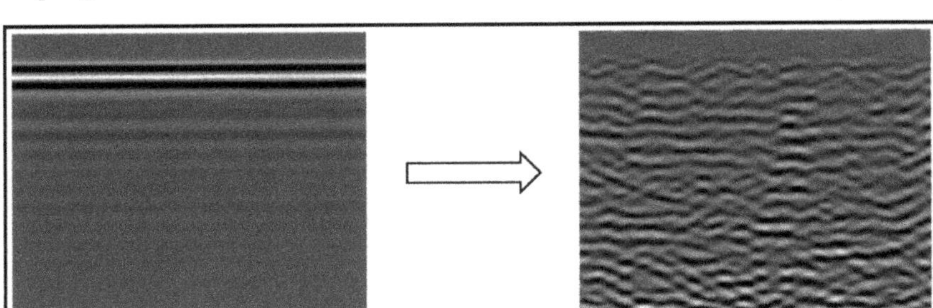

(a)Original image (b)Filtered image

Figure 11. Comparison of GPR images before and after preprocessing; (**a**) displays the original output image, (**b**) shows the filtered image.

(1) Experimental Site 1: Urban Road

The first experimental site is located around the Beijing New Technology Base of the Chinese Academy of Sciences, as shown in Figure 12a. The total length of the urban test trajectory is approximately 7 km. The equipment setup for the road experiment is illustrated in Figure 12b. The GPR antenna is mounted at the rear of the vehicle to ensure that the scanning line is aligned with the vehicle's travel trajectory.

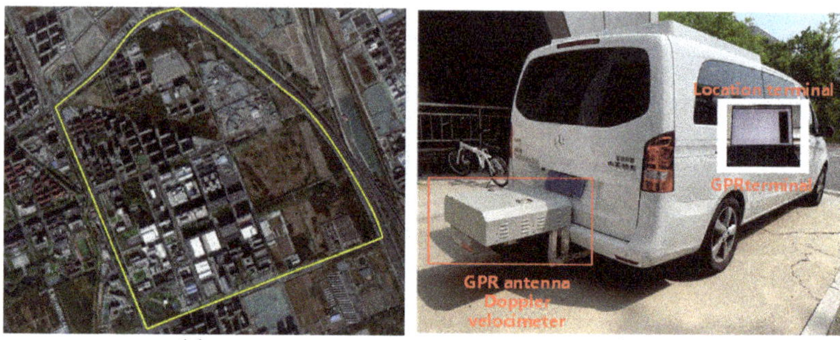

(a) (b)

Figure 12. Urban test trajectory and equipment setup. (**a**) Urban test trajectory, the yellow line represents the test trajectory. (**b**) The equipment setup for the road experiment.

(2) Experimental Site 2: Railway Test Track

The second experimental site is the section of the Shenhua Railway, as shown in Figure 13a. The train test trajectory primarily consists of railway subgrade, including track ballast, subgrade fill, and foundation soil, with a total length of approximately 6.7 km. The equipment setup for the railway experiment is depicted in Figure 13b. The ground-penetrating radar (GPR) antenna is mounted on the side of the train to precisely measure the railway subgrade while minimizing the impact of the rails and sleepers on the imaging.

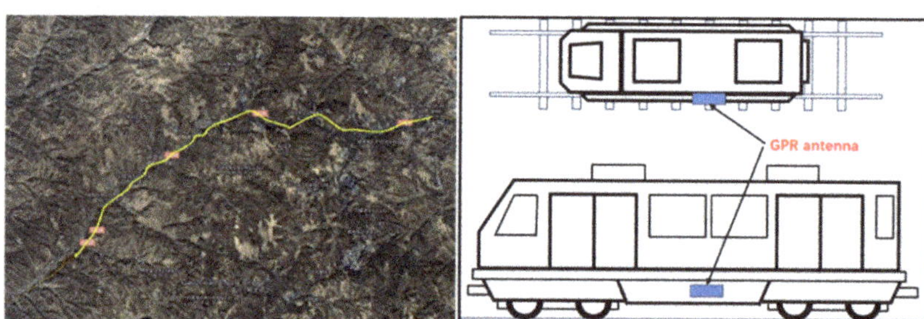

(a) (b)

Figure 13. Train test trajectory and equipment setup. (**a**) Train test trajectory total length is approximately 6.7 km, the yellow line represents the test trajectory. (**b**) The equipment setup for the railway experiment.

4.2. Experimental Evaluation Metrics

In this paper, the Root Mean Square Error (RMSE) is employed as the evaluation metric for assessing both the pseudo-corner point matching accuracy and the GPR image matching localization accuracy. RMSE is defined as follows:

$$
\begin{aligned}
RMSE_{point} &= \sqrt{\frac{1}{N}\sum_{i=1}^{N}\left[(x_{i,img}-x'_{i,img})^2+(y_{i,img}-y'_{i,img})^2\right]} \\
RMSE_{urban} &= \sqrt{\frac{1}{N}\sum_{i=1}^{N}\left[(x_{i,urban}-x'_{i,urban})^2+(y_{i,urban}-y'_{i,urban})^2\right]} \\
RMSE_{raliway} &= \sqrt{\frac{1}{N}\sum_{i=1}^{N}(mile_i-mile'_i)^2}
\end{aligned}
\quad (21)
$$

where $(x_{i,img}, y_{i,img})$ and $(x'_{i,img}, y'_{i,img})$ represent the coordinates of points in the reference image and the points in the image to be matched, respectively, and N denotes the number of matched point pairs. $(x_{i,urban}, y_{i,urban})$ and $(x'_{i,urban}, y'_{i,urban})$ represent the coordinates of the true position and calculated position, respectively. $mile_i$ and $mile'_i$ represent the mileage of the true position and calculated position, respectively.

In this study, recall and precision are utilized to evaluate the performance of the image registration algorithm. Precision describes the usefulness of the search results and serves as a measure of accuracy. Recall, on the other hand, describes the completeness of the matching results, acting as a measure of completeness. The F-measure is a metric that incorporates both recall and precision. The specific definition is as follows:

$$
\begin{cases}
R_{precision} = \frac{N_{correct}}{N_{correct}+N_{false}} \\
R_{recall} = \frac{N_{correct}}{N_{correspondences}}
\end{cases}
\quad (22)
$$

where $N_{correct}$ represents the number of actual correct matching point pairs; N_{false} denotes the number of incorrect matching point pairs; and $N_{correspondences}$ is the theoretical number of correct matching point pairs. The higher the values of $R_{precision}$ and R_{recall}, the stronger the algorithm's capability to correctly match points.

4.3. Experiment and Parameter Design

To comprehensively evaluate the performance of the proposed algorithm, this study is structured into three experimental analyses. First, in Section 4.3.1, the impact of key parameters of the algorithm on localization performance is assessed. Second, Section 4.3.2 compares the proposed CDSC feature descriptor extraction algorithm with existing typical visual image feature extraction algorithms to validate the designed descriptor's registration performance. Finally, in Section 4.3.3, the proposed matching and localization algorithm is compared with conventional visual processing algorithms in terms of accuracy and computational efficiency, providing a comprehensive evaluation of the algorithm's performance in vehicle localization tasks. This demonstrates its superiority and potential for practical application.

4.3.1. Feature Descriptor Parameter Analysis

The proposed descriptor in this paper aims to quantify the strip-like elements in GPR images into feature vectors. Consequently, the parameter settings for generating the descriptor directly impact the results of matching and localization. Therefore, experiments were conducted on both the railway and urban test trajectory to evaluate the effects of different parameters on the matching performance.

As shown in Equation (4), the parameters that influence the performance of the descriptor primarily include the feature statistics range and the number of grid divisions (both distance and angular). From the principles of image matching, it is known that the windows size also affects localization performance—larger windows contain more information that can be utilized for localization. Hence, this paper selected three key parameters: matching window size, feature statistics range, and the number of grid divisions. A series of experiments was performed using the controlled variable method, where the parameter values were gradually adjusted to explore their effects on descriptor performance.

It was necessary to select an appropriate matching window and search range. The size of the matching window is related to the distribution of underground features and includes two dimensions: depth and length. Different depths and lengths affect localization performance, as shown in Table 2. Theoretically, the larger the matching window involved in localization, the higher the accuracy. This is because a larger window extracts more pseudo-corner points for matching and localization. However, in practice, once the image depth exceeds a certain range, localization performance declines. This can be attributed to two factors: firstly, the shallow subsurface structure features are richer and more suitable for matching and localization. Deeper features exhibit less variation, contributing less to localization performance. Secondly, the GPR system used in this study operates at 300 MHz and primarily detects depths of 0–3 m. Beyond this depth, the imaging resolution decreases, making image matching prone to loss of correlation. For the urban road environment, a detection depth of 2 m and a length of 10 m were chosen, while for the railway environment, a depth of 2.5 m and a length of 10 m were selected, to strike a balance between localization accuracy and operational efficiency.

Table 2. The impact of the size of the matching window.

Nums	Matching Window Depth (m)	Matching Window Length (m)	Urban RMSE (m)	Train RMSE (m)
1	1.5	6	1.76	0.11
2	1.5	8	1.51	0.07
3	1.5	10	1.35	0.06
4	2.0	6	1.79	0.09
5	2.0	8	1.52	0.06
6	2.0	10	1.22	0.06
7	2.5	6	1.86	0.09
8	2.5	8	1.58	0.07
9	2.5	10	1.33	0.06

First, the impact of the feature statistics range on localization performance was analyzed. In GPR images, the resolution in the vertical direction is higher than that in the horizontal direction, while the stratified stripe structures in GPR images are distributed horizontally. To ensure the extraction of more information along the horizontal direction, the feature statistics range was set to an elliptical shape, with the minor axis ranging from 0.2 to 0.4 m and the major axis ranging from 0.5 to 0.95 m. The value of the major axis increasing from 0.5 to 0.95 is maintained, with the minor axis increasing proportionally, to access the localization performance of the algorithm.

The experimental results are shown in Table 3. As the feature statistics range increases, the localization error initially decreases and then increases. Ideally, a larger radius should provide more information for describing pseudo-corner points, leading to a corresponding reduction in matching localization error. However, the mutual influence between pseudo-corner points in the GPR image must also be considered. When the feature statistics range exceeds a certain threshold, the statistical regions of adjacent pseudo-corner points start to overlap, leading to redundancy in the feature descriptors, which can, in turn, increase localization errors.

Table 3. The impact of the feature statistical range.

Nums	Feature Statistical Range (m)	Urban RMSE (m)	Train RMSE (m)
1	0.50	3.55	0.43
2	0.65	2.70	0.21
3	0.80	2.64	0.27
4	0.95	3.04	0.28

The test results indicate that, in both urban and railway environments, a major axis range of around 0.8 m strikes an optimal balance between the quantity of neighborhood information and the distance between adjacent pseudo-corner points, ensuring the best localization accuracy.

Next, the influence of the density of grid divisions on localization performance is analyzed, as shown in Table 4. Increasing the number of angle or distance intervals effectively reduces localization errors. For the same feature statistics range, sparser grid divisions result in more pixels being contained within each grid, which weakens the grid's ability to describe the shape. This is because, in larger grids, the variations in the stripe features become more smoothed, leading to more indistinct feature representations. Based on overall performance, dividing the distance into five grids and the angles into eight grids yields the best localization results.

Table 4. The impact of the density of grid divisions.

Nums	Distance Intervals	Angle Intervals	Urban RMSE (m)	Train RMSE (m)
1	4	4	5.44	1.17
2	4	6	3.37	0.46
3	4	8	2.64	0.27
4	5	4	4.38	0.55
5	5	6	3.15	0.38
6	5	8	2.32	0.18

In summary, moderately increasing the matching window size can improve localization accuracy. However, in urban environments, increasing depth may have certain negative effects. An appropriate feature statistics range can effectively balance the amount of information and the relationship between neighboring pseudo-corner points. Reducing angle and distance intervals moderately can improve localization accuracy. The parameters for the algorithm are set as follows:

Railway Test Environment: The feature statistics range is 0.8 m, with angle divisions set to 8 and distance divisions set to 5. The matching search radius is 20 m, and the matching window size is 2.5 m × 10 m.

Urban Test Environment: The feature statistics range is 0.8 m, with angle divisions set to 10 and distance divisions set to 5. The matching search radius is 20 m, and the matching window size is 2 m × 10 m.

4.3.2. Feature Descriptor Matching Analysis

A comparative analysis of the matching performance between the proposed method and three feature descriptors—ORB, SIFT, and SURF—was conducted. The statistical results for the urban test trajectory are shown in Table 5. The CDSC descriptor achieves the highest recall rate and, under the condition of high recall, also maintains a high precision rate. Among the four algorithms, the F-measures from lowest to highest are ORB, SURF, SIFT, and CDSC, with CDSC showing the best overall performance.

Table 5. The statistical results for the urban test trajectory.

Method	Recall Rate	Precision Rate	F-Measure
SIFT	5.3%	30.1%	0.09013
ORB	0.6%	37.7%	0.011812
SURF	2.6%	11.9%	0.042676
CDSC	6.4%	33.7%	0.107571

The statistical results for the railway test trajectory are shown in Table 6. Compared to the urban test results, all four descriptors showed improvements in both recall and precision rates. Notably, the CDSC algorithm exhibited the highest recall and precision rates. These findings demonstrate that the proposed algorithm outperforms conventional descriptor in both urban road and railway test environments, providing more accurate feature descriptions and higher matching precision.

Table 6. The statistical results for the railway test trajectory.

Method	Recall Rate	Precision Rate	F-Measure
SIFT	11.9%	36.9%	0.179963
ORB	0.7%	47.5%	0.013797
SURF	5.1%	22.0%	0.082804
CDSC	21.3%	60.9%	0.315613

4.3.3. Localization Analysis

The proposed method was compared with SIFT, ORB, SURF, and NCC image matching methods in terms of localization accuracy and computational efficiency. The localization scatter plots are shown in Figure 14, where red, green, blue, black, and yellow represent the localization results of SIFT, ORB, SURF, NCC, and CDSC, respectively. The ORB, SIFT, and SURF descriptors showed a large number of mismatches, with maximum errors exceeding 20 m. In contrast, the localization results of CDSC and NCC were largely consistent, with localization errors of less than 1 m. The NCC algorithm, which uses a full matching approach by leveraging all intensity information in the GPR images, ensured high localization accuracy. However, the proposed CDSC algorithm, which derives its features from stable structures in GPR images, also guaranteed high localization accuracy.

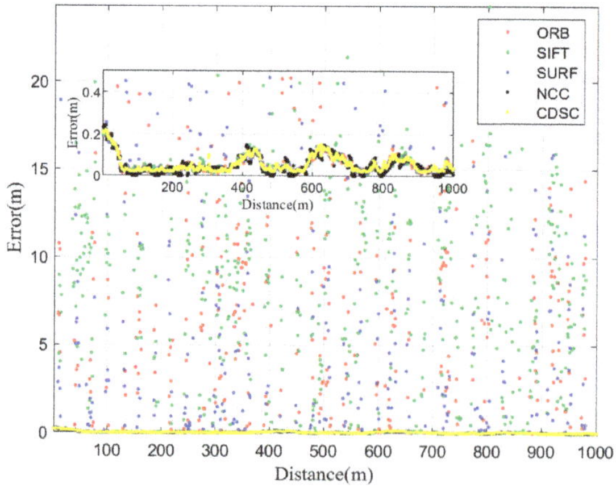

Figure 14. Railway test trajectory positioning error.

A section of GPR imaging with strong interference is shown in Figure 15a. Common pseudo-corner point descriptors like SIFT and ORB failed to identify correct matching point pairs on highly disturbed images, while the SURF algorithm managed to retrieve a small number of matching pairs. However, the proposed method identified a significant number of correct matching point pairs, showing markedly better performance than SIFT, ORB, and SURF descriptors when faced with strongly interfered GPR images, demonstrating superior feature description capabilities. As shown in Figure 15b, when the interference in the GPR image was weak, the number of correct matching pseudo-corner points identified by the CDSC algorithm was far higher than that of the other descriptor.

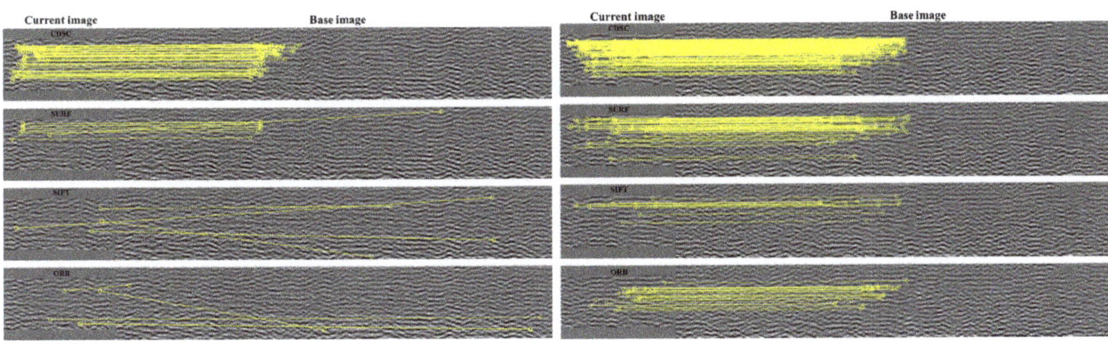

(**a**) (**b**)

Figure 15. Comparison of matching results of different methods in the railway test trajectory. (**a**) Matching GPR image with strong interference; (**b**) matching GPR image without interference.

A statistical analysis of the localization error produced a cumulative probability distribution (CDF), as shown in Figure 16. The SIFT algorithm achieved 55.5% of the results within 1 m accuracy, ORB achieved 74.6%, SURF achieved 80.4%, NCC achieved 86.8% within 0.1 m, and CDSC achieved 85.6% within 0.1 m; CDSC and NCC have comparable performance.

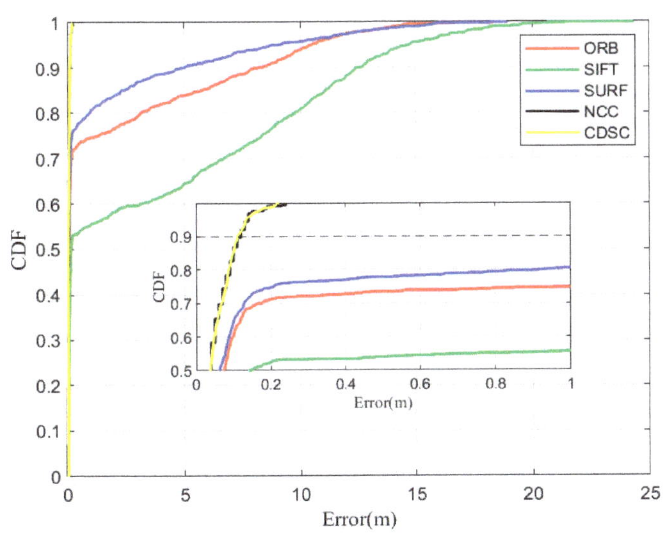

Figure 16. Railway test trajectory positioning error CDF.

As shown in Table 7, the RMSE (Root Mean Square Error) of the proposed algorithm is 0.055 m, representing a significant improvement in accuracy compared to other pseudo-corner point-based descriptors and is comparable to the localization accuracy of the NCC algorithm.

Table 7. Railway test trajectory positioning RMSE.

Methods	SIFT	ORB	SURF	NCC	CDSC
RMSE (m)	4.11	1.84	1.28	0.05	0.06

Due to the more complex road conditions in urban sections compared to railway sections, the probability of image decorrelation caused by GPR image interference is higher. As a result, the localization errors of all algorithms increased. However, the localization errors of CDSC and NCC were similar and significantly better than those of SURF, ORB, and SIFT. In regions with strong interference, the proposed algorithm demonstrated strong robustness. As shown in the subplots of Figure 17, at the mileage between 480 and 530 m, all algorithms except the CDSC algorithm exhibited substantial localization errors. Figure 18 compares the GPR images in this section, showing that clutter interference caused changes in image features. Conventional matching methods resulted in numerous mismatches or even failure to match, while CDSC could still ensure the correct matching of a large number of pseudo-corner points.

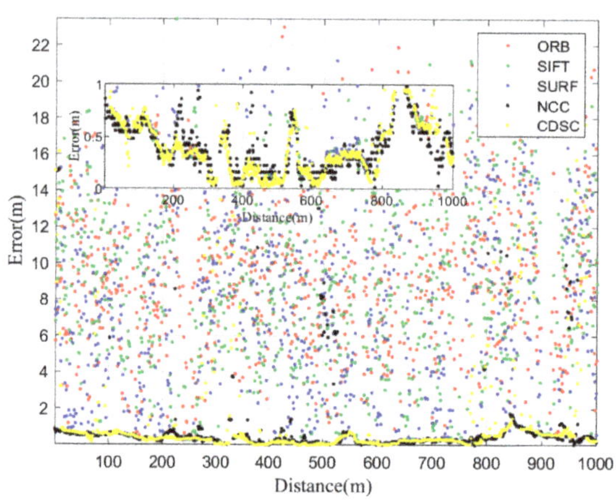

Figure 17. Urban test trajectory positioning error.

Statistical analysis of the localization results produced cumulative probability distributions (CDFs) shown in Figure 19. The SURF, ORB, and SIFT algorithms had 22.4%, 7.1%, and 12.2% of their localization errors within 1 m, respectively. NCC had 89.6% of localization errors within 1 m, while the proposed CDSC algorithm had 86.6% of its localization errors within 1 m, outperforming the other algorithms overall.

The matching and localization results for the urban test trajectory are shown in Table 8. Compared to the railway test results, the localization errors in the urban tests increased. The localization errors for SIFT, ORB, and SURF exceeded 5 m, while NCC had an error of 0.92 m and CDSC had an error of 1.22 m.

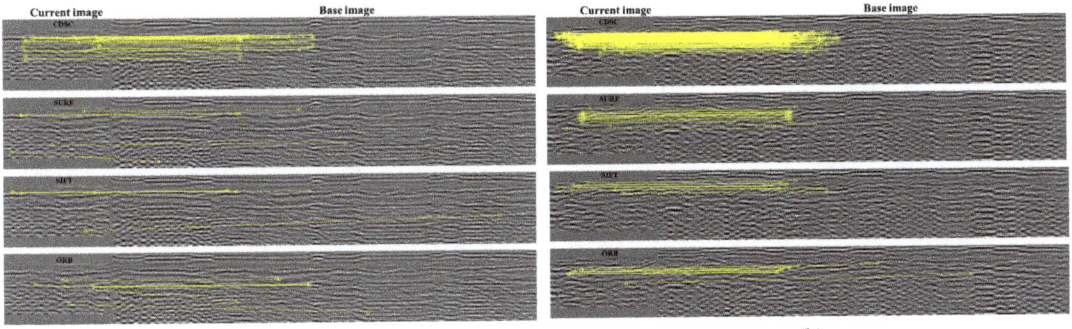

(a) (b)

Figure 18. Comparison of matching results of different methods in the urban test trajectory. (**a**) Matching GPR image with strong interference; (**b**) matching GPR image without interference.

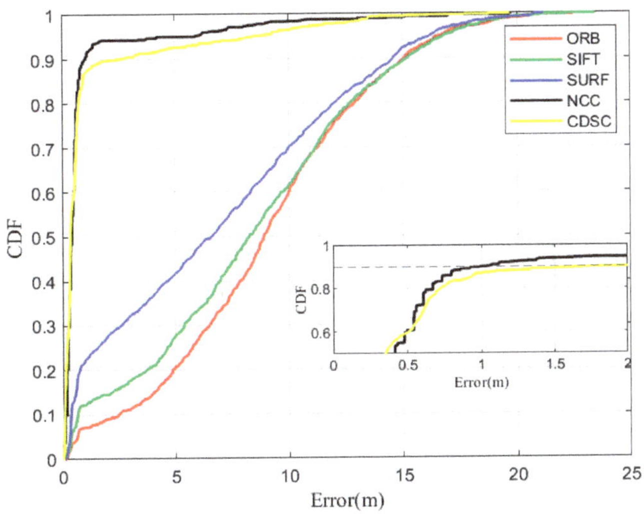

Figure 19. Urban test trajectory positioning error CDF.

Table 8. Urban test trajectory positioning RMSE.

Methods	SIFT	ORB	SURF	NCC	CDSC
RMSE (m)	8.34	8.92	6.90	0.92	1.22

NCC utilizes all pixels in the matching process, effectively making use of all the image information, which results in the best positioning accuracy. However, the drawback of NCC is the large computational load and high storage space usage caused by involving all pixels in the computation. The CDSC algorithm extracts effective local features from the GPR images and uses feature vectors to replace the entire image in the computation. Although its positioning accuracy is slightly lower than that of NCC, it shows significant improvements in terms of positioning computation time and storage space usage. The average time and storage space required for localizing 1 km of data were recorded, as shown in Table 9.

Table 9. Comparison of storage space and runtime.

Methods	SIFT	ORB	SURF	NCC	CDSC
Runtime (s)	292.1	733.0	304.3	2113.9	532.3
Storage (MB)	19	49	164	233	46

As indicated in Table 9, the proposed algorithm has a time advantage in feature description. This is because the CDSC descriptor reduces a large number of partial derivative computations compared to traditional methods, simplifying the calculations. The SIFT algorithm has certain advantages over the CDSC algorithm in terms of runtime and storage space. As shown in Figures 15 and 18, the number of correctly matched point pairs in SIFT is significantly fewer than in CDSC. This is because the SIFT algorithm is affected by the loss of correlation in local regions of the GPR images, leading to fewer feature points being extracted with lower accuracy, which in turn reduces the localization computation time. This limitation results in significantly inferior localization performance compared to the CDSC algorithm proposed in this manuscript, making it unsuitable for practical localization scenarios. Additionally, the proposed algorithm reduced matching time by 74.8% compared to NCC due to the lower dimensionality of the feature descriptor, which reduces the time required for matching. In summary, the proposed algorithm has a time advantage over other algorithms in both feature description and feature matching, which is beneficial for real-time matching and localization. In terms of storage space, the proposed algorithm reduced the required storage by 80.2% compared to NCC, demonstrating a significant improvement.

5. Conclusions

This paper presented a novel method for extracting features from GPR images, namely the central dense structure context (CDSC) features, which takes full advantage of the pseudo-corner and strip-like elements of GPR images. At the same time, a complete and reliable matching and positioning method was designed based on CDSC, which reduces the space size of reference image data and improves the matching calculation efficiency. The proposed vehicle matching method primarily includes steps such as extracting pseudo-corner points using SURF, CDSC descriptor extraction, feature matching, and outlier removal.

The proposed algorithm was evaluated on datasets collected from urban roads and railway tracks, and the performance of the localization method was evaluated. The results show that, compared to traditional methods, the proposed method demonstrates advantages in terms of localization accuracy, the number of matching point pairs, and precision. It significantly improves overall performance. This paper provides new ideas for research in the new technology of ground-penetrating radar navigation and positioning.

Achieving vehicle localization using GPR is a challenging task. Although the proposed algorithm improves localization accuracy and operational efficiency compared to traditional methods, it is still some ways from achieving real-time localization output. Future research will focus on optimizing threshold segmentation and graphical statistical strategies to further enhance the algorithm's real-time performance. Additionally, the suppression of image interference and the assessment of matching availability are also areas of interest for future research.

Author Contributions: Conceptualization, J.X. and Q.L.; formal analysis, X.J. and G.S.; investigation, D.W.; methodology, J.X.; project administration, Q.L. and D.W.; resources, D.W.; supervision, H.Y.; validation, J.X. and Q.L.; writing—original draft, J.X.; writing—review and editing, Q.L. and D.W. All authors will be informed about each step of manuscript processing including submission, revision, revision reminder, etc., via emails from our system or assigned assistant editor. All authors have read and agreed to the published version of the manuscript.

Funding: This research received no external funding.

Data Availability Statement: Restrictions apply to the datasets. The datasets presented in this article are not readily available because the data are part of an ongoing study.

Conflicts of Interest: The authors declare no conflicts of interest.

References

1. Zhang, G.; Wen, W.; Xu, B.; Hsu, L.-T. Extending shadow matching to tightly-coupled GNSS/INS integration system. *IEEE Trans. Veh. Technol.* **2020**, *69*, 4979–4991. [CrossRef]
2. Li, X.; Li, X.; Huang, J.; Shen, Z.; Wang, B.; Yuan, Y.; Zhang, K. Improving PPP–RTK in urban environment by tightly coupled integration of GNSS and INS. *J. Geod.* **2021**, *95*, 132. [CrossRef]
3. Zhang, T.; Zhou, L.; Feng, X.; Shi, J.; Zhang, Q.; Niu, X. INS Aided GNSS Pseudo-range Error Prediction Using Machine Learning for Urban Vehicle Navigation. *IEEE Sens. J.* **2024**, *24*, 6. [CrossRef]
4. Wen, W.W.; Hsu, L.-T. 3D LiDAR aided GNSS NLOS mitigation in urban canyons. *IEEE Trans. Intell. Transp. Syst.* **2022**, *23*, 18224–18236. [CrossRef]
5. Yue, J.; Wen, W.; Han, J.; Hsu, L.-T. 3D point clouds data super resolution-aided LiDAR odometry for vehicular positioning in urban canyons. *IEEE Trans. Veh. Technol.* **2021**, *70*, 4098–4112. [CrossRef]
6. Wan, G.; Yang, X.; Cai, R.; Li, H.; Zhou, Y.; Wang, H.; Song, S. Robust and precise vehicle localization based on multi-sensor fusion in diverse city scenes. In Proceedings of the 2018 IEEE International Conference on Robotics and Automation (ICRA), Brisbane, Australia, 21–25 May 2018; pp. 4670–4677.
7. Su, Y.; Chunlin, H.; Wentai, L. *Ground Penetrating Radar: Theory and Applications*; Science Press: Beijing, China, 2006.
8. Solla, M.; Pérez-Gracia, V.; Fontul, S. A review of GPR application on transport infrastructures: Troubleshooting and best practices. *Remote Sens.* **2021**, *13*, 672. [CrossRef]
9. Bestagini, P.; Lombardi, F.; Lualdi, M.; Picetti, F.; Tubaro, S. Landmine detection using autoencoders on multipolarization GPR volumetric data. *IEEE Trans. Geosci. Remote Sens.* **2020**, *59*, 182–195. [CrossRef]
10. Li, Y.; Ren, Y.; Peng, S.S.; Cheng, H.; Wang, N.; Luo, J. Measurement of overburden failure zones in close-multiple coal seams mining. *Int. J. Min. Sci. Technol.* **2021**, *31*, 43–50. [CrossRef]
11. Cornick, M.; Koechling, J.; Stanley, B.; Zhang, B. Localizing ground penetrating radar: A step toward robust autonomous ground vehicle localization. *J. Field Robot.* **2016**, *33*, 82–102. [CrossRef]
12. Ort, T.; Gilitschenski, I.; Rus, D. Autonomous navigation in inclement weather based on a localizing ground penetrating radar. *IEEE Robot. Autom. Lett.* **2020**, *5*, 3267–3274. [CrossRef]
13. Baikovitz, A.; Sodhi, P.; Dille, M.; Kaess, M. Ground encoding: Learned factor graph-based models for localizing ground penetrating radar. In Proceedings of the 2021 IEEE/RSJ International Conference on Intelligent Robots and Systems (IROS), Prague, Czech Republic, 27 September 2021; pp. 5476–5483.
14. Zhang, P.; Shen, L.; Huang, X.; Xin, Q. Ground penetrating radar image template matching based on symmetrical structure features. *Prog. Geophys.* **2022**, *37*, 2657–2666.
15. Ni, Z.; Ye, S.; Shi, C.; Pan, J.; Zheng, Z.; Fang, G. A deep learning assisted ground penetrating radar localization method. *J. Electron. Inf. Technol.* **2022**, *44*, 1265–1273.
16. Bi, B.; Shen, L.; Zhang, P.; Huang, X.; Xin, Q.; Jin, T. TSVR-Net: An End-to-End Ground-Penetrating Radar Images Registration and Location Network. *Remote Sens.* **2023**, *15*, 3428. [CrossRef]
17. Kim, G.; Kim, A. Scan context: Egocentric spatial descriptor for place recognition within 3d point cloud map. In Proceedings of the 2018 IEEE/RSJ International Conference on Intelligent Robots and Systems (IROS), Madrid, Spain, 1–5 October 2018; pp. 4802–4809.
18. Belongie, S.; Malik, J.; Puzicha, J. Shape matching and object recognition using shape contexts. *IEEE Trans. Pattern Anal. Mach. Intell.* **2002**, *24*, 509–522. [CrossRef]
19. Yang, M.; Cao, M.; Cheng, L.; Jiang, H.; Ai, T.; Yan, X. Classification of urban interchange patterns using a model combining shape context descriptor and graph convolutional neural network. *Geo-Spat. Inf. Sci.* **2024**, *27*, 5. [CrossRef]
20. Sun, C.; Wu, X.; Sun, J.; Qiao, N.; Sun, C. Multi-stage refinement feature matching using adaptive ORB features for robotic vision navigation. *IEEE Sens. J.* **2021**, *22*, 2603–2617. [CrossRef]
21. Zhang, W. Combination of SIFT and Canny edge detection for registration between SAR and optical images. *IEEE Geosci. Remote Sens. Lett.* **2020**, *19*, 1–5. [CrossRef]
22. Bay, H.; Tuytelaars, T.; Van Gool, L. Surf: Speeded up robust features. In Proceedings of the Computer Vision—ECCV 2006: 9th European Conference on Computer Vision, Graz, Austria, 7–13 May 2006; Proceedings, Part I. 9, pp. 404–417.
23. Liu, C.; Cao, H.; Fan, Y. Study on Adaptive DCP Image Optimization Algorithm Combined with OSTU Threshold Method. *Fire Control. Command. Control.* **2022**, 047.
24. Xu, H.; Li, S.; Ji, Y.; Cao, R.; Zhang, M.; Li, H. Panoramic Camera Image Mosaic Method Based on Feature Points. *Trans. Chin. Soc. Agric. Mach.* **2019**, *50*, 9.

25. *JTG D50—2017 (EN)*; Specifications for Design of Highway Asphalt Pavement. Ministry of Transport of the People's Republic of China: Beijing, China, 2017.
26. *TB 10001-2016*; Code for Design of Railway Earth Structure. State Railway Administration: Beijing, China, 2016.
27. Deng, P. Research on Detection of Railway Tunnel Seepage with Train-Mounted GPR. Master's Thesis, Southwest Jiaotong University, Chengdu, China, 2015.

remote sensing

MDPI

Article

What Is Beyond Hyperbola Detection and Characterization in Ground-Penetrating Radar Data?—Implications from the Archaeological Site of Goting, Germany

Tina Wunderlich [1,*], Bente S. Majchczack [2], Dennis Wilken [1], Martin Segschneider [3] and Wolfgang Rabbel [1]

1 Institute of Geosciences, Christian-Albrechts-University of Kiel, Otto-Hahn-Platz 1, 24118 Kiel, Germany; dennis.wilken@ifg.uni-kiel.de (D.W.); wolfgang.rabbel@ifg.uni-kiel.de (W.R.)
2 Cluster of Excellence ROOTS, Christian-Albrechts-University of Kiel, 24118 Kiel, Germany; bmajchczack@roots.uni-kiel.de
3 NihK—Institute for Historical Coastal Research, Viktoriastraße 26/28, 26382 Wilhelmshaven, Germany; segschneider@nihk.de
* Correspondence: tina.wunderlich@ifg.uni-kiel.de; Tel.: +49-431-8803903

Abstract: Hyperbolae in radargrams are caused by a variety of small subsurface objects. The analysis of their curvature enables the determination of propagation velocity in the subsurface, which is important for exact time-to-depth conversion and migration and also yields information on the water content of the soil. Using deep learning methods and fitting (DLF) algorithms, it is possible to automatically detect and analyze large numbers of hyperbola in 3D Ground-Penetrating Radar (GPR) datasets. As a result, a 3D velocity model can be established. Combining the hyperbola locations and the 3D velocity model with reflection depth sections and timeslices leads to improved archaeological interpretation due to (1) correct time-to-depth conversion through migration with the 3D velocity model, (2) creation of depthslices following the topography, (3) evaluation of the spatial distribution of hyperbolae, and (4) derivation of a 3D water content model of the site. In an exemplary study, we applied DLF to a 3D GPR dataset from the multi-phased (2nd to 12th century CE) archaeological site of Goting on the island of Föhr, Northern Germany. Using RetinaNet, we detected 38,490 hyperbolae in an area of 1.76 ha and created a 3D velocity model. The velocities ranged from approximately 0.12 m/ns at the surface to 0.07 m/ns at approx. 3 m depth in the vertical direction; in the lateral direction, the maximum velocity variation was ±0.048 m/ns. The 2D-migrated radargrams and subsequently created depthslices revealed the remains of a longhouse, which was not known beforehand and had not been visible in the unmigrated timeslices. We found hyperbola apex points aligned along linear strong reflections. They can be interpreted as stones contained in ditch fills. The hyperbola points help to differentiate between ditches and processing artifacts that have a similar appearance as the ditches in time-/depthslices. From the derived 3D water content model, we could identify the thickness of the archaeologically relevant layer across the whole site. The layer contains a lot of humus and has a high water retention capability, leading to a higher water content compared to the underlying glacial moraine sand, which is well-drained.

Keywords: GPR; deep learning; hyperbola; object detection; velocity determination; archaeological interpretation

1. Introduction

Hyperbolic diffractions in Ground-Penetrating Radar (GPR) data (radargrams) are caused by a variety of subsurface objects such as pipes, cables, stones, or archeological artifacts. The spatial distribution of their apex points can aid in investigating the spatial distribution of these subsurface structures. An additional piece of information derivable from the hyperbola is the propagation velocity of electromagnetic waves in the subsurface, which is needed to determine the depth of the objects and/or the 3D distribution of the water content of the soil.

With the advent of multi-channel GPR antenna arrays, it has become easy to cover large areas in short time with high spatial sampling, i.e., inline and crossline trace distances of 4 and 8 cm, respectively [1]. This leads to large data volumes, from which often only a small portion of information is used for interpretation. This is partly due to the limited aim of the survey, e.g., delineating structures in timeslices such as pipes or rebar, and partly due to high costs and time of experienced personnel to inspect the data carefully in all aspects. However, the latter work can be supported by machine learning algorithms, which have found widespread use in the last decades and have also been used in the context of GPR processing to automatically detect pipes or rebar (e.g., [2–8]), defects in concrete (e.g., [9]), tree roots (e.g., [10–12]) in radargrams, or archaeological remains of hearths in timeslices [13].

In urban infrastructure detection, the research mostly stops with the localization of pipes or rebar, but further important information can be derived from the detected hyperbolae: the propagation velocity of the subsurface. This is necessary for correct time-to-depth-conversion, migration algorithms, and topographic corrections and can also be used to estimate the water content of the soil or to differentiate between different subsurface layers.

In general, there are different possibilities to derive a velocity model from GPR data: (a) using multi-offset measurements such as common midpoint (CMP) measurements [14], which require preferably horizontal reflectors at different depths (or the need to correct for inclined reflectors); (b) evaluating the shape of diffraction hyperbolae in constant-offset data, which requires the presence of small objects in the subsurface producing a diffraction; (c) using Time Domain Reflectometry (TDR) measurements at the surface [15], which does not provide velocity information at greater depths; or (d) comparing the reflector depths in radargrams with coring and thus deriving a 1D velocity model.

In order to use multi-offset measurements (point (a) above) for a velocity model, manual CMP measurements with high spatial coverage are very time-consuming and are not feasible for larger areas. With some multi-channel GPR systems, this high spatial coverage of multi-offset measurements can be achieved [16,17], but a second measurement run for "normal" constant offset measurements is also necessary for archaeological interpretation with timeslices. Also, the use of TDR or coring (points (c) and (d)) is limited to some points and requires additional equipment and labor. Therefore, it is easier to use the already collected constant-offset data. Then, if enough diffraction hyperbolae are present, their shape can be used to derive velocity information at different spatial locations. The evaluation of the shape can be achieved by, e.g., fitting test hyperbolae to the hyperbolae seen in the data [18], performing test migrations with a range of velocities [19], and evaluating the roundness of the resulting reflections [20], or using measures such as the semblance [21], which is commonly used in seismic data processing. All mentioned approaches require the detection of the diffraction hyperbolae itself, which can of course be performed manually, but which is not feasible for larger datasets. Therefore, an automatic hyperbola detection and velocity extraction workflow has been proposed by [18]. When applying this procedure, we have three datasets containing different information: (1) the constant-offset GPR data with timeslices and radargrams, (2) the locations (x, y, z) of small objects causing the diffraction hyperbola, and (3) RMS velocities of the overlying soil at these locations. So, the further questions that arise are the following:

1. Can we use the RMS velocities to derive a 3D velocity model? Can that model be used for migration of the radargrams? Does it tell us anything about the archaeological site?
2. What can we learn from the spatial distribution of the detected hyperbolae?

We set out to answer these questions using a Malå MIRA (Malå Imaging Radar) GPR dataset collected with 16 constant-offset channels measured on the archeological site of Goting on the Wadden Sea island of Föhr (district of Nordfriesland, Germany). The site is a multi-phased settlement distributed on a slightly elevated sandy moraine around a small wetland, which provided access to the North Sea via a tidal inlet. Magnetic measurements and archaeological excavations established a settlement phase of the Younger Roman Iron Age and Migration Period (2nd to early 6th century CE) consisting of dispersed single farmsteads and a settlement of the Early through High Middle Ages (7th through 11th/12th century CE). During the younger phase, the settlement can be characterized as a harbor site with specialization in trade and craft production [22] (pp: 163–182) or [23]. For the concluding archaeological interpretation, we not only used GPR data and hyperbola information but also a magnetic map acquired for the same area and results from excavations.

After reporting on the data collection, a short description of the DLF algorithm used is given, before we focus on the further parameters derived and the results in Goting. The findings are discussed with respect to geophysical evaluation and archeological interpretation.

2. Materials and Methods

2.1. Magnetic Data Acquisition and Processing

Magnetic data were collected with a 6-sensor Fluxgate array (Förster) mounted on a non-magnetic cart. The internal vertical sensor spacing was 65 cm, and the horizontal sensor spacing on the cart was 50 cm, measuring the vertical difference in the vertical component of the magnetic field. The data were sampled with 10 Hz while driving in straight lines over the area. Accurate positioning was achieved with a Leica RTK-DGPS (real-time kinematic differential global positioning system), yielding an accurate coordinate (cm accuracy) for each sample. Processing consisted of removing a mean value for each sensor on each profile, to remove the effect of metal on the cart. Then, the data were binned on a grid with 20 $\times$ 20 cm cells, and missing values were linearly interpolated.

2.2. GPR

2.2.1. Data Acquisition and Processing

The GPR dataset was measured with a Malå Multichannel Imaging Radar (MIRA) with 16 constant-offset channels and a 400 MHz center frequency. The traces of 100 ns length were recorded with 1024 samples and stacked 4 times. Profile spacing was 8 cm and the inline trace distance 3 cm. Accurate positioning was achieved with a Leica total station TS15.

The data processing was performed in MATLAB using the in-house software MultichannelGPR V1.3.0 [24] (available on GitHub: https://github.com/tinawunderlich/MultichannelGPR, accessed on 30 October 2024), which consisted of (a) DC amplitude removal, (b) time zero correction by shifting channels 2–16 in relation to channel 1 for each profile by means of cross-correlation, (c) adjustment of time zero across all profiles for all channels, (d) bandpass filtering (100–600 MHz), (e) cutting of the time axis at 80 ns, (f) application of a custom gain function with [-20, 0, 10, 20, 30 dB] stretched over the time range of 80 ns, and (g) removal of horizontal reflections with the help of a wavenumber highpass with cutoff wavenumber 0.1 m^{-1}.

The processed radargrams were binned on an 8 cm $\times$ 8 cm grid. Subsequently, the data cube was cut into timeslices by summing the absolute amplitudes in 5 ns intervals and interpolating missing values. Linear stripe noise in the profile direction was eliminated by a wavenumber–wavenumber–pieslice filter (later denoted as kk-filter) [25].

The processed radargrams were used to extract the propagation velocities (see Sections 2.2.2–2.2.4). After generating the 3D velocity model, the velocities along the GPR profiles were extracted and used for 2D migration by means of semicircle superposition, taking the topography and the antenna aperture of about 30° into account [26]. After migrating all profiles, depthslices were generated by summing the absolute amplitudes in 20 cm intervals following the topography, i.e., resulting in curved slices and not horizontal ones. Again, a kk-filter was applied to reduce stripe noise.

2.2.2. Hyperbola Detection and Velocity Determination

In addition to timeslices, we used the processed radargrams for automatic hyperbola detection and velocity determination. For this task, the radargrams were exported as jpg images of 256 × 256 pixel size and fed into the convolutional neural network (CNN) RetinaNet [27,28] in a PyTorch implementation by [29] (Figure 1a). We applied transfer learning by using a pretrained ResNet50 [30] as the backbone. All images of channel 1 of our dataset were then used to train the network. We reached an average precision of 58% in detecting hyperbolae. More details can be found in [18]. The outcomes were bounding boxes around the classified hyperbolae in all radargrams (Figure 1b). In total, 38,490 hyperbolae were detected on the complete area of 1.76 ha.

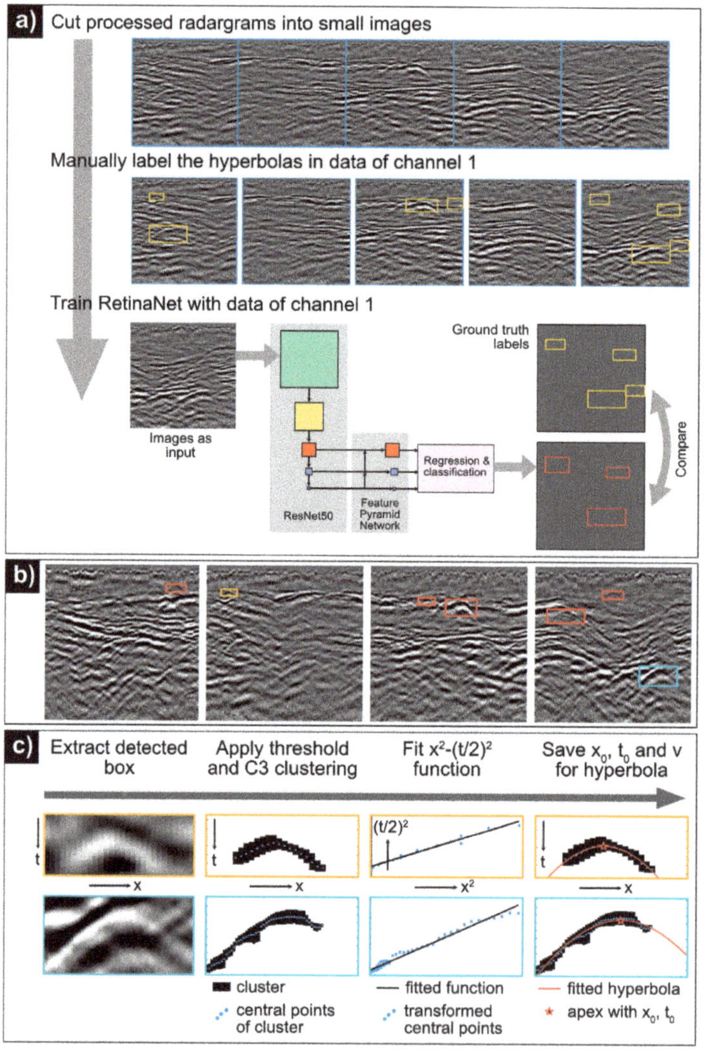

Figure 1. (**a**) The processed radargrams of all 16 channels have been cut into images of 256 × 256 pixels. In all images of the first channel, hyperbolae were labeled manually. These images and ground truth labels were used to train a RetinaNet. (**b**) During inference, all images of all channels were fed into the trained RetinaNet and hyperbolae were automatically detected (red, blue

and yellow boxes), which is shown here for four example images. (**c**) Examples of the velocity determination workflow (see yellow and blue box in (**b**)): After detection of all hyperbolae, the boxes were extracted from the images and subsequently a thresholding and C3 clustering algorithm was applied. This results in central points of a cluster, which are used in a x^2-$(t/2)^2$-diagram for fitting of a linear equation. We can derive the velocity from the slope of this fit, and from the intercept, the apex time t_0.

The detected hyperbolae were cut out from the images, and the pixels of the hyperbolae were extracted using an amplitude threshold and a column connection clustering (C3) algorithm [31]. The points were then fitted with a straight line in a x^2-$(t/2)^2$-plot in order to determine the velocity v, the horizontal apex location x_0, and the temporal apex location t_0 (Figure 1c). Again, more details can be found in [18]. This provided an RMS velocity error of 0.021 m/ns compared to manual velocity determination [18].

2.2.3. Clustering of Apex Points

As a small archaeological object such as a stone in the subsurface produces hyperbolae on adjacent radargrams, they need to be grouped together into one point. Otherwise, one could misinterpret the neighboring hyperbolae as an elongated subsurface structure, because they are arranged perpendicular to the profile direction in a straight line. Because we neither expected pipes or other linear modern infrastructure nor tree roots in the investigated area, this assumption is valid.

In order to group hyperbolae originating from the same subsurface object, we applied Density-Based Spatial Clustering of Applications with Noise (DBSCAN, [32]). A cluster is defined if 2 or more points are neighbors in a radius of 0.25 m. If a point has no neighbors in this radius, it is not assigned to a cluster and stays as single point. Again, more details on the clustering can be found in [18]. This results in a list of apex points (Easting, Northing, depth), velocities, and a flag if the point is a cluster or a single point.

2.2.4. Building a 3D Velocity Model

The complete area was divided into a 2 m × 2 m grid, and for each grid, cell apex points in a radius of 5 m were taken in time intervals of 10 ns in order to derive a mean velocity. Thus, the area was split in three dimensions in 2 m × 2 m × 10 ns cells, and in each cell, a mean velocity was calculated, if more than 2 apex points were found in the valid radius. An exemplary profile along the 6,060,229 m Northing shows the velocity distribution along the profile with some missing values at early times and especially in most of the profile at later times (Figure 2a).

To fix this, the first velocity in each horizontal grid cell was set at time zero and the last also at 80 ns. Then, the cube was interpolated linearly along the time axis onto the 828-sample-long time vector of the radargrams (80 ns) and smoothed with a moving average filter over 150 samples (Figure 2b). In order to fill the other missing cells and to further smooth the model for better migration results, the grid was smoothed in the x and y directions over 7 cells with a box filter (Figure 2c). Then, the velocities for each radargram were extracted from the smoothed 3D cube by means of triangulation. Before migration, the velocity model for each profile was again smoothed over 200 samples (approx. 20 ns) in the time direction and 100 samples (approx. 3 m) in the horizontal profile direction with moving average filters.

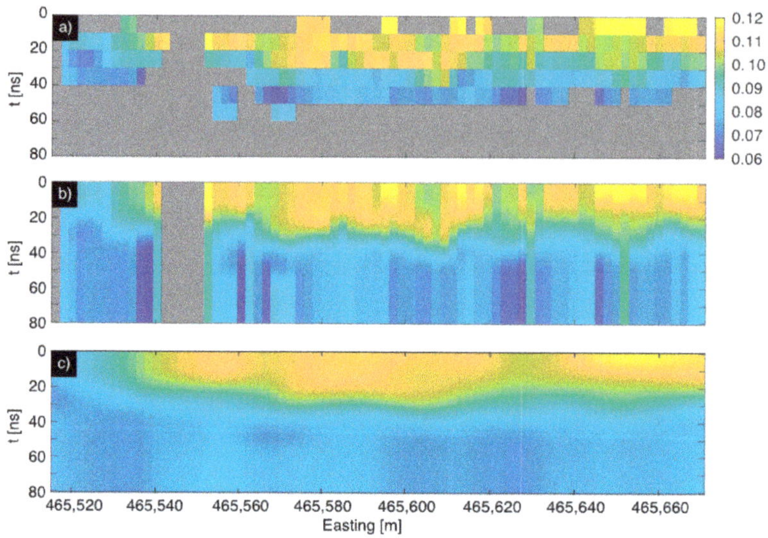

Figure 2. (**a**) Mean velocity distribution along a west–east profile taken from the smoothed 3D data cube. (**b**) Missing velocity values at early or late times are extrapolated constantly from the existing ones and smoothed with a moving average filter over 150 samples. (**c**) Lateral extrapolation and smoothing with a box filter over 7 cells.

2.2.5. Water Content Calculation

Further information that could be taken from the 3D velocity model was an estimate of the volumetric water content. In order to achieve this, the 3D velocity model, which reflected an RMS velocity, was first converted to the interval velocity (v) [33]. Then, the relative permittivity (ε_r) was calculated using the equation

$$\varepsilon_r = \frac{1}{v^2 \varepsilon_0 \mu_0 \mu_r} \quad (1)$$

with $\varepsilon_0 = 8.854 \cdot 10^{-12} \frac{F}{m}$ and $\mu_0 = 4\pi \cdot 10^{-7} \frac{H}{m}$ are the permittivity and magnetic permeability in a vacuum, respectively. For the relative magnetic permeability, we assumed $\mu_r = 1$. From the resulting 3D relative permittivity model, we could estimate the volumetric water content (θ), e.g., by using the equation of [34]:

$$\theta = -5.3 \cdot 10^{-2} + 2.92 \cdot 10^{-2} \cdot \varepsilon_r - 5.5 \cdot 10^{-4} \cdot \varepsilon_r^2 + 4.3 \cdot 10^{-6} \cdot \varepsilon_r^3 \quad (2)$$

Further functions for calculating the water content exist in the literature, e.g., [35,36], but due to the lack of calibration samples for a more sophisticated choice, we chose the easiest one.

In addition, we calculated heatmaps of point density. The area was gridded into 2 m × 2 m cells, and for each cell, the number of hyperbolae in a radius of 5 m was calculated. This was performed for time intervals of 10 ns each.

3. Results

The Goting settlement site is located on the southern edge of the island's moraine core (Figure 3a). The site extends along three gently sloping sides of a depression in the moraine, which harbors a marshy lowland of 8 ha with a tidal inlet. The moraine reaches a height of up to 8 m a.s.l., the lowland 1.8 m a.s.l. (Figure 3d). The moraine is composed of glacial sands and occasionally till [37]. The magnetic map of the measurement areas around the lowland reveals differently structured zones of settlement (numbers in Figure 3b), which require individual interpretations and dating approaches based on archaeological finds, excavations, and comparison with other sites [22] (pp: 165–168):

(1) Dispersed farmsteads with longhouses and yard enclosures, Younger Roman Iron Age and Migration Period.
(2) Possible seasonal pit house settlement, early medieval.
(3) Workshop area with elongated plot division on the slope towards the tidal inlet used as a harbor, early medieval.
(4) Low-density area with pit houses and longhouses, interpreted as an early medieval residential area.
(5) Unclear cluster of dense magnetic anomalies, possibly overlap with early medieval residential area.
(6) Dense cluster of linear and rectangular settlement structures and adjacent early medieval longhouses.
(7) Early to high medieval pit houses and longhouses.

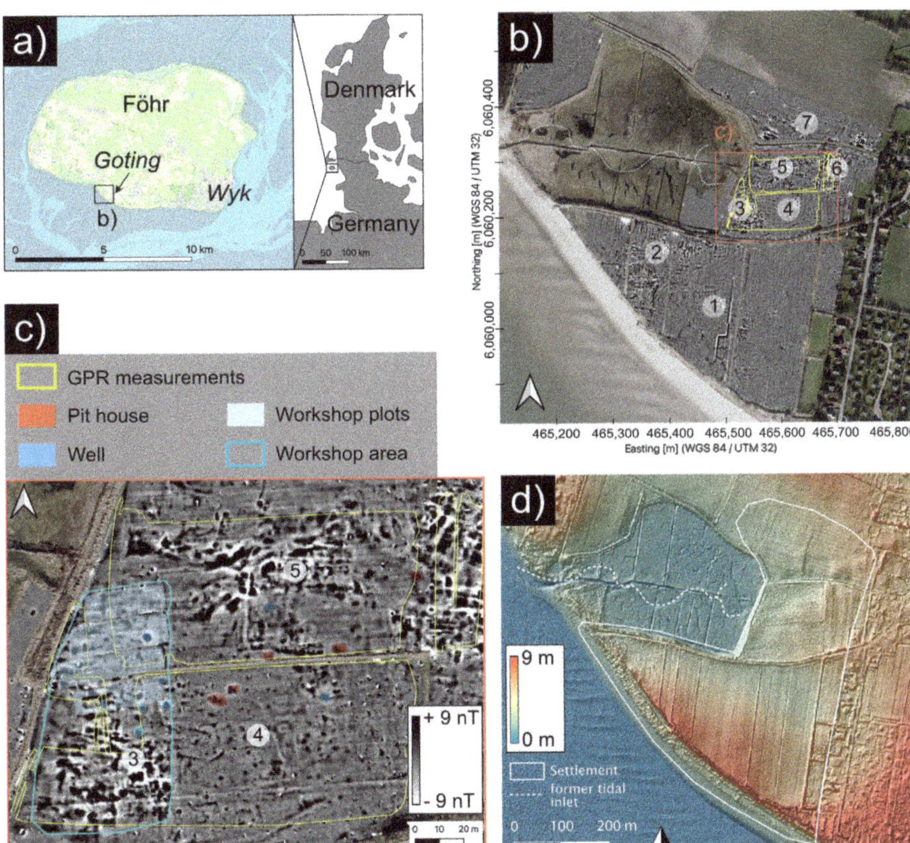

Figure 3. (**a**) Location of the island of Föhr in northern Germany (Basemap: OpenStreetMap); (**b**) magnetic map of Goting, where the numbers refer to different zones of the settlement, which are further explained in the text; (**c**) zoomed in view of the magnetic map with interpretation and the GPR measurement area; (**d**) topographic overview based on a digital elevation model (Basemaps: DGM1, DOP20 ©GeoBasis-DE/LVermGeo SH/CC BY 4.0).

The GPR measurements cover zones 4 and 5 and parts of 3 and 6 (Figure 3c). The magnetic map in this area shows several pit houses and wells, as well as a workshop area on the western side.

Hyperbolae were detected in almost the complete area and in traveltimes between 0 and 70 ns (Figure 4). At the borders of the area, hyperbolae were missing, which was mainly caused by technical problems leading to smearing of GPR data at the beginning and ending of profiles and thus no reliable information. The determined velocities ranged between 0.05 and 0.15 m/ns. Grouping the detected hyperbolae into time intervals and smoothing results in a 3D velocity model that could be visualized, e.g., by timeslices (Figure 5), provided a clearer picture. The velocity range was slightly decreased due to smoothing. The highest velocities of about 0.12 cm/ns can be found in the southern part of the area between 0 and 10 ns. The lowest velocities of about 0.06 m/ns are found in deeper timeslices in the northern part of the area. The velocity is decreasing from top to bottom, with the highest variation in the upper 30 ns.

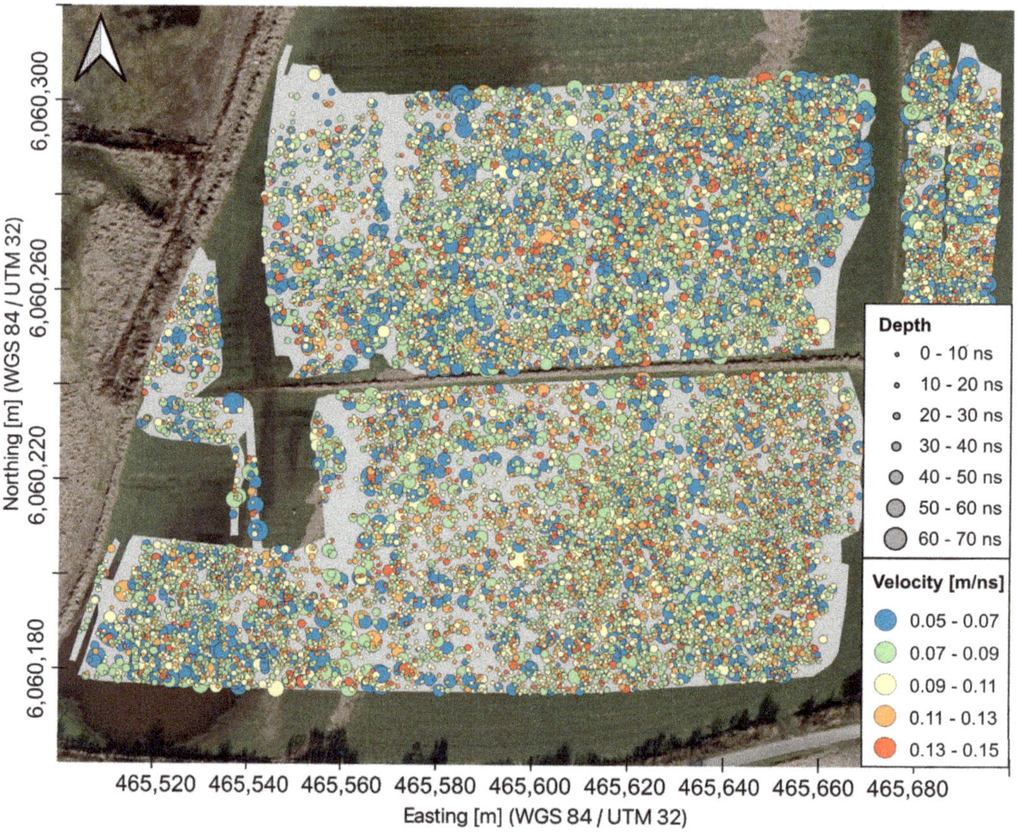

Figure 4. Detected hyperbola in different time ranges on the complete area (Basemap: ©GeoBasis-DE/LVermGeo SH/CC BY 4.0).

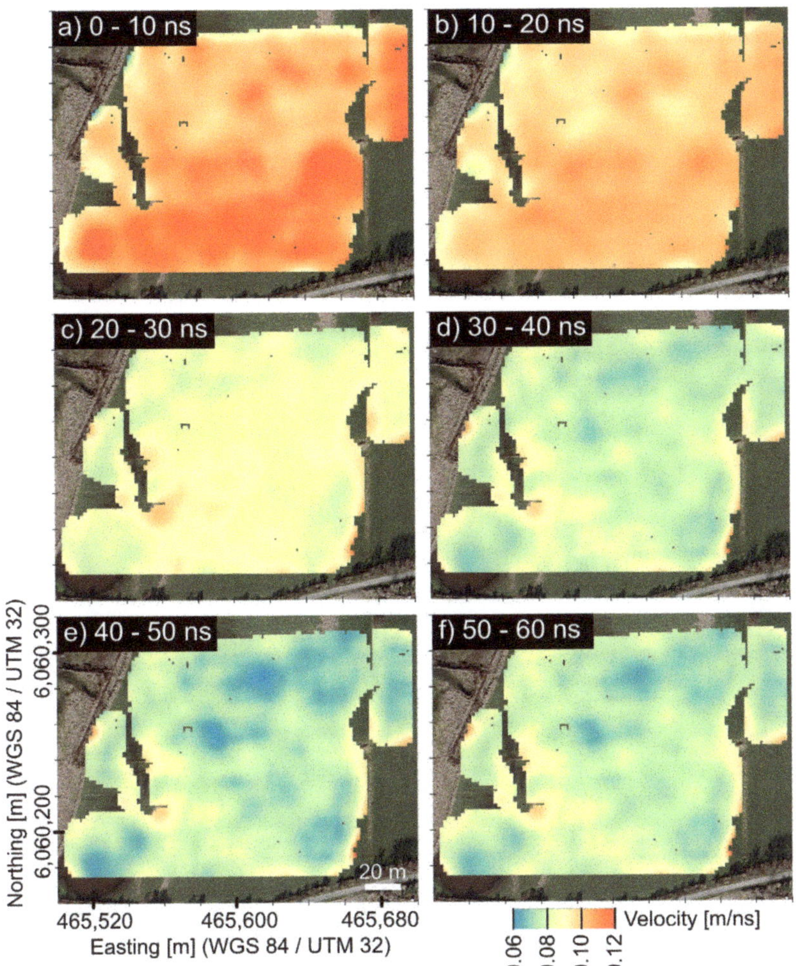

Figure 5. Timeslices of interpolated RMS velocities from the 3D velocity model (Basemap: ©GeoBasis-DE/LVermGeo SH/CC BY 4.0).

The spatial distribution of hyperbola points varies strongly over the area and in different time ranges (Figure 6). Most hyperbolae are found in the range 10 to 20 ns, with up to 68 points in a radius of 5 m in the southeastern part. In the uppermost layer, most hyperbolae are found in the southern area. In the time intervals 20–30 ns and 30–40 ns, the eastern part contains more hyperbolae and is delimited from the western part by an almost straight north–south-running line of higher density.

Timeslices and depthslices, respectively, were calculated for the complete GPR area. The border areas were hard to interpret, especially in the western part close to the depression, due to technical positioning problems and high attenuation. The northern part of the area is characterized by strong geological variations that cannot be related to any archaeological structures, so we can neither explain the magnetic anomalies nor the GPR features in this part. The southern part of the area is shown in Figure 7, together with the location of exemplary profile 109 (Figure 8). The magnetic map shows two almost rectangular anomalies in the northwestern part, which can be interpreted as the remains of pit houses. The strong round anomaly in the northeast is a well. Many more pit-like anomalies are visible as well as a round structure in the middle. The timeslices between 10 and 15 ns (Figure 7c) and between 15 and 20 ns (Figure 7d) show a variety of linear features, interpreted as ditches. The deeper timeslice also shows the southern pit house and the well as an absorbing area. In the depthslices, the ditches are also visible, as well as both pit houses and the well. In addition, the deeper slice between 80 and 100 cm depth below the surface (Figure 7f) shows an approximately southwest–northeast-oriented elongated shape of stronger reflections that might be interpreted as the remains of a longhouse with a size of 27 × 6 m (Figure 8), probably from the Younger Roman Iron Age or the Migration Period. The roundish magnetic anomaly in the middle of the area can neither be seen in the timeslices nor in the depthslices. The northern pit house is only clearly visible in the magnetic map, because it was excavated before the GPR measurements took place.

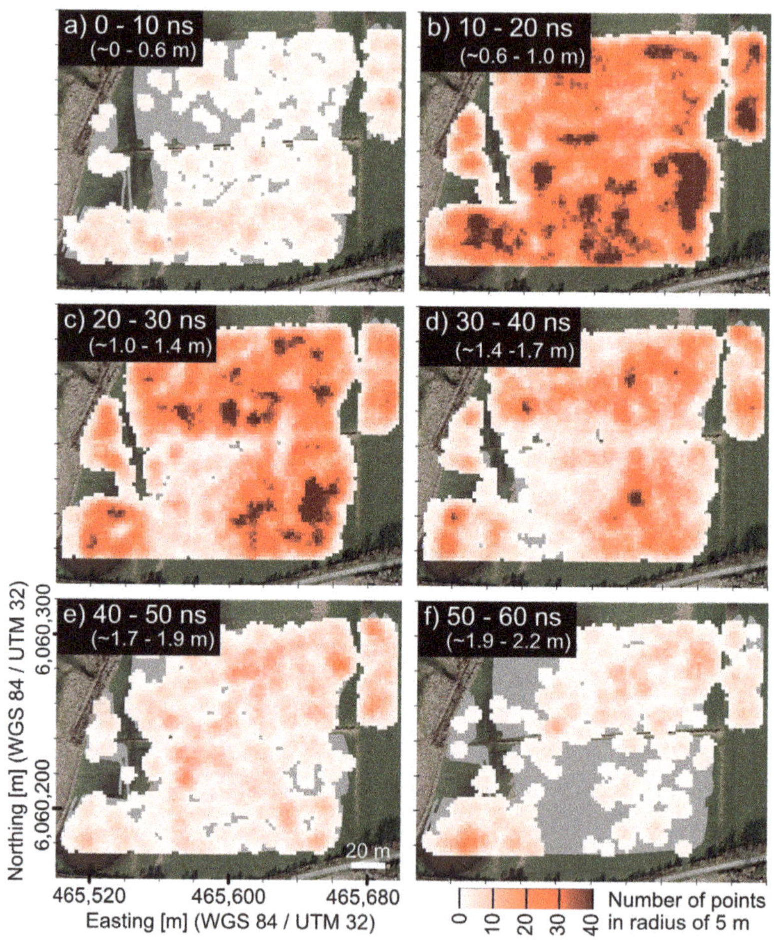

Figure 6. Density maps of hyperbola points in different time intervals for grid cells of 2 m × 2 m in a radius of 5 m. The approximate depth ranges were calculated with a mean 1D velocity function, which was derived by calculating the average velocity in each time interval over the complete area (Basemap: ©GeoBasis-DE/LVermGeo SH/CC BY 4.0).

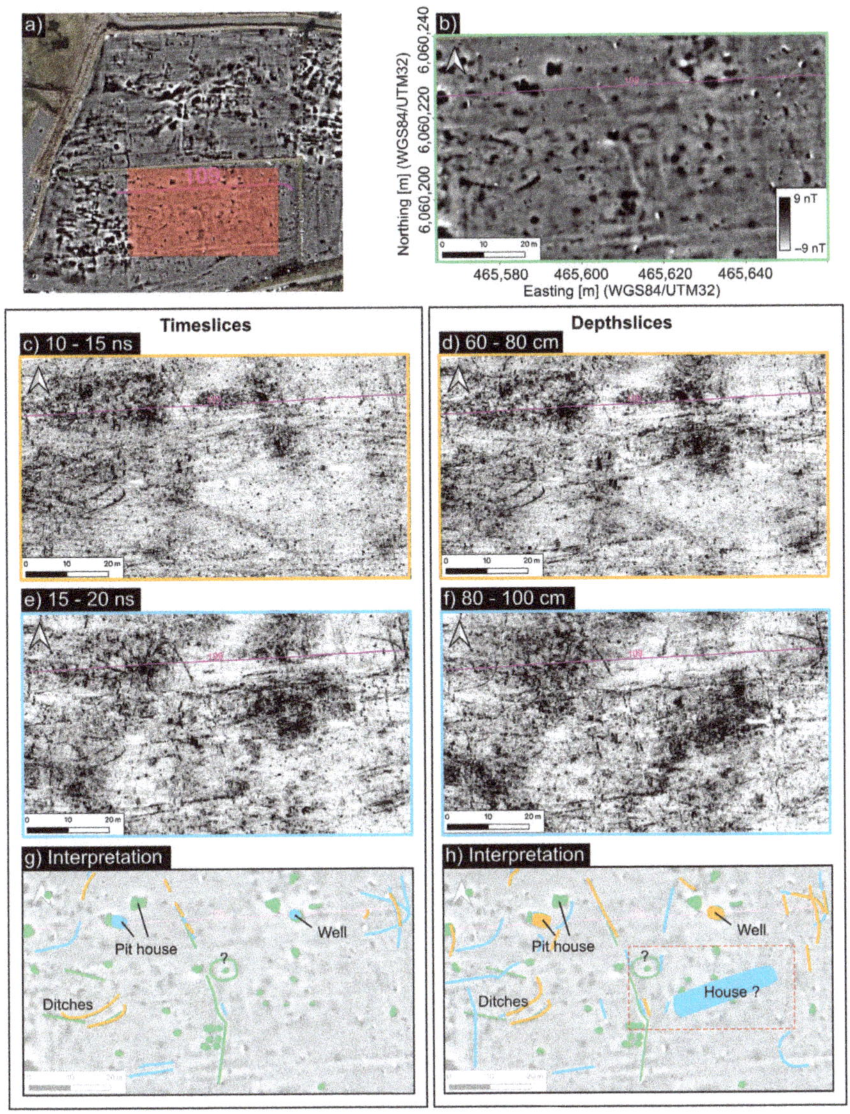

Figure 7. (**a**) Magnetic map with location of smaller area (red) shown in (**b**). The purple line shows the position of the example profile in Figure 9. (**b**) Cut-out from the magnetic map. (**c**,**e**) Timeslices. (**d**,**f**) Depthslices following the topography with a given depth below the surface. (**g**) Interpretation of magnetics (green), timeslice 10–15 ns (yellow), and timeslice 15–20 ns (blue) (main features). (**h**) Interpretation of magnetics (green), depthlice 60–80 cm (yellow), and depthslice 80–100 cm (blue) (main features). The red dotted box marks the location of the cut-out shown in Figure 8 (Basemap: ©GeoBasis-DE/LVermGeo SH/CC BY 4.0).

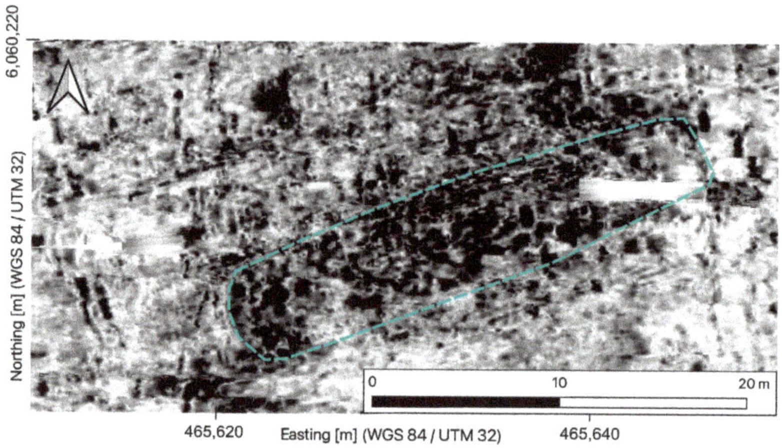

Figure 8. Depthslice between 80 and 100 cm (for location, see red dotted box in Figure 7h) and marked outline of the longhouse (blue dotted line).

Exemplary profile 109 (channel 1) is running from west to east crossing the southern pit house, the well, and ditches (Figure 9). In both the time-domain radargram (Figure 9b) and depth-migrated radargram (Figure 9e), the pit house is visible as a 3.5 m wide depression reaching a depth of up to 1.2 m below the surface. In contrast, the well has a diameter of only 3 m and reaches to a depth of about 2 m below the surface. Another differentiation point between both anomalies is their shapes: the pit house is rectangular and the well is round. The RMS velocity model along the profiles shows a higher-velocity layer of about 0.12 m/ns in the first 15 ns, which then decreases rapidly to below 0.08 m/ns (Figure 9c,g). For times greater than about 40 ns, there are some lateral variations between 0.06 and 0.08 m/ns. If the velocity model is used to calculate the approximate volumetric water content, we can observe a relatively dry layer (10–15 vol%) of 1 m thickness followed by a wet layer (>30 vol%) of about the same thickness (Figure 9f,g). Three exemplary 1D velocity functions extracted from the profile (Figure 9g) show a surface layer of a slightly higher interval velocity in the eastern part of the profile. All interval velocity functions decrease in the high-water-content layer and increase to a similar value below.

The linear features in the time-/depthslices, which we interpret as ditches, are originating from a line of hyperbolae (Figure 10a). These are probably caused by stones, which are located in the infill of some of the ditches, as was exemplarily found during excavations in the western part of the area (Figure 10b; [22], pp. 543–578). Unfortunately, due to the above-mentioned problems, no GPR data are available for comparison with the excavation.

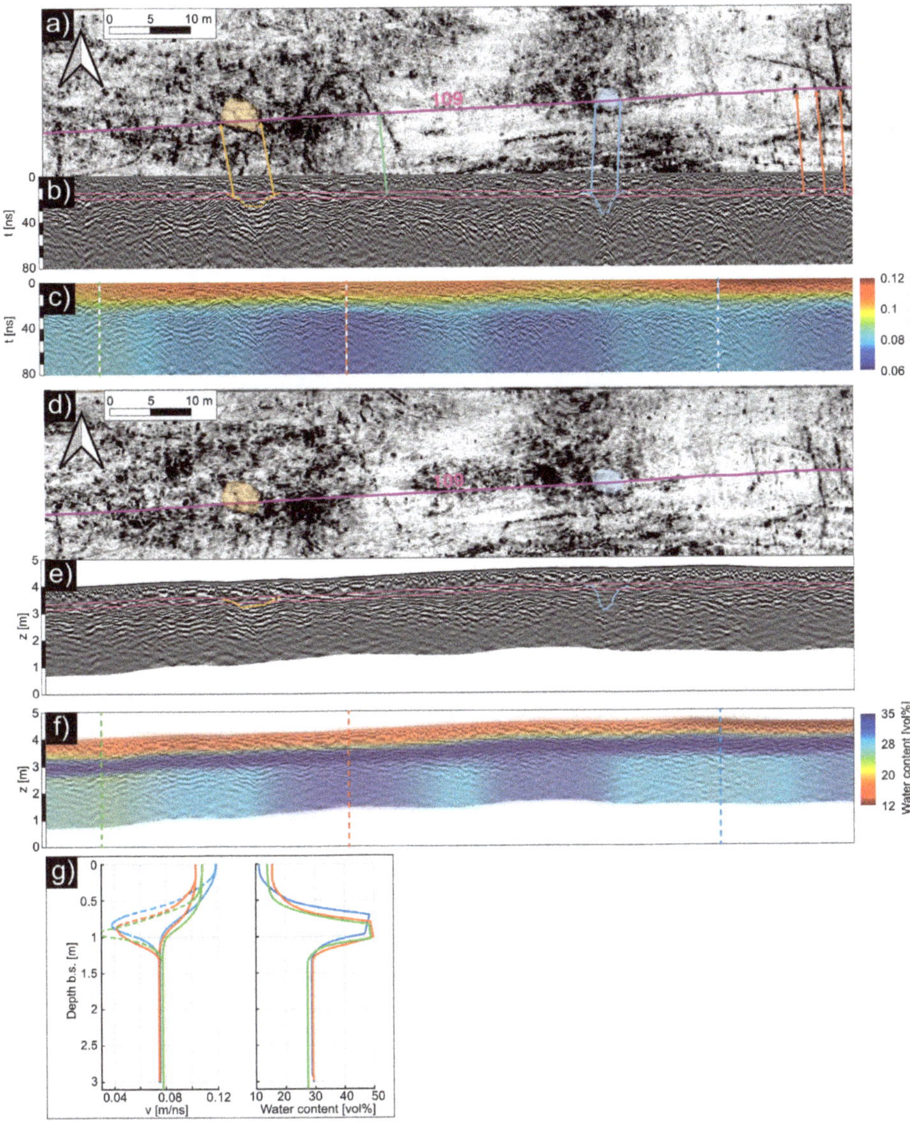

Figure 9. Comparison of slices and radargrams: (**a**) Timeslices between 15 and 20 ns with location of exemplary radargram, (**b**) processed radargram (Profile 109 of channel 1), (**c**) smoothed velocity model along the profile as an overlay on the radargram, (**d**) depthslice between 60 and 80 cm below the surface, following the topography, (**e**) migrated radargram using the velocity model in (**c**) and taking the topography into account, (**f**) water content distribution as an overlay on the migrated radargram. The time or depth ranges of the slices are shown as purple lines in the radargrams, respectively. The yellow color marks the remains of a pit house, whereas the blue anomaly marks a well. Green and orange arrows in (**a**,**b**) indicate the locations of isolated hyperbolae forming linear features in the time-/depthslices, interpreted as ditches. (**g**) Exemplary 1D velocity and water content functions. Their locations are marked by the blue/reg/green lines in (**c**,**f**). Dashed lines represent interval velocities, bold lines RMS velocity functions.

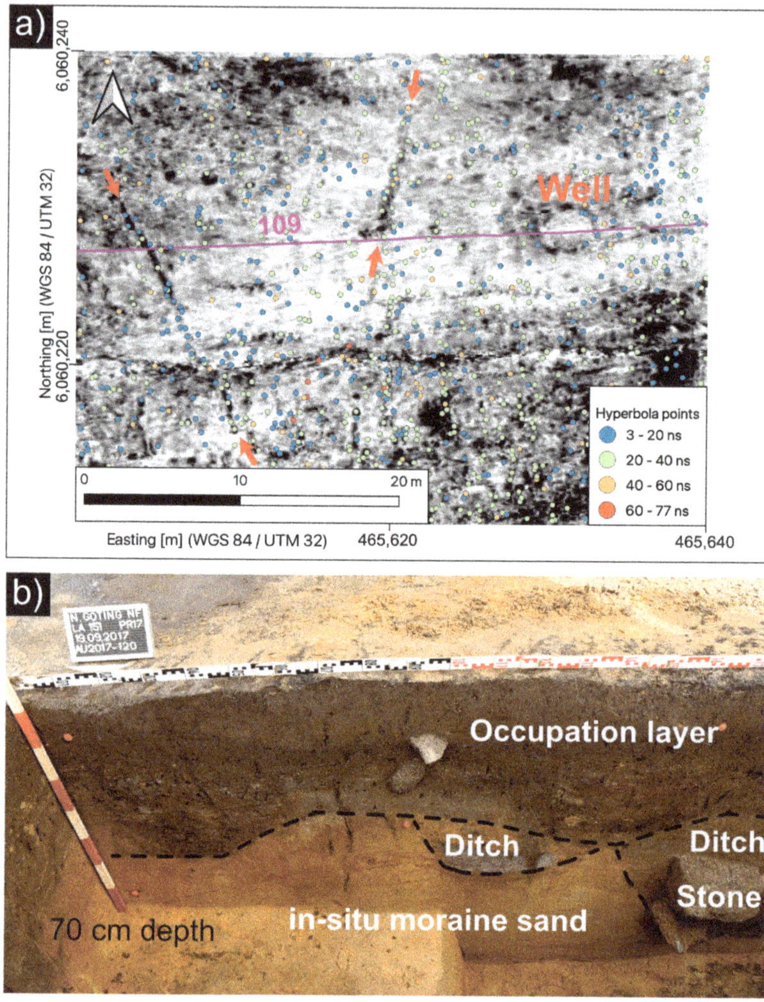

Figure 10. (**a**) Depthslice between 80 and 100 cm with hyperbola locations. Red arrows indicate the alignment of points in ditches. (**b**) Photography of an excavation profile in the eastern part of the area with a simplified interpretation. A stone is located in a ditch filling (Photograph by J. Gebühr, NIhK Wilhelmshaven; full description in [22], pp. 550–551, Profile 17).

The 3D smoothed velocity model was used to calculate the volumetric water contents for different depths (Figure 11). The water content was relatively homogenous laterally, but it varied strongly with depth. The highest values (>30 vol%) were reached between 80 and 120 cm. This is approximately the occupation layer, as seen in excavations. The in situ moraine sand below again had lower water contents, as it drains the water very well.

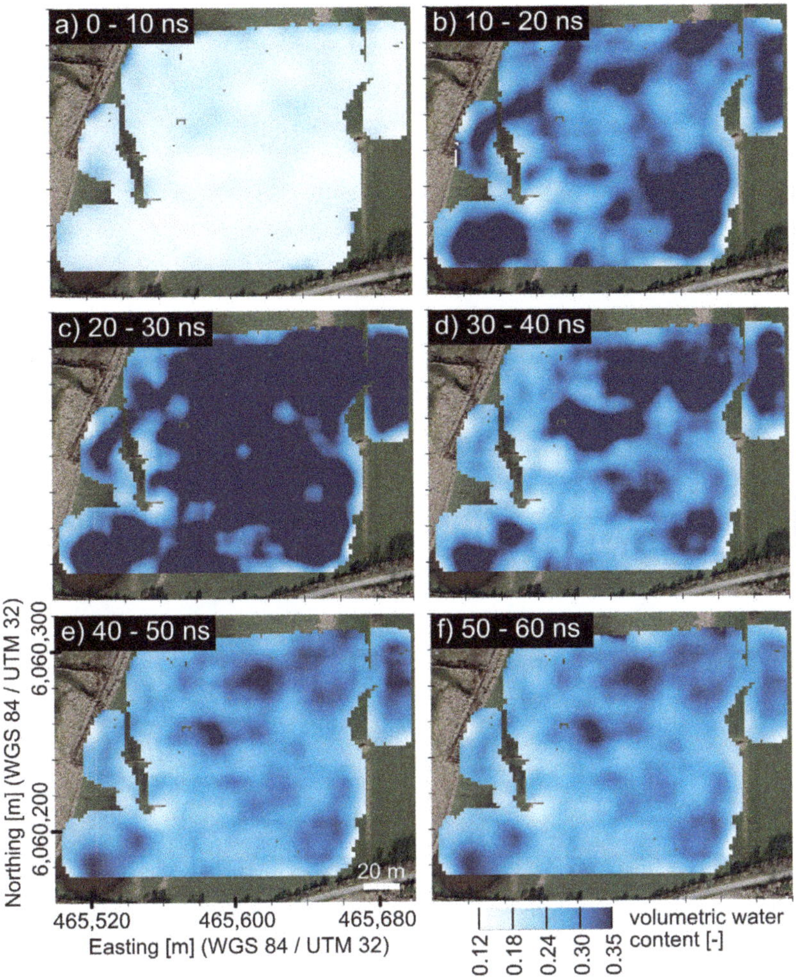

Figure 11. Volumetric water content maps at different depths (Basemap: ©GeoBasis-DE/LVermGeo SH/CC BY 4.0).

4. Discussion

4.1. Archaeological Interpretation

The GPR measurements in areas 3 and 6 (Figure 3b) cannot be interpreted properly due to high attenuation and technical problems. Still, some hyperbolae were detected here, which enabled us to determine a velocity and thus a water content model for these places. In the aerial photographs, one can also see a line of seaweed in area 3, indicating that salty North Sea water is sometimes overflowing up to this point through the wetland depression in the west. This causes higher electrical conductivity of the soils and thus higher attenuation of the GPR signals.

The topographic 2D migration with the derived 3D velocity model yields depthslices showing clearer results compared to the timeslices. In area 4, different linear features are visible, which can be interpreted as ditches. In combination with the locations of hyperbolae that align here, we can surmise that there are probably stones in the ditches, as was also found in an excavation. These ditches are partly visible as negative anomalies also on the magnetic map, but they are more pronounced in the depthslices. Traces of a previously unknown longhouse appear in the slice between 80 and 100 cm, which was not visible in either the magnetic map or timeslices.

An advantage of interpreting the depthslices together with the locations of hyperbolae is the differentiation between ditches and artifacts: In Figure 10a, the approximately north–south-running ditches are seen as high-reflectivity linear features in the depthslice, and that is supported by the alignment of hyperbolae, whereas the west–east-running (slightly zigzag) line has no alignment of hyperbolae and can be interpreted as an artifact. This is caused by the west–east profile direction of measurement and not fully adjusted antenna signals between channels in this location.

In area 5, the magnetic map as well as the GPR results are difficult to interpret in an archaeological sense. The depthslices mainly show geological structures and could hint at a surficial occurrence of till in this location.

The derived water content shows a strong maximum between about 60 and 120 cm depths below the surface. This is the zone of anthropogenic activity we are interested in, consisting of applied soil layers and occupation layers with a high humus content and high water-holding capacity. In contrast, the underlying Pleistocene moraine sand is well drained and drier. In the days before the measurements, there was heavy rainfall, which could explain the high water contents in the archaeological layer.

4.2. Further Considerations

With the proposed deep learning approach [18], 38,490 hyperbolae were detected in an area of 1.76 ha, with a maximum of 9 hyperbola per square meter between 20 and 30 ns. The velocities of neighboring hyperbolae vary quite strongly, which we do not attribute to real velocity changes in the subsurface, but rather to misdetected features (false positive), distorted hyperbola, and/or skewed automatic velocity determination. The application of strong smoothing in all directions yields a 3D velocity still capable of explaining moderate local velocity variations. For comparison, a mean 1D velocity model can be calculated by averaging all velocities over each time interval from the complete area. The maximum difference between this mean 1D velocity model and the 3D velocity model is 0.02 m/ns at the surface and increases to a maximum of 0.048 m/ns at later times. These pronounced local velocity variations lead to local differences in depth determination compared to a 1D or constant velocity model. In order to estimate the error made by using a 1D or constant velocity model compared to the (true) 3D model, we calculated the corresponding depths for all traveltimes for different models (Figure 12). The depths derived by a mean 1D velocity model (dark blue curve) and constant velocities (approximate mean velocity is 0.08 m/ns, violet curve; green and light blue curves for adjacent velocities) are similar, with maximum differences of about 0.5 m for long traveltimes. But if we assume a higher or lower mean 1D velocity function by 0.05 m/ns (which is comparable to the maximum variation in the 3D model compared to the mean 1D model), the differences are greater: about 0.5 m at 20 ns and up to 2 m at 80 ns. These assumed-wrong depth determinations are non-negligible when interpreting the archaeological structures and when planning targeted excavations. Thus, the 3D velocity model has a high impact on the correct depths determined from the GPR data.

The 3D velocity model also enables us to estimate the water contents of the site at different depths. At first glance, this might not be interesting with respect to archaeological interpretation, but at least in the geological setting on the island of Föhr, we can use this proxy to estimate the thickness of archaeologically relevant layers. These are normally rich in humus and can retain the water better compared to the more drained in situ Pleistocene sand below. Although the derivation of water content from GPR velocity is dependent on the material and several different transfer functions exist (e.g., [35,36]), we derive relative changes in water content independent from the model used. Thus, the easiest Topp equation [34] is sufficient.

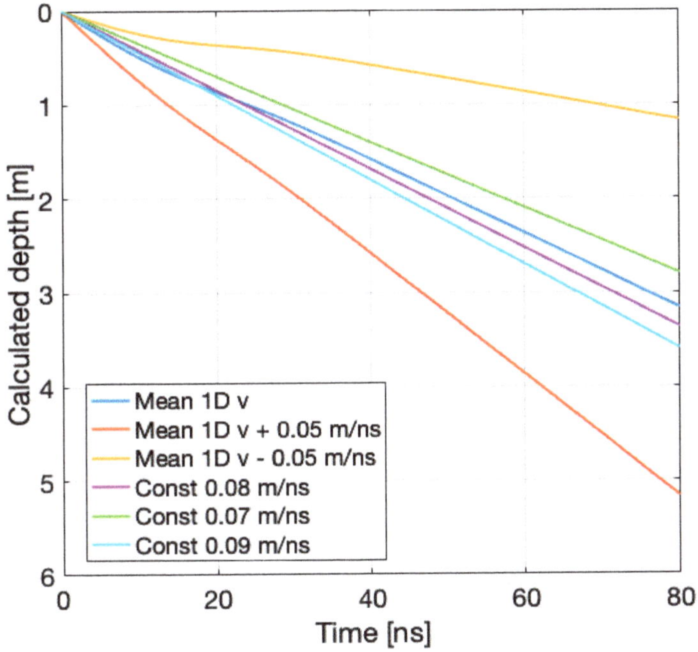

Figure 12. Comparison of depth calculations using a mean 1D velocity model, a mean 1D velocity model varied by 0.05 m/ns, or constant velocity models.

The automatic detection of hyperbola had a mean average precision of 0.58, a recall of 0.70, and a precision of 0.63 for the test data [18]. This means that about 70% of all hyperbolae were detected, but that only 63% of all detections were real hyperbola. This can be discussed with respect to several aspects:

(a) We can derive a smooth 3D velocity model from the detected hyperbolae in combination with automatic velocity determination. As we need a relatively smooth model for the migration and as we assume that we cannot resolve sharp velocity changes, the number of detected hyperbolae is sufficient.

(b) During automatic velocity determination on the detected hyperbolae, the only quality control of the hyperbolae is carried out by defining a valid velocity range between 0.05 and 0.15 m/ns. If a determined velocity is outside this range, it will be filtered out, which occurred for 20% of the detected hyperbolae in this study. As an improvement, further quality criteria might be useful in order to exclude false positive detections. These could be metrics such as the skewness of the detected objects or the amplitude with respect to the background. The first is probably universally applicable, whereas the second is surely site dependent (and probably also depends on the specific location on site). Since the shapes of the hyperbolae not only depend on depth (indirectly given by t_0) and the velocity but also on the dimension of the object, one could also try to include this parameter in the classification scheme and thus also improve the resulting velocity.

(c) The heterogeneity of the subsurface is very high on this site, which strongly influences the GPR signal, leading to overlapping and interfering reflections. Still, the automatic hyperbola detection is satisfactorily working on these data and will probably increase in performance if the amount of training data increases.

(d) Before the start of this study, one hypothesis was that we can use the distribution pattern of the detected hyperbolae as an indicator for archaeological settlement activity: A higher density of hyperbolae in some places might indicate higher settlement activity, with more small objects, such as post holes or stones, "disturbing" the homogeneous background subsurface. On the site considered in this article, we can see from the data that there are a lot of hyperbolae caused by geological structures (stones) not related to archaeology. Thus, the assumption of an otherwise homogeneous background soil is violated here, except if the geological structures were distributed homogeneously. But this cannot be proven. On other sites, the hypothesis is valid: One example is the site Hunte 1 [38], where a high spatial density of hyperbolae caused by partly worked piles and timbers indicated settlement activity and thus extended the formerly known settlement area, which had been excavated previously. On the site Hunte 1, the hyperbolae were found by visual inspection and picked manually. If we assumed an average precision of 0.58 in this case, the picture of spatial hyperbola distribution would look different. In order to achieve a realistic hyperbola distribution, the recall of 0.70 is not satisfying and needs to be improved. Still, at the Goting site, we could identify the alignment of detected hyperbolae in the ditches, and we can use this additional information to find and exclude artifacts from the interpretation.

5. Conclusions

We have shown that the automatic detection of diffraction hyperbolae in radargrams from an archaeological site can aid the archaeological interpretation, although the automatic detection is not perfect. In particular, we found following:

1. It is feasible to use automatically determined velocities to create a 3D velocity model. This is essential for correct depth calculations and migration, leading to improved visualization using depthslices that follow the topography. These depthslices revealed a possible new longhouse structure at between 80 and 100 cm depth, which was not visible in the time-domain data. Strong linear reflections were attributed to ditches, partly including stones, which were seen in both the time-domain and depth-domain data. The relatively low precision of detection and thus the number of detected hyperbolae were sufficient for this task.
2. At our site, the geological setting contained a lot of stones not attributed to archaeology, but which were equally detected as hyperbolae in the data. Thus, the spatial distribution of hyperbolae could not be used for differentiation of settlement activities. On other sites with "homogeneous" backgrounds and hyperbolae purely caused by archaeologically relevant features, the spatial distribution of hyperbolae can be interpreted in an archaeological way, but for this task, the recall of 0.70 of detections needs to be improved.
3. An aspect normally not included in archaeological interpretation is the 3D water content distribution, which can be derived through the 3D velocity model. In our geological setting, with well-drained glacial sand below the humus, representing a rich and water-retaining archaeological layer, differentiation between the two layers is possible based on the water content. This can be used to estimate the depths required for excavations.

A question for further research is the applicability to new sites, both with respect to (a) the data processing and (b) the archaeological interpretation. We are sure that the training data used here will not be representative of all cases. Therefore, re-training with new data is needed. It should be evaluated whether, at some point, the collected training data are sufficient for most applications, or if a continuous replenishment of training data is needed. For the second point, researchers should assess how different geological settings alter the interpretation scheme proposed in this study.

Author Contributions: Conceptualization, T.W.; methodology, T.W.; software, T.W.; validation, T.W.; formal analysis, T.W.; investigation, D.W. and B.S.M.; resources, D.W. and B.S.M.; data curation, D.W. and T.W.; writing—original draft preparation, T.W.; writing—review and editing, T.W., D.W., B.S.M., M.S. and W.R.; visualization, T.W.; supervision, D.W., M.S. and W.R.; project administration, D.W.,

M.S. and W.R.; funding acquisition, D.W., M.S. and W.R. All authors have read and agreed to the published version of the manuscript.

Funding: The work presented here was funded by the Deutsche Forschungsgemeinschaft (DFG, German Research Foundation) in two projects: Project-ID 290391021 (SFB 1266) and RA 496/26-2 (within SPP1630).

Data Availability Statement: The original contributions presented in the study are included in the article, further inquiries can be directed to the corresponding author.

Acknowledgments: The computations were performed on the NEC-HPC Linux-Cluster at the computation center of Kiel University. We acknowledge the helpful comments of four reviewers.

Conflicts of Interest: The authors declare no conflicts of interest.

References

1. Trinks, I.; Hinterleitner, A.; Neubauer, W.; Nau, E.; Löcker, K.; Wallner, M.; Gabler, M.; Filzwieser, R.; Wilding, J.; Schiel, H.; et al. Large-Area High-Resolution Ground-Penetrating Radar Measurements for Archaeological Prospection. *Archaeol. Prospect.* **2018**, *25*, 171–195. [CrossRef]
2. Gamba, P.; Lossani, S. Neural Detection of Pipe Signatures in Ground Penetrating Radar Images. *IEEE Trans. Geosci. Remote Sens.* **2000**, *38*, 790–797. [CrossRef]
3. Shaw, M.R.; Millard, S.G.; Molyneaux, T.C.K.; Taylor, M.J.; Bungey, J.H. Location of Steel Reinforcement in Concrete Using Ground Penetrating Radar and Neural Networks. *NDT E Int.* **2005**, *38*, 203–212. [CrossRef]
4. Kang, M.-S.; Kim, N.; Im, S.B.; Lee, J.-J.; An, Y.-K. 3D GPR Image-Based UcNet for Enhancing Underground Cavity Detectability. *Remote Sens.* **2019**, *11*, 2545. [CrossRef]
5. Ishitsuka, K.; Iso, S.; Onishi, K.; Matsuoka, T. Object Detection in Ground-Penetrating Radar Images Using a Deep Convolutional Neural Network and Image Set Preparation by Migration. *Int. J. Geophys.* **2018**, *2018*, 1–8. [CrossRef]
6. Lei, W.; Hou, F.; Xi, J.; Tan, Q.; Xu, M.; Jiang, X.; Liu, G.; Gu, Q. Automatic Hyperbola Detection and Fitting in GPR B-Scan Image. *Autom. Constr.* **2019**, *106*, 102839. [CrossRef]
7. Feng, J.; Yang, L.; Wang, H.; Song, Y.; Xiao, J. GPR-Based Subsurface Object Detection and Reconstruction Using Random Motion and DepthNet. In Proceedings of the 2020 IEEE International Conference on Robotics and Automation (ICRA), Paris, France, 31 May–31 August 2020; IEEE: Paris, France, 2020; pp. 7035–7041.
8. Liu, H.; Lin, C.; Cui, J.; Fan, L.; Xie, X.; Spencer, B.F. Detection and Localization of Rebar in Concrete by Deep Learning Using Ground Penetrating Radar. *Autom. Constr.* **2020**, *118*, 103279. [CrossRef]
9. Liu, Z.; Wu, W.; Gu, X.; Li, S.; Wang, L.; Zhang, T. Application of Combining YOLO Models and 3D GPR Images in Road Detection and Maintenance. *Remote Sens.* **2021**, *13*, 1081. [CrossRef]
10. Hou, F.; Lei, W.; Li, S.; Xi, J. Deep Learning-Based Subsurface Target Detection from GPR Scans. *IEEE Sens. J.* **2021**, *21*, 8161–8171. [CrossRef]
11. Li, W.; Cui, X.; Guo, L.; Chen, J.; Chen, X.; Cao, X. Tree Root Automatic Recognition in Ground Penetrating Radar Profiles Based on Randomized Hough Transform. *Remote Sens.* **2016**, *8*, 430. [CrossRef]
12. Mertens, L.; Persico, R.; Matera, L.; Lambot, S. Automated Detection of Reflection Hyperbolas in Complex GPR Images With No A Priori Knowledge on the Medium. *IEEE Trans. Geosci. Remote Sens.* **2016**, *54*, 580–596. [CrossRef]
13. Cornett, R.L.; Ernenwein, E.G. Object-Based Image Analysis of Ground-Penetrating Radar Data for Archaic Hearths. *Remote Sens.* **2020**, *12*, 2539. [CrossRef]
14. Allroggen, N.; Tronicke, J.; Delock, M.; Böniger, U. Topographic Migration of 2D and 3D Ground-penetrating Radar Data Considering Variable Velocities. *Near Surf. Geophys.* **2015**, *13*, 253–259. [CrossRef]
15. Verdonck, L.; Simpson, D.; Cornelis, W.; Plyson, A.; Bourgeois, J. Analysing the Velocity of Ground-Penetrating Radar Waves: A Case Study from Koekelare (Belgium). In *Proceedings of the 1st International EARSeL Workshop "Advances on Remote Sensing for Archaeology and Cultural Heritage Management"*; Lasaponara, R., Masini, N., Eds.; Aracne editrice SRL: Rome, Italy, 2008; pp. 143–146, ISBN 978-88-548-2030-2.
16. Angelis, D.; Warren, C.; Diamanti, N.; Martin, J.; Peter Annan, A. Development of a Workflow for Processing Ground-Penetrating Radar Data from Multiconcurrent Receivers. *Geophysics* **2022**, *87*, WB9–WB18. [CrossRef]
17. Kaufmann, M.S.; Klotzsche, A.; Vereecken, H.; Van Der Kruk, J. Simultaneous Multichannel Multi-offset Ground-penetrating Radar Measurements for Soil Characterization. *Vadose Zone J.* **2020**, *19*, e20017. [CrossRef]
18. Wunderlich, T.; Wilken, D.; Majchczack, B.S.; Segschneider, M.; Rabbel, W. Hyperbola Detection with RetinaNet and Comparison of Hyperbola Fitting Methods in GPR Data from an Archaeological Site. *Remote Sens.* **2022**, *14*, 3665. [CrossRef]
19. Leckebusch, J. Two- and Three-Dimensional Ground-Penetrating Radar Surveys across a Medieval Choir: A Case Study in Archaeology. *Archaeol. Prospect.* **2000**, *7*, 189–200. [CrossRef]
20. Linford, N. A Vector-Based Approach to Assessing the Results from Constant Velocity Migration of GPR Data. *Archeosciences* **2021**, *45-1*, 263–266. [CrossRef]

21. Taner, M.T.; Koehler, F. Velocity spectra—Digital computer derivation applications of velocity functions. *Geophysics* **1969**, *34*, 859–881. [CrossRef]
22. Majchczack, B.S. *Die Rolle der Nordfriesischen Inseln im Frühmittelalterlichen Kommunikationsnetzwerk*; Studien zur Landschafts- und Siedlungsgeschichte im südlichen Nordseeraum; Verlag Marie Leidorf: Rahden/Westfahlen, Germany, 2020; ISBN 978-3-86757-341-2.
23. Majchczack, B.S.; Schneider, S.; Wunderlich, T.; Wilken, D.; Rabbel, W.; Segschneider, M. Early Medieval Trading Sites on the North-Frisian Island of Föhr. First Results of Fieldwork in Witsum and Goting. In *Harbours as Objects of Interdisciplinary Research – Archaeology + History + Geosciences*; von Carnap-Bornheim, C., Daim, F., Ettel, P., Warnke, U., Eds.; RGZM Tagungen; RGZM: Mainz, Germany, 2018; Volume 34, pp. 311–328.
24. Wunderlich, T. MultichannelGPR—A New MATLAB-Tool for the Processing of GPR Data. *Archeosciences* **2021**, *45*, 279–283. [CrossRef]
25. Wilken, D.; Feldens, P.; Wunderlich, T.; Heinrich, C. Application of 2D Fourier Filtering for Elimination of Stripe Noise in Side-Scan Sonar Mosaics. *Geo-Mar. Lett.* **2012**, *32*, 337–347. [CrossRef]
26. Wilken, D.; Wunderlich, T.; Zori, D.; Kalmring, S.; Rabbel, W.; Byock, J. Integrated GPR and Archaeological Investigations Reveal Internal Structure of Man-Made Skiphóll Mound in Leiruvogur, Iceland. *J. Archaeol. Sci. Rep.* **2016**, *9*, 64–72. [CrossRef]
27. Lin, T.-Y.; Dollar, P.; Girshick, R.; He, K.; Hariharan, B.; Belongie, S. Feature Pyramid Networks for Object Detection. In Proceedings of the 2017 IEEE Conference on Computer Vision and Pattern Recognition (CVPR), Honolulu, HI, USA, 21–26 July 2017; IEEE: Honolulu, HI, USA, 2017; pp. 936–944.
28. Lin, T.-Y.; Goyal, P.; Girshick, R.; He, K.; Dollár, P. Focal Loss for Dense Object Detection. *arXiv* **2017**, arXiv:1708.02002. [CrossRef]
29. Henon, Y. *Pytorch-Retinanet 2021*; GitHub: San Francisco, CA, USA, 2021.
30. He, K.; Zhang, X.; Ren, S.; Sun, J. Deep Residual Learning for Image Recognition. In Proceedings of the 2016 IEEE Conference on Computer Vision and Pattern Recognition (CVPR), Las Vegas, NV, USA, 27–30 June 2016; IEEE: Las Vegas, NV, USA, 2016; pp. 770–778.
31. Dou, Q.; Wei, L.; Magee, D.R.; Cohn, A.G. Real-Time Hyperbola Recognition and Fitting in GPR Data. *IEEE Trans. Geosci. Remote Sens.* **2017**, *55*, 51–62. [CrossRef]
32. Ester, M.; Kriegel, H.-P.; Sander, J.; Xiaowei, X. A Density-Based Algorithm for Discovering Clusters in Large Spatial Databases with Noise. In Proceedings of the Proceedings of the Second International Conference on Knowledge Discovery in Databases and Data Mining, Portland, OR, USA, 2–4 August 1996; AAAI Press: Portland, OR, USA, 1996; pp. 226–231.
33. Dix, C.H. Seismic velocities from surface measurements. *Geophysics* **1955**, *20*, 68–86. [CrossRef]
34. Topp, G.C.; Davis, J.L.; Annan, A.P. Electromagnetic Determination of Soil Water Content: Measurements in Coaxial Transmission Lines. *Water Resour. Res.* **1980**, *16*, 574–582. [CrossRef]
35. Steelman, C.M.; Endres, A.L. Comparison of Petrophysical Relationships for Soil Moisture Estimation Using GPR Ground Waves. *Vadose Zone J.* **2011**, *10*, 270–285. [CrossRef]
36. Wunderlich, T.; Petersen, H.; Attia Al Hagrey, S.; Rabbel, W. Pedophysical Models for Resistivity and Permittivity of Partially Water-Saturated Soils. *Vadose Zone J.* **2013**, *12*, 1–14. [CrossRef]
37. Burschil, T.; Scheer, W.; Kirsch, R.; Wiederhold, H. Compiling Geophysical and Geological Information into a 3-D Model of the Glacially-Affected Island of Föhr. *Hydrol. Earth Syst. Sci.* **2012**, *16*, 3485–3498. [CrossRef]
38. Wunderlich, T.; Brozio, J.P.; Feeser, I.; Heumüller, M.; Mohr, C. Hunte 1 Reloaded—Combining Ground Penetrating Radar, Electrical Resistivity Tomography, Corings and Excavations at the Neolithic Domestic Site Hunte 1, Germany. In *Advances in On- and Offshore Archaeological Prospection*; Wunderlich, T., Hadler, H., Blankenfeldt, R., Eds.; Universitätsverlag Kiel | Kiel University Publishing: Kiel, Germany, 2023; pp. 305–309, ISBN 978-3-928794-83-1.

remote sensing

MDPI

Article

Study on the Identification Method of Planar Geological Structures in Coal Mines Using Ground-Penetrating Radar

Jialin Liu [1,*], Xiaosong Tang [1,†], Feng Yang [1], Xu Qiao [1], Fanruo Li [1], Suping Peng [2], Xinxin Huang [1], Yuanjin Fang [1] and Maoxuan Xu [1]

[1] School of Artificial Intelligence, China University of Mining and Technology (Beijing), Beijing 100083, China; bqt2310405005@student.cumtb.edu.cn (X.T.); yangf@cumtb.edu.cn (F.Y.); qiaoxu@cumtb.edu.cn (X.Q.); bqt2200405019@student.cumtb.edu.cn (F.L.); zqt2200405128@student.cumtb.edu.cn (X.H.); bqt1900405017@student.cumtb.edu.cn (Y.F.); xmx@cumtb.edu.cn (M.X.)

[2] State Key Laboratory for Fine Exploration and Intelligent Development of Coal Resources, China University of Mining & Technology (Beijing), Beijing 100083, China; psp@cumtb.edu.cn

* Correspondence: bqt2000405023@student.cumtb.edu.cn; Tel.: +86-132-0632-2893

† This author is the first author for this manuscript.

Abstract: The underground detection environment in coal mines is complex, with numerous interference sources. Traditional ground-penetrating radar (GPR) methods suffer from limited detection range, high noise levels, and weak deep signals, making it extremely difficult to accurately identify geological structures without stable feature feedback. During research, it was found that the detection energy of the same target significantly changes with the antenna direction. Based on this phenomenon, this paper proposes a geological radar advanced detection method using spatial scanning. This method overcomes constraints imposed by the underground coal mine environment on detection equipment, enhancing both detection range and accuracy compared to traditional approaches. Experiments using this method revealed pea-shaped response characteristics of planar geological structures in radar images, and the mechanisms behind their formation were analyzed. Additionally, this paper studied the changes in response characteristics under changes in target inclination, providing a basis for understanding the spatial distribution of geological structures. Finally, application experiments in underground coal mine environments explored the practical potential of this method. Results indicate that, compared to drilling data, this method achieves identification accuracies of 91.88%, 90.42%, and 78.72% for the depth and spatial extent of geological structures, providing effective technical support for coal mining operations.

Keywords: coal mine; advanced detection; ground-penetrating radar; spatial scanning; planar geological structure; feature analysis

1. Introduction

China's energy structure is characterized by being 'rich in coal, poor in oil, and scarce in gas', with coal playing a dominant role in energy production and consumption. According to relevant studies, coal will continue to account for over 50% of China's primary energy consumption from 2030 to 2050 [1]. In February 2020, China released *Guiding Opinions on Accelerating the Development of Intelligent Coal Mines*, emphasizing the need to achieve key technological breakthroughs in precise geological detection for coal mines, intelligent perception of major hazard sources, and early warning systems. The plan aims to establish a comprehensive set of technical standards for intelligent coal mining by 2025, thereby promoting high-quality development in the coal industry and accelerating its transformation and upgrading [2–4]. The fundamental premise of intelligent coal mine construction is a comprehensive and in-depth understanding of the geological environment of the mine, ensuring full transparency of the geological conditions at the working face [5,6]. In current coal mining operations, geological transparency primarily relies on the collection and analysis of geological data. Through various exploration techniques such as drilling and physical and chemical prospecting, geological data are deeply explored and thoroughly revealed to achieve a comprehensive understanding of the working face's geological conditions, particularly in relation to key information on stratigraphy, geological structures, and geological anomalies [7,8]. However, due to various factors, different exploration techniques exhibit issues such as inaccurate detection results, low precision, and insufficient depth to meet the demands of mining operations, leading to significant differences in the transparency capabilities for the coal mine working face [9,10]. The main reasons for these issues include: (1) the complex underground environment in coal mines, especially at the tunneling face, which often faces harsh conditions such as high temperature, high humidity, high dust, high noise, and high gas levels; (2) difficult underground transportation and narrow working spaces impose strict constraints on the layout of geophysical systems and the size and weight of detection equipment; (3) the numerous interference sources at the working face, such as rail tracks, anchor nets, industrial electricity, and mechanical vibrations, coupled with the requirement for explosion-proof detection equipment and limited transmission power, resulting in weakened signals from deeper layers. In the absence of stable and reliable geological structural features, it becomes extremely challenging to effectively identify geological information from interference signals [11–13]; (4) the quasi-full-space conditions at the working face make it difficult to directly apply surface detection theories, thereby increasing the complexity of data processing and interpretation [14,15]. These challenges pose significant obstacles to achieving full geological transparency at the coal mine working face. Meanwhile, as deep mining accelerates, the demand for transparency in the coal mine geological environment continues to grow.

To achieve geological transparency at the coal mine working face, refined geological surveys of the working face conditions are necessary. Among various detection technologies, ground-penetrating radar (GPR) exhibits significant advantages due to its high resolution, strong anti-interference capabilities, and non-destructive detection characteristics [16–18]. In recent years, both domestic and international scholars have conducted studies on the application of GPR in the coal mining field. Domestic scholars, such as Peng S.P. and Feng Y. [19,20], have developed deep-penetration GPR systems and equipment to meet the requirements for explosion-proof operations and ultra-deep detection in mines, increasing the detection depth from several meters to tens of meters. Xu X.L. and colleagues improved the performance of antenna systems through a new microwave front-end design, optimizing the size of the antenna's radiation surface and shielding cavity, making GPR systems more suitable for detecting geological hazards in coal mine tunnels [21]. In experiments on detecting coal–rock interfaces in fully mechanized mining faces, the detection error was within 1.2 cm, with an average error of 8.6%, demonstrating the superiority of the equipment in complex geological environments [22]. Additionally, a coal–rock stratigraphy positioning algorithm was proposed based on a two-dimensional mapping of relative coal thickness errors within the antenna's main lobe extremum range, reducing the shallow coal–rock interface identification error to less than 7.6% [23]. Due to the lack of explosion-proof radar equipment abroad, surface radar systems are unsuitable for underground coal mine environments, with most foreign research focusing on detecting hazards in near-surface coal mines. For example, Strange et al. proposed a method for detecting thin coal thickness near the surface, and validated it in a laboratory environment using a 1.4 GHz radar antenna, with an effective detection range for coal seam thickness within 5 cm [24]. Jecny and others used GPR systems to scan the thickness of coal seams in open-pit mines, achieving detection results with errors within 5% of actual measurements [25]. Other scholars have also conducted research in areas such as subsurface geological structure surveys [26,27] and surface mineral resource assessments [28]. In summary, research on GPR detection in mine environments remains relatively scarce, with most studies focusing on the development and optimization of technical equipment. Given the limited working face space and extreme operating conditions in underground coal mines, accurately identifying geological structures and effectively obtaining spatial distribution information about geological structures continues to pose significant challenges for existing GPR technology and detection methods.

To address challenges associated with traditional GPR detection methods in underground coal mine environments, this study presents an innovative spatial scanning detection approach. Throughout the research process, significant planar geological structures, such as fault lines and coal seam thinning zones, were selected as subjects of investigation. Experiments were designed to analyze the response characteristics of these planar geological structures in GPR detection results. Additionally, detection experiments targeting different angles of geological structures were conducted to examine the spatial variation of target response characteristics. Finally, application experiments were carried out in a coal mining recovery area in Hebei Province, China. Through comparative analysis of detection results and drilling data, the feasibility of this method in actual production environments was validated.

2. Materials and Methods

The process of advanced detection using GPR in coal mines (Figure 1) involves directing high-frequency pulsed electromagnetic waves from a transmitting antenna (TX in Figure 1) towards the mining face. The reflected signals are then captured by a receiving antenna (TR in Figure 1), aiming to conduct a detailed investigation of the geological structures in the unknown areas ahead of the working face. During this process, electromagnetic waves penetrate through the coal mine medium and encounter various physical phenomena such as reflection, transmission, and refraction upon encountering significant differences in electrical characteristics of anomalous geological structures like aquifers, goafs, faults, fractured zones, and collapsed pillars [29]. Through thorough analysis of received electromagnetic signals, spatial attributes including types and shapes of anomalous geological structures within the coal mine can be estimated. The distance L between geological structures and observation points can be determined using Equations (1) and (2):

$$v = c/\sqrt{\varepsilon_r} \quad (1)$$

$$L = \frac{\sqrt{(vt)^2 - d^2}}{2} \quad (2)$$

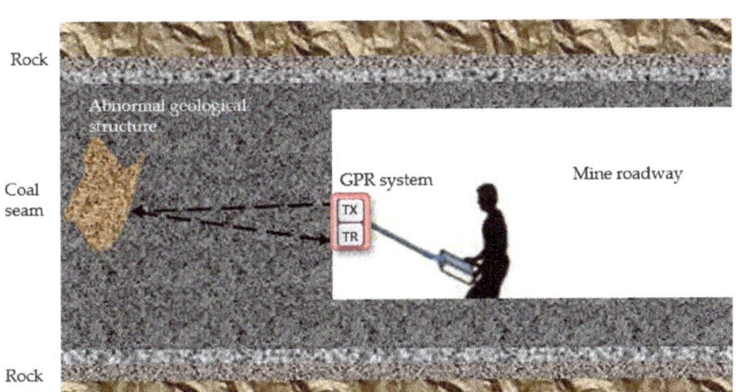

Figure 1. Advanced detection schematic of the coal mine working face.

In the equation, v represents the propagation velocity of electromagnetic waves in a medium with a dielectric constant ε_r (m/ns); t denotes the bidirectional travel time of the electromagnetic waves (ns); d is the distance between the transmitting and receiving antennas (m) (depth detection can be ignored); and c is the propagation velocity of electromagnetic waves in a vacuum (0.3 m/ns) [30].

The traditional advanced detection method of GPR in coal mine environments (Figure 2b) typically involves laying survey lines on the mining face to detect and acquire geological information about the unknown areas ahead. However, this traditional approach presents several limitations:

1. Limited detection accuracy: As shown in Figure 2a, the coal mining face is a fully enclosed and confined space, with typical height and width dimensions of 3 m and 5 m, respectively. In extreme geological conditions, this space becomes even narrower, severely restricting the layout of survey lines. Due to these spatial constraints, only one or two survey lines can be arranged on the mining face, leading to highly limited spatial coverage and data acquisition (Figure 2c). This makes precise analysis of the unknown geological conditions ahead extremely difficult.
2. Limited detection range: As shown in Figure 2b, the survey lines are positioned on the mining face, allowing for detection only in the direction of mining operations. However, it is ineffective in detecting potential geological hazards in areas beyond the observation range, such as the roof of the coal mine roadway (Figure 2a).
3. Limited detection depth: While the theoretical detection depth of GPR in mine environments can reach tens of meters, the actual effective depth is limited to only a few meters to over ten meters. This is primarily due to the exponential attenuation of high-frequency electromagnetic waves as they propagate through the medium, weakening the strength of deep signals. Additionally, signals reflected by steel supports, metal mesh, and electromagnetic interference from mining equipment, along with scattered signals in confined spaces, are all captured by the receiving antenna, further reducing the signal-to-noise ratio of deep detections. In the absence of stable geological structure features and prior geological information, effectively distinguishing geological information from the interference signals becomes extremely difficult, presenting additional challenges for the processing, analysis, and interpretation of deep data.
4. Limited spatial information on geological structures: GPR is based on the principle of electromagnetic wave reflection and detects geological structures according to differences in dielectric properties. When geological structures develop in horizontal or vertical directions and are either parallel or at steep angles to the mining direction, their reflected signals may be directed into unknown areas and cannot be captured by the receiving antenna. This can result in suboptimal detection of geological structures, making it challenging to obtain critical geological information such as spatial distribution and extension direction.

In response to the limitations of traditional advanced detection methods using GPR in underground coal mine environments, this study proposes a novel GPR-based advanced detection method utilizing spatial scanning, tailored to the actual working conditions of coal mining faces. As illustrated in Figure 3b, this method is implemented through a GPR spatial scanning system. During the detection process, the spatial scanning system is deployed at the coal mine working face (as shown in Figure 3a), and the antenna's horizontal and vertical rotation angles, θ and ψ, are adjusted to control the antenna's orientation. This adjustment modifies the transmission and reception directions of the antenna, thereby acquiring geological information from unknown spatial areas (Figures 3c and 4e). Throughout the detection process, the antenna's orientation and its transmission and reception directions are continuously changing. Let $\vec{p}$ and $\vec{p}_b$ represent the antenna's positions in the inertial and spatial coordinate systems, respectively, with the origins of both coordinate systems coinciding but differing in angular orientations [31]. In the initial state, the relationship between the spatial and inertial coordinate systems is $\vec{p}_b^{\,0} = \vec{p}$. The antenna performs horizontal scanning by rotating θ degrees around the y-axis (Figure 4a,b) and vertical scanning by rotating ψ degrees around the x-axis (Figure 4c,d). The rotation matrices for these two operations are $\mathcal{R}_y(\theta)$ and $\mathcal{R}_x(\psi)$, respectively. After rotation, the antenna's posture in space, $\vec{p}_b$, can be expressed through matrix multiplication as shown in Equation (3):

$$\begin{aligned}\vec{p}_b &= \mathcal{R}_x(\psi)\mathcal{R}_y(\theta)\vec{p} \\ &= \begin{bmatrix} \cos\theta & 0 & -\sin\theta \\ \sin\psi\sin\theta & \cos\psi & \sin\psi\cos\theta \\ \cos\psi\sin\theta & -\sin\psi & \cos\psi\sin\theta \end{bmatrix}\vec{p} \\ &= \mathcal{R}_{yx}(\theta,\psi)\,\vec{p}\end{aligned} \tag{3}$$

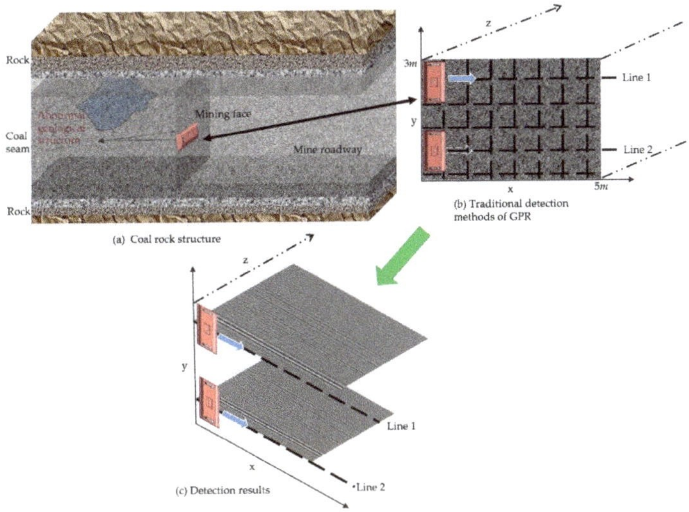

Figure 2. Schematic diagram of the application of the traditional GPR advanced detection method in the coal mine working face.

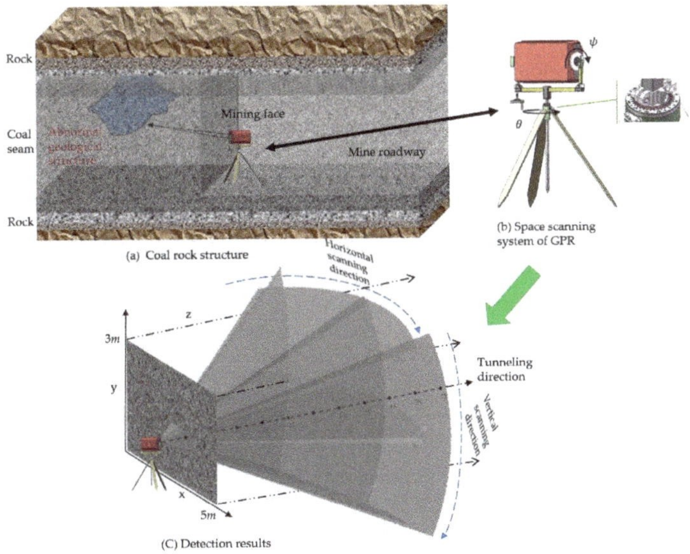

Figure 3. Schematic diagram of spatial scanning detection using GPR at the coal mine working face.

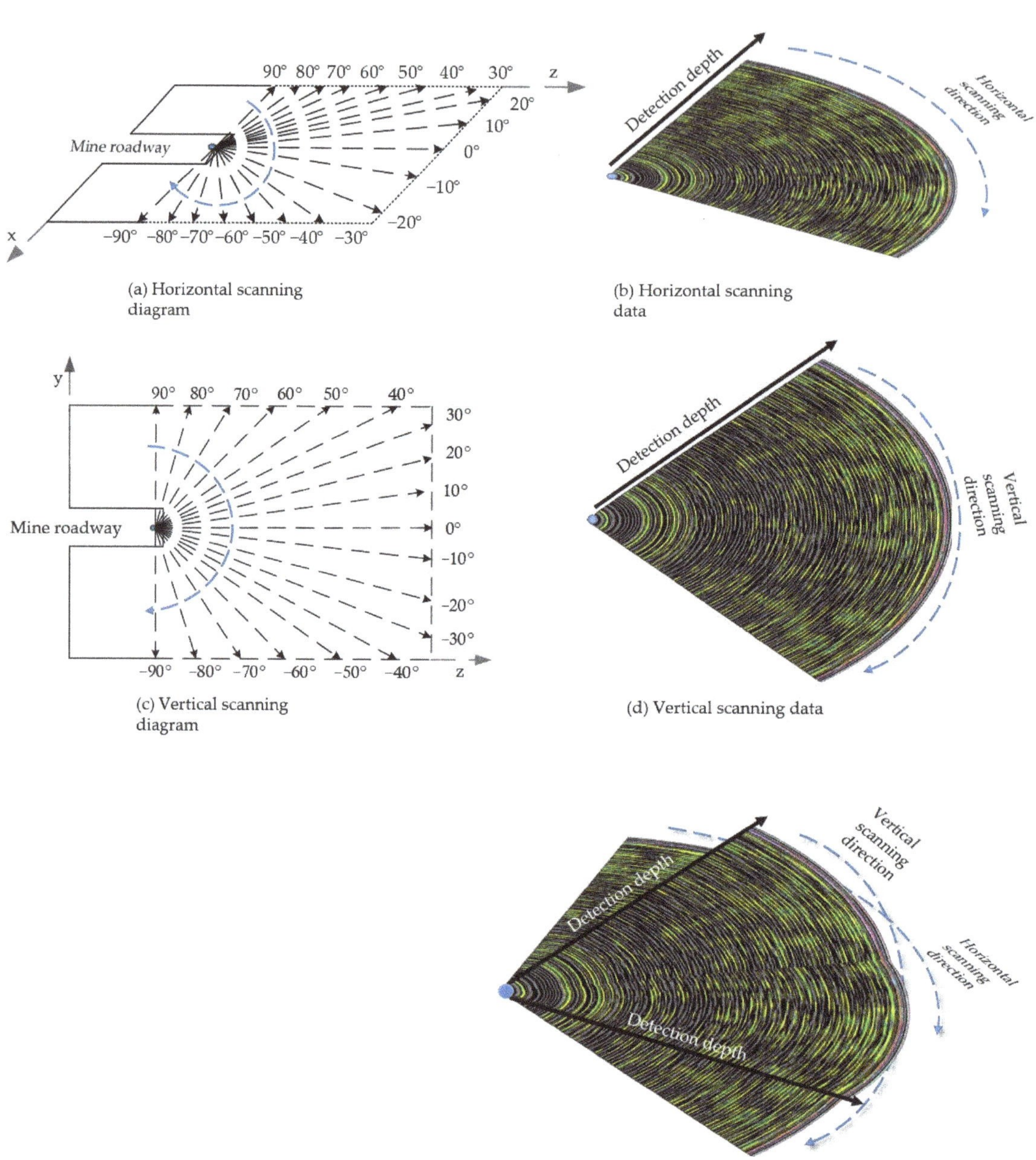

(a) Horizontal scanning diagram

(b) Horizontal scanning data

(c) Vertical scanning diagram

(d) Vertical scanning data

(e) Spatial scanning data

Figure 4. Schematic diagram of GPR horizontal and vertical scanning detection of the coal mine working face and the corresponding detection results.

In the equation, $\mathcal{R}_{yx}(\theta, \psi)$ represents the rotation matrix after the antenna undergoes horizontal and vertical scanning, which is influenced by the order of rotation. Due to the limitations of the detection equipment (as shown in Figure 4), the angles θ and ψ are restricted to the range $[-\pi/2,\pi/2]$. Therefore, the rotation matrix is expressed as shown in Equation (4):

$$\mathcal{R}_{yx}(\theta,\psi)=\begin{bmatrix}\cos\theta & 0 & -\sin\theta\\ \sin\psi\sin\theta & \cos\psi & \sin\psi\cos\theta\\ \cos\psi\sin\theta & -\sin\psi & \cos\psi\sin\theta\end{bmatrix}\vec{p} \tag{4}$$

In the process of the antenna scanning in space, its movement can be expressed by Equation (5):

$$\vec{p}=\vec{p}_0+\mathcal{R}_{yx}(\theta,\psi)^T\vec{p}_b \tag{5}$$

In the equation, $\vec{p}_0$ represents the initial orientation of the antenna. Given that the rotation axes are mutually orthogonal, the inverse of the rotation matrix $\mathcal{R}_{yx}(\theta,\psi)$ is equal to its transpose $\mathcal{R}_{yx}(\theta,\psi)^T$. Therefore, the reception of energy during the antenna's spatial scanning process, including the reflection and refraction of electromagnetic waves, can be described by Equation (6):

$$P_{L(\theta,\psi)}=Re\left\{\frac{P_tG^2\lambda^2}{64\pi^2L^2}e^{-2aL}\left(1-R_{air,coal}\right)^2R_{coal,rock}\right\}\mathcal{R}_{yx}(\theta,\psi) \tag{6}$$

In the equation, $P_{L(\theta,\psi)}$ represents the received power of the electromagnetic wave reflected by the geological structure at a depth L. P_t is the transmitted power of the antenna; G is the gain of the antenna's main lobe in the direction of detection; λ is the wavelength of the electromagnetic wave in the medium, which is related to the antenna's center frequency and the dielectric parameters of the medium; a is the attenuation factor of the electromagnetic wave in the lossy medium; and $R_{air,coal}$ and $R_{coal,rock}$ are the reflection coefficients at the air–coal and coal–rock interfaces, respectively.

In the underground coal mine environment, the aforementioned spatial scanning detection method offers significant advantages over traditional forward-looking detection techniques of GPR:

- Higher detection accuracy: During the spatial scanning process, the detection area is divided into $k \times k$ sector scanning units, with the antenna transmitting and receiving signals at the center of each unit. As the rotation angles θ and ψ are adjusted, the transmission points *Tri*, where $I = \theta_1\psi_1, \theta_1\psi_2,\ldots, \theta_1\psi_k,\ldots, \theta_k\psi_1, \theta_k\psi_2,\ldots, \theta_k\psi_k$, and the corresponding reception points *Rei*, change accordingly. This method allows for fine-grained management of the scanning space, significantly increasing the amount of spatial detection data (as shown in Figure 3c) and enhancing detection accuracy. Additionally, this detection approach eliminates the need for laying detection lines along the working face, thus freeing the detection equipment from the constraints of the working surface.
- Significant expansion of detection range: As illustrated in Figure 3a,b, there is no need to deploy survey lines above the mining working face for wall-attached detection, thereby removing spatial constraints imposed by the working face. By utilizing the spatial scanning system, comprehensive coverage from $-90°$ to $90°$ in both horizontal and vertical directions (as shown in Figure 4a,d) can be achieved, facilitating the exploration of nearly all potential detection areas along the mining direction.
- Capacity for obtaining spatial distribution information of geological structures: This method is especially suitable for planar geological structures (such as faults, thin coal layers, etc.), which exhibit stable response characteristics. Through detailed analysis of energy variation characteristics, geological structures can be accurately identified amidst numerous interference signals. Moreover, crucial geological information, including spatial distribution, can be obtained based on changes in rotation angles during the detection process. The subsequent sections of this paper will delve into how this method significantly enhances detection accuracy and retrieval of spatial distribution information.

3. Simulation Experiment and Result Analysis

3.1. Experimental Detection of Target Response Characteristics

3.1.1. Detection Process and Phenomena

To investigate the response characteristics of the proposed method and the imaging principles, simulated detection experiments were conducted at the South Plaza of China University of Mining and Technology (Beijing) Shahe Campus. In this experiment, as shown in Figure 5, two detection targets, f_1 and f_2, were set up to simulate the spatial distribution of mining faces and planar geological structures. Specifically, f_1 was a solid wall, made of solid brick and concrete with dimensions of 6 m height, 12 m width, and 0.5 m thickness; while f_2 was tin foil with dimensions of 1 m height, 1 m width, and 0.002 m thickness. The targets f_1 and f_2 were arranged in parallel with a separation of 6.5 m between them. The antenna was positioned at point p_0, 1 m away from target f_1, with its radiation plane initially forming a perpendicular relationship with the planes of targets f_1 and f_2. During the experiment, the antenna rotates clockwise (as shown in Figure 4a) with the vertical rotation angle fixed at 0°. The horizontal rotation occurs at intervals of 0.2°, covering a horizontal space from −90° to 90°. Additionally, the radar antenna is a 100 MHz shielded enhanced antenna from China University of Mining and Technology (Beijing), designed with explosion-proof capabilities for underground coal mines. The antenna operates at a central frequency of 100 MHz with a sampling frequency of 100 kHz, utilizing 1024 sampling points and a minimum sampling interval of 5 ps. The antenna's preamplifier provides a gain of 48 dB. Notably, the radar antenna supports both single-point and time-domain sampling modes. For this experiment, the single-point sampling mode was chosen with a sampling window set to 150 ns. During the detection process, the horizontal rotation angle θ is adjusted manually using the dial (as shown in Figure 3b), triggering data sampling with each adjustment, thereby completing the collection of horizontal spatial data from −90° to 90°.

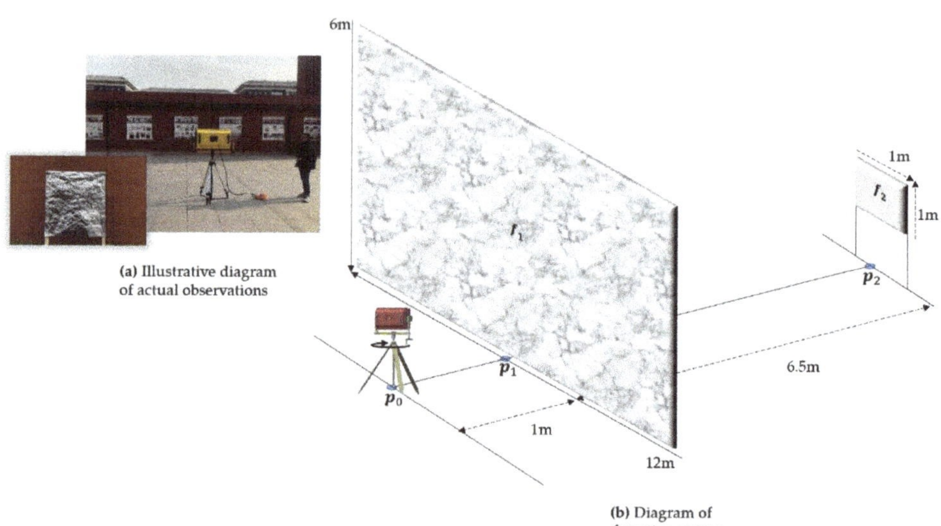

Figure 5. Diagram of spatial scan detection experiment using GPR.

The detection data were processed using conventional methods such as filtering, background noise reduction, and gain adjustment to obtain detection results (Figure 6). In the detection results, the bidirectional travel time difference between the peak energy positions of the direct wave and the anomalous reflection region R_1 was approximately 7 ns. Considering the speed of electromagnetic waves in air as 0.3 m/ns, R_1 was calculated to be approximately 1.05 m from the antenna. Given the 1.0 m distance between target f_1 and the antenna, and ignoring errors during the detection process, R_1 was confirmed as the electromagnetic signal response from f_1. Similarly, the bidirectional travel time difference between R_1 and R_2 was approximately 44 ns, estimating the distance between R_1 and R_2 to be around 6.6 m. Considering the 6.5 m distance between targets f_1 and f_2, R_2 was identified as the electromagnetic signal response from f_2. The characteristics of R_1 and R_2 differ from the traditional hyperbolic response features of GPR (Figure 6a). Specifically, the travel times of R_1 and R_2 remain stable regardless of the antenna's rotation angle, and their responses exhibit a linear distribution overall. In the radar data image, R_1 and R_2 appear brighter at their central positions, gradually becoming darker towards both sides at the same travel time, displaying smooth and regular variations. The overall contour of R_1 and R_2 shows higher brightness at the center, lower brightness at the sides, a thicker center, narrower sides, and a symmetrical distribution resembling a pea-shaped pattern, highlighting their regular distribution characteristics.

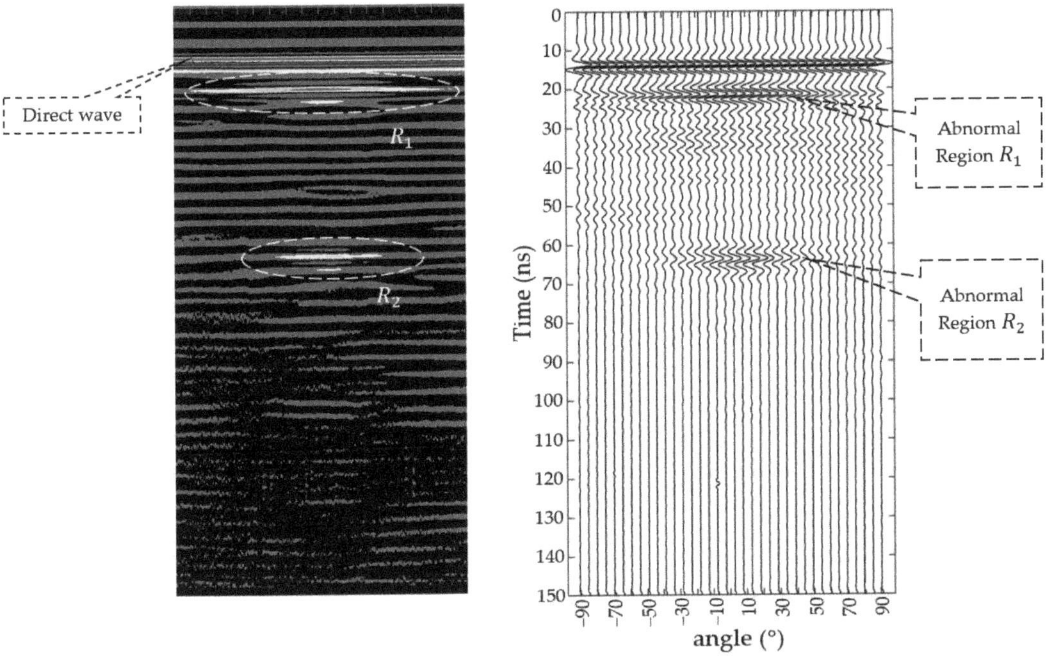

(**a**) Horizontal scanning data radar map (**b**) Horizontal scanning data waveform plot

Figure 6. Schematic diagram of GPR horizontal scanning detection of the coal mine working face and the corresponding detection results.

Figure 6b depicts the vertical axis as bidirectional travel time and the horizontal axis as horizontal scanning angle. Within the reflected wave groups R_1 and R_2, each component's energy exhibits a trend of initial enhancement followed by attenuation over time, resulting in significantly higher energy peaks at the central positions of the reflected sub-wavelets, while the energy peaks at the upper and lower sides are relatively weaker. The reflected wave groups display a symmetric state at the central position. During horizontal rotation from $-90°$ to $90°$, the energy of the reflected wave groups demonstrates a spatial pattern of initial enhancement followed by decay, gradually diminishing from the central position towards both sides. Throughout this attenuation process, the energy of the wavelet peaks at the upper and lower sides of the reflected wave groups decays more prominently due to their initial weaker intensity, whereas the energy of the wavelet at the central position exhibits less noticeable decay despite its stronger initial intensity. This results in R_1 and R_2 displaying a pea-shaped pattern with a thicker center, narrower sides, brighter middle, and darker sides.

3.1.2. Analysis of Detection Results

The formation of target response characteristics is related to antenna motion. During the spatial scan process of the antenna, its radiation direction continuously changes, leading to variations in the antenna radiation field with movement. Since commonly used GPR antennas excite electric fields, the far-field radiation field strength $E(r, \alpha, \beta)$ distributed at a distance r from the antenna can be expressed as Equation (7):

$$E(r, \alpha, \beta) = E_{max}(r, I_A) f(\alpha, \beta) e^{-jkr} \hat{\alpha} \tag{7}$$

In the equation, α and β represent the horizontal and vertical angles between the observation point and the radiation direction, respectively; $\hat{\alpha}$ is the directional unit vector; e^{-jkr} denotes the phase change of electromagnetic waves, varying with distance; E_{max} is the maximum electric field strength in the antenna's main radiation direction; I_A is the current amplitude; and $f(\alpha, \beta)$ is the antenna radiation pattern function, indicating the relative distribution of radiation intensity in the radiation space [32]. As depicted in Figure 7, the antenna directional radiation pattern was constructed using $f(\alpha, \beta)$ [33]; this illustrates the non-uniform distribution of electric field intensity in the radiation space. In the radiation pattern, the electric or magnetic field consists of multiple lobes, with the main lobe reflecting the antenna's strongest radiation direction E_{max}. The angle $2\delta_{0.5}$ on either side of the main lobe peak represents the half-power attenuation region, where the power density decreases to half (electric field strength of $0.707E_{max}$). For GPR antennas, the range of half-power attenuation region typically spans approximately 60° to 90°.

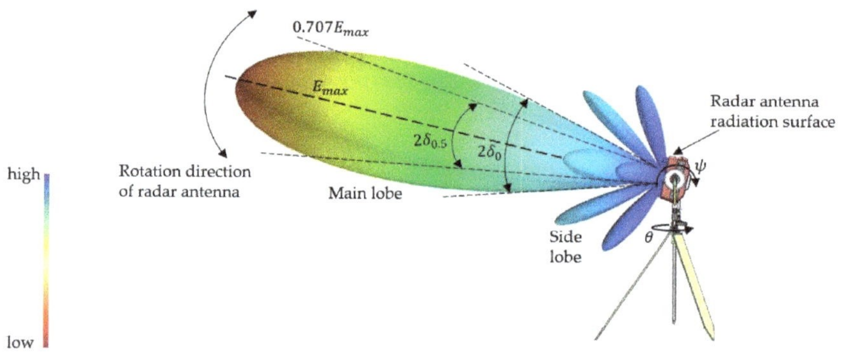

Figure 7. Antenna directional radiation pattern diagram.

Therefore, it can be inferred that as the antenna rotates (Figure 7), the radiation direction of the antenna also rotates in space around the coordinate axes, leading to a continuous change in the direction of the main lobe radiation. During this process, the central position of the antenna remains stable, and its relative position to the detection target does not change. Furthermore, the reflection wave groups from R_1 and R_2 exhibit a linear distribution in space (Figure 6b), indicating that the distance from the reflection points to the antenna has not changed significantly. The horizontal rotation angle of the energy peak at R_1 is −3.0°, and when the peak energy attenuates to 0.707 times, the range of horizontal rotation angles is −31.8° to 33.6°, which falls within the antenna's half-power reduction range. Within this range, the reflection wave groups demonstrate a symmetrical distribution characteristic on either side of the energy peak position. Further analysis shows that in the R_1 and R_2 regions, the average correlation coefficient of the sub-waves at adjacent detection angles reaches 98.97%, with an average spectral similarity of 98.78%. In summary, the formation of the target response characteristics is attributed to the responses generated by wave lobes of varying energy intensities reflecting off the same vertical reflection area during the antenna's rotation. Since these response characteristics originate from the same reflection area, the reflected sub-waves at different detection angles in the R_1 and R_2 regions exhibit significant similarity. Moreover, the formation mechanism of R_1 and R_2 is primarily induced by the antenna's motion relationships, which also demonstrate an intrinsic correlation in the space of detection angles.

3.2. Experiment and Analysis of Target Spatial Distribution

To further analyze the variation in response characteristics of geological structures with different spatial distributions, experiments were conducted using targets with varying tilt angles (Figure 8). Tin foil was selected as target f_3, with dimensions of 1.0 m in height, 5.0 m in length, and 0.002 m in thickness. The antenna was positioned 5.0 m away from target f_3, with its radiation plane initially perpendicular to the plane of f_3. During the experiment, target f_3 was tilted at horizontal angles of 0°, 15°, 30°, 45°, 60°, and 75°. The antenna rotated clockwise with a fixed vertical rotation angle of 0°, and horizontal rotation angles were adjusted in 0.2° increments to cover the range from −90° to 90° of horizontal rotation angles.

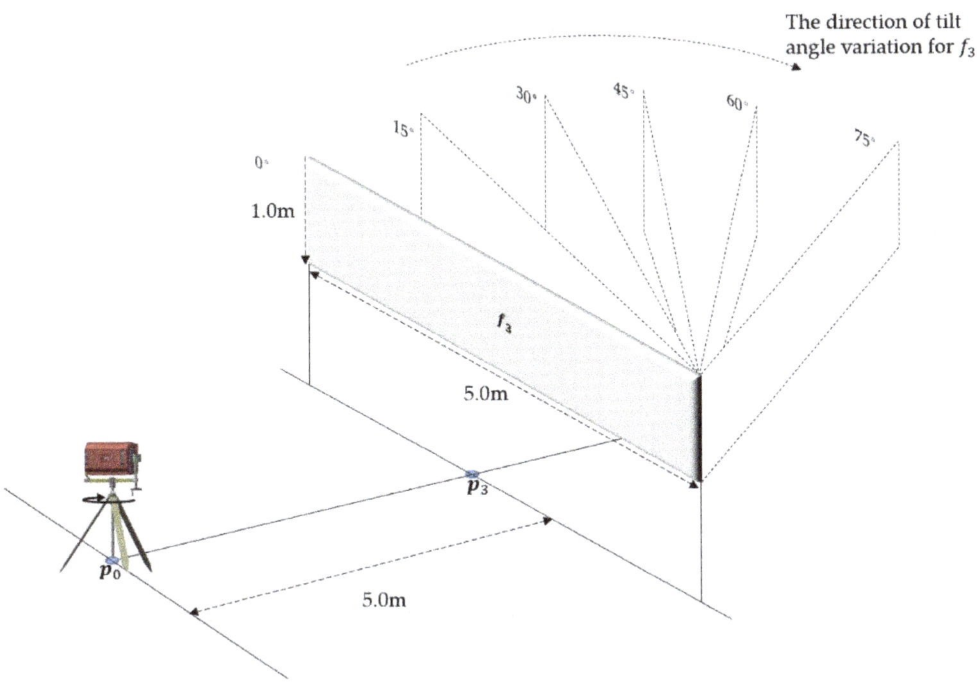

Figure 8. Experimental schematic of tilted target detection in GPR.

Figure 9a–f respectively show the waveform results when the tilt angle of target f_3 varies from 0° to 75°. The data are processed to remove direct waves. Observing the information in the figures, it can be noted that as the tilt angle of target f_3 changes from 0° to 75°, the corresponding response area of the target in the waveform also shifts. When the tilt angles are 0°, 15°, 30°, and 45°, the reflection energy is stronger, and the characteristics of the target response in the waveform are more pronounced. However, as the tilt angle increases to 60° and 75°, the reflection energy weakens, resulting in a more blurred response feature. According to the principle of electromagnetic wave reflection, when the receiving surface of the antenna is perpendicular to the reflecting surface of the target, the antenna can receive the reflection energy to the greatest extent [34]. When the tilt angle of the detection target is 0°, its reflecting surface is located at the center of the target. As the tilt angle of the detection target changes, the reflecting surface gradually shifts from the left to the right.

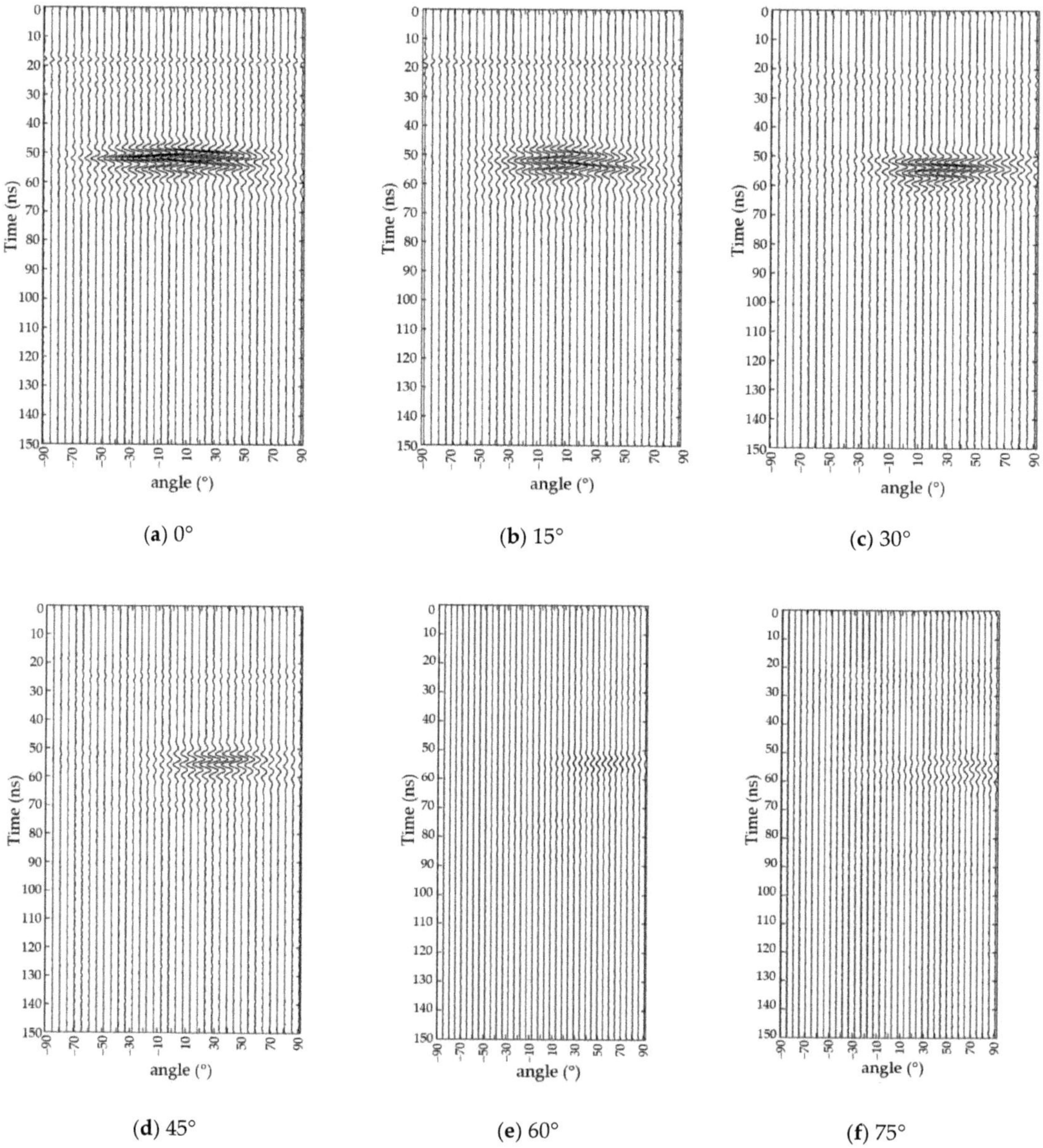

Figure 9. Target detection data waveform plot.

As shown in Table 1, the energy peak values and half-power decay angles of the target response regions in Figure 9 were statistically analyzed. The results indicate that with adjustments in the tilt angle of the detection target, the position of the energy peak in the horizontal scanning angle also correspondingly shifted. As the tilt angle of the detection target varied from 0° to 75°, the horizontal scanning angles where the energy peaks were located were −2°, 14.4°, 27.2°, 41.8°, 55.8°, and 72°, respectively. This indicates that the position of the energy peak has shifted spatially from left to right. There exists a linear relationship between the tilt angle of the target and the horizontal scanning angle where the energy peak is located, showing consistency between the two, disregarding potential errors in the detection process. This finding can provide a basis for determining the orientation of geological structures in the observed space. Furthermore, the half-power decay angles of the target response were 62.2°, 61.8°, 45°, 42.6°, 37.4°, and 41.2°, respectively, indicating that as the tilt angle of the detection target increases, the reflection of electromagnetic waves gradually weakens. In addition, based on practical detection data from Figure 9e,f, when the tilt angle of the target is too large, detection results may exhibit deviations, highlighting an aspect that deserves attention in future research.

Table 1. Information on energy peak and half-power decay range in detection results of targets with different tilt angles.

No.	Tilt Angle	Peak Energy Position	Half-Power Decay Range
1	0°	−2°	−31.4°~30.8°
2	15°	14.4°	−18.8°~43°
3	30°	27.2°	2.6°~47.6°
4	45°	41.8°	19.4°~62°
5	60°	55.8°	31.2°~68.6°
6	75°	72°	48.8°~90°

4. Experimental in Underground Coal Mine Environment

4.1. Detection Process and Phenomena

To investigate the method's application effectiveness in underground coal mine production environments, a retreat mining area in a Hebei Province, China coal mine was selected as the experimental subject (Figure 10). Characterized by roadway roof and floor strata of sandy mudstone and siltstone, with an average coal seam thickness of 3.1 m, primarily consisting of bright coal within a medium-thick coal layer, this area presents a unique geological setting. Figure 10 illustrates that point P represents the detection location at a depth of −676.2 m, which is 18 m and 25.5 m from marked points 173 and 175 in the roadway, respectively. The black dashed line area indicates the detection range. The red solid line, orange solid line, and green solid line represent boundaries of mining areas at different depths. The green arc indicates the direction of antenna rotation, while the gray solid line represents depth contours. Experimentally, the antenna's center frequency is set to 100 MHz, with a 1000 ns sampling time window and sampling number 1024, resulting in an effective detection depth of approximately 50 to 80 m. The horizontal and vertical scanning ranges are −80° to 80° and −50° to 50°, respectively, with intervals of 5° and 0.5°. During the detection process, the horizontal and vertical rotation angles θ and ψ are manually adjusted using the dial (as shown in Figure 3b). Specifically, the detection begins by manually adjusting the horizontal angle for one rotation interval. Subsequently, the vertical rotation angle is sequentially adjusted from −50° to 50°, with data collection performed after each adjustment of the vertical rotation angle. After completing one cycle of data collection, the process repeats, covering the entire spatial range of −80° to 80° for horizontal rotation and −50° to 50° for vertical rotation, thereby completing the spatial data acquisition.

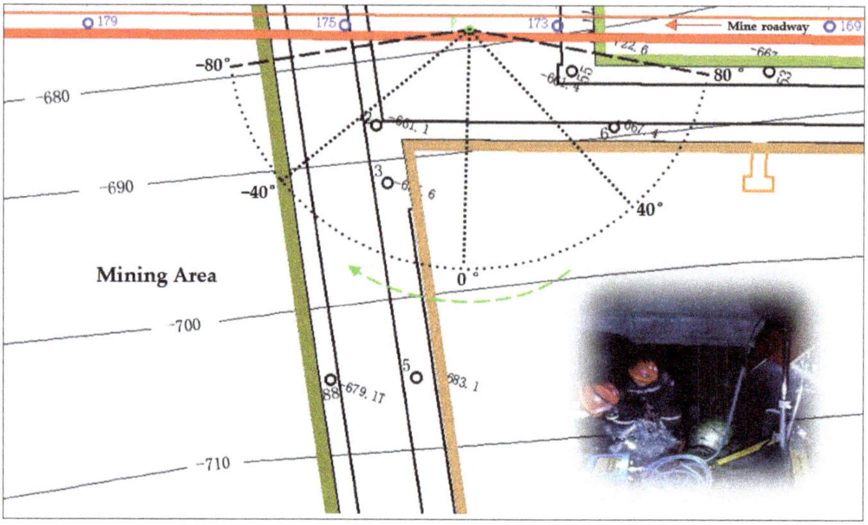

Figure 10. Schematic diagram of the application experiment in the retreat mining area of a coal mine in Hebei Province, China.

In the processing of raw data, the following steps were taken: (1) zero-point calibration was performed to correct any potential baseline offsets; (2) one-dimensional filtering and Wiener predictive filtering techniques were applied to suppress noise and enhance signal quality; (3) background noise elimination was conducted to further clean non-target signals in the data; (4) wavelet transform was employed to analyze signals at different scales, crucial for identifying and separating useful signal components; (5) considering the relatively weak nature of deep signals and significant differences in gain scales, a power-law gain curve was applied to enhance data accordingly, ensuring signal clarity and resolution [35,36]. It is noteworthy that during filtering, given the antenna's central frequency of 100 MHz and bandwidth of 100 MHz, cutoff frequencies were set at 50 MHz and 200 MHz to meet frequency requirements for signal processing. As shown in Figure 11, this paper selects the vertically scanned data corresponding to a specific horizontal rotation angle at 10° intervals from the horizontal scanning range for in-depth analysis. Figure 11a presents the bidirectional travel time on the vertical axis and the horizontal rotation angle on the horizontal axis, corresponding to the vertically scanned data at various horizontal angles. The data were conventionally processed, removing direct waves. In Figure 11a, vertical scan data for each group are arranged from $-50°$ to $50°$ in vertical angles. For each vertical scan angle, 1024 samples within 1000 ns are collected, and these samples are mapped to the RGB color space [37]. As shown in Figure 11, within the color spectrum, colors closer to the center represent lower reflection energy, while colors closer to the ends of the spectrum indicate stronger reflection energy. As the horizontal rotation angle changes from $-80°$ to $-10°$, a prominent strong reflection zone R_3 appears around the 300 ns position, with the reflection energy showing an overall increasing trend. Figure 11b analyzes six groups of vertically scanned data corresponding to horizontal angles from $-80°$ to $-30°$, with two-way travel times ranging between 290 ns and 320 ns. Notably, within the $-40°$ to $40°$ vertical scanning range, the anomalous area R_3 exhibits energy significantly higher than the background, with uneven energy distribution. Localized strong reflections manifest as fine bright stripes, gradually attenuating from the center to the sides, displaying an energy distribution pattern of "strong in the center, weak on the sides." The energy variation within the R_3 area shows significant regularity, forming high-amplitude, high-frequency reflection wave groups, characterized by uniform waveforms without chaotic reflections or multiple waves. The radar image reveals clear parallel reflective co-phases, which, although slightly inclined, exhibit good continuity. In the −80° to −50° horizontal angle range, the corresponding data response characteristics are pronounced, while features at −40° and −30° weaken. In the R_3 region, the average correlation coefficients of strong reflection sub-waves across adjacent detection angles are 98.97%, 99.67%, 97.34%, 89.35%, 81.67%, and 85.35%; the average frequency spectrum similarity is 98.67%, 98.40%, 93.53%, 88.32%, 79.41%, and 74.78%, indicating high similarity of response sub-waves. Furthermore, the characteristics in the R_3 area exhibit a linear distribution trend with varying detection angles, forming an overall structure resembling a pea shape, thicker in the center and narrower on the sides. Additionally, the vertical rotation angles of the energy peaks for each group are −1.5°, −2.5°, −5.5°, −10°, −5.0°, and −8.5°.

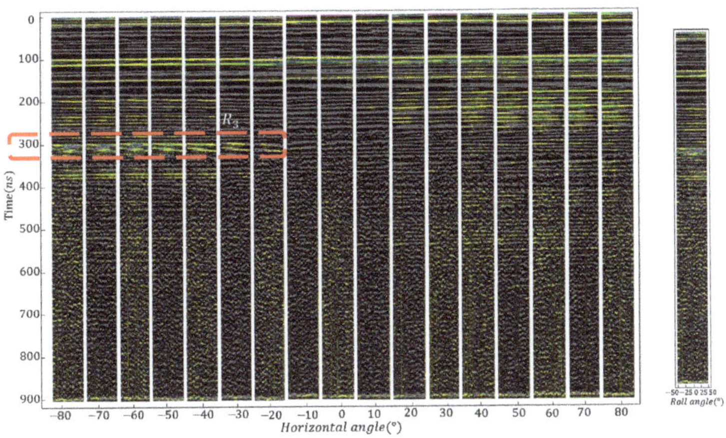

(**a**) Vertical scanning data corresponding to different horizontal rotation angles.

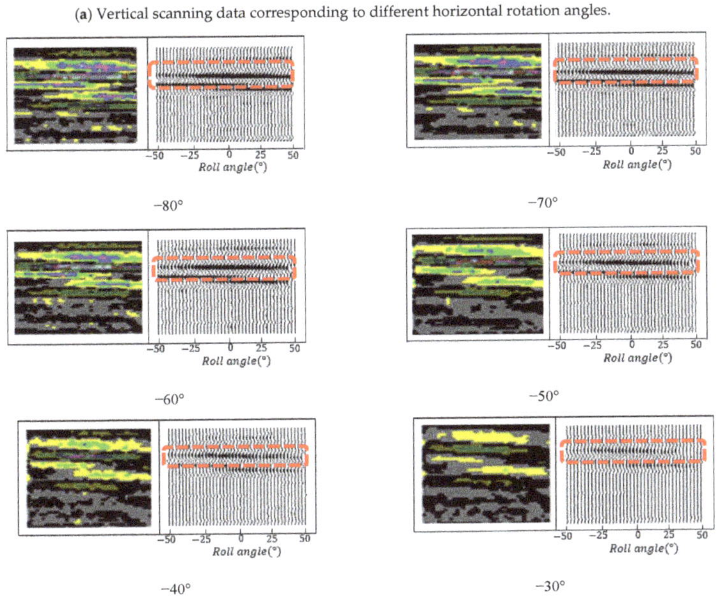

(**b**) Vertical scanning data corresponding to horizontal rotation angles of −80°, −70°, −60°, −50°, −40°, and −30°.

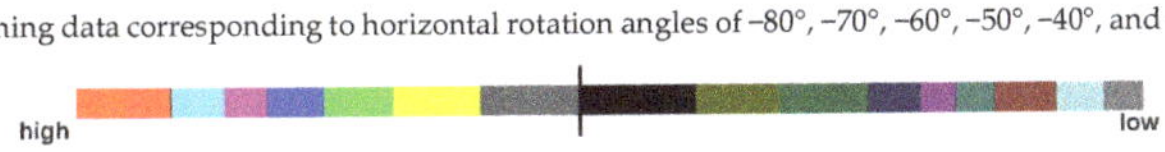

Figure 11. Schematic diagram of the application experiment detection results in an underground coal mine environment (the locations of data anomalies are marked with red dotted box).

Based on the above analysis, anomalous area R_3 exhibits distinct reflective interface characteristics, indicating the presence of two media with significant electrical differences. The absence of chaotic reflections and scattering phenomena suggests that the reflective interface between the two media is relatively smooth. Furthermore, as shown in Figure 12, horizontal scanning data corresponding to a vertical rotation angle of 0° indicate that the horizontal rotation angle range for anomalous area R_3 is approximately −80° to −15°, with the energy peak located around −44.5°.

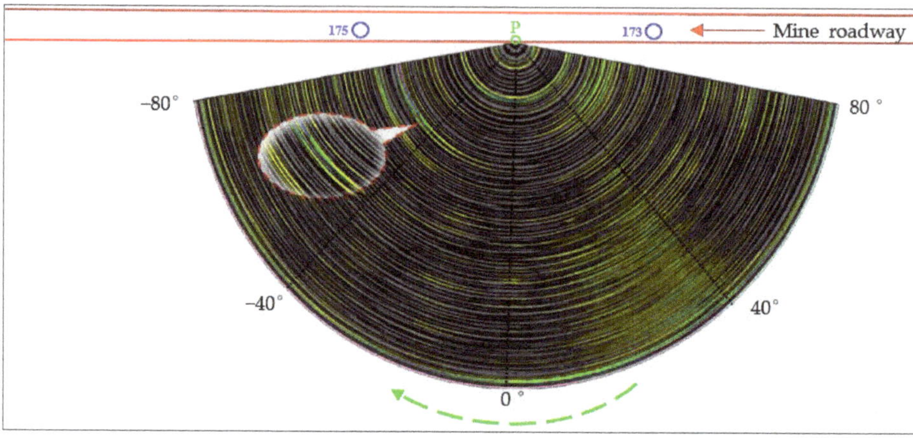

Figure 12. Schematic diagram of the horizontal scanning results corresponding to a vertical rotation angle of 0°.

In summary, it can be inferred that there is a distribution of coal–rock interfaces in the detection area, and the rock layers are overall uniformly developed, with intact, dry rock masses unaffected by groundwater. The strata of the roof and floor of the roadway in the detection area are primarily composed of sandy mudstone and siltstone. During the detection process, exposed coal seams were observed near the detection location, characterized primarily by bright coal, which is a mixture primarily composed of coking coal and lean coal. Based on Table 2, the electromagnetic wave velocity in bright coal is approximately 0.156 m/ns, allowing us to estimate the distance from the detection point to the center of the rock layer as 23.4 m. The development direction of the rock layer forms an angle of approximately 44.5° with the direction of the tunnel. Considering errors in the detection process and localized non-uniform coal seam distribution, the average peak energy change variation during vertical scanning indicates an inclination of the rock layer in the vertical direction of approximately 5.5°.

Table 2. Electrical characteristics of common coal and rock media at a frequency of 100 MHz.

No.	Medium	Conductivity (S/m)	Relative Permittivity (F/m)	Electromagnetic Wave Speed (m/ns)
1	Sand stone	4×10^{-5}	4.6	0.14
2	Coking coal	2.7×10^{-5}	2.8	0.179
3	Lean coal	2.21×10^{-5}	2.6	0.186
4	Mud stone	1×10^{-4}	6.5	0.118

4.2. Analysis of Detection Results

To validate detection results and investigate coal seam distribution in the detection area, drilling operations were conducted to accurately depict coal seam distribution. As shown in Figure 13a, the drilling area measures 150 m in length and 80 m in width, with points 1# to 23# indicating drilling lines on the detection area's surface. Dense drilling points are found in regions with significant coal seam undulations. Drilling results indicate uneven coal seam thickness distribution in the detection area. Specifically, as illustrated in Figure 13a, brown solid, blue dashed, and gray dashed lines represent contour lines for coal seam thicknesses of 1.5 m, 1.0 m, and 0.8 m, respectively. The contour distribution clearly shows that coal seam thickness within the contours is significantly lower than the average thickness of 3.1 m in other regions, with contours indicating that coal seam thickness decreases from point 1# to point 23#. Figure 13b shows the representative drilling lines 2#, 5#, and 7#, selected to analyze vertical coal seam distribution, with the vertical axis denoting drilling depth and the horizontal axis indicating horizontal distance. It can be observed that rock layers intrude into coal seams from below. On drilling line 2#, coal seam thickness variation is relatively small. However, on drilling lines 5# and 7#, significant coal seam thickness changes occur within the horizontal distance range of 10 m to 60 m. Notably, on drilling line 7#, severe rock layer erosion occurs in the region from 30 m to 55 m, leading to near-extinct coal seams and ultra-thin coal layers. Additionally, along drilling lines 7# to 23#, local coal seam thickness is less than 0.8 m, indicating severe rock layer erosion and the presence of coal–rock boundary regions and coal-free zones.

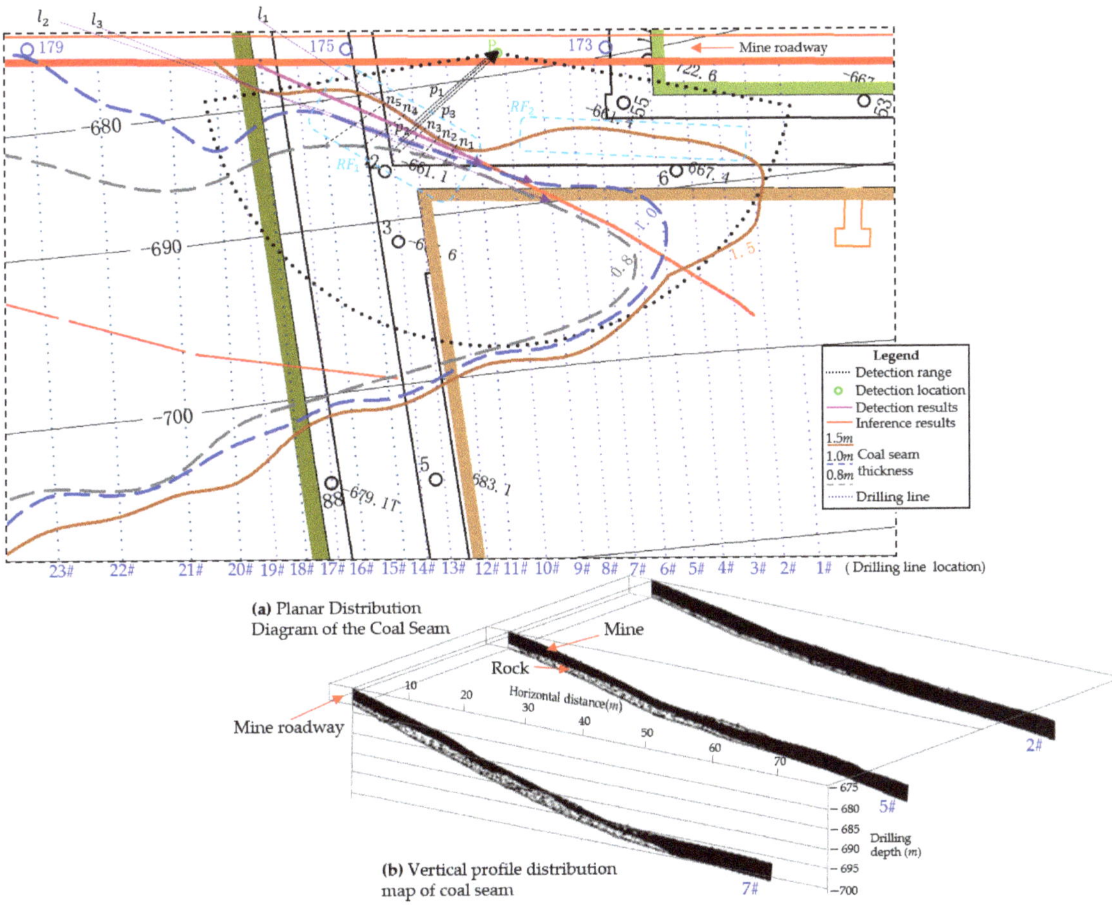

Figure 13. Schematic diagram of coal seam distribution in the retreat mining area.

In summary, drilling results for lines 1# to 23# in Figure 13 indicate that, as drilling progresses, the total thickness of the coal seam decreases from 1.5m to 1.0m and 0.8m. This suggests that rock strata invade the coal seam vertically from bottom to top and horizontally from 23# to 1#, forming a downward sloping intrusion surface, which is consistent with vertical detection results. Figure 13a shows two distinct coal–rock boundary regions, RF_1 and RF_2, within the detection range. Region RF_1 has significant coal thickness variation, resulting in a relatively complete reflection interface. The detection results (solid pink line in Figure 13a) accurately reflect spatial coal thickness variation. To analyze errors in detection results, tangents l_3, l_2, l_1 for 0.8 m, 1.0 m, and 1.5 m coal seam contour lines at RF_1 were drawn. The angles between tangents and tunnel are 33.63°, 31.29°, and 55.81°, respectively, with an average angle of 40.24°, deviating 4.26° from the detection horizontal inclination. At the same time, the distance p_1, p_2, p_3 of the detection position P and the contours of 0.8 m, 1.0 m, and 1.5 m coal seams in the direction of the horizontal rotation angle of 44.5° are 16.5 m, 22.4 m and 25.6 m, respectively. The distance from the mean value is 21.5 m, and the error with the detection depth is 1.2 m. In addition, in the RF_1 region, n_1, n_2, n_3, n_2, n_5 are selected to calculate the vertical dip angle for the specific position of the coal seam drop of 0.7 m. The lengths of n_1~n_5 are 3.4 m, 4.9 m, 6.4 m, 9.3 m, and 12.7 m, respectively. Therefore, the vertical dip angles are 11.63°, 8.12°, 6.24°, 4.3°, and 3.15° respectively, and the average dip angle is 6.67°, which gives a 1.17° error with the detection vertical dip angle. Considering localized non-uniform coal seam distribution and potential detection errors, the accuracy of recognizing depth and spatial extension in detection results for RF_1 is 91.88%, 90.42%, and 78.72%, respectively. In contrast, coal seam variations in RF_2 are more gradual, with a steeper reflection angle and weaker reflection energy, resulting in a less distinct response in the radar spectrum. Future adjustments to the antenna's horizontal position and use of multi-point data energy superposition will enable clearer revelations of spatial information in RF_2.

5. Discussion

This paper presents an innovative forward-looking detection method for coal mining working faces. Through a series of detection experiments, the response characteristics and imaging patterns of targets under this method are thoroughly investigated, and the mechanisms behind the formation of these response characteristics are analyzed. Detection experiments targeting different inclination angles are designed to examine the energy variation patterns of detected targets as the inclination changes, providing a basis for acquiring spatial distribution information of geological structures. Application experiments conducted underground in a coal mine validate the feasibility of this method for forward-looking detection in underground environments. The experimental results robustly demonstrate that this technology can provide strong technical support for tunneling and coal mining operations, showcasing its practical application value. In response to issues encountered during actual detection, future research will focus on the following aspects: building on existing findings, further exploring the response characteristics of geological radar signals related to different types, scales, and spatial distribution patterns of geological structures and their intrinsic relationships. Additionally, an intelligent extraction algorithm for target response characteristics will be developed to enhance the efficiency of geological radar signal feature analysis.

Author Contributions: Conceptualization, J.L., F.Y. and X.Q.; methodology, J.L., F.Y., X.T. and X.Q.; software, F.Y. and X.Q.; validation, J.L. and F.Y.; formal analysis, J.L. and F.Y.; investigation, J.L., F.Y., X.T. and X.Q.; resources, F.Y. and S.P.; data curation, J.L., F.Y. and X.Q.; writing—original draft preparation, J.L.; writing—review and editing, J.L., F.L., F.Y. and S.P.; visualization, J.L. and X.Q.; supervision, F.Y., X.Q., M.X., Y.F. and X.H.; project administration, J.L.; funding acquisition, S.P., F.Y. and X.Q. All authors have read and agreed to the published version of the manuscript.

Funding: This research was funded by the National Key Research and Development Program of China (grant number 2021YFC3090304) and the National Key Laboratory of Coal Resources and Safe Mining (grant number SKLCRSM19KFA07).

Data Availability Statement: Restrictions apply to the availability of these data. Data were obtained from the China University of Minning and Technology (Beijing) and are available from the authors with the permission of the China University of Minning and Technology (Beijing).

Conflicts of Interest: The authors declare no conflicts of interest.

References

1. Song, D.; Hu, Z.; Yu, Y.; Zhang, F.; Sun, H. Assessing Ecological Impacts and Recovery in Coal Mining Areas: A Remote Sensing and Field Data Analysis in Northwest China. *Remote Sens.* **2024**, *16*, 2236. [CrossRef]
2. Yuan, L.; Zhang, P.J. Key technology and path thinking of dynamic reconstruction of mine transparent geological model. *J. China Coal Soc.* **2023**, *48*, 1–14.
3. Zhang, Q.; Zhang, R.; Liu, J.; Wang, C.; Zhang, H.; Tian, Y. Review on coal and rock identification technology for intelligent mining in coal mines. *Coal Sci. Technol.* **2022**, *50*, 1–26.
4. Wang, G.; Xu, Y.; Ren, H. Intelligent and ecological coal mining as well as clean utilization technology in China: Review and prospects. *Int. J. Min. Sci. Technol.* **2019**, *29*, 161–169. [CrossRef]
5. Cheng, J.; Li, F.; Peng, S. Research progress and development direction on advanced detection in mine roadway working face using geophysical methods. *J. China Coal Soc.* **2014**, *39*, 1742–1750.
6. Siddiqui, M.A.H.; Akhtar, S.; Chattopadhyaya, S.; Sharma, S.; Kumar, A.; Abbas, M. 611 Universal Drilling Machine Reliability Modeling and Performance Evaluation in Subterranean Coal Mines. *Rock Mech. Rock Eng.* **2024**, *57*, 3559–3575. [CrossRef]
7. Shuning, D.; Zaibin, L.; Jianyuan, C.; Baohui, C.; Zhenhua, D.; Dan, L. Technologies and prospect of geological guarantee for intelligent coal mining. *Coal Geol. Explor.* **2021**, *49*, 4.
8. Peng, S. Current status and prospects of research on geological assurance system for coal mine safe and high efficient mining. *J. China Coal Soc.* **2020**, *45*, 2331–2345.
9. Su, B.; Liu, S.; Deng, L.; Gardoni, P.; Krolczyk, G.M.; Li, Z. Monitoring direct current resistivity during coal mining process for underground water detection: An experimental case study. *IEEE Trans. Geosci. Remote Sens.* **2022**, *60*, 1–8. [CrossRef]
10. Cheng, J.-L.; Song, G.-D.; Liu, T.-Y.; Hu, B.-X.; Wang, J.-Q.; Wang, J.-Y. High precision location of micro-seismic source in underground coal mine. *Chin. J. Geophys.* **2016**, *59*, 734–743.
11. Anbazhagan, P.; Dixit, P.N.; Bharatha, T. Identification of type and degree of railway ballast fouling using ground coupled GPR antennas. *J. Appl. Geophys.* **2016**, *126*, 183–190. [CrossRef]
12. Liu, C.; Song, C.; Lu, Q. Random noise de-noising and direct wave eliminating based on SVD method for ground penetrating radar signals. *J. Appl. Geophys.* **2017**, *144*, 125–133. [CrossRef]
13. Song, J.; Wang, L.J. Study on the high precision detection of buried drill pipe by ground penetrating radar in the coal face. *J. China Coal Soc.* **2014**, *39*, 537–542.
14. Wang, D.J.G.E. Ground penetrating radar and engineering geological exploration. *Geophys. Equip.* **2001**, *11*, 188–194.
15. Li, S.C.; Zhou, Z.Q.; Ye, Z.H.; Li, L.P.; Zhang, Q.Q.; Xu, Z.H. Comprehensive geophysical prediction and treatment measures of karst caves in deep buried tunnel. *J. Appl. Geophys.* **2015**, *116*, 247–257. [CrossRef]
16. Orlando, L.; Cardarelli, E.; Cercato, M.; De Donno, G.; Di Giambattista, L. Pavement testing by integrated geophysical methods: Feasibility, resolution and diagnostic potential. *J. Appl. Geophys.* **2017**, *136*, 462–473. [CrossRef]
17. Cheng, J.; Lu, Z.; Xue, B.; Li, P.; Wang, P. The "long excavation and long exploration" technology of rapid excavation of coal mine roadway. *J. China Coal Soc.* **2022**, *47*, 404–412.
18. Fernández-Álvarez, J.-P.; Rubio-Melendi, D.; Castillo, J.A.Q.; González-Quirós, A.; Cimadevilla-Fuente, D. Combined GPR and ERT exploratory geophysical survey of the Medieval Village of Pancorbo Castle (Burgos, Spain). *J. Appl. Geophys.* **2017**, *144*, 86–93. [CrossRef]
19. Xu, X.; Peng, S.; Yang, F. Development of a ground penetrating radar system for large-depth disaster detection in coal mine. *J. Appl. Geophys.* **2018**, *158*, 41–47. [CrossRef]
20. Xu, X.; Ma, Z.; Sun, T. A Whole closed space intrinsically safe GPR system for detection of geological hazard sources in coal mines. *IEEE Geosci. Remote Sens. Lett.* **2021**, *19*, 1–5. [CrossRef]
21. Ma, Z.; Xu, X.; Peng, S.; Zhu, P.; Wang, Y. Development and application of air-coupled GPR antenna for coal-rock interface detection in mines. *J. Appl. Geophys.* **2023**, *219*, 105238. [CrossRef]
22. Zhu, P.; Xu, X.; Ma, Z.; Wang, Y. Research on accuracy and error analysis of coal and rock strata detection based on air-coupled GPR. *IEEE Geosci. Remote Sens. Lett.* **2022**, *19*, 1–5. [CrossRef]
23. Strange, A.D. Robust Thin Layer Coal Thickness Estimation Using Ground Penetrating Radar. Ph.D. Thesis, Queensland University of Technology, Brisbane, Australia, 2007.
24. Strange, A.D.; Jecny, Z. Coal subsurface mapping for selective mining. In Proceedings of the 2018 17th International Conference on Ground Penetrating Radar (GPR), Rapperswil, Switzerland, 18–21 June 2018; pp. 1–4.

25. Ralston, J.C.; Strange, A.D. An industrial application of ground penetrating radar for coal mining horizon sensing. In Proceedings of the 2015 International Symposium on Antennas and Propagation (ISAP), Hobart, Australia, 9–12 November 2015; pp. 1–4.
26. Ralston, J.; Strange, A. 3D robotic imaging of coal seams using ground penetrating radar technology. In Proceedings of the 2016 16th International Conference on Ground Penetrating Radar (GPR), Hong Kong, China, 13–16 June 2016; pp. 1–6.
27. Francke, J. Applications of GPR in mineral resource evaluations. In Proceedings of the XIII International Conference on Ground Penetrating Radar, Lecce, Italy, 21–25 June 2010; pp. 1–5.
28. Ralston, J.C.; Hainsworth, D.W.; McPhee, R.J. Application of ground penetrating radar for coal thickness measurement. In *TENCON'97 Brisbane-Australia. Proceedings of IEEE TENCON'97. IEEE Region 10 Annual Conference. Speech and Image Technologies for Computing and Telecommunications (Cat. No.97CH36162), Brisbane, QLD, Australia, 4 December 1997*; IEEE: Piscataway, NJ, USA, 2002.
29. Lai, W.W.; Chang, R.K.; Sham, J.F. A blind test of nondestructive underground void detection by ground penetrating radar (GPR). *J. Appl. Geophys.* **2018**, *149*, 10–17. [CrossRef]
30. Yuan, W.; Liu, S.; Zhao, Q.; Deng, L.; Lu, Q.; Pan, L.; Li, Z. Application of Ground-Penetrating Radar with the Logging Data Constraint in the Detection of Fractured Rock Mass in Dazu Rock Carvings, Chongqing, China. *Remote Sens.* **2023**, *15*, 4452. [CrossRef]
31. Zhang, J.; Wang, Y.; Zhang, T.; Yang, K.; Zhang, J.; Wang, X.J.S. A Study on Refraction Error Compensation Method for Underwater Spinning Laser Scanning Three-Dimensional Imaging. *Sensors* **2024**, *24*, 343. [CrossRef]
32. Poljak, D.; Sesnic, S.; Susnjara, A.; Paric, D.; Drissi, K.E.K.; Lallechere, S. Direct time domain evaluation of the transient field transmitted into a lossy ground due to GPR antenna radiation. *Eng. Anal. Bound. Elem.* **2017**, *82*, 27–31. [CrossRef]
33. Jiao, Y.; A McMechan, G.; Pettinelli, E. In situ 2-D and 3-D measurements of radiation patterns of half-wave dipole GPR antennas. *J. Appl. Geophys.* **2000**, *43*, 69–89. [CrossRef]
34. Dhande, P. Antennas and Its Applications. *DRDO Sci. Spectr.* **2009**, 66–78.
35. Wang, W.; Hao, L.; Li, X.; Sun, J. Signal processing technology and its application research of ground penetrating radar (GPR). *Proc. IOP Conf. Ser. Earth Environ. Sci.* **2017**, *69*, 012077. [CrossRef]
36. Abujarad, F. *Ground Penetrating Radar Signal Processing for Landmine Detection*; Otto-von-Guericke-Universität Magdeburg, Universitätsbibliothek: Magdeburg, Germany, 2007.
37. Süsstrunk, S.; Buckley, R.; Swen, S. Standard RGB color spaces. In Proceedings of the Color and Imaging Conference, Scottsdale, AZ, USA, 16–19 November 1999; pp. 127–134.

remote sensing

MDPI

Article

Development of a Background Filtering Algorithm to Improve the Accuracy of Determining Underground Cavities Using Multi-Channel Ground-Penetrating Radar and Deep Learning

Dae Wook Park [1], Han Eung Kim [2], Kicheol Lee [3] and Jeongjun Park [4,*]

1 Department of Civil Engineering, Kunsan National University, 558 Daehak-ro, Gunsan-si 54150, Republic of Korea; dpark@kunsan.ac.kr
2 Geotechnical Korea Engineering Co., Ltd., 91 LS-ro, Dongan-gu, Anyang-si 14119, Republic of Korea; he-kim@gk-eng.co.kr
3 Corporate Affiliated Research Institute, UCI Tech, 313 Inha-ro, Michuhol-gu, Incheon 22012, Republic of Korea; leeki3539@gmail.com
4 Incheon Disaster Prevention Research Center, Incheon National University, 119 Academy-ro, Yeonsu-gu, Incheon 22012, Republic of Korea
* Correspondence: jjpark72@inu.ac.kr; Tel.: +82-10-4722-2971

Abstract: In the process of using multi-channel ground-penetrating radar (GPR) for underground cavity exploration, the acquired 3D data include reflection data from underground cavities or various underground objects (structures). Reflection data from unspecified structures can interfere with the identification process of underground cavities. This study aims to identify underground cavities using a C-GAN model with an applied ResBlock technique. This deep learning model demonstrates excellent performance in the image domain and can automatically classify the presence of cavities by analyzing 3D GPR data, including reflection waveforms (A-scan), cross-sectional views (B-scan), and plan views (C-scan) measured from the ground under roads. To maximize the performance of the C-GAN model, a background filtering algorithm (BFA) was developed and applied to enhance the visibility and clarity of underground cavities. To verify the performance of the developed BFA, 3D data collected from roads in Seoul, Republic of Korea, using 3D GPR equipment were transformed, and the C-GAN model was applied. As a result, it was confirmed that the recall, an indicator of cavity prediction, improved by approximately 1.15 times compared to when the BFA was not applied. This signifies the verification of the effectiveness of the BFA. This study developed a special algorithm to distinguish underground cavities. This means that in the future, not only the advancement of separate equipment and systems but also the development of specific algorithms can contribute to the cavity exploration process.

Keywords: GPR; underground cavity; 3D data; C-GAN; BFA

1. Introduction

With the increase in aging infrastructure, redevelopment projects are being actively discussed, particularly in urban areas. However, due to the saturation of above-ground spaces, there are limitations to new and redevelopment projects. This has led to the active development of underground spaces, but the frequency of underground cavities is rapidly increasing. Ground subsidence resulting from these cavities is one of the most severe disasters that can occur in urban areas.

Underground cavities are formed due to changes in groundwater extraction and drainage patterns, damage to water supply connections, and aging sewer pipes. Since these underground cavities can appear without warning in urban areas, there is an increasing demand for early detection [1–3]. Unexpected ground subsidence can lead not only to road closures and economic losses but also to casualties [4,5].

Recently, ground-penetrating radar (GPR) has been used as a highly reliable technique and is evaluated as the most suitable technology for responding to ground subsidence [6–9]. GPR is a fully non-contact, non-destructive inspection technology that has gained popularity in the field of structural condition monitoring due to rapid scanning speed and 3D imaging capabilities [10]. GPR emits high-frequency electromagnetic waves into the ground and receives the reflected waves from underground objects or empty space. By analyzing the received signals, it is possible to identify the boundaries of these anomalies if there are changes in the electromagnetic properties, such as the relative permittivity of the medium or the composition of the soil layers [11,12]. Thus, GPR has been applied in archaeological and geological contexts for mapping underground features and has recently been expanded to derive the engineering properties of underground cavities and soil [13–17].

However, raw GPR signals are generally dominated by reflections from upper layers, such as paved roads, making analysis extremely difficult. Additionally, the complex geometry, changes in water content, and unexpected contaminants associated with urban roads make data interpretation even more challenging [18]. Additionally, the analysis of data obtained through GPR surveys requires specialized expertise due to the complexity and specificity of the equipment, necessitating the involvement of expert analysts. However, in reality, there is a significant shortage of specialized equipment and personnel. And there is no separate interpretation method for the 3D data obtained from GPR surveys, so experts must manually inspect and assess the data, relying heavily on subjective judgment and effort. As a result, there is a risk of missing signals indicating cavities if the expert's concentration wanes or subjective judgment excessively interferes. Moreover, the lengthy process from the start of the survey to the completion of the analysis poses a challenge for urgently responding to sudden road subsidence disasters [19,20].

To address these problems, neural network classifiers have been developed to detect buried utilities and solid objects in GPR images, and artificial neural networks have been introduced to detect pipe signatures in GPR images [21,22]. Also, machine learning techniques have been studied to extract amplitude and time delay features from GPR longitudinal scan images [23]. While traditional machine learning techniques have been successfully used to detect dominant or clear features in GPR signals, they are not well-suited for classifying multiple underground objects in complex urban areas. Furthermore, the manual feature extraction process is extremely difficult and cumbersome when interpreting GPR data in complex field conditions, which include measurement noise.

More recently, there has been active exploration of ground cavity detection using neural network models, including deep learning models. Studies have focused on traditional methods and deep learning approaches for GPR data detection [24], methods for predicting the permittivity of targets using deep learning models [25], and improving speed and accuracy through neural network models [26]. Additionally, there is ongoing research on verifying the performance of newly developed GPR systems [27]. In other words, new methods are continuously being explored. However, this study aims to focus on improving accuracy through data quality enhancement rather than developing a new prediction model. While this approach may not be as groundbreaking as innovative prediction technologies, it is expected to contribute to the overall advancement of ground cavity prediction models.

The deep learning used is focused on automation, incorporating automatic feature extraction, unsupervised learning for pre-training, and utilizing images directly in the training and testing process without image reconstruction. This paper describes a method for quickly and automatically detecting underground cavities using artificial intelligence cavity analysis technology. Specifically, it focuses on developing and validating an algorithm to remove noise in the direction of travel. This algorithm, known as the background filtering algorithm (BFA), combines a signal amplification filter and a local average subtraction technique to enhance the visibility and clarity of underground cavities beneath urban road pavements. The developed BFA was applied to a C-GAN (conditional generative adversarial network) model incorporating the ResBlock technique, which has shown excellent performance in the image domain. Finally, the C-GAN model was trained using 3D GPR data obtained from urban roads. The effectiveness of the BFA was validated by comparing the data processed with and without the BFA, confirming the improvement in detecting underground cavities.

2. Automated Technique for the Detection of Underground Cavities Based on Deep Learning

The deep-learning-based technology developed for the automated detection of underground cavities can be detailed in three stages. First, an explanation of the operating principle of multi-channel GPR (ground-penetrating radar) will present the process of data acquisition and the various types of data. Next, the details of developing a basic tracking-based background filtering algorithm that can enhance the visibility of underground structures in GPR images will be explained. Finally, the C-GAN model with ResBlock techniques applied for the automatic detection of underground cavities using filtered images will be described.

2.1. Process and Principles for the Detection of Underground Cavities Using Multi-Channel GPR

Figure 1 illustrates the underground visualization process for acquiring 3D survey data using multi-channel GPR in a road site. The 3D GPR system consists of multiple transmitters and receiver antennas and a system for the acquisition of data. The transmitters emit high-frequency electromagnetic waves, and the receivers collect the returning waves. The typical frequency range of these electromagnetic waves spans from a few MHz to GHz. The appropriate frequency range should be selected based on the size and depth of the target object. Generally, high frequencies are used for shallower and smaller targets, while low frequencies are used for deeper and larger targets.

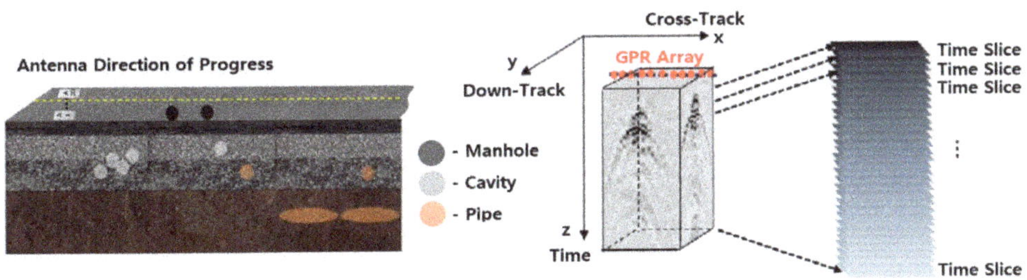

Figure 1. Schematic of data acquisition.

Multi-channel GPR typically acquires three different types of data when scanning a target area: reflection waveform (A-scan), longitudinal cross-section (B-scan), and horizontal cross-section (C-scan).

In GPR surveys, data are acquired and analyzed through three types of scans to identify cavities. First, to analyze cavity signals, the changes in signals are examined three-dimensionally (by depth variation) on planar images (C-scan). During this process, the signals are analyzed to determine whether they form a continuous linear pattern (indicating buried pipes) or an independently separated area (indicating a cavity). In the second stage, the preliminary candidate cavity images identified as independent shapes in the planar C-scan images are further analyzed in a two-dimensional B-scan cross-section. Here, strong signal intensity, phase changes, and the pattern of parabolic shapes are examined. Finally, in the third stage, considering the significant difference in dielectric constants between air and the surrounding medium, the A-scan data of the survey points are analyzed. Additional analyses of waveform phase changes, amplitude, and signal intensity are conducted to confirm the presence of cavities.

Additionally, when there are changes in the electromagnetic properties of the scanned area, such as variations in the relative permittivity of the ground layer or the presence of anomalous objects with different properties, a portion of the electromagnetic waves is reflected back to the receiving antenna. These reflected waves are then collected by the data acquisition system and used for underground object detection. In this study, various data and parameters acquired through multi-channel GPR are comprehensively analyzed. Based on the criteria for cavity selection, the presence and characteristics of cavities are determined.

2.1.1. Images of A-Scan

The A-scan provides precise time-domain information at a specific spatial point (z-axis in Figure 1). By considering the significant difference in permittivity between air and the medium, the reflection wave exhibits a waveform that progresses from negative (-) to positive (+) amplitude when reflected from the upper low-velocity layer (pavement layer) and the lower high-velocity layer (air). Therefore, the A-scan in multi-channel GPR involves analyzing the phase changes, amplitude, signal strength, and intensity of the waveform to identify cavities. Figure 2 illustrates the A-scan waveforms for two cavities and non-cavities, such as loose gravel layers and buried pipes.

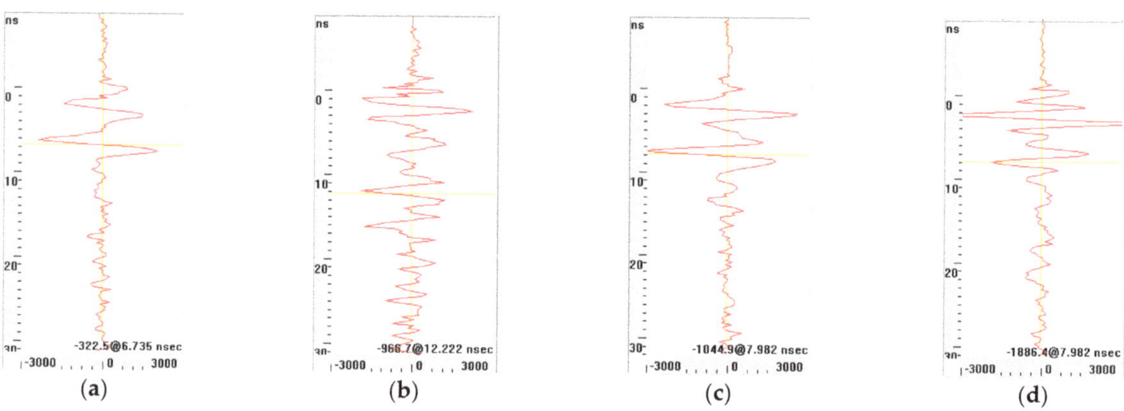

Figure 2. A-scan images: (**a**) underground cavity −1; (**b**) underground cavity −2; (**c**) loose gravel layer; (**d**) buried pipes.

2.1.2. Images of B-Scan

The B-scan in the x-z domain is comprised based on various A-scan data along the scanning path (x-axis in Figure 1). For the cavities selected by independent signal shapes in the A-scan plane image, a two-dimensional B-scan is generated, displaying longitudinal and cross-sectional images as shown in Figure 3. After, in the B-scan, the signal strength, phase changes, and patterns of parabolic shapes are analyzed and verified. Typically, boundary surfaces reflecting the received signal are represented as curves or straight lines. The boundaries of underground objects or cavities appear in a parabolic shape. The upper point of a cavity is indicated by the topmost part of the parabola formed beneath the survey point, shown as white and black sections.

Additionally, A-scan and B-scan images can be used to determine whether an underground pipe is made of metal or PVC material. In the analysis of A-scan and B-scan images, if the start of the embedded image appears as (+) or white, it indicates a metallic material, such as a cast iron or steel pipe used for gas or water lines. Conversely, if the image appears as (-) or black, it indicates a PVC material, commonly used for communication conduits.

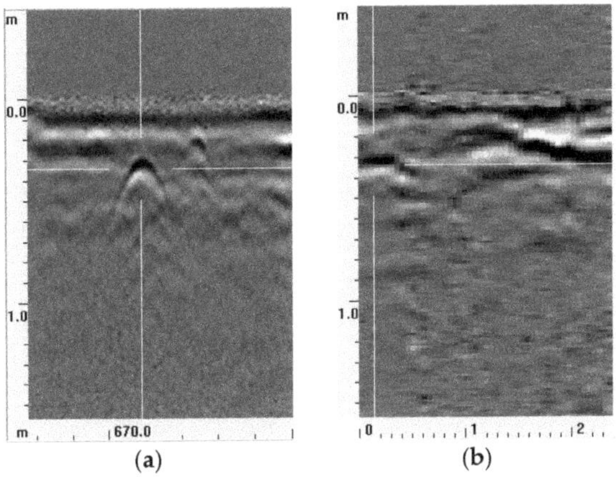

Figure 3. B-scan images: (**a**) longitudinal section; (**b**) cross-section.

2.1.3. Images of C-Scan

Due to strong surface reflections, B-scan images often display dominant linear features that appear nearly straight. These surface reflections typically overwhelm underground reflections via the sudden change in permittivity from air (1 F/m^{-1}) to paved road (5 F/m^{-1}). These linear features in B-scan images can often interfere with the detection of underground objects. Therefore, B-scan images obtained using multiple antennas are combined to generate C-scan images that are visible in the x-y domain. The various C-scan images can be examined along the z-axis, representing depth, and these cross-sectional C-scan images are used to further characterize the detected features.

To analyze cavity signals, the changes in signals are examined three-dimensionally in the planar image, and shape changes with depth are confirmed. As shown in Figure 4, buried pipelines exhibit a linear pattern, while cavities form a distinct area with abrupt amplitude changes along the boundary, showing an independently separated pattern. Therefore, cavities need to be identified by comprehensively analyzing planar, longitudinal, and cross-sectional images based on the independent signal shapes.

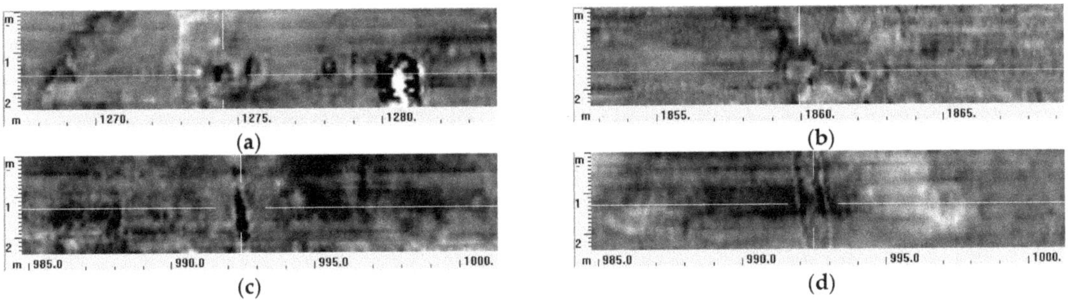

Figure 4. C-scan images with depth from the ground surface: (**a**) underground cavity at a depth of 32 cm; (**b**) underground cavity at a depth of 42 cm; (**c**) buried pipes at a depth of 26 cm; (**d**) buried pipes at a depth of 36 cm.

2.2. Background Filtering Algorithm (BFA)

In this study, a filtering algorithm of noise removal was developed to enhance the visibility of underground objects by removing dominant linear features and unwanted pattern noise from B-scan images. It is a background filtering algorithm (BFA). This algorithm consists of six steps. The filtered images can significantly enhance the visibility of underground cavities and are effectively used in the subsequent process involving the

ResBlock technique applied to the C-GAN model. The workflow diagram for the BFA is illustrated in Figure 5.

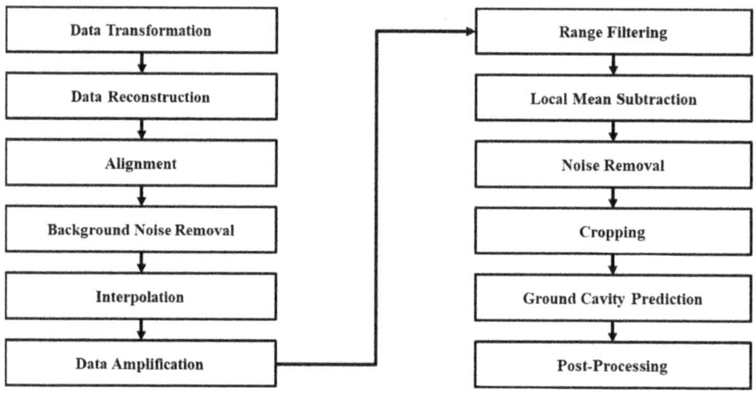

Figure 5. Workflow of the BFA.

2.2.1. Collection of 3D GPR Data

3D GPR data are collected using multi-channel antennas to receive raw data from the underground space, as illustrated in Figure 1. This setup allows for the acquisition of raw data corresponding to 25 channels, providing 25 longitudinal, planar, and cross-sectional views of the underground space. It is important to note that the measured GPR signals generally attenuate as they propagate through specific media. This attenuation must be considered during the data analysis process.

2.2.2. Conversion and Restructuring of Data

The raw data obtained from 3D GPR must be converted into a format that can be processed by deep learning models. The raw data are in an RD3 format, consisting of a combination of texts and numbers, and are not directly analyzable. Therefore, a decoding process was performed to convert these data into a one-dimensional array (1D-array) format, with numbers listed in a single column. This process utilized the Numpy library and involved converting the RD3 format raw data into the NPY format (*.npy). Subsequently, a data restructuring step was carried out to convert the one-dimensional array data in the NPY format into a three-dimensional array format, making them easier to use in deep learning applications. This conversion allows the data to be visualized and analyzed based on each antenna channel, the depth of the underground space, and the distance between the antenna and the underground cross-section.

As shown in Figure 6, the reshape function of the Numpy library was used to reconstruct the one-dimensional array data into three-dimensional array data. The three-dimensional array data consist of elements arranged as follows:

- [Antenna Channels (width of the survey vehicle)]
- [Depth of the Underground Space]
- [Distance between the Antenna and the Underground Cross-Section (distance along the survey vehicle's driving direction)]
- Etc.

```
array([[[-190,  -190,  -190, ...,   -59,   -18,    -4],
        [-3071, -3071, -3071, ..., -2640, -2661, -2939],
        [-9244, -9244, -9244, ..., -8670, -7999, -8983],
        ...,
        [  988,   988,   988, ...,   881,   840,   818],
        [ 1018,  1018,  1018, ...,   983,   960,   777],
        [ 1043,  1043,  1043, ...,  1002,   909,  1257]]])
```

Figure 6. Example of reconstructing data in the form of a 3-dimensional matrix.

2.2.3. Alignment of Data

A ground alignment process is performed to correct the time differences using the maximum and minimum values of the data for each channel. In the case of a 3D GPR, the reflected waves received by the antenna with 25 channels are not received at exactly the same time, so they need to be aligned by tracking the ground surface. Specifically, as shown in Figure 7, the first alignment value is calculated as the average of the maximum and minimum values of the data for each channel, and the second alignment value is calculated as the median of the maximum and minimum values of the data for each channel. By considering the pixel point corresponding to the average of the first and second alignment values as the ground surface corresponding to the underground space, the time difference for each channel can be corrected based on this pixel point.

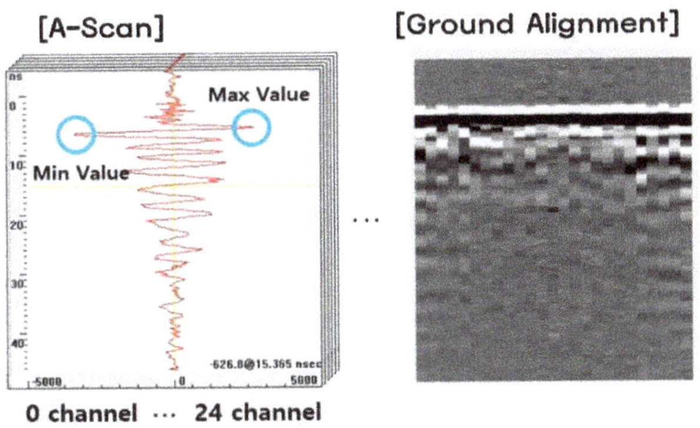

Figure 7. Process of ground alignment.

2.2.4. Background Removal and Data Interpolation

The surface portion of the data, which has undergone the data alignment process, generates strong signals unrelated to cavity signals, interfering with the detection of cavity signals. Additionally, some channels occur bias. Therefore, noise in the surface portion must be removed and bias values adjusted by subtracting the average pixel values at specific depths in the underground space. This process is known as background removal. Subsequently, data interpolation is performed. This step ensures that the data structure and format of the pre-processed data match those of the training data used for the deep learning model. The intervals between pixels in the data are adjusted to a set value.

2.2.5. Data Amplification

When the transmitted waves from the antenna pass through the boundaries of different media, reflected waves occur based on the reflectivity of the media, and only the transmitted waves continue. As the depth of the underground space increases, the strength of the transmitted waves weakens, and the strength of the reflected waves also diminishes. Consequently, signals below a specific depth become difficult to observe during the process of visualizing and analyzing the data. Previously, signal amplification was performed using linear functions or exponential functions, but this approach often resulted in excessive noise.

Therefore, this study aims to achieve an optimal data amplification appropriate for different depths in the underground space by varying the degree of amplification for the relatively shallow and relatively deep parts of the data. Specifically, a signal amplification filter, represented by a function (*erf*), is applied to the data to dynamically adjust the amplification weights for different depth intervals: the shallowest part (Zone 1), a deeper part than Zone 1 (Zone 2), and a deeper part than Zone 2 (Zone 3). The amplification filter function (*erf*) is a Sigmoid error function, as shown in Equation (1) and Figure 8a.

$$erf = \frac{2}{\sqrt{\pi}} \int_0^x e^{-t^2} dt \quad (1)$$

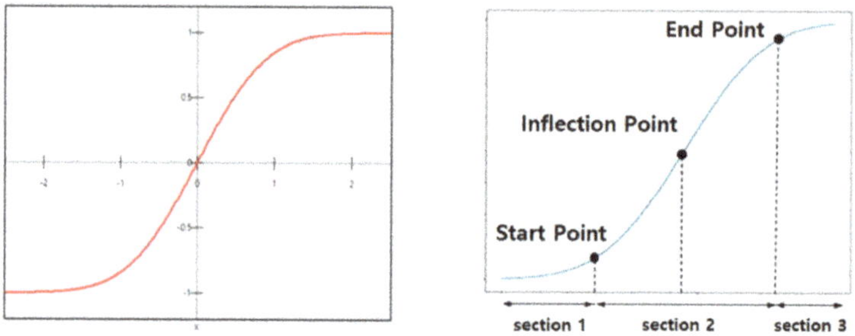

(a) (b)

Figure 8. Data amplification function: (**a**) error function; (**b**) improved error correction function.

When applying this error function directly to amplify data with underground depth, it proved difficult to accurately detect cavities due to mismatches with the depths, widths, and sizes where cavities occur. Therefore, the error function was improved to create a new function (F), and this signal amplification filter was applied to amplify the data. Specifically, based on the error function expressed in Equation (1), additional variables were set, including the inflection range (the width between the start and end points of the inflection), the inflection midpoint (the midpoint between the start and end points), the slope of the function (*erf*) within the inflection range, and the y-axis intercept (bias) of the function (*erf*).

The final corrected error function (F) can be expressed as shown in Equation (2). Figure 8b illustrates the graph of the improved signal amplification filter function (F) based on Equation (2). Here, F represents the function of the signal amplification filter, x denotes the pixel value according to the depth of the underground space, a is the inflection midpoint (=140), b is the inflection range (=70), K is the slope of the function (F) within the inflection range (=1), C is the y-axis intercept (bias) of the function (F) (=1.5), and t is time. The x-axis represents the pixel values according to the depth of the underground space, which can be in units of pixels or cm. The y-axis represents the weight for the degree of data amplification. Additionally, the default values may vary depending on the characteristics of the underground space.

$$F = \left(erf\left(\left(\frac{x-a}{b} \right) + 1 \right) + K \right) + C \tag{2}$$

Figure 9 shows the data images before and after the application of the signal amplification filter. When the signal amplification filter is applied, it is evident that the data near the cavities are relatively more prominently amplified. This indicates that the applied improved error correction function is appropriate.

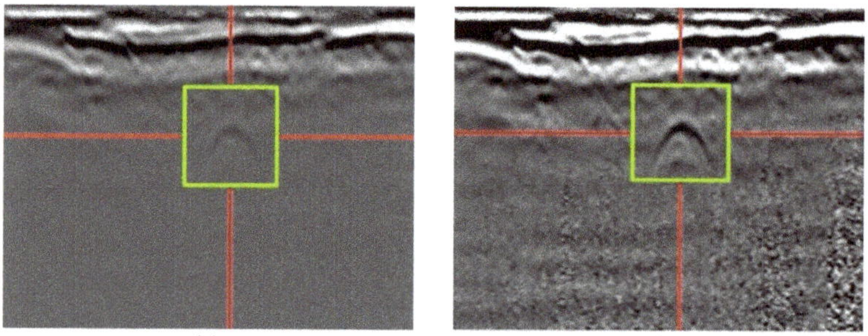

(a) (b)

Figure 9. Data images: (**a**) non-application of the improved error correction function; (**b**) application of the improved error correction function.

2.2.6. Noise Removal Using Local Average Subtraction

In data where errors have been corrected using traditional Kalman filters, signals that appear as straight lines in the direction of the survey vehicle's travel (e.g., pipe signals, surface signals) are not effectively removed. As a result, these straight-line signals frequently overlap with the signals from the cavity regions. This means that the cavity signals are obscured by the straight-line signals in the direction of travel.

In this study, a Gaussian function was used to remove noise in the direction of the survey vehicle's travel. However, applying a typical Gaussian filter would remove not only the noise but also the signals from the cavity regions. Therefore, in this study, instead of directly applying the Gaussian filter, a method is used in which the operation value of the Gaussian function is subtracted from the pixel value of the target pixel. This helps to selectively remove only the noise according to the driving direction of the survey vehicle.

The C-GAN model is considered the most suitable model for ground cavity prediction. The traditional deep learning method of a CNN (convolutional neural network) was initially used to extract image features for ground cavity detection. A total of 1396 image files were created and applied for training the CNN model, and 236 images were prepared for testing, resulting in an accuracy rate of 80.2%. In comparison to the C-GAN model, which achieved a 100% accuracy rate, the CNN model failed to reflect the entire dataset. Thus, the suitability of the C-GAN model for prediction after ground cavity detection was confirmed.

Figure 10 shows the data images before and after removing the straight-line noise in the direction of travel using local average subtraction. It is evident that noise near the surface and in the direction of travel, such as from pipes, is significantly reduced. Specifically, pixel values within a defined range centered on the direction of travel are selected, and the calculated value of the Gaussian function $G(x)$, expressed by Equation (3), is obtained based on the central pixel value xxx within the defined range. By subtracting this calculated value $G(x)$ from the pixel value (x), noise in the direction of travel can be removed. Here, σ is a predefined standard deviation value.

$$G(\chi) = \frac{1}{\sqrt{2\pi\sigma^2}} e^{-\frac{x^2}{2\sigma^2}} \tag{3}$$

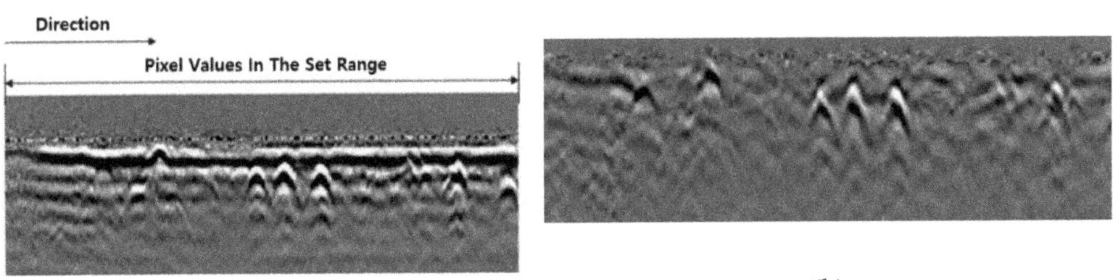

(a) (b)

Figure 10. Noise removal using local average subtraction: (**a**) non-application; (**b**) application.

2.3. *C-GAN-Based Automated Underground Cavity Detection*

By inputting preprocessed data into the deep learning model, it is possible to predict cavities within underground spaces. The deep learning model used in this study is a C-GAN (conditional generative adversarial network), which is an improved form of the traditional GAN. In a GAN, a generator and a discriminator compete during training, but in C-GAN, conditional data are input to obtain the desired images. There are two representative types of C-GANs: pix2pix and cycle GAN. This study uses pix2pix, which, unlike traditional GANs, inputs images instead of noise for training, and the generator learns to produce images as a result. To enhance image generation performance, the model structure employs a U-Net form, incorporating the residual block technique with skip connections to prevent vanishing gradients.

Figure 11 illustrates the data processing flow of the deep learning model utilized in this study. Figure 12 depicts the C-GAN architecture that can output cavity sections within the underground space based on pre-processed input data and pre-trained training data from 3D GPR data. Using this model eliminates the need for the reliance on expert concentration or subjective judgment in cavity surveys, allowing for real-time cavity detection and a rapid response to sudden road subsidence incidents. The predicted cavity sections in the deep learning model can be adjusted to display as images highlighted with specific colors.

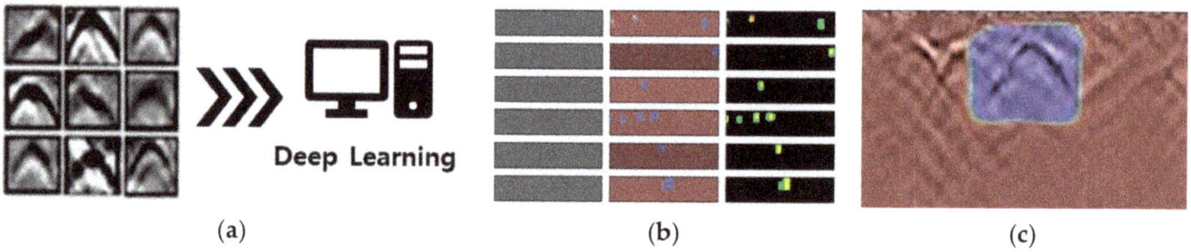

Figure 11. Data processing images of the deep learning model: (**a**) input data image; (**b**) data processing; (**c**) output of the cavity section.

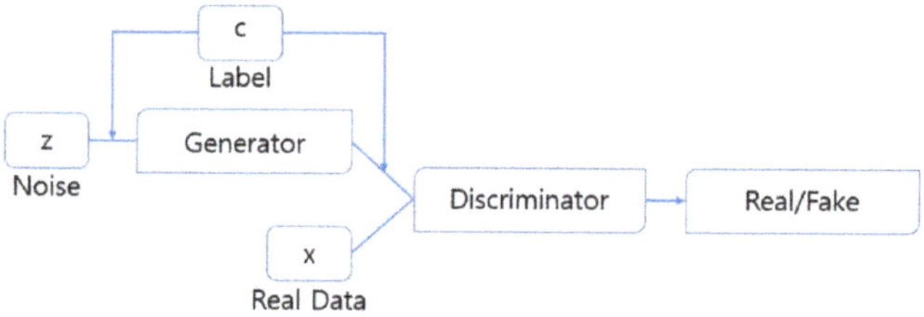

Figure 12. C-GAN architecture tailored for 3D GPR data classification.

3. Field Validation Tests

The deep-learning-based underground cavity automatic detection technique derived in the previous section was applied to the roads in Seoul, Republic of Korea. The validity of this technique was verified using the collected 3D GPR data.

3.1. Experimental Setup

Figure 13 shows the vehicle-type multi-channel GPR used in this field investigation. The GPR has a frequency of 400 MHz, and its effective investigation depth and width are up to 2.0 m from the surface of flexible pavement and 2.4 m, respectively. Moreover, the multi-channel GPR can perform 3-dimensional analysis; therefore, it can detect the interface of different materials and abnormal signals owing to the differences in the relative permittivity, even in identical materials. The GPR system comprises GPR antennas, a camera recording surface conditions, a global positioning system (GPS), and positioning cameras [16].

This system consists of 14 transmitting antennas and 14 receiving antennas, totaling 25 channels, and extends the survey width to 2.4 m, allowing data acquisition without survey shadow areas in a single pass. Additionally, the vehicle-based GPR survey software is divided into software for recording and controlling in the field and software for analyzing and recording data indoors. In the field, MIRA Software v.3.57.01 from MALA was used to record and control the survey equipment. For survey data analysis, a combination of MALA's rSlicer v1.0.0 and the GPRIS System developed in the Republic of Korea was utilized.

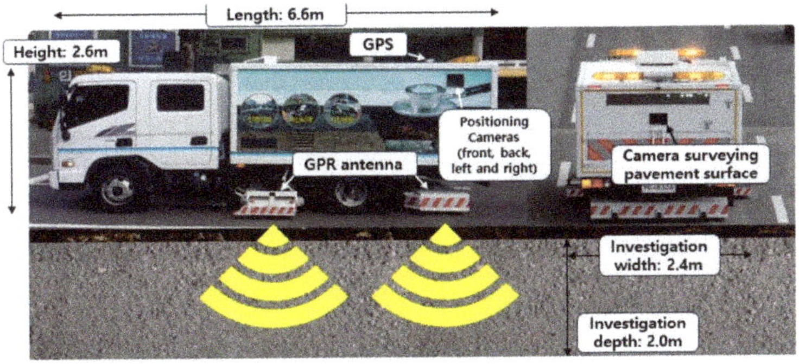

Figure 13. Field validation tests using a van with a mounted 3D GPR.

Figure 14a shows the radargram of a cross-section in the longitudinal direction from the GPR scanning, and the yellow circle represents that an underground cavity may exist. The horizontal axis indicates the distance in the longitudinal direction, and the vertical axis indicates the depth from the pavement surface. The depth from the top surface to the center of the parabolic black strip is defined as an overburden thickness measured as 0.33 m in this study. To determine the size of the cavity from the image, the boundary of the white strip above the black parabola that indicated an amplitude change in the GPR signal was set as the boundary of the cavity, and the length between the boundaries was defined as the size of the cavity. Figure 14b shows the radargram of a planar image at a depth of 0.67 m that represents the overburden thickness of the cavity.

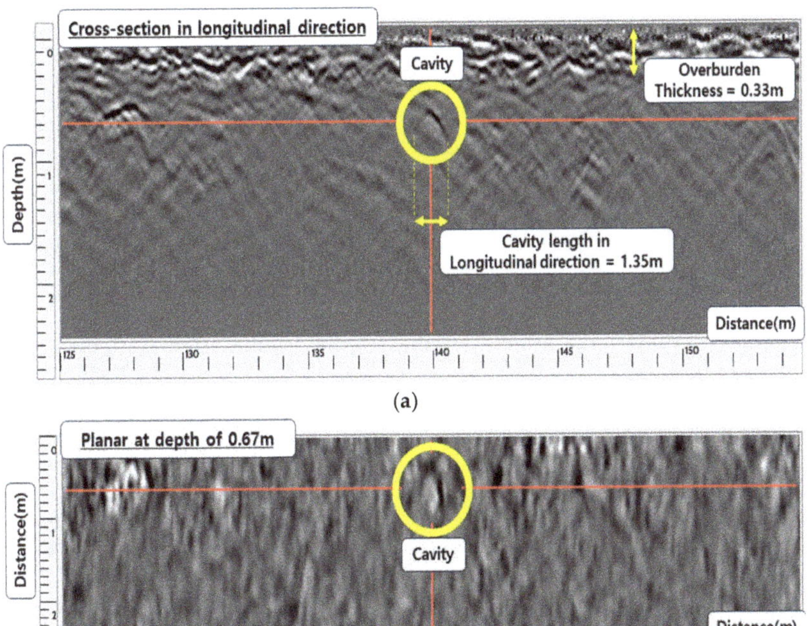

Figure 14. GPR radargrams: (**a**) cross-section in the longitudinal direction; (**b**) planar at a depth of 0.67 m.

The data in Figure 14a are non-migration data. The objective of this study is not to determine the size of the cavities but to evaluate the performance of cavity pre-processing using the BFA, making migration unnecessary in this context. Of course, during data acquisition, the apparent structure (hyperbola) reflected in the images is converted into the actual underground structure using correction methods to measure the scale of the cavity. Migration is typically used to reposition inclined surfaces to their actual underground locations and to reduce diffractions. According to the literature [16], approximately 70% of cavity waveforms detected through GPR surveys are considered to represent the actual size of the cavity, allowing for an approximate estimation of its size. However, as previously mentioned, migration is not a mandatory requirement for this study, so it was not performed.

3.2. Results of Appling the BFA

Figure 14 shows the longitudinal and cross-sectional views at locations with cavities, comparing images detected with and without the application of the BFA. In the longitudinal section, as shown in Figure 15a, the cavity appears in the form of a curve known as a parabola or hyperbola. In the upper image of Figure 15a, where the BFA is not applied, the boundary feature information of the target (underground cavity) has a weak intensity, and the geometric structure of the hyperbola is not distinct. However, in the lower image of Figure 15a, where the BFA is applied, signals that appear as black and white bands, indicating the boundary characteristics of the asphalt pavement and soil medium near the surface, are removed, making the features of the underground cavity in the longitudinal section more clearly identifiable.

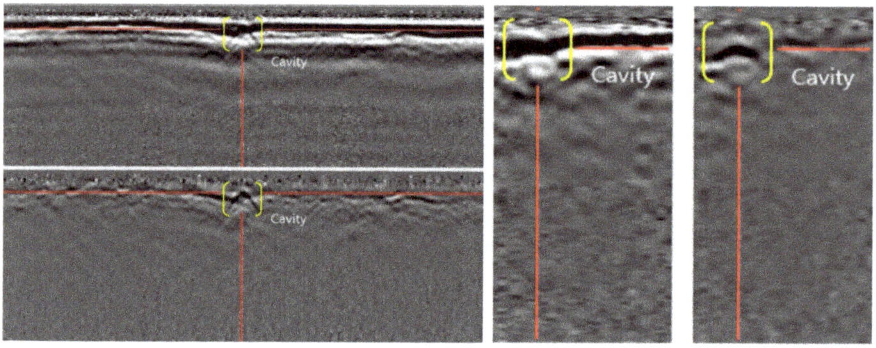

(a) (b)

Figure 15. Comparison of B-Scan data with the BFA applied: (**a**) longitudinal section; (**b**) cross-section.

Similarly, the underground cavity in the cross-sectional view shown in Figure 15b also exhibits the characteristic parabolic or hyperbola curve shape. In the image on the right side of Figure 15b, where the BFA is applied, the black and white bands near the surface are removed, making the boundary of the cavity's curved shape more prominently defined.

Figure 16 shows the plan view (C-scan) at locations with cavities, comparing images with and without the application of the BFA. When examining the plan view at the same depth, it is observed that the local area is represented in black before the BFA is applied. In contrast, after applying the BFA, the black areas in the image are clearly distinguishable as underground cavities, confirming the effectiveness of the BFA in enhancing the visibility of these features.

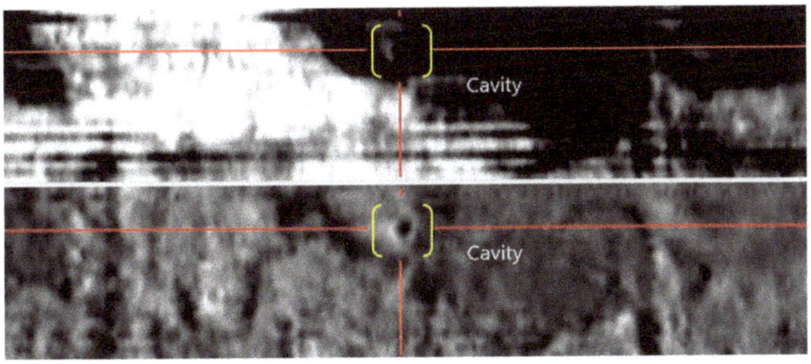

Figure 16. Comparison of plan view (C-Scan) data with the BFA applied.

Figures 17 and 18 present the analysis of underground cavity images at specific depths in the plan view, comparing cases with and without the application of the BFA. The data with the BFA applied, obtained at 3 cm depth intervals, showed a reduced influence from surrounding signals at all depths. As a result, the closed-loop boundaries characteristic of the underground cavities in the plan view became more distinct, and independent region images were formed more clearly.

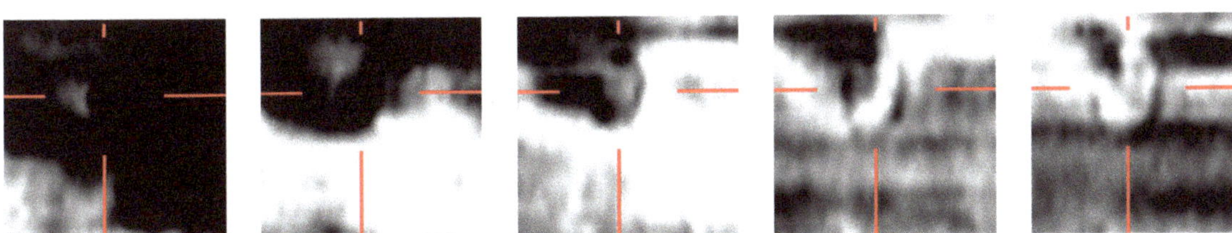

(a) (b) (c) (d) (e)

Figure 17. Comparison of the plan view of cavities at different depths without the BFA: (**a**) 15 cm; (**b**) 18 cm; (**c**) 21 cm; (**d**) 24 cm; (**e**) 27 cm.

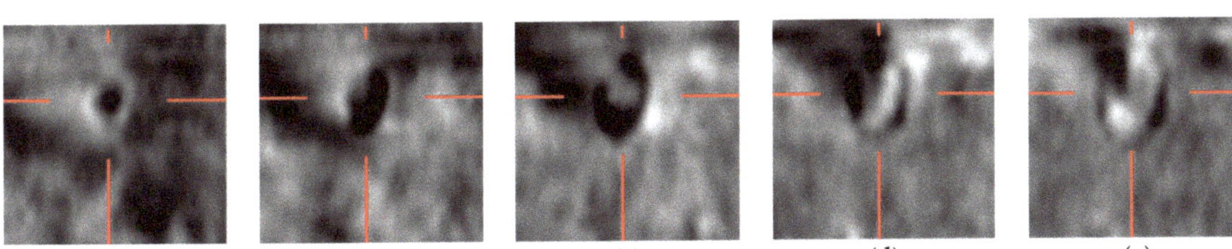

(a) (b) (c) (d) (e)

Figure 18. Comparison of the plan view of cavities at different depths with the BFA: (**a**) 15 cm; (**b**) 18 cm; (**c**) 21 cm; (**d**) 24 cm; (**e**) 27 cm.

Figures 19 and 20 show the images of data from five channels where the parabolic shape characteristic of cavities in the longitudinal section is well-visible. These images compare the data without the BFA applied and with the BFA applied. In the five channels, the data without the BFA areis obscured by the black and white straight-line signals representing the boundary characteristics of the asphalt pavement and soil medium. In contrast, the data with the BFA applied clearly show the upper and lower boundary characteristics of the parabolic cavity shape in the longitudinal section, demonstrating that the cavity features are more distinctly detected.

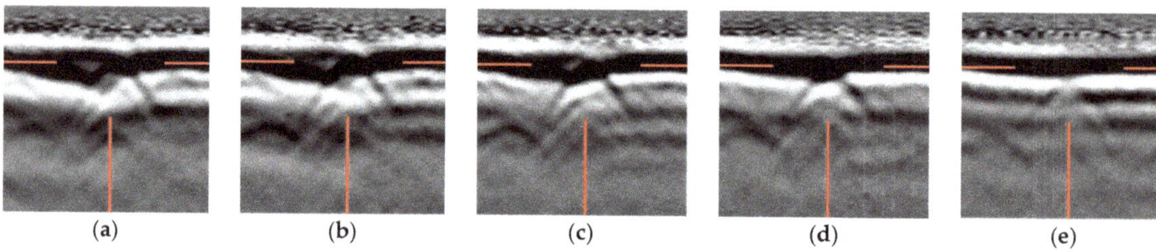

Figure 19. Comparison of the longitudinal view of cavities at cChannels without the BFA: (**a**) Channel 5; (**b**) Channel 6; (**c**) Channel 7; (**d**) Channel 8; (**e**) Channel 9.

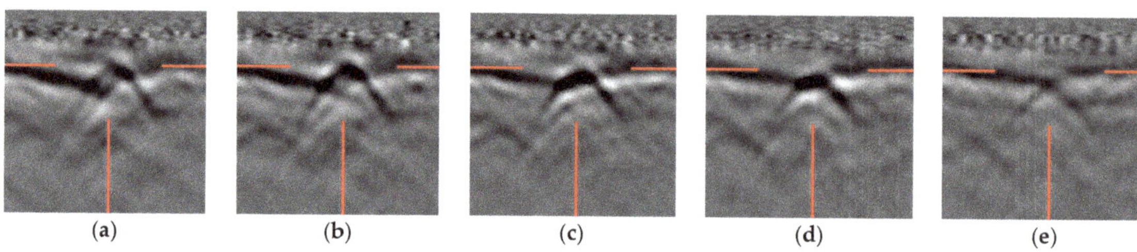

Figure 20. Comparison of the longitudinal view of cavities at cChannels with the BFA: (**a**) Channel 5; (**b**) Channel 6; (**c**) Channel 7; (**d**) Channel 8; (**e**) Channel 9.

3.3. Verification of the BFA Effect Using the Pre-Trained C-GAN Model

The 3D GPR data used in the C-GAN model include a total of 12 cavities, with 25 channels, a depth of 2.5 m, and a survey distance of 6215 m. To verify the effect of applying the BFA, tests were conducted and analyzed in two cases: without the BFA and with the BFA applied. Each dataset was input into the pre-trained C-GAN, and it took approximately 22 min to analyze a total distance of 12.4 km.

Assuming the analysis depth is around 80 cm, the 3D GPR data can be viewed as a rectangular prism with dimensions of 25 channels × 0.8 m (depth) × 6215 m (distance). Dividing this rectangular prism into smaller prisms of 5 channels × 0.4 m (depth) × 1 m (distance) results in 62,150 smaller rectangular prisms. The classification results of these 62,150 prisms, when using a confusion matrix, are shown in Figure 21. The x-axis represents the actual values, and the y-axis represents the predicted values. In detail, in the actual value, T is True (Cavity), and F is False (Non-Cavity), while in the predicted value, P is positive (Cavity), and N is Negative (Non-Cavity).

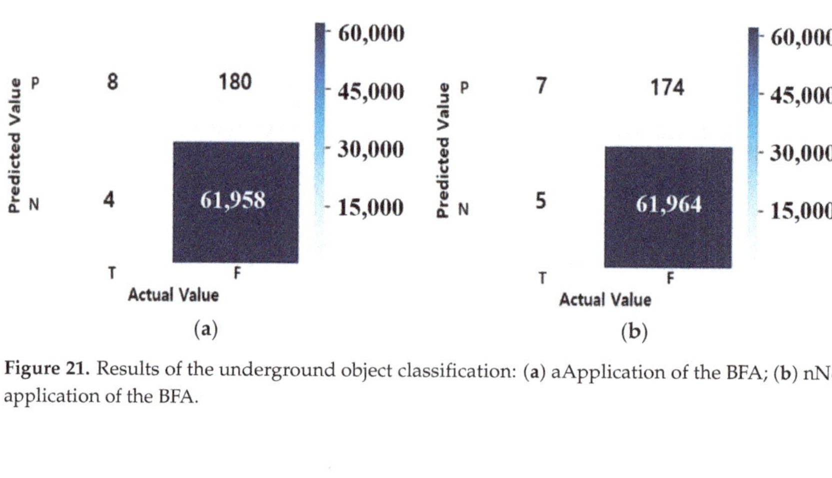

Figure 21. Results of the underground object classification: (**a**) aApplication of the BFA; (**b**) nNon-application of the BFA.

Thus, TP and FN indicate the accuracy of the C-GAN model, while TN or FP indicates the prediction failures by the C-GAN model. As an effect of applying BFA, TP and FN increased slightly, but these values are negligible compared to the overall data.

To quantitatively analyze the accuracy of the underground cavity detection results, a comprehensive evaluation index (*F1 Score*) for the fitted characteristic hyperbola was used. This can be calculated as shown in Equations (4)–(6). Here, *P* is precision, and *R* is recall. TP, FP, and FN represent true positives, false positives, and false negatives.

$$F1\ Score = \frac{2PR}{P+R} \tag{4}$$

$$P = \frac{TP}{TP+FP} \tag{5}$$

$$R = \frac{TP}{TP+FN} \tag{6}$$

Figure 22 shows the results of the model classification applied to Equations (4)–(6). These results illustrate the impact of the proposed BFA technique on performance through three indicators: F1 score, Precision, and Recall. The high value of false positives affects the F1 score and Precision, resulting in generally some lower values for these metrics. However, both metrics showed higher values when the BFA was applied. Additionally, in a cavity survey, the most crucial aspect is ensuring safety by accurately detecting all cavities. Therefore, Recall, the indicator of how well the C-GAN model predicts actual cavities, is vital. Comparing the Recall values with and without the BFA application, the model with the BFA applied showed higher Recall values by about 15%.

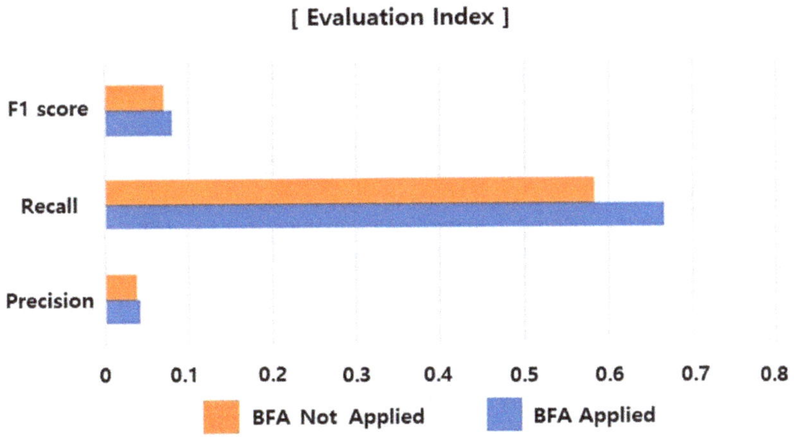

Figure 22. Evaluation index with and without BFA aApplication.

4. Discussion

In this study, a background filtering algorithm (BFA) was developed to enhance the visibility and clarity of underground cavities in a C-GAN model capable of autonomously detecting underground spaces under urban roads using 3D GPR data. A field test was performed by converting 3D GPR data collected in Seoul through the C-GAN model (a deep-learning-based automatic underground cavity detection technique) to which the developed BFA was applied, and the following discussions were derived.

(1) In this study, six steps were performed for the development of the BFA: collecting and converting 3D GPR data, data reconstruction, data alignment, background removal and data interpolation, data amplification, and noise removal using local average

subtraction. Subsequently, the BFA was proposed to enhance the visibility of underground cavities by eliminating dominant linear features and unwanted pattern noise in the 3D GPR data. The application of the BFA to the plan view, longitudinal, and cross-sectional views of the cavity locations in the 3D data revealed a clearer detection of the upper and lower boundary characteristics of the parabolic cavity images, which were previously obscured by the boundary characteristics of the asphalt pavement and soil medium.

(2) The C-GAN model, which outputs cavity sections within the underground space based on pre-processed input data and pre-trained learning data from 3D GPR data, eliminates the need for expert concentration or subjective judgment in cavity surveys. Additionally, real-time cavity detection allows for an immediate response to sudden road collapse incidents. The C-GAN model is designed to extract high-level semantic features from scan images, which can be used to train classification networks. Because high-level semantic features contain less interference information, the network model is easier to train and achieves higher recognition rates. Furthermore, the C-GAN-based data enhancement strategy used in this study demonstrates potential for improving the detection accuracy of field data by the network. The C-GAN model demonstrates a higher accuracy rate compared to the CNN model, and its application will significantly aid in predicting ground cavities. Furthermore, by collecting more high-quality training data to enhance the model's learning process and building a comprehensive GPR data library, an improvement in the accuracy of detecting underground structures is also anticipated.

(3) This technology surpasses existing GPR techniques by achieving fast and fully automated data interpretation. The proposed method was experimentally validated using actual 3D GPR data from roads in Seoul. Such data provided a unique opportunity to test the methodology. Although the proposed method demonstrated a satisfactory accuracy in detecting underground cavities, results may vary in different field conditions. Therefore, it is essential to conduct enhanced training processes based on diverse field data, and technological improvements need to be made.

(4) GPR signals reflected from gravel that resemble underground cavities or rocks often cannot be distinguished from actual cavity signals. To overcome these technical limitations, it is necessary to collect higher-quality training data verified through boreholes, enhance learning, and build a robust GPR data library. Additionally, developing further signal processing techniques, such as the phase analysis of reflection waveforms (A-scan), could combine with the current BFA-applied C-GAN architecture to improve the automatic classification accuracy for various types of underground objects, including cavities.

5. Conclusions

This study developed a special algorithm, called a background filtering algorithm (BFA), which can be used in ground cavity detection processes utilizing 3D GPR data. This algorithm enhances the clarity and resolution of GPR data, contributing to the improvement of the quality of data used in the C-GAN model, which is commonly applied in ground cavity detection. It is not a new equipment development or a systemic improvement for finding ground cavities. Rather, it is an economically advantageous technology that can upgrade the existing technology and is considered an excellent innovation due to its shift in approach.

Author Contributions: Conceptualization, D.W.P., H.E.K., K.L. and J.P.; methodology, H.E.K. and J.P.; software, H.E.K.; validation, H.E.K., D.W.P., K.L. and J.P.; formal analysis, K.L. and J.P.; investigation, H.E.K. and J.P.; resources, K.L. and J.P.; data curation, D.W.P. and H.E.K.; writing—original draft preparation, H.E.K. and J.P.; writing—review and editing, D.W.P., K.L. and J.P.; visualization, H.E.K. and K.L.; supervision, J.P.; project administration, K.L. and J.P. All authors have read and agreed to the published version of the manuscript.

Funding: This research received no external funding.

Institutional Review Board Statement: Not applicable.

Informed Consent Statement: Not applicable.

Data Availability Statement: No new data were created or analyzed in this study. Data sharing is not applicable to this article.

Conflicts of Interest: Han Eung Kim was employed by the company Geotechnical Korea Engineering Co., Ltd. The remaining authors declare that the research was conducted in the absence of any commercial or financial relationships that could be construed as a potential conflict of interest.

References

1. Kim, S.M.; Choi, H.; Yoon, J.S.; Park, J.J. A study on the change of cavity area through groundwater injection test under pavement cavity. *J. Soc. Disaster Inf.* **2020**, *16*, 267–275.
2. Park, J.J.; Chung, Y.; Hong, G. A method for cavity scale estimation based on ground-penetrating radar (GPR) explorations: An experimental study. *Adv. Civ. Eng.* **2019**, *1*, 1365759. [CrossRef]
3. Park, J.J.; Kim, I.D. Analysis of the under pavement cavity growth rate using multi-channel GPR equipment. *J. Soc. Disaster Inf.* **2020**, *16*, 60–69.
4. Lee, K.; Kim, D.; Park, J.J. Study on management system of ground sinking based on underground cavity grade. *J. Korean Geosynth. Soc.* **2017**, *16*, 23–33.
5. Han, J.G.; Ryu, Y.S.; Kim, D.; Park, J.J.; Hong, G. A study on expansion and strength characteristics of material for emergency restoration in ground cavity. *J. Korean Geosynth. Soc.* **2017**, *16*, 131–138.
6. Endres, A.L.; Clement, W.P.; Rudolph, D.L. Ground penetrating radar imaging of an aquifer during a pumping test. *Groundwater* **2000**, *38*, 566–576. [CrossRef]
7. Hagrey, A. GPR study of pore water content and salinity in sand. *Geophys. Prospect.* **2000**, *48*, 63–85. [CrossRef]
8. Pyke, K.; Eyuboglu, S.; Daniels, J.J.; Vendl, M. A controlled experiment to determine the water table response using ground penetrating radar. *J. Environ. Eng. Geophys.* **2008**, *13*, 335–342. [CrossRef]
9. Chang, P.; Flatau, A.; Liu, S. Review paper: Health monitoring of civil infrastructure. *Struct. Health Monit.* **2003**, *2*, 257–267. [CrossRef]
10. Nuzzo, L.; Leucci, G.; Negri, S.; Carrozzo, M.T.; Quarta, T. Application of 3D visualization techniques in the analysis of GPR data for archaeology. *Ann. Geophys.* **2002**, *45*, 321–337. [CrossRef]
11. Batayneh, A.; Abdelruhman, A.; Moumani, K. Use of ground-penetrating radar for assessment of potential sinkhole conditions: An example from Ghor al Haditha area. *Environ. Geol.* **2002**, *41*, 977–983.
12. Rodriguez, V.; Gutie'rrez, F.; Green, A.; Carbonel, D.; Horstmeyer, H.; Schmelzbach, C. Characterizing sagging and collapse sinkholes in a Mantled Karst by means of ground penetrating radar (GPR). *Environ. Eng. Geosci.* **2014**, *20*, 109–132. [CrossRef]
13. Conyers, L.B. Analysis and Interpretation of GPR Datasets for Integrated Archaeological Mapping. *Near Surf. Geophys.* **2015**, *13*, 645–651. [CrossRef]
14. Daniels, D.J. *Ground Penetrating Radar*, 2nd ed.; IEE Radar Sonar Navigation and Avionics Series; The Institution of Engineering and Technology: London, UK, 2004.
15. Cassidy, N.J.; Eddies, R.; Dods, S. Void detection beneath reinforced concrete sections: The practical application of ground penetrating radar and ultrasonic techniques. *J. Appl. Geophys.* **2011**, *74*, 263–273. [CrossRef]
16. Lee, S.; Park, J.J.; Cho, B.H. Management of cavities under flexible pavement road network in metropolitan area: Detection, evaluation, and rehabilitation. *Dev. Built Environ.* **2022**, *12*, 100091. [CrossRef]
17. Kim, H.E.; Kim, C.H.; Kim, T.G.; Park, J.J. Verification of ground subsidence risk map based on underground cavity data using DNN technique. *J. Soc. Disaster Inf.* **2023**, *19*, 334–343.
18. Grote, K.; Hubbard, S.; Harvey, J.; Rubin, Y. Evaluation of infiltration in layered pavements using surface GPR reflection techniques. *J. Appl. Geophys.* **2005**, *57*, 129–153. [CrossRef]
19. Annan, A. *Practical Processing of GPR Data*; Sensors & Software: Mississauga, ON, Canada, 1999.
20. Park, B.; Kim, J.; Lee, J.; Kang, M.S.; An, Y.K. Underground object classification for urban roads using instantaneous phase analysis of ground-penetrating radar (GPR) data. *Remote Sens.* **2018**, *10*, 1417. [CrossRef]
21. Al-Nuaimy, W.; Huang, Y.; Nakhkash, M.; Fang, M.T.C.; Nguyen, V.T.; Eriksen, A. Automatic detection of buried utilities and solid objects with GPR using neural networks and pattern recognition. *J. Appl. Geophys.* **2000**, *43*, 157–165. [CrossRef]
22. Gamba, P. Neural detection of pipe signatures in ground penetrating radar images. *IEEE Trans Geosci. Remote Sens.* **2000**, *38*, 790–797. [CrossRef]
23. Zhang, Y.; Huston, D.; Xia, T. Underground object characterization based on neural networks for ground penetrating radar data. In Proceedings of the SPIE, Las Vegas, NV, USA, 21–24 March 2016.
24. Bai, X.; Yang, Y.; Wei, S.; Chen, G.; Li, H.; Li, Y.; Tian, H.; Zhang, T.; Cui, H. A Comprehensive review of conventional and deep learning approaches for Ground-Penetrating Radar detection of raw data. *Appl. Sci.* **2023**, *13*, 7992. [CrossRef]

25. Wang, H.; Ouyang, S.; Liu, Q.; Liao, K.; Zhou, L. Deep-learning-based method for estimating permittivity of Ground-Penetrating Radar targets. *Remote Sens.* **2022**, *14*, 4293. [CrossRef]
26. Xue, W.; Chen, K.; Li, T.; Liu, L.; Zhang, J. Efficient underground target detection of urban roads in Ground-Penetrating Radar images based on neural networks. *Remote Sens.* **2023**, *15*, 1346. [CrossRef]
27. Fang, L.; Yang, F.; Xu, M.; Liu, F. Research on development 3D Ground Penetrating Radar acquisition and control technology for road underground diseases with dual-band antenna arrays. *Sensors* **2023**, *23*, 8301. [CrossRef]

MDPI AG
Grosspeteranlage 5
4052 Basel
Switzerland
Tel.: +41 61 683 77 34

Remote Sensing Editorial Office
E-mail: remotesensing@mdpi.com
www.mdpi.com/journal/remotesensing

www.ingramcontent.com/pod-product-compliance
Lightning Source LLC
Chambersburg PA
CBHW061551220926
51850CB00005B/311
9783725873968